Rectangular Solid

Volume: $V = lwh$

Right Circular Cone

Volume: $V = \frac{1}{3}\pi r^2 h$

Lateral Surface Area: $L = \pi rs$

Total Surface Area: $S = \pi r^2 + \pi rs$

Slant Height: $s = \sqrt{r^2 + h^2}$

Right Circular Cylinder

Volume: $V = \pi r^2 h$

Lateral Surface Area: $L = 2\pi rh$

Total Surface Area:
$S = 2\pi rh + 2\pi r^2$

Cube

Volume: $V = s^3$

Sphere

Volume: $V = \frac{4}{3}\pi r^3$

Surface Area: $S = 4\pi r^2$

Fundamental Algebra and Trigonometry
Second Edition

Mervin L. Keedy
PURDUE UNIVERSITY

Marvin L. Bittinger
INDIANA UNIVERSITY—
PURDUE UNIVERSITY AT
INDIANAPOLIS

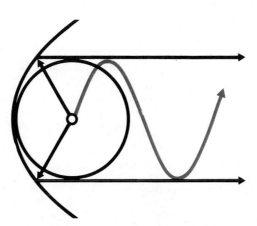

▲▼ ADDISON-WESLEY
PUBLISHING COMPANY

Reading, Massachusetts
Menlo Park, California
London □ Amsterdam
Don Mills, Ontario □ Sydney

Sponsoring Editor: Patricia Mallion
Production Editor: Martha Morong
Designer: Vanessa Piñeiro
Illustrator: VAP International Communications Ltd.
Cover Design: Vanessa Piñeiro
Cover Illustrator: Bob Trevor

Library of Congress Cataloging in Publication Data

Keedy, Mervin L.
   Fundamental algebra and trigonometry.

   Includes index.
   1. Algebra. 2. Trigonometry. I. Bittinger,
Marvin L., joint author. II. Title.
QA154.2.K433   1981   512'.13   80-19331
ISBN 0-201-03839-0

Reprinted with corrections, May 1982

Copyright © 1981 by Addison-Wesley Publishing Company, Inc. Philippines copyright 1981 by Addison-Wesley Publishing Company, Inc.

All rights reserved. No part of this publication may be reproduced, stored in a retrieval system, or transmitted, in any form or by any means, electronic, mechanical, photocopying, recording, or otherwise, without the prior written permission of the publisher. Printed in the United States of America. Published simultaneously in Canada. Library of Congress Catalog Card No. 80-19331.

ISBN 0-201-03839-0
EFGHIJK-HA-898765

Preface

There are substantial differences between the first edition of this text and the present second edition. Following is a list of features that characterize this edition.

1. *Completeness.* Topics have been added to the first edition and the paperback version of this text in an effort to satisfy the needs of various users. As a result, this book has become "topic-wise" very complete. It is hoped that it will now meet the needs of any user, besides being a valuable reference source.

Naturally, no class can cover this entire book in an ordinary freshman course. Topics must be chosen judiciously and some topics will have to be omitted. The following feature is therefore very important.

2. *Flexibility.* There are many ways in which this book can be used. Many topics are optional, and there are numerous paths that one can take through various topics, teaching them in various orders. Some of the possibilities are detailed on pp. vi–vii.

3. *Exercises.* This edition contains many new exercises. In response to comments from users, the authors have added exercises that require something of the student other than an understanding of the immediate objectives of the lesson at hand, yet are not necessarily highly challenging. The challenge exercises of the first edition have been augmented here. Thus, the first exercises in an exercise set are very much like the examples in the text for that section. These exercises are graded in difficulty and are paired. That is, each even-numbered exercise is very much like the one that precedes it. The next exercises (marked ☆) require the student to go beyond the immediate objectives. For the first two types of exercises, answers to the odd-numbered ones are given at the back of the book. The instructor can therefore easily make an assignment that is varied in terms of availability of answers. The challenge exercises are marked ★, and some of them are quite difficult. Answers to these exercises are not in the text, but are given in the answer booklet.

4. *Calculators.* Many exercises, as well as some parts of the development, are designed with the calculator in mind. Although it is perfectly feasible to use this text without a calculator, we have indicated those exercises or sections in which the use of a calculator is recommended by the symbol ▦. A calculator will be found useful in many other places as well.

Most of the calculator exercises in the first edition were much like the other exercises, except that the numbers were more complicated. In this edition the use of the calculator has been made much more comprehensive.

5. *Readability and understandability.* Although readability and understandability are related, they are actually separate features of a text. It is easy to write prose that is easy to read but impossible to understand. Therefore, we discuss these items separately.

With respect to *readability,* we have striven to say what we feel needs to be said, but without excess verbiage. The goal was to make the reading level of this text quite low, without sounding condescending. This book is written *to the student.* Theorems, principles, and procedures are stated for maximum student understanding, yet the tone of the book is still mature.

With respect to *understandability,* the goal was to produce a sequence in which each topic is developed, step by carefully-described step. At each appropriate point, examples are given, sufficient in coverage that the routine part of the homework exercises is thoroughly covered. *Cautions* are frequently given to the student; for example, "Don't make the mistake of thinking that $\sqrt{a^2 + b^2} = a + b$."

6. *Functions and transformations.* This text applies the concepts of function, relation, and transformation quite thoroughly. The idea of transformations makes Chapter 3 unique and sets up the study of later material.

Chapter 3 may be a bit long. Users have given us valuable feedback with respect to this chapter. Some have recommended that it be broken up, with some of the topics imbedded in other chapters, while others feel quite positive about teaching the entire chapter. After careful consideration, we have decided to leave the chapter intact. Although the topics can be taught in various orders, there is no *best* order. Moreover, for reference purposes, it is desirable to leave all the material together in its own chapter. Some possible reorderings and omissions of the topics are given on pp. vi–vii.

The concept of transformation helps to simplify topics related to quadratic functions and to graphing. It provides a new approach to solving inequalities with absolute value, and many of the trigonometric identities are rendered very simple. Instructors tell us of cases in which students are afraid that they may not be understanding the material because they grasp the ideas so readily.

7. *Accuracy.* In this edition the text, examples, and answers have been checked thoroughly by several different people, in a conscientious effort to make the book error-free. Although we cannot guarantee that nothing has "slipped through the cracks," we have certainly tried hard to minimize errors and inaccuracies. We will be most grateful to any user who calls our attention to any error that somehow went undetected.

8. *Supplementary materials.* The following supplementary materials are available.

- An Answer Booklet containing the answers to the even-numbered exercises and to the more challenging exercises.
- A Student Solutions Booklet containing worked-out solutions to selected odd-numbered exercises.
- A Test Booklet containing five classroom-ready tests for each chapter and five final examinations.

TOPIC SELECTION AND ORDERING

There are numerous topics that can be omitted, depending on the nature of a particular class and/or course. Chapter 6 can easily be omitted entirely, for example, as can several of the later chapters. We will not attempt to list all the topics that can be omitted. Such a list would be unduly long and probably not very helpful. Rather, we shall point out some possibilities that might not be so easy to see on cursory inspection of the book.

Functions and Transformations

Some users consider Chapter 3 a bit long. The following may help such users see how to break it up or omit parts of it.

Sections 3.1, 3.2, and 3.3, on Cartesian products, graphs, and functions, respectively, are the parts of the chapter that are essential in sequence. They should be taught before moving on.

Sections 3.4 and 3.5, on symmetry and transformations, can be postponed until Chapter 4, prior to the material on quadratic functions. Section 3.4 can be partly or entirely omitted without causing much difficulty at this point.

Section 3.7, on inverses of relations, is needed for the material on logarithmic and exponential functions in Chapter 6, and can be postponed until the beginning of that chapter.

Section 3.6, on special classes of functions (including periodic functions), is needed for Chapter 8 on the circular functions and can be taught along with that chapter.

Trigonometry

The trigonometry occurs in successive chapters, but need not necessarily be taught in a block. There are now several different possible tracks through the trigonometry. We detail three such tracks, one with analytic emphasis, the others with triangle emphasis.

Track I (Analytic emphasis)
Proceed through Chapters 8–11 as written, omitting Section 9.6. This track gives a thorough grounding in trigonometry with an initial emphasis on the analytic aspects.

Section 8.1, which is an introduction to triangle trigonometry, could be postponed until the beginning of Chapter 9. Certain topics, such as vectors and polar coordinates, can be omitted.

Track II (Triangle emphasis)
Section 8.1
Chapter 9 (omitting 9.6)   } This is triangle trigonometry.
Chapter 11
Remainder of Chapter 8   } This is analytic trigonometry.
Chapter 10

Track III (Minimal course, triangle emphasis)
Section 8.1
Chapter 9 (omitting 9.6)
Chapter 11 (omitting certain   } This is triangle trigonometry.
  optional sections
  if desired)
Section 9.6
Perhaps Section 8.7   } This is analytic trigonometry.
Chapter 10

Note that in the minimal course, most of Chapter 8 is omitted. There are other variations possible. One could teach Chapters 8, 9, 11, and then 10, for example.

The authors wish to thank Stephen H. Brown, Auburn University; Rodney Chase, Oakland Community College–Orchard Ridge Campus; Gary Cornell, the University of Connecticut; Mark P. Hale, Jr., the University of Florida; Lloyd A. Iverson, University of Oklahoma; Stanley M. Lukawecki, Clemson University; Mohammed G. Rajah, Mira Costa College; Eugene Spiegel, the University of Connecticut; and Jerry F. Tate, San Jacinto College; who helped with the development of this new edition.

January 1981

M. L. K.
M. L. B.

Contents

1

1.1	The Real-Number System	2
1.2	Exponential Notation	11
1.3	Addition and Subtraction of Algebraic Expressions	15
1.4	Multiplication of Polynomials	19
1.5	Factoring	22
1.6	Solving Equations and Inequalities	24
1.7	Fractional Expressions	29
1.8	Radical Notation and Absolute Value	36
1.9	Further Calculations with Radical Notation	43
1.10	Rational Exponents	46
1.11	Handling Dimension Symbols	48
	Review	51

2

2.1	Solving Equations	56
2.2	Formulas and Applied Problems	61
2.3	Quadratic Equations	71
2.4	Formulas and Applied Problems	77
2.5	Radical Equations	82
2.6	Equations Reducible to Quadratic	84
2.7	Variation	88
	Review	93

3 RELATIONS, FUNCTIONS, AND TRANSFORMATIONS

3.1	Relations and Ordered Pairs	96
3.2	Graphs of Equations	98
3.3	Functions	104
3.4	Symmetry	112
3.5	Transformations	119
3.6	Some Special Classes of Functions	126
3.7	Inverses	136
	Review	143

4 LINEAR AND QUADRATIC FUNCTIONS AND INEQUALITIES

4.1	Lines and Linear Functions	148
4.2	Parallel and Perpendicular Lines; the Distance Formula	154
4.3	Quadratic Functions	159
4.4	Mathematical Models	166
4.5	Sets, Sentences, and Inequalities	171
4.6	Equations and Inequalities with Absolute Value	177
4.7	Quadratic, Polynomial, and Rational Inequalities	181
	Review	185

5 SYSTEMS OF LINEAR EQUATIONS AND INEQUALITIES

5.1	Systems of Equations in Two Variables	190
5.2	Inconsistent and Dependent Equations; Applied Problems	195
5.3	Systems of Equations in Three or More Variables	198
5.4	Matrices and Systems of Equations	206

5.5	Determinants and Cramer's Rule	**209**
5.6	Systems of Inequalities	**214**
5.7	Linear Programming	**217**
	Review	**221**

6 MATRICES AND DETERMINANTS

6.1	Operations on Matrices	**224**
6.2	Determinants of Higher Order	**231**
6.3	Properties of Determinants	**235**
6.4	Inverses of Matrices	**241**
	Review	**249**

7 EXPONENTIAL AND LOGARITHMIC FUNCTIONS

7.1	Exponential and Logarithmic Functions	**252**
7.2	Properties of Logarithmic Functions	**257**
7.3	Common Logarithms	**262**
7.4	Interpolation	**269**
7.5	Exponential and Logarithmic Equations	**273**
7.6	The Number e; Natural Logarithms; Change of Base	**279**
	Review	**288**

8 THE CIRCULAR FUNCTIONS

8.1	Triangles	**292**
8.2	The Unit Circle	**297**
8.3	The Sine and Cosine Functions	**301**
8.4	The Other Circular Functions	**307**
8.5	Some Relations Among the Circular Functions	**315**
8.6	Graphs	**320**
8.7	Algebraic and Trigonometric Manipulations	**324**
	Review	**329**

9

9.1	Rotations and Angles	332
9.2	Angular Speed	336
9.3	Trigonometric Functions of Angles or Rotations	339
9.4	Relationships Among Function Values	346
9.5	Tables of Trigonometric Functions	348
9.6	Further Relationships Among the Functions	354
	Review	360

10

10.1	Sum and Difference Formulas	364
10.2	Some Important Identities	371
10.3	Proving Identities	378
10.4	Inverses of the Trigonometric Functions	385
10.5	Combinations of Trigonometric Functions and Their Inverses	390
10.6	Trigonometric Equations	394
10.7	Identities in Solving Equations	398
	Review	402

11

11.1	Solving Right Triangles	406
11.2	The Law of Sines	409
11.3	The Law of Cosines	414
11.4	Vectors	418
11.5	Vectors and Coordinates	420
11.6	Polar Coordinates	425
11.7	Forces in Equilibrium	428
	Review	432

12

12.1	Introduction to Imaginary and Complex Numbers	434
12.2	Conjugates and Division	438
12.3	Equations and Complex Numbers	443
12.4	Graphical Representation and Polar Notation	447
12.5	DeMoivre's Theorem	452
	Review	455

13

13.1	Polynomials and Polynomial Functions	458
13.2	The Remainder and Factor Theorems	461
13.3	Theorems About Roots	466
13.4	Rational Roots	471
13.5	Further Helps in Finding Roots	476
13.6	Graphs of Polynomial Functions	485
13.7	Rational Functions	491
13.8	Partial Fractions	499
	Review	503

14

14.1	Conic Sections	506
14.2	Ellipses	512
14.3	Hyperbolas	517
14.4	Parabolas	523
14.5	Systems of First-degree and Second-degree Equations	528
14.6	Systems of Second-degree Equations	531
	Review	535

15 SEQUENCES, SERIES, AND MATHEMATICAL INDUCTION

15.1	Sequences	538
15.2	Arithmetic Sequences and Series	542
15.3	Geometric Sequences and Series	547
15.4	Infinite Geometric Sequences and Series	552
15.5	Mathematical Induction	557
	Review	563

16 COMBINATORICS AND PROBABILITY

16.1	Permutations	566
16.2	Combinations	573
16.3	Binomial Series	576
16.4	Probability	580
	Review	586

1	Powers, Roots, and Reciprocals	589
2	Common Logarithms	590
3	Values of Trigonometric Functions	592
4	Exponential Functions	597
5	Factorials and Large Powers of 2	598
6	Tables of Measures	598
7	Natural Logarithms (ln x)	599

ANSWERS TO SELECTED ODD-NUMBERED EXERCISES   A-1

INDEX   I-1

Basic Concepts of Algebra

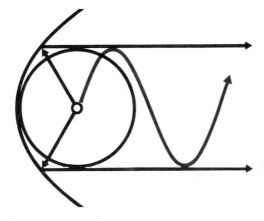

1.1 THE REAL-NUMBER SYSTEM

Real Numbers

There are various kinds of numbers. Those most used in elementary algebra are the *real numbers*. Later we consider a more comprehensive number system called the *complex numbers*. The real numbers are often pictured in one-to-one correspondence with the points of a line, as in Fig. 1. The positive numbers are shown to the right of 0 and the negative numbers to the left. Zero itself is neither positive nor negative.

There are several subsystems of the real numbers. They are as follows.

Natural numbers. Those numbers used for counting: 1, 2, 3, ...

Whole numbers. The natural numbers and 0: 0, 1, 2, 3, ...

Integers. The whole numbers and their additive inverses: 0, 1, -1, 2, -2, 3, -3, ...

Rational numbers. The integers and all quotients of integers (excluding division by 0): $\frac{4}{5}, \frac{-4}{7}, \frac{9}{1}, 6, -4, 0, \frac{78}{-5}, -\frac{2}{3}$ (can also be named $\frac{-2}{3}$, or $\frac{2}{-3}$).

Any real number that is not rational is called *irrational*. The rational numbers and the irrational numbers can be described in several ways.

The *rational numbers* are:

1. Those numbers that can be named with fractional notation a/b, where a and b are integers and $b \neq 0$ (definition);
2. Those numbers for which decimal notation either ends or repeats.

Examples All of these are rational.

1. $\frac{5}{16} = 0.3125$ Ending (terminating) decimal.
2. $-\frac{8}{7} = -1.142857142857\ldots = -1.\overline{142857}$ Repeating decimal. The bar indicates the repeating part.
3. $\frac{3}{11} = 0.2727\ldots = 0.\overline{27}$ Repeating decimal.

The *irrational numbers* are:

1. Those real numbers that are not rational (definition);
2. Those real numbers that cannot be named with fractional notation a/b, where a and b are integers and $b \neq 0$;

Figure 1

3. **Those real numbers for which decimal notation does not end and does not repeat.**

There are many irrational numbers. For example, $\sqrt{2}$ is irrational. We can find rational numbers a/b for which $(a/b) \cdot (a/b)$ is close to 2, but we cannot find such a number a/b for which $(a/b) \cdot (a/b)$ is exactly 2.

Unless a whole number is a perfect square its square root is irrational. For example, $\sqrt{9}$ and $\sqrt{25}$ are rational, but all of the following are irrational:

$$\sqrt{3}, \quad -\sqrt{14}, \quad \sqrt{45}.$$

There are also many irrational numbers that cannot be obtained by taking square roots. The number π is an example.* Decimal notation for π does not end and does not repeat.

Examples All of these are irrational.

4. $\pi = 3.1415926535\ldots$ Numeral does not repeat.
5. $-1.101001000100001000001\ldots$ Numeral does not repeat.
6. $\sqrt[3]{2} = 1.25992105\ldots$ Numeral does not repeat.

In Example 5 there is a pattern, but it is not a repeating pattern.

Algebra and Properties of Real Numbers

In arithmetic we use numbers, performing calculations to obtain certain answers. In algebra we use arithmetic symbolism, but in addition we use symbols to represent unknown numbers. We do calculations and manipulate symbols, on the basis of the properties of numbers, which we review now. Algebra is thus an extension of arithmetic and a more powerful tool for solving problems.

Addition

Assuming that addition of nonnegative real numbers poses no problem, let us review how the definition of addition is extended to include the negative numbers. Recall first that the absolute value of a nonnegative number is that number itself. To get the absolute value of a negative number, change its sign (make it positive). The absolute value of a number a is denoted $|a|$. Thus, $|3| = 3$ and $|-7.4| = 7.4$.

---

*$\frac{22}{7}$ and 3.14 are only rational approximations to the irrational number π.

1. To add two negative numbers, add their absolute values (the sum is negative).
2. To add a negative number and a positive number, find the difference of their absolute values. The result will have the sign of the summand with the larger absolute value. If the absolute values are the same, the sum is 0.

For example,

$$-5 + (-6) = -11, \quad -\frac{3}{4} + \left(-\frac{7}{8}\right) = -\frac{13}{8}, \quad 8 + (-5) = 3,$$

$$-5 + 3 = -2, \quad 8.6 + (-4.2) = 4.4, \quad \pi + (-\pi) = 0,$$

$$-\frac{9}{5} + \frac{3}{5} = -\frac{6}{5}, \quad -\sqrt{3} + (-4\sqrt{3}) = -5\sqrt{3}.$$

Properties of Real Numbers Under Addition

We now list, for review, the fundamental properties of real numbers under addition. These are properties on which algebraic manipulations are based, especially when symbols for unknown numbers are used.

Commutativity. For any real numbers a and b, $a + b = b + a$. (The *order* in which numbers are added does not affect the sum.)

Associativity. For any real numbers $a, b,$ and c, $a + (b + c) = (a + b) + c$. (When only additions are involved, parentheses for grouping purposes may be placed as we please without affecting the sum.)

Identity. There exists a unique real number 0 such that for any real number a, $a + 0 = 0 + a = a$. (Adding 0 to any number gives that same number.)

Inverses. For every real number a, there exists a unique number, denoted $-a$, for which $-a + a = a + (-a) = 0$.

In connection with additive inverses, we caution the reader about one of the most misunderstood ideas in elementary algebra: It is common to read an expression such as $-x$ as "negative x." This can be confusing, because $-x$ may be positive, negative, or zero, depending on the value of x. The symbol "$-$," used in this way, indicates an *additive inverse;* somewhat unfortunately, the same symbol may also indicate a negative number, as in -5, or it may indicate subtraction, as in $3 - x$.

Caution: An initial $-$ sign, as in $-x$ or $-(x^2 + 3x - 2)$ should always be interpreted as meaning "the additive inverse of." The entire expression may be positive, negative, or zero, depending on the value of the part of the expression that follows the − sign.*

Examples Find the additive inverse.

7. $-x$, when $x = 5$ $-(5) = -5$ (negative 5)
8. $-x$, when $x = -3$ $-(-3) = 3$
9. $-x$, when $x = 0$ $-(0) = 0$
10. $-(x^2 + 4x + 2)$, when $x = 1$ $-(1^2 + 4 \cdot 1 + 2) = -7$
    $-(x^2 + 4x + 2)$, when $x = -1$ $-((-1)^2 + 4(-1) + 2) = 1$

It is easily shown that $-1 \cdot x = -x$ for any number x. That is, multiplying a number x by negative 1 produces the additive inverse of x. We list this as a theorem.

THEOREM 1

For any real number x, $-1 \cdot x = -x$. (Multiplying negative 1 times x produces the additive inverse of x.)

Multiplication

Assuming that multiplication of nonnegative real numbers poses no problem, let us review how the definition of multiplication is extended to include the negative numbers.

1. **To multiply two negative numbers, multiply their absolute values (the product is positive).**
2. **To multiply a positive number and a negative number, multiply their absolute values and take the additive inverse of the result (the product is negative).**

For example,

$$3 \cdot (-4) = -12, \qquad 1.5 \cdot (-3.8) = -5.7,$$

$$-5 \cdot (-4) = 20, \qquad -\frac{2}{3} \cdot \left(-\frac{4}{5}\right) = \frac{8}{15},$$

$$-3 \cdot (-2) \cdot (-4) = -24.$$

We now list the properties of the real numbers under multiplication.

---

*Taking the additive inverse is sometimes called "changing the sign."

Commutativity. For any real numbers a and b, $ab = ba$. (The *order* in which numbers are multiplied does not affect the product.)

Associativity. For any real numbers a, b, and c, $a(bc) = (ab)c$. (When only multiplications are involved, parentheses for grouping purposes may be placed as we please without affecting the sum.)

Identity. There exists a unique real number 1 such that for any real number a, $a \cdot 1 = 1 \cdot a = a$. (Multiplying any number by 1 gives that same number.)

Inverses. For each nonzero real number a, there exists a unique number, denoted $1/a$ or a^{-1}, for which $a(1/a) = (1/a)(a) = 1$.

Multiplicative inverses are also called *reciprocals*.

Examples

11. The multiplicative inverse, or reciprocal, of 2 is $\frac{1}{2}$.

12. The multiplicative inverse of $-\frac{2}{3}$ is $-\frac{3}{2}$.

13. The reciprocal of 0.16 is 6.25.

There is a very special property that connects addition and multiplication, as follows.

Distributivity. For any real numbers a, b, and c,

$$a(b + c) = ab + ac.*$$

(This is also called the distributive law of multiplication over addition.)

Any number system having the preceding properties for addition and multiplication is called a *field*. Thus we refer to this list of properties as the *field properties*.

Many other properties, important in algebraic manipulations, can be proved from the field properties. We list some of these as theorems.

Subtraction

Subtraction is the operation opposite to addition, as given in the following definition.

---

*The expression $ab + ac$ means $(a \cdot b) + (a \cdot c)$. By agreement, we can omit parentheses around multiplications. Because of this agreement, multiplications are to be done before additions or subtractions.

DEFINITION

Subtraction. **For any real numbers a and b, $a - b = c$ if and only if $b + c = a$ ($a - b$ is the number which when added to b gives a).**

In any field, we actually subtract by adding an inverse, according to the following theorem.

THEOREM 2*

For any real numbers a and b, $a - b = a + (-b)$.

Theorem 2 follows immediately from the definitions of subtraction and additive inverse. It says that to subtract a number we can add its additive inverse. For example,

$$8 - 5 = 8 + (-5) = 3,$$
$$3 - 7 = 3 + (-7) = -4,$$
$$10 - (-4) = 10 + 4 = 14,$$
$$8.6 - (-2.3) = 8.6 + 2.3 = 10.9,$$
$$-15 - (-5) = -15 + 5 = -10,$$
$$\frac{5}{9} - \frac{2}{9} = \frac{5}{9} + \left(-\frac{2}{9}\right) = \frac{1}{3}.$$

Multiplication is distributive over subtraction in the real-number system, as the following theorem states.

THEOREM 3

Distributivity. **For any real numbers a, b, and c,**

$$a(b - c) = ab - ac.$$

(This is the distributive law of multiplication over subtraction.)

Theorem 3 follows easily from Theorem 2 and the other distributive laws.

---

*Theorem 2 is often used as a *definition* of subtraction. The definition of subtraction used here is more general, since it does not depend on the existence of inverses. Our definition is valid in the system of natural numbers, for example, where Theorem 2 would not even make sense since additive inverses do not exist.

Division

Division is the operation opposite to multiplication, as given in the following definition.

DEFINITION

Division. **For any number a and any nonzero number b, $a \div b = c$ if and only if $b \cdot c = a$ ($a \div b$ is the number which when multiplied by b gives a).**

In any field, we usually divide by multiplying by a reciprocal, according to the following theorem.

THEOREM 4

For any real number a and nonzero number b, $a \div b = a(1/b)$.

The definition of division parallels that for subtraction, and Theorems 2 and 4 are also parallels.

Examples Divide by multiplying by a reciprocal.

14. $\frac{3}{4} \div \left(-\frac{2}{3}\right) = \frac{3}{4}\left(-\frac{3}{2}\right) = -\frac{9}{8}$
15. $-\frac{6}{7} \div \left(-\frac{3}{5}\right) = -\frac{6}{7}\left(-\frac{5}{3}\right) = \frac{10}{7}$

From Theorem 4 it follows easily that the quotient of two negative numbers is positive and that the quotient of a positive number and a negative number is negative.

Order

The order of the real numbers is shown intuitively by a number line. If a number a is pictured to the left of a number b, then a is less than b ($a < b$). In this event, if we subtract a from b, the answer will be a positive number. This idea motivates the definition of $<$.

DEFINITION

For any real numbers a and b, $a < b$ if and only if $b - a$ is positive.

Examples Verify the inequality using the definition of $<$.

16. $2 < 8 \qquad 8 - 2 = 6$ and 6 is positive.

17. $-4 < 9$ $9 - (-4) = 13$ and 13 is positive.
18. $-7 < -5$ $-5 - (-7) = 2$ and 2 is positive.

The symbol for "greater than" ($>$) is defined in terms of $<$, as follows.

DEFINITION

$a > b$ means $b < a$.

Thus to show that $a > b$ we show that $a - b$ is positive.

The Use of Calculators

In the exercise set that follows you will find problems designed for the use of a calculator. These, and other examples and discussions, will be denoted by the symbol 🔳. Although a calculator is not essential for working problems in this text, you will find one useful.

Keep in mind that there can be differences in answers found on the calculator because of rounding error differences. For example, suppose you were asked to find $\sqrt{18}$, precise to seven decimal places. On a certain kind of calculator with an eight-digit readout, the display would show

$$\sqrt{18} = 4.2426406.$$

On another kind of calculator with a ten-digit readout, the display would show

$$\sqrt{18} = 4.242640687.$$

The answer would be found by rounding back to the seventh decimal place and would be given as

$$\sqrt{18} = 4.2426407.$$

Thus there can be variance in the seventh decimal place.

EXERCISE SET 1.1

Given the numbers

$$-6,\ 0,\ 3,\ -\frac{1}{2},\ \sqrt{3},\ -2,\ -\sqrt{7},\ \sqrt[3]{2},\ \frac{5}{8},\ 14,\ -\frac{9}{4},\ 8.53,\ 9\frac{1}{2},$$

which are:

1. natural numbers?
2. whole numbers?

3. irrational numbers?

4. rational numbers?

5. integers?

6. real numbers?

Which of the following are rational? irrational?

7. $-\dfrac{6}{5}$ **8.** $-\dfrac{3}{7}$ **9.** -9.032 **10.** 3.14

11. $4.51\overline{6516}$ (Numeral repeats)   **12.** $-7.32\overline{3232}$ (Numeral repeats)

13. $4.303003000300003\ldots$ (Numeral does not repeat)

14. $6.414114111411114\ldots$ (Numeral does not repeat)

15. $\sqrt{6}$ **16.** $\sqrt{7}$ **17.** $-\sqrt{14}$ **18.** $-\sqrt{12}$

19. $\sqrt{49}$ **20.** $-\sqrt{16}$ **21.** $\sqrt[3]{5}$ **22.** $\sqrt[4]{10}$

Simplify.

23. $|12|$ **24.** $|-2.56|$ **25.** $|-47|$ **26.** $|0|$

Find $-x$ and $-1 \cdot x$, when:

27. $x = -7$.   **28.** $x = -\dfrac{10}{3}$.   **29.** $x = 57$.   **30.** $x = \dfrac{13}{14}$.

Find $-(x^2 - 5x + 3)$, when: Find $-(7 - y)$, when:

31. $x = 12$.   **32.** $x = -8$.   **33.** $y = -9$.   **34.** $y = 19$.

Compute.

35. $-3.1 + (-7.2)$ **36.** $-735 + 319$ **37.** $\dfrac{9}{2} + \left(-\dfrac{3}{5}\right)$ **38.** $-6 + (-4) + (-10)$

39. $-7(-4)$ **40.** $-\dfrac{8}{3}\left(-\dfrac{9}{2}\right)$ **41.** $(-8.2) \times 6$ **42.** $-6(-2)(-4)$

43. $-7(-2)(-3)(-5)$ **44.** $(-7.1)(-2.3)$ **45.** $-\dfrac{14}{3}\left(-\dfrac{17}{5}\right)\left(-\dfrac{21}{2}\right)$ **46.** $-\dfrac{13}{4}\left(-\dfrac{16}{5}\right)\left(\dfrac{23}{2}\right)$

47. $\dfrac{-20}{-4}$ **48.** $\dfrac{49}{-7}$ **49.** $\dfrac{-10}{70}$ **50.** $\dfrac{-40}{8}$

51. $\dfrac{2}{7} \div \left(-\dfrac{14}{3}\right)$ **52.** $-\dfrac{3}{5} \div \left(-\dfrac{6}{7}\right)$ **53.** $-\dfrac{10}{3} \div \left(-\dfrac{2}{15}\right)$ **54.** $-\dfrac{12}{5} \div (-0.3)$

55. $11 - 15$ **56.** $-12 - 17$ **57.** $12 - (-6)$ **58.** $-13 - (-4)$

59. $15.8 - 27.4$ **60.** $-19.04 - 15.76$ **61.** $-\dfrac{21}{4} - \left(-\dfrac{7}{4}\right)$ **62.** $\dfrac{10}{3} - \left(-\dfrac{17}{3}\right)$

Calculate. Round to six decimal places. The symbol 🖩 indicates an exercise meant to be done with a calculator.

63. 🖩 a) $(1.4)^2$ $(1.4142)^2$
        $(1.41)^2$ $(1.41421)^2$
        $(1.414)^2$

   b) What number does the sequence of numbers 1.4, 1.41, 1.414, and so on, seem to approach as a limit?

64. 🖩 a) $(2.1)^3$ $(2.1544)^3$
        $(2.15)^3$ $(2.15443)^3$
        $(2.154)^3$

   b) What number does the sequence of numbers 2.1, 2.15, 2.154, and so on, seem to approach as a limit?

In Exercises 65–74, what property is illustrated by each sentence?

65. $k + 0 = k$
66. $ax = xa$
67. $-1(x + y) = (-1x) + (-1y)$
68. $4 + (t + 6) = (4 + t) + 6$
69. $c + d = d + c$
70. $-67 \cdot 1 = -67$
71. $4(xy) = (4x)y$
72. $5(a + t) = 5a + 5t$
73. $y\left(\dfrac{1}{y}\right) = 1, \quad y \neq 0$
74. $-x + x = 0$

75. Show that subtraction is not commutative. That is, find real numbers a and b such that $a - b \neq b - a$.

76. Show that division is not commutative.

77. Show that division is not associative.

78. Show that subtraction is not associative.

Use the field properties to prove:

79. Theorem 1. **80.** Theorem 2. **81.** Theorem 3. **82.** Theorem 4.

Prove the following for any real numbers a, b, c, and d.

83. $(b + c)a = ba + ca$
84. $a(b + c + d) = ab + ac + ad$
85. $a(b + c - d) = ab + ac - ad$
86. Prove that any positive number is greater than 0.

1.2 EXPONENTIAL NOTATION

Integers as Exponents

When an integer greater than 1 is used as an exponent, the integer gives the number of times the base is used as a factor. For example, 5^3 means $5 \cdot 5 \cdot 5$. An exponent of 1 does not change the meaning of an expression. For example, $(-3)^1 = -3$. When 0 occurs as the exponent of a nonzero expression, we agree that the expression is equal to 1. For example, $37^0 = 1$. When a minus sign occurs with exponential notation, a certain caution is in order. For example, $(-4)^2$ means that -4 is to be raised to the second power. Hence $(-4)^2 = (-4)(-4) = 16$. On the other hand, -4^2 represents the additive inverse of 4^2. Thus $-4^2 = -16$. It may help to think of $-x^2$ as $-1 \cdot x^2$, according to Theorem 1.

Negative integers as exponents have meaning as follows.

DEFINITION

If n is any positive integer, then a^{-n} means $1/a^n$ for $a \neq 0$. In other words, a^n and a^{-n} are reciprocals of each other.

Examples

1. $\dfrac{1}{5^2} = 5^{-2}$ 2. $7^{-3} = \dfrac{1}{7^3}$ 3. $5^{-4} = \dfrac{1}{5^4} = \dfrac{1}{5 \cdot 5 \cdot 5 \cdot 5} = \dfrac{1}{625}$

Properties of Exponents

Let us consider an example involving multiplication:

$$b^5 \cdot b^{-2} = (b \cdot b \cdot b \cdot b \cdot b) \cdot \dfrac{1}{b \cdot b} = \dfrac{b \cdot b}{b \cdot b} \cdot (b \cdot b \cdot b)$$

$$= 1 \cdot (b \cdot b \cdot b) = b^3.$$

Note that the result can be obtained by adding the exponents. This is true in general.

THEOREM 5

For any number a, and any integers m and n, $a^m \cdot a^n = a^{m+n}$.

Examples

4. $x^4 \cdot x^{-2} = x^{4+(-2)} = x^2$ 5. $5^4 \cdot 5^6 = 5^{10}$
6. $c^{-3} \cdot c^{-2} = c^{-5}$ 7. $a^4 \cdot a^3 = a^7$

Let us consider an example involving division:

$$\dfrac{8^5}{8^3} = 8^5 \cdot \dfrac{1}{8^3} = 8^5 \cdot 8^{-3} = 8^{5-3} = 8^2.$$

Note that the result could also be obtained by subtracting the exponents. Here is another example:

$$\dfrac{7^{-2}}{7^3} = 7^{-2} \cdot 7^{-3} = 7^{-2-3} = 7^{-5}.$$

Again, the result could be obtained by subtracting the exponents. This is true in general.

THEOREM 6

For any nonzero a, and any integers m and n, $a^m/a^n = a^{m-n}$.

Examples

8. $\dfrac{9^{-2}}{9^5} = 9^{-2-5} = 9^{-7}$ 9. $\dfrac{7^{-4}}{7^{-5}} = 7^{-4-(-5)} = 7^1 = 7$

This property can be used to show why a^0 is not defined when $a = 0$. Consider the following:

$$a^0 = a^{3-3} = \frac{a^3}{a^3}.$$

If a were 0, we would then have 0/0, which is meaningless.

Raising a Power to a Power

Consider this example:

$$(5^2)^4 = 5^2 \cdot 5^2 \cdot 5^2 \cdot 5^2 = 5^8.$$

The result can be obtained by multiplying the exponents. This is true in general.

THEOREM 7

For any number a, and any integers m and n, $(a^m)^n = a^{m \cdot n}$.

Examples

10. $(8^{-2})^3 = 8^{-2} \cdot 8^{-2} \cdot 8^{-2} = 8^{-6}$
11. $(x^{-5})^4 = x^{-20}$
12. $(3x^2y^{-2})^3 = 3^3(x^2)^3(y^{-2})^3 = 3^3 x^6 y^{-6} = 27 x^6 y^{-6}$
13. $(5x^3y^{-5}z^2)^4 = 5^4 x^{12} y^{-20} z^8 = 625 x^{12} y^{-20} z^8$
14. $(4x^2y^{-3})^{-4} = 4^{-4} x^{-8} y^{12} = \frac{1}{256} x^{-8} y^{12}$

Scientific Notation

Scientific notation is particularly useful for naming very large or very small numbers. It also has uses for our later study of logarithms. Some examples are:

$$7.8 \times 10^{13} = 78{,}000{,}000{,}000{,}000,$$
$$5.64 \times 10^{-8} = 0.0000000564.$$

DEFINITION

Scientific notation for a number consists of exponential notation for a power of 10, and, if needed, decimal notation for a number a between 1 and 10 and a multiplication sign. The following are both scientific notation: $a \times 10^b$; 10^b.

We can convert to scientific notation by multiplying by 1, choosing an appropriate symbol $10^k/10^k$ for 1.

Example 15 Convert to scientific notation.

$96{,}000 = 96{,}000 \times \dfrac{10^4}{10^4}$ We use 10^4 in order to "move the decimal point" between 9 and 6.

$\phantom{96{,}000} = \dfrac{96{,}000}{10^4} \times 10^4$

$\phantom{96{,}000} = 9.6 \times 10^4$

With practice such conversions can be done mentally, and you should try to do this as much as possible.

Example 16 Convert to scientific notation.

$0.00000478 = 0.00000478 \times \dfrac{10^6}{10^6}$ We use 10^6 in order to "move the decimal point" between 4 and 7.

$ = (0.00000478 \times 10^6) \times 10^{-6}$
$ = 4.78 \times 10^{-6}$

Again you should try to make conversions mentally as much as possible.

Examples Convert to decimal notation.

17. $6.043 \times 10^5 = 604{,}300$
18. $4.7 \times 10^{-8} = 0.000000047$

EXERCISE SET 1.2

Simplify.

1. $2^3 \cdot 2^{-4}$
2. $3^4 \cdot 3^{-5}$
3. $b^2 \cdot b^{-2}$
4. $c^3 \cdot c^{-3}$
5. $4^2 \cdot 4^{-5} \cdot 4^6$
6. $5^2 \cdot 5^{-4} \cdot 5^5$
7. $2x^3 \cdot 3x^2$
8. $3y^4 \cdot 4y^3$
9. $(5a^2b)(3a^{-3}b^4)$
10. $(4xy^2)(3x^{-4}y^5)$
11. $(2x)^3(3x)^2$
12. $(4y)^2(3y)^3$
13. $(6x^5y^{-2}z^3)(-3x^2y^3z^{-2})$
14. $(5x^4y^{-3}z^2)(-2x^2y^4z^{-1})$
15. $\dfrac{b^{40}}{b^{37}}$
16. $\dfrac{a^{39}}{a^{32}}$
17. $\dfrac{x^2y^{-2}}{x^{-1}y}$
18. $\dfrac{x^3y^{-3}}{x^{-1}y^2}$
19. $\dfrac{9a^2}{(-3a)^2}$
20. $\dfrac{16y^2}{(-4y)^2}$
21. $\dfrac{24a^5b^3}{8a^4b}$
22. $\dfrac{30x^6y^4}{5x^3y^2}$
23. $\dfrac{12x^2y^3z^{-2}}{21xy^2z^3}$
24. $\dfrac{15x^3y^4z^{-3}}{45xyz^5}$

25. $(2ab^2)^3$
26. $(4xy^3)^2$
27. $(-2x^3)^4$
28. $(-3x^2)^4$
29. $-(2x^3)^4$
30. $-(3x^2)^4$
31. $(6a^2b^3c)^2$
32. $(5x^3y^2z)^2$
33. $(-5c^{-1}d^{-2})^{-2}$
34. $(-4x^{-1}z^{-2})^{-2}$
35. $\dfrac{4^{-2}+2^{-4}}{8^{-1}}$
36. $\dfrac{3^{-2}+2^{-3}}{7^{-1}}$
37. $\dfrac{(-2)^4+(-4)^2}{(-1)^8}$
38. $\dfrac{(-3)^2+(-2)^4}{(-1)^6}$
39. $\dfrac{(3a^2b^{-2}c^4)^3}{(2a^{-1}b^2c^{-3})^2}$
40. $\dfrac{(2a^3b^{-3}c^3)^3}{(3a^{-1}b^{-3}c^{-5})^2}$
41. $\dfrac{6^{-2}x^{-3}y^2}{3^{-3}x^{-4}y}$
42. $\dfrac{5^{-2}x^{-4}y^3}{2^{-3}x^{-5}y}$

Convert to scientific notation.

43. 58,000,000
44. 27,000
45. 365,000
46. 3645
47. 0.0000027
48. 0.0000658
49. 0.027
50. 0.0038
51. $910,000,000,000 (A recent national debt)
52. 93,000,000 (The distance, in miles, from the earth to the sun)

Convert to decimal notation.

53. 4×10^5
54. 5×10^{-4}
55. 6.2×10^{-3}
56. 7.8×10^6
57. 7.69×10^{12}
58. 8.54×10^{-7}
59. 9.46×10^{12} (The distance, in kilometers, that light travels in one year)
60. 1.7×10^{-24} (The mass, in grams, of a hydrogen atom)

Find $-x^2$ and $(-x)^2$, when:

61. $x = 5$.
62. $x = -7$.
63. $x = -1.08$.
64. $x = \sqrt{3}$.

The formula

$$M = P\left[\dfrac{\dfrac{i}{12}\left(1+\dfrac{i}{12}\right)^n}{\left(1+\dfrac{i}{12}\right)^n - 1}\right]$$

gives the monthly mortgage payment M on a home loan of P dollars at interest rate i, and n is the total number of payments (12 times the number of years).

65. ▪ The cost of a house is $92,000. The down payment is $14,000. The interest rate is $15\tfrac{3}{4}\%$. The loan period is 25 years. What is the monthly payment?

66. ▪ Repeat Exercise 65 for loan periods of 20 years and 30 years.

1.3 ADDITION AND SUBTRACTION OF ALGEBRAIC EXPRESSIONS

Polynomials

Expressions like these are called *polynomials in one variable*:

$-7x + 5$, $3y^3 - 5y^2 + 7y - 4$, 0, $-5t^4$, $x^5 - 9$.

BASIC CONCEPTS OF ALGEBRA

DEFINITION

A *polynomial in one variable* is any expression of the type
$$a_n x^n + a_{n-1} x^{n-1} + \cdots a_2 x^2 + a_1 x + a_0,$$
where n is a nonnegative integer and $a_n, \ldots, a_0$ are real numbers, called *coefficients*. Some or all of the coefficients may be 0. Each $a_i x^i$ is called a *term*.

Expressions like these are called *polynomials in several variables*:
$$5x^2 y^3 + 17 x^2 y - 2, \quad 14 a^2 b, \quad \pi r^2 + 2\pi r h.$$

Example 1 The polynomial $5x^3 y - 7xy^2 + 2$ has three terms. They are:
$$5x^3 y, \quad -7xy^2, \quad \text{and} \quad 2.$$
The coefficients of the terms are 5, -7, and 2.

The *degree of a term* is the sum of the exponents of the variables. The *degree* of a nonzero polynomial is the degree of the term of highest degree. The polynomial consisting only of the number 0 has no degree.

Example 2 In the polynomial $5x^3 y - 7xy^2 + 2$, the degrees of the terms are 4, 3, and 0. The polynomial is of degree 4.

A polynomial with just one term is called a *monomial*. If there are just two terms, a polynomial is called a *binomial*. If there are just three terms, it is called a *trinomial*.

Much of the algebraic manipulation we do with polynomials can also be done with expressions that are not polynomials. Here are some examples of expressions that are not polynomials.

Examples

3. $3\sqrt{x} + 4y$ 4. $\dfrac{3x^2 + 2}{x - 1}$ 5. $4x^{1/2} - 5y^{3/2}$

If two terms of an expression have the same letters raised to the same powers, the terms are called *similar*. This is true even if the expression is not a polynomial. Similar terms can be "combined" using the distributive laws.

Examples

6. $3x^2 - 4y + 2x^2 = 3x^2 + 2x^2 - 4y$ Rearranging
$ = (3 + 2)x^2 - 4y$ Using the distributive laws
$ = 5x^2 - 4y$

7. $4x^{1/2}y + 7x^{1/2}y = 11x^{1/2}y$
8. $-2x^2\sqrt{y^3} + 5x^2\sqrt{y^3} = 3x^2\sqrt{y^3}$

Addition

The sum of two polynomials can be found by writing a plus sign between them and then combining similar terms.

Example 9 Add: $-3x^3 + 2x - 4$ and $4x^3 + 3x^2 + 2$.

$$(-3x^3 + 2x - 4) + (4x^3 + 3x^2 + 2) = x^3 + 3x^2 + 2x - 2$$

Additive Inverses

The additive inverse of a polynomial can be found by changing the sign of every term.

THEOREM 8

The additive inverse of a polynomial can be found by replacing every term by its additive inverse.

Example 10 The additive inverse of $-3xy^2 + 4x^2y - 5x - 3$ is
$$3xy^2 - 4x^2y + 5x + 3.$$

Example 11 The additive inverse of $7xy^2 - 6xy - 4y + 3$ can be symbolized as
$$-(7xy^2 - 6xy - 4y + 3).$$
Thus
$$-(7xy^2 - 6xy - 4y + 3) = -7xy^2 + 6xy + 4y - 3.$$

The preceding example may recall the following rule: To remove parentheses preceded by an additive inverse sign, we change the sign of every term inside the parentheses.

Subtraction

By Theorem 2 we can subtract by adding an inverse. Thus to subtract one polynomial from another, we add its additive inverse. We change the sign of each term of the subtrahend and then add.

Example 12 Subtract.

$(-9x^5 - x^3 + 2x^2 + 4) - (2x^5 - x^4 + 4x^3 - 3x^2)$
$= (-9x^5 - x^3 + 2x^2 + 4) + [-(2x^5 - x^4 + 4x^3 - 3x^2)]$
$= (-9x^5 - x^3 + 2x^2 + 4) + (-2x^5 + x^4 - 4x^3 + 3x^2)$
$= -11x^5 + x^4 - 5x^3 + 5x^2 + 4$

On occasion, it may be helpful to write polynomials to be subtracted with similar terms in columns.

Example 13 Subtract the second polynomial from the first.

$$\begin{array}{l} 4x^2y - 6x^3y^2 + x^2y^2 - 5y \\ \underline{4x^2y + x^3y^2 + 3x^2y^3 + 6y} \\ -7x^3y^2 - 3x^2y^3 + x^2y^2 - 11y \end{array}$$ Mentally, change signs and add.

EXERCISE SET 1.3

Determine the degree of each term and the degree of the polynomial.

1. $-11x^4 - x^3 + x^2 + 3x - 9$
2. $t^3 - 3t^2 + t + 1$
3. $y^3 + 2y^6 + x^2y^4 - 8$
4. $u^2 + 3v^5 - u^3v^4 - 7$
5. $a^5 + 4a^2b^4 + 6ab + 4a - 3$
6. $8p^6 + 2p^4t^4 - 7p^3t + 5p^2 - 14$

Add.

7. $5x^2y - 2xy^2 + 3xy - 5$ and
$-2x^2y - 3xy^2 + 4xy + 7$

8. $6x^2y - 3xy^2 + 5xy - 3$ and
$-4x^2y - 4xy^2 + 3xy + 8$

9. $-3pq^2 - 5p^2q + 4pq + 3$ and
$-7pq^2 + 3pq - 4p + 2q$

10. $-5pq^2 - 3p^2q + 6pq + 5$ and
$-4pq^2 + 5pq - 6p + 4q$

11. $2x + 3y + z - 7$ and
$4x - 2y - z + 8$ and
$-3x + y - 2z - 4$

12. $2x^2 + 12xy - 11$ and
$6x^2 - 2x + 4$ and
$-x^2 - y - 2$

13. $7x\sqrt{y} - 3y\sqrt{x} + \frac{1}{5}$ and
$-2x\sqrt{y} - y\sqrt{x} - \frac{3}{5}$

14. $10x\sqrt{y} - 4y\sqrt{x} + \frac{4}{3}$ and
$-3x\sqrt{y} - y\sqrt{x} - \frac{1}{3}$

Rename each additive inverse without parentheses.

15. $-(5x^3 - 7x^2 + 3x - 6)$
16. $-(-4y^4 + 7y^2 - 2y - 1)$

Subtract.

17. $(3x^2 - 2x - x^3 + 2)$
$- (5x^2 - 8x - x^3 + 4)$

18. $(5x^2 + 4xy - 3y^2 + 2)$
$- (9x^2 - 4xy + 2y^2 - 1)$

19. $(4a - 2b - c + 3d)$
    $- (-2a + 3b + c - d)$

20. $(5a - 3b - c + 4d)$
    $- (-3a + 5b + c - 2d)$

21. $(x^4 - 3x^2 + 4x)$
    $- (3x^3 + x^2 - 5x + 3)$

22. $(2x^4 - 5x^2 + 7x)$
    $- (5x^3 + 2x^2 - 3x + 5)$

23. $(7x\sqrt{y} - 4y\sqrt{x} + 7.5)$
    $- (-2x\sqrt{y} - y\sqrt{x} - 1.6)$

24. $\left(10x\sqrt{y} - 4y\sqrt{x} + \frac{4}{3}\right) - \left(-3x\sqrt{y} + y\sqrt{x} - \frac{1}{3}\right)$

Simplify.

25. ▦ $(0.565p^2q - 2.167pq^2 + 16.02pq - 17.1)$
    $+ (-1.612p^2q - 0.312pq^2 - 7.141pq - 87.044)$

26. ▦ $(5003.2xy^{-2} + 3102.4\sqrt{xy} - 5280)$
    $- (2143.6xy^{-2} + 6153.8xy - 4141\sqrt{xy} + 4979.12)$

1.4 MULTIPLICATION OF POLYNOMIALS

Multiplication of polynomials is based on the distributive laws. To multiply two polynomials, we multiply each term of one by every term of the other and then add the results.

Example 1 Multiply $4x^4y - 7x^2y + 3y$ by $2y - 3x^2y$.

$$
\begin{array}{r}
4x^4y - 7x^2y + 3y \\
2y - 3x^2y \\
\hline
8x^4y^2 - 14x^2y^2 + 6y^2 \\
-12x^6y^2 + 21x^4y^2 - 9x^2y^2 \\
\hline
-12x^6y^2 + 29x^4y^2 - 23x^2y^2 + 6y^2
\end{array}
$$

Multiplying by $2y$
Multiplying by $-3x^2y$
Adding

Products of Two Binomials

We can find a product of two binomials mentally. We multiply the first terms, then the outside terms, then the inside terms, then the last terms (this procedure is sometimes abbreviated FOIL), and then add the results. This also works for expressions that are not polynomials.

Examples Multiply.

                      F   O   I   L

2. $(3xy + 2x)(x^2 + 2xy^2) = 3x^3y + 6x^2y^3 + 2x^3 + 4x^2y^2$

3. $(x + \sqrt{2})(y - \sqrt{2}) = xy - \sqrt{2}x + \sqrt{2}y - 2$
4. $(2x - \sqrt{3})(y + 2) = 2xy + 4x - \sqrt{3}y - 2\sqrt{3}$
5. $(2x + 3y)(x - 4y) = 2x^2 - 5xy - 12y^2$

Squares of Binomials

Multiplying a binomial by itself, we obtain the following:
$$(a + b)^2 = a^2 + 2ab + b^2$$
and
$$(a - b)^2 = a^2 - 2ab + b^2.$$

Thus, to square a binomial we square the first term, add twice the product of the terms, and then add the square of the second term.

Examples Multiply.

6. $(2x + 9y^2)^2 = (2x)^2 + 2(2x)(9y^2) + (9y^2)^2$
$= 4x^2 + 36xy^2 + 81y^4$

7. $(3x^2 - 5xy^2)^2 = (3x^2)^2 - 2(3x^2)(5xy^2) + (5xy^2)^2$
$= 9x^4 - 30x^3y^2 + 25x^2y^4$

Products of Sums and Differences

The following multiplication gives a result to be remembered:
$$(a + b)(a - b) = (a + b) \cdot a - (a + b) \cdot b$$
$$= a^2 + ab - ab - b^2$$
$$= a^2 - b^2;$$
in other words,
$$(a + b)(a - b) = a^2 - b^2.$$

The product of a sum and a difference of the same two terms is the difference of their squares. Thus to find such a product, we square the first term, square the second term, and write a minus sign between the results.

Examples Multiply.

8. $(y + 5)(y - 5) = y^2 - 25$
9. $(3x + 2)(3x - 2) = (3x)^2 - 2^2$
$= 9x^2 - 4$
10. $(2xy^2 + 3x)(2xy^2 - 3x) = (2xy^2)^2 - (3x)^2$
$= 4x^2y^4 - 9x^2$
11. $(5x + \sqrt{2})(5x - \sqrt{2}) = (5x)^2 - (\sqrt{2})^2$
$= 25x^2 - 2$

12. $(5y + 4 + 3x)(5y + 4 - 3x) = (5y + 4)^2 - (3x)^2$
$= 25y^2 + 40y + 16 - 9x^2$

13. $(3xy^2 + 4y)(-3xy^2 + 4y) = -(3xy^2)^2 + (4y)^2$
$= 16y^2 - 9x^2y^4$

In the following exercises you should do as much of the calculating mentally as you can. If possible, write only the answer. Work for speed with accuracy.

EXERCISE SET 1.4

Multiply.

1. $2x^2 + 4x + 16$ and $3x - 4$
2. $3y^2 - 3y + 9$ and $2y + 3$
3. $4a^2b - 2ab + 3b^2$ and $ab - 2b + 1$
4. $2x^2 + y^2 - 2xy$ and $x^2 - 2y^2 - xy$
5. $(a - b)(a^2 + ab + b^2)$
6. $(t + 1)(t^2 - t + 1)$
7. $(2x + 3y)(2x + y)$
8. $(2a - 3b)(2a - b)$
9. $\left(4x^2 - \frac{1}{2}y\right)\left(3x + \frac{1}{4}y\right)$
10. $\left(2y^3 + \frac{1}{5}x\right)\left(3y - \frac{1}{4}x\right)$
11. $(\sqrt{2}x^2 - y^2)(\sqrt{2}x - 2y)$
12. $(\sqrt{3}y^2 - 2)(\sqrt{3}y - x)$
13. $(2x + 3y)^2$
14. $(5x + 2y)^2$
15. $(2x^2 - 3y)^2$
16. $(4x^2 - 5y)^2$
17. $(2x^3 + 3y^2)^2$
18. $(5x^3 + 2y^2)^2$
19. $\left(\frac{1}{2}x^2 - \frac{3}{5}y\right)^2$
20. $\left(\frac{1}{4}x^2 - \frac{2}{3}y\right)^2$
21. $(0.5x + 0.7y^2)^2$
22. $(0.3x + 0.8y^2)^2$
23. $(3x - 2y)(3x + 2y)$
24. $(3x + 5y)(3x - 5y)$
25. $(x^2 + yz)(x^2 - yz)$
26. $(2x^2 + 5xy)(2x^2 - 5xy)$
27. $(3x^2 - \sqrt{2})(3x^2 + \sqrt{2})$
28. $(5x^2 - \sqrt{3})(5x^2 + \sqrt{3})$
29. $(2x + 3y + 4)(2x + 3y - 4)$
30. $(5x + 2y + 3)(5x + 2y - 3)$
31. $(x^2 + 3y + y^2)(x^2 + 3y - y^2)$
32. $(2x^2 + y + y^2)(2x^2 + y - y^2)$
33. $(x + 1)(x - 1)(x^2 + 1)$
34. $(y - 2)(y + 2)(y^2 + 4)$
35. $(2x + y)(2x - y)(4x^2 + y^2)$
36. $(5x + y)(5x - y)(25x^2 + y^2)$
37. ▦ $(0.051x + 0.04y)^2$
38. ▦ $(1.032x - 2.512y)^2$
39. ▦ $(37.86x + 1.42)(65.03x - 27.4)$
40. ▦ $(3.601x - 17.5)(47.105x + 31.23)$

☆
41. Find a formula for $(a + b)^3$.
42. Find a formula for $(a - b)^3$.
43. Multiply: $P(1 + i)^3$.
44. Multiply: $(3t^2 - 4)^3$.

1.5 FACTORING

To factor a polynomial we do the reverse of multiplying; that is, we find an expression that is a product. Facility in factoring is an important algebraic skill.

Terms with Common Factors

When an expression is to be factored, we should always look first for a possible factor that is common to all terms. We then "factor it out" using the distributive laws.

Examples Factor.

1. $4x^2 + 8 = 4(x^2 + 2)$
2. $12x^2y - 20x^3y = 4x^2y(3 - 5x)$
3. $7x\sqrt{y} + 14x^2\sqrt{y} - 21\sqrt{y} = 7\sqrt{y}(x + 2x^2 - 3)$
4. $(a - b)(x + 5) + (a - b)(x - y^2) = (a - b)[(x + 5) + (x - y^2)]$
$= (a - b)(2x + 5 - y^2)$

In some polynomials pairs of terms have a common factor that can be removed, as in the following examples. This process is called *factoring by grouping*, and uses the distributive laws repeatedly.

Examples Factor.

5. $y^2 + 3y + 4y + 12 = y(y + 3) + 4(y + 3)$
$= (y + 4)(y + 3)$
6. $ax^2 + ay + bx^2 + by = a(x^2 + y) + b(x^2 + y)$
$= (a + b)(x^2 + y)$

Differences of Squares

Recall that $(a + b)(a - b) = a^2 - b^2$. We can use this equation in reverse to factor an expression that is the difference of two squares.

Examples Factor.

7. $x^2 - 9 = (x + 3)(x - 3)$
8. $y^2 - 2 = (y + \sqrt{2})(y - \sqrt{2})$
9. $9a^2 - 16x^4 = (3a + 4x^2)(3a - 4x^2)$
10. $9y^4 - 9x^4 = 9(y^4 - x^4)$
$= 9(y^2 + x^2)(y^2 - x^2)$
$= 9(y^2 + x^2)(y + x)(y - x)$

Trinomial Squares

You should recall that $a^2 + 2ab + b^2 = (a + b)^2$ and $a^2 - 2ab + b^2 = (a - b)^2$. We can use these equations to factor trinomials that are squares. To factor a trinomial, you should check to see if it is a square. For this to be the case, two of the terms must be squares and the other term must be twice the product of the square roots.

Examples Factor.

11. $x^2 - 10x + 25 = (x - 5)^2$
12. $16y^2 + 56y + 49 = (4y + 7)^2$
13. $-4y^2 - 144y^8 + 48y^5 = -4y^2(1 + 36y^6 - 12y^3)$ We first
    $ = -4y^2(1 - 12y^3 + 36y^6)$ removed the
    $ = -4y^2(1 - 6y^3)^2$ common factor.

Trinomials That Are Not Squares

Certain trinomials that are not squares can be factored into two binomials. To do this, we can use the equation $acx^2 + (ad + bc)x + bd = (ax + b) \times (cx + d)$.

Example 14 Factor: $2x^2 + 11x + 12$.

We look for binomials $(ax + b)$ and $(cx + d)$. The product of the first terms must be $2x^2$. The product of the last terms must be 12. When we multiply the inside terms, then the outside terms, and add, we must get $11x$. By trial, we thus determine the factors to be as follows: $(x + 4)(2x + 3)$.

Sums or Differences of Cubes

We can use the following equations to factor a sum or a difference of two cubes:

$$a^3 + b^3 = (a + b)(a^2 - ab + b^2),$$
$$a^3 - b^3 = (a - b)(a^2 + ab + b^2).$$

Examples Factor.

15. $27x^3 + y^3 = (3x + y)(9x^2 - 3xy + y^2)$
16. $64y^6 - 125x^6 = (4y^2 - 5x^2)(16y^4 + 20y^2x^2 + 25x^4)$

EXERCISE SET 1.5

Factor.

1. $18a^2b - 15ab^2$
2. $4x^2y + 12xy^2$
3. $a(b - 2) + c(b - 2)$
4. $a(x^2 - 3) - 2(x^2 - 3)$
5. $x^2 + 3x + 6x + 18$
6. $3x^3 + x^2 - 18x - 6$
7. $9x^2 - 25$
8. $16x^2 - 9$
9. $4xy^4 - 4xz^2$
10. $5xy^4 - 5xz^4$
11. $y^2 - 6y + 9$
12. $x^2 + 8x + 16$
13. $1 - 8x + 16x^2$
14. $1 + 10x + 25x^2$
15. $4x^2 - 5$
16. $16x^2 - 7$
17. $x^2y^2 - 14xy + 49$
18. $x^2y^2 - 16xy + 64$
19. $4ax^2 + 20ax - 56a$
20. $21x^2y + 2xy - 8y$
21. $a^2 + 2ab + b^2 - c^2$
22. $x^2 - 2xy + y^2 - z^2$
23. $x^2 + 2xy + y^2 - a^2 - 2ab - b^2$
24. $r^2 + 2rs + s^2 - t^2 + 2tv - v^2$
25. $5y^4 - 80x^4$
26. $6y^4 - 96x^4$
27. $x^3 + 8$
28. $y^3 - 64$
29. $3x^3 - \dfrac{3}{8}$
30. $5y^3 + \dfrac{5}{27}$
31. $x^3 + 0.001$
32. $y^3 - 0.125$
33. $3z^3 - 24$
34. $4t^3 + 108$
35. $a^6 - t^6$
36. $64m^6 + y^6$
37. $16a^7b + 54ab^7$
38. $24a^2x^4 - 375a^8x$
39. ▦ $x^2 - 17.6$
40. ▦ $x^2 - 8.03$
41. ▦ $37x^2 - 14.5y^2$
42. ▦ $1.96x^2 - 17.4y^2$
    (*Hint:* First remove the common factor 37.)
    (*Hint:* First remove the common factor 1.96.)

Factor.

43. $(x + h)^3 - x^3$
44. $(x + 0.01)^2 - x^2$

1.6 SOLVING EQUATIONS AND INEQUALITIES

A *solution* of an equation is any number that makes the equation true when that number is substituted for the variable. The set of all solutions of an equation is called its *solution set*. When we find all the solutions of an equation (find its solution set), we say we have *solved* it.

1.6 SOLVING EQUATIONS AND INEQUALITIES

The Addition and Multiplication Principles

Two simple principles* allow us to solve many equations. The first of these is as follows.

The addition principle. **For any real numbers a, b, and c, if an equation $a = b$ is true, then $a + c = b + c$ is true.**

The second principle is similar to the first.

The multiplication principle. **For any real numbers a, b, and c, if an equation $a = b$ is true, then $ac = bc$ is true.**

These principles may be used together as needed to solve an equation.

Example 1 Solve: $3x + 4 = 15$.

$3x + 4 + (-4) = 15 + (-4)$ Here we used the addition principle, adding -4.

$3x = 11$ Here we simplified.

$\frac{1}{3} \cdot 3x = \frac{1}{3} \cdot 11$ Here we used the multiplication principle, multiplying by $\frac{1}{3}$.

$x = \frac{11}{3}$ Here we simplified.

Check:

$$\begin{array}{c|c} 3x + 4 = 15 & \\ \hline 3 \cdot \frac{11}{3} + 4 & 15 \\ 11 + 4 & \\ 15 & \end{array}$$

The only solution is the number $\frac{11}{3}$.

The solution set can be indicated $\{\frac{11}{3}\}$, but for brevity we shall most often omit the braces.

In Example 1 we used the addition and multiplication principles. Thus we know that if $3x + 4 = 15$ is true, then $x = \frac{11}{3}$ is true. This does

---

*These "principles" are actually very easy theorems. Suppose $a = b$ is true. Then a and b are the same number. If we add c to this number, the result is $a + c$. It is also $b + c$.

not guarantee that if $x = \frac{11}{3}$ is true, then $3x + 4 = 15$ is true. Thus it is important to check by substituting $\frac{11}{3}$ in the original equation.

Example 2 Solve: $3(7 - 2x) = 14 - 8(x - 1)$.

$21 - 6x = 14 - 8x + 8$	Here we multiplied to remove parentheses.
$21 - 6x = 22 - 8x$	Here we simplified.
$8x - 6x = 22 - 21$	Here we added -21 and also $8x$.
$2x = 1$	Here we combined like terms and simplified.
$x = \frac{1}{2}$	Here we multiplied by $\frac{1}{2}$.

Check:

$$\begin{array}{c|c} 3(7 - 2x) = 14 - 8(x - 1) \\ \hline 3\left(7 - 2 \cdot \frac{1}{2}\right) & 14 - 8\left(\frac{1}{2} - 1\right) \\ 3(7 - 1) & 14 - 8\left(-\frac{1}{2}\right) \\ 3 \cdot 6 & 14 + 4 \\ 18 & 18 \end{array}$$

The number $\frac{1}{2}$ checks. Thus the solution is $\frac{1}{2}$.

Example 3 Solve: $x + 3 = x$.

$-x + x + 3 = -x + x$	Here we added $-x$.
$3 = 0$	Here we simplified.

In Example 3 we get a false equation. No replacement for x will make the equation true. Thus there are no solutions. The solution set is the *empty set*, denoted $\varnothing$.

A third principle for solving equations is called the *principle of zero products*. It is as follows:

The principle of zero products. **For any numbers *a* and *b*, if $ab = 0$, then $a = 0$ or $b = 0$; and if $a = 0$ or $b = 0$, then $ab = 0$.**

To solve an equation using this principle, there must be 0 on one side of the equation and a product on the other. The solutions are then obtained by setting the factors equal to 0 separately.

Example 4 Solve: $x^2 + x - 12 = 0$.

$\quad\quad (x + 4)(x - 3) = 0 \quad\quad$ Here we factored.
$\quad\quad x + 4 = 0 \quad$ or $\quad x - 3 = 0 \quad\quad$ Here we used the principle of zero products.
$\quad\quad x = -4 \quad$ or $\quad\quad x = 3$

The solutions are -4 and 3. The solution set is $\{-4, 3\}$.

Solving Inequalities

Principles for solving inequalities are similar to those for solving equations. We can add the same number to both sides of an inequality. We can also multiply both sides by the same nonzero number, but if that number is negative, we must reverse the inequality sign.

Example 5 Solve: $3x < 11 - 2x$.

$\quad\quad 3x + 2x < 11 \quad\quad$ Here we added $2x$.
$\quad\quad 5x < 11 \quad\quad$ Here we combined similar terms.
$\quad\quad x < \dfrac{11}{5} \quad\quad$ Here we multiplied by $\frac{1}{5}$.

Any number less than $\frac{11}{5}$ is a solution. The solution set is the set of all x such that $x < \frac{11}{5}$. We abbreviate this using *set-builder* notation as follows:

$$\left\{x \,\middle|\, x < \frac{11}{5}\right\}.$$

For brevity we often write merely $x < \frac{11}{5}$.

Example 6 Solve: $16 - 7y \geq 10y - 4$.

$\quad\quad -16 + 16 - 7y \geq -16 + 10y - 4 \quad\quad$ We added -16.
$\quad\quad\quad -7y \geq 10y - 20 \quad\quad$ We have simplified.
$\quad\quad\quad -17y \geq -20 \quad\quad$ We added $-10y$ and simplified.
$\quad\quad\quad y \leq \dfrac{20}{17} \quad\quad$ We multiplied by $-\frac{1}{17}$ and reversed the inequality sign.

Any number less than or equal to $\frac{20}{17}$ is a solution. The solution set is $\{y \mid y \leq \frac{20}{17}\}$.

EXERCISE SET 1.6

Solve.

1. $4x + 12 = 60$
2. $2y - 11 = 37$
3. $4 + \frac{1}{2}x = 1$
4. $4.1 - 0.2y = 1.3$
5. $y + 1 = 2y - 7$
6. $5 - 4x = x - 13$
7. $5x - 2 + 3x = 2x + 6 - 4x$
8. $5x - 17 - 2x = 6x - 1 - x$
9. $1.9x - 7.8 + 5.3x = 3.0 + 1.8x$
10. $2.2y - 5 + 4.5y = 1.7y - 20$
11. $7(3x + 6) = 11 - (x + 2)$
12. $4(5y + 3) = 3(2y - 5)$
13. $2x - (5 + 7x) = 4 - [x - (2x + 3)]$
14. $y - (9y - 8) = [5 - 2y - 3(2y - 3)] + 29$
15. $(2x - 3)(3x - 2) = 0$
16. $(5x - 2)(2x + 3) = 0$
17. $x(x - 1)(x + 2) = 0$
18. $x(x + 2)(x - 3) = 0$
19. $3x^2 + x - 2 = 0$
20. $10x^2 - 16x + 6 = 0$
21. $(x - 1)(x + 1) = 5(x - 1)$
22. $6(y - 3) = (y - 3)(y - 2)$
23. $x[4(x - 2) - 5(x - 1)] = 2$
24. $14\left[(x - 4) - \frac{1}{14}(x + 2)\right] = (x + 2)(x - 4)$
25. $(3x^2 - 7x - 20)(2x - 5) = 0$
26. $(8x + 11)(12x^2 - 5x - 2) = 0$
27. $16x^3 = x$
28. $9x^3 = x$
29. $x + 6 < 5x - 6$
30. $3 - x < 4x + 7$
31. $3x - 3 + 2x \geq 1 - 7x - 9$
32. $5y - 5 + y \leq 2 - 6y - 8$
33. $17 - 5y \leq 8y - 5$
34. $12x - 6 < 10x + 4$
35. $-\frac{3}{4}x \geq -\frac{5}{8} + \frac{2}{3}x$
36. $-\frac{5}{6}x \leq \frac{3}{4} + \frac{8}{3}x$
37. $4x(x - 2) < 2(2x - 1)(x - 3)$
38. $(x + 1)(x + 2) > x(x + 1)$

Write set-builder notation for each set.

39. The set of all x such that $x > 2.5$
40. The set of all y such that $y \leq -7$
41. The set of all t such that $t^2 = 5$
42. The set of all m such that $m^3 + 3 = m^2 - 2$

☆

Solve.

43. ▦ $2.905x - 3.214 + 6.789x = 3.012 + 1.805x$
44. ▦ $(13.14x + 17.152)(15.15 - 7.616x) = 0$
45. ▦ $1.52(6.51x + 7.3) < 11.2 - (7.2x + 13.52)$
46. ▦ $4.73(5.16y + 3.62) \geq 3.005(2.75y - 6.31)$

★

Solve.

47. $(x + 1)^3 = (x - 1)^3 + 26$
48. $(x - 2)^3 = x^3 - 2$

1.7 FRACTIONAL EXPRESSIONS

Expressions like these are called *fractional expressions*:

$$\frac{8}{5}, \quad \frac{5}{x-2}, \quad \frac{3x^2 + 5\sqrt{x} - 2}{x^2 - y^2}.$$

Fractional expressions represent division. Certain substitutions are not sensible in such expressions. Since division by 0 is not defined, any number that makes a denominator zero is not a sensible replacement.

Multiplication and Division

To multiply two fractional expressions, we multiply their numerators and also their denominators. By Theorem 4, when we divide, we multiply by the reciprocal of the divisor. The latter can be obtained by inverting the divisor.

Examples

1. Multiply.

$$\frac{x+3}{y-4} \cdot \frac{x^3}{y+5} = \frac{(x+3)x^3}{(y-4)(y+5)}$$

$$= \frac{x^4 + 3x^3}{y^2 + y - 20}$$

2. Divide.

$$\frac{x-2}{x+1} \div \frac{x+5}{x-3} = \frac{x-2}{x+1} \cdot \frac{x-3}{x+5} \qquad \text{Inverting}$$

$$= \frac{(x-2)(x-3)}{(x+1)(x+5)} \qquad \text{Multiplying}$$

$$= \frac{x^2 - 5x + 6}{x^2 + 6x + 5}$$

Simplifying

The basis of simplifying fractional expressions rests on the fact that certain expressions are equal to 1 for all sensible replacements. Such expressions have the same numerator and denominator.* Here

---

*By Theorem 4, $a \div a = a/a = a(1/a)$, and since a and $1/a$ are reciprocals, their product is 1.

are some examples:

$$\frac{x-2}{x-2}, \quad \frac{3x^2-4x+2}{3x^2-4x+2}, \quad \frac{4x-5}{4x-5}.$$

When we multiply by such an expression we obtain an equivalent expression. This means that the new expression will name the same number as the first for all sensible replacements. The set of sensible replacements may not be the same for the two expressions.

Example 3 Multiply.

$$\frac{y+4}{y-3} \cdot \frac{y-2}{y-2} = \frac{(y+4)(y-2)}{(y-3)(y-2)}$$

$$= \frac{y^2+2y-8}{y^2-5y+6}$$

The expressions $(y+4)/(y-3)$ and $(y^2+2y-8)/(y^2-5y+6)$ are equivalent. That is, they will name the same number for all sensible replacements. The only nonsensible replacement in the first expression is 3. For the second expression the nonsensible replacements are 2 and 3.

Simplification can be accomplished by reversing the procedure in the above example; that is, we try to factor the fractional expression in such a way that one of the factors is equal to 1 and then "remove" that factor.

Examples Simplify.

4. $\dfrac{15x^3y^2}{20x^2y} = \dfrac{(5x^2y)3xy}{(5x^2y)4}$ Factoring numerator and denominator

   $= \dfrac{5x^2y}{5x^2y} \cdot \dfrac{3xy}{4}$ Factoring the expression

   $= \dfrac{3xy}{4}$ "Removing" a factor of 1

Note that in the original expression neither x nor y can be 0. In the simplified expression, however, all replacements are sensible.

5. $\dfrac{x^2-1}{2x^2-x-1} = \dfrac{(x-1)(x+1)}{(2x+1)(x-1)} = \dfrac{x-1}{x-1} \cdot \dfrac{x+1}{2x+1}$

   $= \dfrac{x+1}{2x+1}$

In the original expression the sensible replacements are all real numbers except 1 and $-\frac{1}{2}$. In the simplified expression the sensible replacements are all real numbers except $-\frac{1}{2}$.

Canceling

Canceling is a shortcut for part of the procedure in the preceding examples. Canceling gives rise to a great many errors, particularly when it is not well understood. It should therefore be used with caution.

Example 6 Simplify.

$$\frac{x^3 - 27}{x^2 + x - 12} = \frac{\cancel{(x - 3)}(x^2 + 3x + 9)}{(x + 4)\cancel{(x - 3)}}$$

$$= \frac{x^2 + 3x + 9}{x + 4}$$

Note that the canceling is a shortcut for "removing" a factor of 1. When fractional expressions are multiplied or divided, they should be simplified when possible.

Examples

7. Multiply and simplify.

$$\frac{x + 2}{x - 2} \cdot \frac{x^2 - 4}{x^2 + x - 2} = \frac{(x + 2)(x^2 - 4)}{(x - 2)(x^2 + x - 2)} \quad \text{Multiplying}$$

$$= \frac{(x + 2)(x + 2)(x - 2)}{(x - 2)(x + 2)(x - 1)} \quad \text{Factoring}$$

$$= \frac{x + 2}{x - 1} \quad \text{"Removing" a factor of 1}$$

8. Divide and simplify.

$$\frac{a^2 - 1}{a + 1} \div \frac{a^2 - 2a + 1}{a + 1} = \frac{a^2 - 1}{a + 1} \cdot \frac{a + 1}{a^2 - 2a + 1}$$

$$= \frac{(a + 1)(a - 1)(a + 1)}{(a + 1)(a - 1)(a - 1)}$$

$$= \frac{a + 1}{a - 1}$$

Addition and Subtraction

When fractional expressions have the same denominator, we can add or subtract them by adding or subtracting the numerators and retaining the common denominator. If denominators are not the same, we then find equivalent expressions with the same denominator and add. If one denominator is the additive inverse of another, we can find a common denominator by multiplying by $-1/-1$.

Examples Add.

9. $\dfrac{3x^2 + 4x - 8}{x^2 + y^2} + \dfrac{-5x^2 + 5x + 7}{x^2 + y^2} = \dfrac{-2x^2 + 9x - 1}{x^2 + y^2}$

In the following example, one denominator is the additive inverse of the other.

10. $\dfrac{3x^2 + 4}{x - y} + \dfrac{5x^2 - 11}{y - x} = \dfrac{3x^2 + 4}{x - y} + \dfrac{-1}{-1} \cdot \dfrac{5x^2 - 11}{y - x}$

$= \dfrac{3x^2 + 4}{x - y} + \dfrac{-1(5x^2 - 11)}{-1(y - x)}$

$= \dfrac{3x^2 + 4}{x - y} + \dfrac{11 - 5x^2}{x - y}$ $\quad -1(y - x) = -y + x$
$\qquad\qquad\qquad\qquad\qquad\qquad\qquad\qquad = x - y$

$= \dfrac{-2x^2 + 15}{x - y}$

When denominators are different, but not additive inverses of each other, we find a common denominator by factoring the denominators. Then we multiply by 1 to get the common denominator in each expression.

Example 11 Add: $\dfrac{1}{2x} + \dfrac{5x}{x^2 - 1} + \dfrac{3}{x + 1}$.

We first find the *Least Common Multiple* (LCM) of the denominators. The denominators, when factored, are

$$2x, \quad (x + 1)(x - 1), \quad x + 1.$$

The LCM is $2x(x + 1)(x - 1)$.

Now we multiply each fractional expression by 1 appropriately:

$$\frac{1}{2x} \cdot \frac{(x+1)(x-1)}{(x+1)(x-1)} + \frac{5x}{(x+1)(x-1)} \cdot \frac{2x}{2x} + \frac{3}{(x+1)} \cdot \frac{2x(x-1)}{2x(x-1)}$$

$$= \frac{(x+1)(x-1) + 10x^2 + 6x(x-1)}{2x(x+1)(x-1)}$$

$$= \frac{17x^2 - 6x - 1}{2x(x+1)(x-1)} \quad \text{or} \quad \frac{17x^2 - 6x - 1}{2x^3 - 2x}.$$

Example 12 Subtract: $\dfrac{x}{x^2 + 5x + 6} - \dfrac{2}{x^2 + 3x + 2}$.

$$\frac{x}{x^2 + 5x + 6} - \frac{2}{x^2 + 3x + 2}$$

$$= \frac{x}{(x+2)(x+3)} - \frac{2}{(x+1)(x+2)} \quad \text{The LCM is } (x+1)(x+2)(x+3).$$

$$= \frac{x}{(x+2)(x+3)} \cdot \frac{x+1}{x+1} - \frac{2}{(x+1)(x+2)} \cdot \frac{x+3}{x+3}$$

$$= \frac{x(x+1) - [2(x+3)]}{(x+1)(x+2)(x+3)} \quad \text{We use the colored brackets here to emphasize that it is important to subtract the } \textit{entire} \text{ numerator, not just part of it.}$$

$$= \frac{x^2 + x - [2x + 6]}{(x+1)(x+2)(x+3)}$$

$$= \frac{x^2 + x - 2x - 6}{(x+1)(x+2)(x+3)}$$

$$= \frac{x^2 - x - 6}{(x+1)(x+2)(x+3)}$$

$$= \frac{(x-3)(x+2)}{(x+1)(x+2)(x+3)}$$

$$= \frac{x-3}{(x+1)(x+3)} \quad \text{Always simplify at the end if possible.}$$

Complex Fractional Expressions

A complex fractional expression is one that has a fractional expression in its numerator or denominator or both. To simplify such an expression, we can combine as necessary in numerator and denominator in order to obtain a single fractional expression for each. Then we divide the numerator by the denominator.

Example 13 Simplify.

$$\frac{x + \frac{1}{5}}{x - \frac{1}{3}} = \frac{x \cdot \frac{5}{5} + \frac{1}{5}}{x \cdot \frac{3}{3} - \frac{1}{3}}$$

$$= \frac{\frac{5x + 1}{5}}{\frac{3x - 1}{3}} \quad \text{Now we have a single fractional expression for both numerator and denominator.}$$

$$= \frac{5x + 1}{5} \cdot \frac{3}{3x - 1} \quad \text{Here we divided by multiplying by the reciprocal of the denominator.}$$

$$= \frac{15x + 3}{15x - 5}$$

Example 14 Simplify.

$$\frac{a^{-2} - b^{-2}}{a^{-1} + b^{-1}} = \frac{\frac{1}{a^2} - \frac{1}{b^2}}{\frac{1}{a} + \frac{1}{b}} = \frac{\frac{b^2}{b^2} \cdot \frac{1}{a^2} - \frac{a^2}{a^2} \cdot \frac{1}{b^2}}{\frac{b}{b} \cdot \frac{1}{a} + \frac{a}{a} \cdot \frac{1}{b}} = \frac{\frac{b^2 - a^2}{a^2 b^2}}{\frac{b + a}{ab}}$$

$$= \frac{b^2 - a^2}{a^2 b^2} \cdot \frac{ab}{b + a}$$

$$= \frac{(b - a)(b + a)ab}{(b + a)a^2 b^2}$$

$$= \frac{ab(b + a)}{ab(b + a)} \cdot \frac{b - a}{ab} = \frac{b - a}{ab}$$

EXERCISE SET 1.7

Determine the sensible replacements.

1. $\dfrac{3x - 2}{x(x - 1)}$
2. $\dfrac{(x^2 - 4)(x + 1)}{(x + 2)(x^2 - 1)}$
3. $\dfrac{7y^2 - 2y + 4}{x(x^2 - x - 6)}$

Simplify. Then determine the replacements that are sensible in the simplified expression.

4. $\dfrac{25x^2 y^2}{10xy^2}$
5. $\dfrac{x^2 - 4}{x^2 + 5x + 6}$
6. $\dfrac{x^2 - 3x + 2}{x^2 + x - 2}$

Multiply or divide, and simplify.

7. $\dfrac{x^2 - y^2}{(x - y)^2} \cdot \dfrac{1}{x + y}$

8. $\dfrac{r - s}{r + s} \cdot \dfrac{r^2 - s^2}{(r - s)^2}$

9. $\dfrac{x^2 - 2x - 35}{2x^3 - 3x^2} \cdot \dfrac{4x^3 - 9x}{7x - 49}$

10. $\dfrac{x^2 + 2x - 35}{3x^3 - 2x^2} \cdot \dfrac{9x^3 - 4x}{7x + 49}$

11. $\dfrac{a^2 - a - 6}{a^2 - 7a + 12} \cdot \dfrac{a^2 - 2a - 8}{a^2 - 3a - 10}$

12. $\dfrac{a^2 - a - 12}{a^2 - 6a + 8} \cdot \dfrac{a^2 + a - 6}{a^2 - 2a - 24}$

13. $\dfrac{m^2 - n^2}{r + s} \div \dfrac{m - n}{r + s}$

14. $\dfrac{a^2 - b^2}{x - y} \div \dfrac{a + b}{x - y}$

15. $\dfrac{3x + 12}{2x - 8} \div \dfrac{(x + 4)^2}{(x - 4)^2}$

16. $\dfrac{a^2 - a - 2}{a^2 - a - 6} \div \dfrac{a^2 - 2a}{2a + a^2}$

17. $\dfrac{x^2 - y^2}{x^3 - y^3} \cdot \dfrac{x^2 + xy + y^2}{x^2 + 2xy + y^2}$

18. $\dfrac{c^3 + 8}{c^2 - 4} \div \dfrac{c^2 - 2c + 4}{c^2 - 4c + 4}$

19. $\dfrac{(x - y)^2 - z^2}{(x + y)^2 - z^2} \div \dfrac{x - y + z}{x + y - z}$

20. $\dfrac{(a + b)^2 - 9}{(a - b)^2 - 9} \cdot \dfrac{a - b - 3}{a + b + 3}$

Add or subtract, and simplify.

21. $\dfrac{3}{2a + 3} + \dfrac{2a}{2a + 3}$

22. $\dfrac{a - 3b}{a + b} + \dfrac{a + 5b}{a + b}$

23. $\dfrac{y}{y - 1} + \dfrac{2}{1 - y}$

24. $\dfrac{a}{a - b} + \dfrac{b}{b - a}$

25. $\dfrac{x}{2x - 3y} - \dfrac{y}{3y - 2x}$

26. $\dfrac{3a}{3a - 2b} - \dfrac{2a}{2b - 3a}$

27. $\dfrac{3}{x + 2} + \dfrac{2}{x^2 - 4}$

28. $\dfrac{5}{a - 3} - \dfrac{2}{a^2 - 9}$

29. $\dfrac{y}{y^2 - y - 20} + \dfrac{2}{y + 4}$

30. $\dfrac{6}{y^2 + 6y + 9} - \dfrac{5}{y + 3}$

31. $\dfrac{3}{x + y} + \dfrac{x - 5y}{x^2 - y^2}$

32. $\dfrac{a^2 + 1}{a^2 - 1} - \dfrac{a - 1}{a + 1}$

33. $\dfrac{9x + 2}{3x^2 - 2x - 8} + \dfrac{7}{3x^2 + x - 4}$

34. $\dfrac{3y}{y^2 - 7y + 10} - \dfrac{2y}{y^2 - 8y + 15}$

35. $\dfrac{5a}{a - b} + \dfrac{ab}{a^2 - b^2} + \dfrac{4b}{a + b}$

36. $\dfrac{6a}{a - b} - \dfrac{3b}{b - a} + \dfrac{5}{a^2 - b^2}$

37. $\dfrac{7}{x + 2} - \dfrac{x + 8}{4 - x^2} + \dfrac{3x - 2}{4 - 4x + x^2}$

38. $\dfrac{6}{x + 3} - \dfrac{x + 4}{9 - x^2} + \dfrac{2x - 3}{9 - 6x + x^2}$

39. $\dfrac{1}{x + 1} - \dfrac{x}{x - 2} + \dfrac{x^2 + 2}{x^2 - x - 2}$

40. $\dfrac{x - 1}{x - 2} - \dfrac{x + 1}{x + 2} + \dfrac{x - 6}{x^2 - 4}$

Simplify.

41. $\dfrac{\dfrac{x^2 - y^2}{xy}}{\dfrac{x - y}{y}}$

42. $\dfrac{\dfrac{a - b}{b}}{\dfrac{a^2 - b^2}{ab}}$

43. $\dfrac{a - a^{-1}}{a + a^{-1}}$

44. $\dfrac{a - \dfrac{a}{b}}{b - \dfrac{b}{a}}$

45. $\dfrac{c + \dfrac{8}{c^2}}{1 + \dfrac{2}{c}}$ 46. $\dfrac{x^{-1} + y^{-1}}{x^{-3} + y^{-3}}$ 47. $\dfrac{x^2 + xy + y^2}{\dfrac{x^2}{y} - \dfrac{y^2}{x}}$ 48. $\dfrac{\dfrac{a^2}{b} + \dfrac{b^2}{a}}{a^2 - ab + b^2}$

49. $\dfrac{\dfrac{x}{y} - \dfrac{y}{x}}{\dfrac{1}{y} + \dfrac{1}{x}}$ 50. $\dfrac{\dfrac{a}{b} - \dfrac{b}{a}}{\dfrac{1}{a} - \dfrac{1}{b}}$ 51. $\dfrac{x^2 y^{-2} - y^2 x^{-2}}{xy^{-1} + yx^{-1}}$ 52. $\dfrac{a^2 b^{-2} - b^2 a^{-2}}{ab^{-1} - ba^{-1}}$

53. $\dfrac{\dfrac{a}{1-a} + \dfrac{1+a}{a}}{\dfrac{1-a}{a} + \dfrac{a}{1+a}}$ 54. $\dfrac{\dfrac{1-x}{x} + \dfrac{x}{1+x}}{\dfrac{1+x}{x} + \dfrac{x}{1-x}}$ 55. $\dfrac{\dfrac{1}{a^2} + \dfrac{2}{ab} + \dfrac{1}{b^2}}{\dfrac{1}{a^2} - \dfrac{1}{b^2}}$ 56. $\dfrac{\dfrac{1}{x^2} - \dfrac{1}{y^2}}{\dfrac{1}{x^2} - \dfrac{2}{xy} + \dfrac{1}{y^2}}$,

☆ ──────────────────────────

Simplify.

57. $\dfrac{(x+h)^2 - x^2}{h}$ 58. $\dfrac{\dfrac{1}{x+h} - \dfrac{1}{x}}{h}$

59. $\dfrac{(x+h)^3 - x^3}{h}$ 60. $\dfrac{\dfrac{1}{(x+h)^2} - \dfrac{1}{x^2}}{h}$

61. $\left[\dfrac{\dfrac{x+1}{x-1} + 1}{\dfrac{x+1}{x-1} - 1}\right]^5$ 62. $1 + \dfrac{1}{1 + \dfrac{1}{1 + \dfrac{1}{1 + \dfrac{1}{x}}}}$

1.8 RADICAL NOTATION AND ABSOLUTE VALUE

The symbol $\sqrt{a}$ denotes the nonnegative square root of the number a. The symbol $\sqrt[3]{a}$ denotes the cube root of a, and $\sqrt[4]{a}$ denotes the nonnegative fourth root of a. In general, $\sqrt[n]{a}$ denotes the nth root of a, that is, a number whose nth power is a. The symbol $\sqrt[n]{}$ is called a *radical* and the symbol under the radical is called the *radicand*. The number n (which is omitted when it is 2) is called the *index*.

Odd and Even Roots

Any positive real number has two square roots, one positive and one negative. The same is true for fourth roots, or roots of any even index. The positive root is called the *principal* root. When a radical such as $\sqrt{4}$ or $\sqrt[4]{18}$ is used, it is understood to represent the principal (nonnegative) root. To denote a nonpositive root we use $-\sqrt{2}$, $-\sqrt[4]{18}$, and so on.

DEFINITION

A radical expression $\sqrt[n]{a}$, where n is even, represents the principal (nonnegative) nth root of a. The nonpositive root is denoted $-\sqrt[n]{a}$.

Since negative numbers do not have even roots in the system of real numbers, any replacement that makes a radicand negative when the index is even is nonsensible.

Every real number, positive, negative, or zero, has just one cube root, and the same is true for any odd root. Thus $\sqrt[n]{a}$, where n is odd, represents the (only) nth root of a. In this case all numbers are sensible replacements in the radicand.

Example 1 What are the sensible replacements in $\sqrt{5x-4}$?

The sensible replacements are those that make the radicand nonnegative, that is, numbers x for which

$$5x - 4 \geq 0$$
$$5x \geq 4$$
$$x \geq \frac{4}{5}.$$

Thus the sensible replacements are any numbers x for which $x \geq \frac{4}{5}$. This is the set $\{x \mid x \geq \frac{4}{5}\}$.

Absolute Value

The absolute value of a nonnegative real number is that number itself. The absolute value of a negative number is its additive inverse. We make our definition as follows.

DEFINITION

For any real number x,
$$|x| = x \quad \text{if} \quad x \geq 0,$$
and
$$|x| = -x \quad \text{if} \quad x < 0.$$

It may help you to understand this definition and avoid confusion if we state it in more ordinary language: the absolute value of a number x is (a) the number x itself, if x is not negative; (b) the additive inverse of x, if x is negative. Confusion commonly arises when $-x$ is interpreted as meaning something negative, rather than "the additive inverse of x."

Certain properties of absolute value notation follow at once. For example, the absolute value of a product is the product of absolute values.

Examples

2. $|-3 \cdot 5| = |-15| = 15$ and $|-3| \cdot |5| = 3 \cdot 5 = 15$, so $|-3 \cdot 5| = |-3| \cdot |5|$.

Similarly,

3. $|-4 \cdot (-3)| = |12| = 12$ and $|-4| \cdot |-3| = 4 \cdot 3 = 12$, so $|-4 \cdot (-3)| = |-4| \cdot |-3|$.

The absolute value of a quotient is similarly the quotient of the absolute values.

Example 4

$$\left|\frac{25}{-5}\right| = |-5| = 5 \quad \text{and} \quad \frac{|25|}{|-5|} = \frac{25}{5} = 5, \quad \text{so} \quad \left|\frac{25}{-5}\right| = \frac{|25|}{|-5|}.$$

The absolute value of an even power can be simplified by leaving off the absolute value signs, because no even power can be negative.

The absolute value of the additive inverse of a number is the same as the absolute value of the number. Their distances from 0 are the same on the number line.

Examples

5. $|(-3)^2| = |9| = 9$ and $(-3)^2 = 9$, so $|(-3)^2| = (-3)^2$.
6. $|-3| = 3$ and $|3| = 3$, so $|-3| = |3|$.

THEOREM 9

For any real numbers a and b and any nonzero number c,
1. $|ab| = |a| \cdot |b|$,
2. $\left|\dfrac{a}{c}\right| = \dfrac{|a|}{|c|}$,
3. $|a^n| = a^n$ if n is an even integer,
4. $|-a| = |a|$.

Examples Simplify, leaving as little as possible inside absolute value signs.

7. $|3x| = |3| \cdot |x| = 3|x|$
8. $|x^2| = x^2$
9. $|x^2 y^3| = |x^2| \cdot |y^3| = x^2 |y^3| = x^2 y^2 |y|$
10. $\left| \dfrac{x^2}{y} \right| = \dfrac{|x^2|}{|y|} = \dfrac{x^2}{|y|}$
11. $|-3x| = 3|x|$

Properties of Radicals

Consider the expression $\sqrt{(-3)^2}$. This is equivalent to $\sqrt{9}$, which simplifies to 3. Similarly, $\sqrt{3^2} = 3$. This illustrates an important general principle for simplifying radicals of even index.

THEOREM 10

For any radicand R, $\sqrt{R^2} = |R|$. Similarly, for any even index n, $\sqrt[n]{R^n} = |R|$.

Examples Simplify.

12. $\sqrt{x^2} = |x|$
13. $\sqrt{x^2 - 2ax + a^2} = \sqrt{(x-a)^2} = |x - a|$
14. $\sqrt{x^2 y^6} = \sqrt{(xy^3)^2} = |xy^3| = y^2 |xy|$

If an index is odd, no absolute value signs are necessary.

THEOREM 11

For any radicand R and any odd index n, $\sqrt[n]{R^n} = R$.

A second property of radicals enables us to multiply. We illustrate it with an example.

Example 15 Compare $\sqrt{4} \cdot \sqrt{9}$ and $\sqrt{4 \cdot 9}$.

$\sqrt{4} \cdot \sqrt{9} = 2 \cdot 3 = 6$ and $\sqrt{4 \cdot 9} = \sqrt{36} = 6$, so $\sqrt{4} \cdot \sqrt{9} = \sqrt{4 \cdot 9}$.

THEOREM 12

For any nonnegative real numbers a and b and any index n,
$\sqrt[n]{a} \cdot \sqrt[n]{b} = \sqrt[n]{a \cdot b}$.

Examples Multiply.

16. $\sqrt{3} \cdot \sqrt{5} = \sqrt{3 \cdot 5} = \sqrt{15}$
17. $\sqrt{x+2} \cdot \sqrt{x-2} = \sqrt{(x+2)(x-2)} = \sqrt{x^2 - 4}$
18. $\sqrt[3]{4} \cdot \sqrt[3]{5} = \sqrt[3]{4 \cdot 5} = \sqrt[3]{20}$

The foregoing property also enables us to simplify radical expressions. The idea is to factor the radicand, obtaining factors that are perfect nth powers.

Examples Simplify.

19. $\sqrt{50} = \sqrt{25 \cdot 2} = \sqrt{25} \cdot \sqrt{2} = 5\sqrt{2}$
20. $\sqrt{5x^2} = \sqrt{x^2 \cdot 5} = \sqrt{x^2} \cdot \sqrt{5} = |x|\sqrt{5}$
21. $\sqrt[3]{32} = \sqrt[3]{8 \cdot 4} = \sqrt[3]{8} \cdot \sqrt[3]{4} = 2\sqrt[3]{4}$
22. $\sqrt{216x^5y^3} = \sqrt{36 \cdot 6 \cdot x^4 \cdot x \cdot y^2 \cdot y} = |6x^2y|\sqrt{6xy} = 6x^2|y|\sqrt{6xy}$
23. $\sqrt{2x^2 - 4x + 2} = \sqrt{2(x-1)^2} = |x-1|\sqrt{2}$

A third fundamental property of radicals is as follows.

THEOREM 13

For any nonnegative number a and any positive number b, and any index n,

$$\sqrt[n]{\frac{a}{b}} = \frac{\sqrt[n]{a}}{\sqrt[n]{b}}.$$

This property can be used to divide and to simplify radical expressions.

Examples Simplify.

24. $\sqrt{16x^3y^{-4}} = \sqrt{\frac{16x^3}{y^4}} = \frac{\sqrt{16x^3}}{\sqrt{y^4}} = \frac{\sqrt{16x^2 \cdot x}}{\sqrt{y^4}} = \frac{4|x|\sqrt{x}}{y^2}$

25. $\sqrt[3]{\frac{27y^5}{343x^3}} = \frac{\sqrt[3]{27y^5}}{\sqrt[3]{343x^3}} = \frac{\sqrt[3]{27y^3 \cdot y^2}}{\sqrt[3]{343x^3}} = \frac{3y\sqrt[3]{y^2}}{7x}$

Fractional expressions are often considered simpler when the denominator is free of radicals. Thus in simplifying, it is usual to remove the radicals in a denominator. This is called *rationalizing the denominator*, and we do it by multiplying by 1.

Examples Simplify.

26. $\sqrt{\dfrac{1}{2}} = \sqrt{\dfrac{1}{2} \cdot \dfrac{2}{2}} = \sqrt{\dfrac{2}{4}} = \dfrac{\sqrt{2}}{\sqrt{4}} = \dfrac{\sqrt{2}}{2}$

27. $\sqrt[3]{\dfrac{7}{9}} = \sqrt[3]{\dfrac{7}{9} \cdot \dfrac{3}{3}} = \sqrt[3]{\dfrac{21}{27}} = \dfrac{\sqrt[3]{21}}{\sqrt[3]{27}} = \dfrac{\sqrt[3]{21}}{3}$

Examples Divide and simplify.

28. $\dfrac{18\sqrt{72}}{6\sqrt{6}} = 3\sqrt{\dfrac{72}{6}} = 3\sqrt{12} = 3\sqrt{4 \cdot 3} = 3 \cdot 2\sqrt{3} = 6\sqrt{3}$

29. $\dfrac{\sqrt[3]{32}}{\sqrt[3]{2}} = \sqrt[3]{\dfrac{32}{2}} = \sqrt[3]{16} = \sqrt[3]{8 \cdot 2} = \sqrt[3]{8} \cdot \sqrt[3]{2} = 2\sqrt[3]{2}$

A fourth fundamental principle of radicals involves an exponent under the radical. We illustrate with an example.

Example 30 Compare $\sqrt{3^4}$ and $(\sqrt{3})^4$.

$$\sqrt{3^4} = \sqrt{81} = 9;$$
$$(\sqrt{3})^4 = \sqrt{3} \cdot \sqrt{3} \cdot \sqrt{3} \cdot \sqrt{3} = 3 \cdot 3 = 9$$

Thus $\sqrt{3^4} = (\sqrt{3})^4$.

The general principle is as follows.

THEOREM 14

For any nonnegative number a and any index n and any natural number m, $\sqrt[n]{a^m} = (\sqrt[n]{a})^m$.

This principle sometimes facilitates radical simplification.

Examples Simplify.

31. $\sqrt[3]{8^5} = (\sqrt[3]{8})^5 = 2^5 = 32$
32. $(\sqrt{2})^6 = \sqrt{2^6} = 2^3 = 8$

EXERCISE SET 1.8

What are the sensible replacements in each of the following?

1. $\sqrt{x-3}$
2. $\sqrt{2x-5}$
3. $\sqrt{3-4x}$
4. $\sqrt{x^2+3}$

Simplify.

5. $|9xy|$
6. $|y^4|$
7. $|3a^2b|$
8. $\left|\dfrac{4a}{b^2}\right|$
9. $\sqrt{(-11)^2}$
10. $\sqrt{(-1)^2}$
11. $\sqrt{16x^2}$
12. $\sqrt{36t^2}$
13. $\sqrt{(b+1)^2}$
14. $\sqrt{(2c-3)^2}$
15. $\sqrt[3]{-27x^3}$
16. $\sqrt[3]{-8y^3}$
17. $\sqrt{x^2-4x+4}$
18. $\sqrt{y^2+16y+64}$
19. $\sqrt[5]{32}$
20. $\sqrt[5]{-32}$
21. $\sqrt{180}$
22. $\sqrt{48}$
23. $\sqrt[3]{54}$
24. $\sqrt[3]{135}$
25. $\sqrt{128c^2d^{-4}}$
26. $\sqrt{162c^4d^{-6}}$
27. $\sqrt{3}\cdot\sqrt{6}$
28. $\sqrt{6}\cdot\sqrt{8}$

In the following exercises simplify, assuming that all letters represent positive numbers and that all radicands are positive. Thus no absolute value signs will be needed.

29. $\sqrt{2x^3y}\sqrt{12xy}$
30. $\sqrt{3y^4z}\sqrt{20z}$
31. $\sqrt[3]{3x^2y}\sqrt[3]{36x}$
32. $\sqrt[5]{8x^3y^4}\sqrt[5]{4x^4y}$
33. $\sqrt[3]{2(x+4)}\sqrt[3]{4(x+4)^4}$
34. $\sqrt[3]{4(x+1)^2}\sqrt[3]{18(x+1)^2}$
35. $\dfrac{\sqrt{21ab^2}}{\sqrt{3ab}}$
36. $\dfrac{\sqrt{128ab^2}}{\sqrt{16a^2b}}$
37. $\dfrac{\sqrt[3]{40m}}{\sqrt[3]{5m}}$
38. $\dfrac{\sqrt{40xy}}{\sqrt{8x}}$
39. $\dfrac{\sqrt[3]{3x^2}}{\sqrt[3]{24x^5}}$
40. $\dfrac{\sqrt[3]{40xy^3}}{\sqrt[3]{8x}}$
41. $\dfrac{\sqrt{a^2-b^2}}{\sqrt{a-b}}$
42. $\dfrac{\sqrt{x^3-y^3}}{\sqrt{x-y}}$
43. $\sqrt{\dfrac{9a^2}{8b}}$
44. $\sqrt{\dfrac{5b^2}{12a}}$
45. $\sqrt[3]{\dfrac{2x^2 2y^3}{25z^4}}$
46. $\sqrt[3]{\dfrac{24x^3y}{3y^4}}$
47. $\dfrac{(\sqrt[3]{32x^4y})^2}{(\sqrt[3]{xy})^2}$
48. $\dfrac{(\sqrt[3]{16x^2y})^2}{(\sqrt[3]{xy})^2}$
49. $\dfrac{3\sqrt{a^2b^2}\sqrt{4xy}}{2\sqrt{a^{-1}b^{-2}}\sqrt{9x^{-3}y^{-1}}}$
50. $\dfrac{4\sqrt{xy^2}\sqrt{9ab}}{3\sqrt{x^{-1}y^{-2}}\sqrt{16a^{-5}b^{-1}}}$

Simplify, assuming that all letters represent positive numbers.

51. $\sqrt{8.2x^3y}\sqrt{12.5xy}$
52. $\sqrt{0.012y^4z}\sqrt{1.305z}$
53. $\sqrt{\dfrac{6.03a^2}{17.13b}}$
54. $\sqrt{\dfrac{3.2b^2}{82.1a}}$

★

55. For what integer values of n is $|a^n|=|a|^n$ for all a?
56. For what values of x and y is $|x|=|y|$?
57. For what values of x and y is $|x+y|=|x|+|y|$?

1.9 FURTHER CALCULATIONS WITH RADICAL NOTATION

Various calculations with radicals can be carried out using the properties of radicals and the properties of numbers, such as the distributive property. The following examples illustrate.

Examples Simplify.

1. $3\sqrt{8} - 5\sqrt{2} = 3\sqrt{4 \cdot 2} - 5\sqrt{2} = 3 \cdot 2\sqrt{2} - 5\sqrt{2}$
$= (6 - 5)\sqrt{2}$ Here we used a distributive law.
$= \sqrt{2}$

2. $(4\sqrt{3} + \sqrt{2})(\sqrt{3} - 5\sqrt{2}) = 4(\sqrt{3})^2 - 20\sqrt{3}\sqrt{2} + \sqrt{2}\sqrt{3} - 5(\sqrt{2})^2$
$= 4 \cdot 3 - 20\sqrt{6} + \sqrt{6} - 5 \cdot 2$
$= 12 - 19\sqrt{6} - 10$
$= 2 - 19\sqrt{6}$

Rationalizing Denominators or Numerators

When a fractional symbol contains radicals, we ordinarily rationalize the denominator, but on occasion we prefer to rationalize the numerator. In either case, we can accomplish the rationalization by multiplying by 1, as in the following examples.

Examples Rationalize the denominator.

3. $\dfrac{\sqrt{7}}{\sqrt{5}} = \dfrac{\sqrt{7}}{\sqrt{5}} \cdot \dfrac{\sqrt{5}}{\sqrt{5}} = \dfrac{\sqrt{35}}{\sqrt{25}} = \dfrac{\sqrt{35}}{5}$

4. $\dfrac{\sqrt{2a}}{\sqrt{5b}} = \dfrac{\sqrt{2a}}{\sqrt{5b}} \cdot \dfrac{\sqrt{5b}}{\sqrt{5b}} = \dfrac{\sqrt{10ab}}{\sqrt{(5b)^2}}$
$= \dfrac{\sqrt{10ab}}{|5b|} = \dfrac{\sqrt{10ab}}{5b}$ The absolute value sign in the denominator is not necessary since $\sqrt{5b}$ would not exist at the outset unless $b \geq 0$.

5. $\dfrac{\sqrt[3]{54x^3}}{\sqrt[3]{4y^5}} = \sqrt[3]{\dfrac{54x^3}{4y^5} \cdot \dfrac{2y}{2y}} = \sqrt[3]{\dfrac{54x^3 \cdot 2y}{8y^6}}$
$= \dfrac{\sqrt[3]{27x^3} \cdot \sqrt[3]{4y}}{\sqrt[3]{8y^6}} = \dfrac{3x \cdot \sqrt[3]{4y}}{2y^2}$

When a numerator or denominator to be rationalized has two terms, we choose the symbol for 1 a little differently. The symbol for 1 will have two terms in its numerator and denominator. The following examples illustrate.

Examples Rationalize the denominator. Assume all letters represent positive numbers.

6. $\dfrac{1}{\sqrt{2} + \sqrt{3}} = \dfrac{1}{\sqrt{2} + \sqrt{3}} \cdot \dfrac{\sqrt{2} - \sqrt{3}}{\sqrt{2} - \sqrt{3}}$

$= \dfrac{\sqrt{2} - \sqrt{3}}{(\sqrt{2} + \sqrt{3})(\sqrt{2} - \sqrt{3})}$

$= \dfrac{\sqrt{2} - \sqrt{3}}{(\sqrt{2})^2 - (\sqrt{3})^2} = \dfrac{\sqrt{2} - \sqrt{3}}{2 - 3}$

$= \dfrac{\sqrt{2} - \sqrt{3}}{-1} = \sqrt{3} - \sqrt{2}$

7. $\dfrac{\sqrt{x} + \sqrt{y}}{\sqrt{x} - \sqrt{y}} = \dfrac{\sqrt{x} + \sqrt{y}}{\sqrt{x} - \sqrt{y}} \cdot \dfrac{\sqrt{x} + \sqrt{y}}{\sqrt{x} + \sqrt{y}}$

$= \dfrac{(\sqrt{x} + \sqrt{y})^2}{(\sqrt{x})^2 - (\sqrt{y})^2}$

$= \dfrac{x + 2\sqrt{x}\sqrt{y} + y}{x - y}$

Examples Rationalize the numerator. Assume all letters represent positive numbers.

8. $\dfrac{1 - \sqrt{2}}{5} = \dfrac{1 - \sqrt{2}}{5} \cdot \dfrac{1 + \sqrt{2}}{1 + \sqrt{2}} = \dfrac{(1 - \sqrt{2})(1 + \sqrt{2})}{5(1 + \sqrt{2})}$

$= \dfrac{1 - 2}{5(1 + \sqrt{2})} = \dfrac{-1}{5 + 5\sqrt{2}}$

9. $\dfrac{\sqrt{x + h} - \sqrt{x}}{h} = \dfrac{\sqrt{x + h} - \sqrt{x}}{h} \cdot \dfrac{\sqrt{x + h} + \sqrt{x}}{\sqrt{x + h} + \sqrt{x}}$

$= \dfrac{(x + h) - x}{h(\sqrt{x + h} + \sqrt{x})} = \dfrac{h}{h(\sqrt{x + h} + \sqrt{x})}$

$= \dfrac{1}{\sqrt{x + h} + \sqrt{x}}$

EXERCISE SET 1.9

In this exercise set assume that all letters represent positive numbers and that all radicands are positive. Thus, absolute value signs will not be necessary.

Simplify.

1. $8\sqrt{2} - 6\sqrt{20} - 5\sqrt{8}$
2. $\sqrt{12} - \sqrt{27} + \sqrt{75}$
3. $2\sqrt[3]{8x^2} + 5\sqrt[3]{27x^2} - 3\sqrt[3]{x^3}$
4. $5a\sqrt{(a+b)^3} - 2ab\sqrt{a+b} - 3b\sqrt{(a+b)^3}$
5. $3\sqrt{3y^2} - \dfrac{y\sqrt{48}}{\sqrt{2}} + \sqrt{\dfrac{12}{4y^{-2}}}$
6. $\sqrt[3]{x^5} - \dfrac{2\sqrt[3]{x}}{\sqrt[3]{x^{-1}}} + \sqrt[3]{\dfrac{8}{x^{-5}}}$
7. $(\sqrt{3} - \sqrt{2})(\sqrt{3} + \sqrt{2})$
8. $(\sqrt{8} + 2\sqrt{5})(\sqrt{8} - 2\sqrt{5})$
9. $(\sqrt{t} - x)^2$
10. $\left(\sqrt{a} + \dfrac{1}{\sqrt{a}}\right)^2$
11. $5\sqrt{7} + \dfrac{35}{\sqrt{7}}$
12. $(\sqrt{a^2 b} + 3\sqrt{y})(2a\sqrt{b} - \sqrt{y})$
13. $(\sqrt{x+3} - \sqrt{3})(\sqrt{x+3} + \sqrt{3})$
14. $(\sqrt{x+h} - \sqrt{x})(\sqrt{x+h} + \sqrt{x})$

Rationalize the denominator.

15. $\dfrac{6}{3 + \sqrt{5}}$
16. $\dfrac{2}{\sqrt{3} - 1}$
17. $\sqrt[3]{\dfrac{16}{9}}$
18. $\dfrac{\sqrt[3]{3}}{\sqrt[3]{6}}$
19. $\dfrac{4\sqrt{x} - 3\sqrt{xy}}{2\sqrt{x} + 5\sqrt{y}}$
20. $\dfrac{5\sqrt{x} + 2\sqrt{xy}}{3\sqrt{x} - 2\sqrt{y}}$

Rationalize the numerator.

21. $\dfrac{\sqrt{2} + \sqrt{5a}}{6}$
22. $\dfrac{\sqrt{3} + \sqrt{5y}}{4}$
23. $\dfrac{\sqrt{x+1} + 1}{\sqrt{x+1} - 1}$
24. $\dfrac{\sqrt{x+4} - 2}{\sqrt{x+4} + 2}$
25. $\dfrac{\sqrt{a+3} - \sqrt{3}}{3}$
26. $\dfrac{\sqrt{a+h} - \sqrt{a}}{h}$

Simplify.

27. $\sqrt{1 + x^2} + \dfrac{1}{\sqrt{1 + x^2}}$
28. $\sqrt{1 - x^2} - \dfrac{x^2}{2\sqrt{1 - x^2}}$
29. Show that $\sqrt{a + b} = \sqrt{a} + \sqrt{b}$ is *false* for positive real numbers a and b by finding two positive numbers a and b for which $\sqrt{a + b} \neq \sqrt{a} + \sqrt{b}$.
30. Show that $(\sqrt{5} + \sqrt{24})^2 = (\sqrt{2} + \sqrt{3})^2$.

★

Prove the following.

31. For any positive real numbers a and b, there exist positive numbers c and d for which $\sqrt{a + b} = \sqrt{c} + \sqrt{d}$. Under what conditions does $a = c$ and $b = d$?

1.10 RATIONAL EXPONENTS

We are motivated to define fractional exponents so that the same rules, or laws, hold for them as for integer exponents. For example, if the laws of exponents are to hold, we would have

$$a^{1/2} \cdot a^{1/2} = a^{1/2+1/2} = a^1 = a.$$

Thus we are led to define $a^{1/2}$ to mean $\sqrt{a}$. Similarly, $a^{1/n}$ would mean $\sqrt[n]{a}$. Again, if the usual laws of exponents are to hold, we would have

$$(a^{1/n})^m = (a^m)^{1/n} = a^{m/n}.$$

Thus we are led to define $a^{m/n}$ to mean $(\sqrt[n]{a})^m$ or $\sqrt[n]{a^m}$.

DEFINITION

An expression $a^{m/n}$, where a is positive and m and n are natural numbers, is defined to mean $(\sqrt[n]{a})^m$ or $\sqrt[n]{a^m}$. An expression $a^{-m/n}$ is defined to mean $1/a^{m/n}$.

Note that in this definition we require a to be positive. Thus in manipulations with fractional exponents we assume that all letters represent positive numbers and that all radicands are positive. No absolute value signs need be used.

Once the definition of rational exponents is made, the question arises whether the usual laws of exponents actually do hold. We shall not prove it, but the answer is that they do. Thus we can simplify or otherwise manipulate expressions containing rational exponents using those laws and the usual arithmetic of rational numbers.

Examples Convert to exponential notation and simplify.

1. $(\sqrt[4]{7xy})^5 = (7xy)^{5/4}$
2. $\sqrt[3]{8^4} = 8^{4/3} = (8^{1/3})^4 = 2^4 = 16$
3. $\sqrt[6]{x^3} = x^{3/6} = x^{1/2}$ (or $\sqrt{x}$)
4. $\sqrt[6]{4} = 4^{1/6} = (4^{1/2})^{1/3} = 2^{1/3}$ (or $\sqrt[3]{2}$)
5. $\sqrt[3]{\sqrt{7}} = \sqrt[3]{7^{1/2}} = (7^{1/2})^{1/3} = 7^{1/6}$ (or $\sqrt[6]{7}$)
6. $\sqrt{6}\sqrt[3]{6} = 6^{1/2} \cdot 6^{1/3} = 6^{1/2+1/3} = 6^{5/6}$ (or $\sqrt[6]{6^5}$)

Examples Simplify and then write radical notation.

7. $x^{5/6} \cdot x^{2/3} = x^{5/6+2/3} = x^{9/6} = x^{3/2} = \sqrt{x^3} = x\sqrt{x}$
8. $(a^5)^{-2/3} = a^{-10/3} = \dfrac{1}{a^{10/3}} = \dfrac{1}{\sqrt[3]{a^{10}}} = \dfrac{1}{a^3 \cdot \sqrt[3]{a}}$

9. $(5^{1/3} - 5^{-5/3}) \cdot 5^{1/3} = 5^{1/3} \cdot 5^{1/3} - 5^{-5/3} \cdot 5^{1/3}$

$\qquad = 5^{2/3} - 5^{-4/3}$

$\qquad = \sqrt[3]{5^2} - \dfrac{1}{\sqrt[3]{5^4}} = \sqrt[3]{25} - \dfrac{1}{5\sqrt[3]{5}}$

In certain expressions containing radicals or fractional exponents, it is possible to simplify in such a way that there is a single radical.

Examples Write an expression containing a single radical.

10. $a^{1/2}b^{-1/2}c^{5/6} = a^{3/6}b^{-3/6}c^{5/6} = (a^3 b^{-3} c^5)^{1/6} = \sqrt[6]{a^3 b^{-3} c^5}$

11. $\dfrac{a^{1/4}b^{3/8}}{a^{1/2}b^{1/8}} = a^{-1/4}b^{1/4} = (a^{-1}b)^{1/4} = \sqrt[4]{a^{-1}b}$

12. $\sqrt[4]{7}\sqrt{3} = 7^{1/4} \cdot 3^{1/2} = 7^{1/4} \cdot 3^{2/4} = (7 \cdot 3^2)^{1/4} = \sqrt[4]{63}$

13. $\dfrac{\sqrt[4]{(x+2)^3}\sqrt[5]{x+2}}{\sqrt{x+2}} = \dfrac{(x+2)^{3/4}(x+2)^{1/5}}{(x+2)^{1/2}}$

$\qquad = (x+2)^{3/4 + 1/5 - 1/2}$

$\qquad = (x+2)^{9/20} = \sqrt[20]{(x+2)^9}$

EXERCISE SET 1.10

Convert to radical notation and simplify.

1. $x^{3/4}$
2. $y^{2/5}$
3. $16^{3/4}$
4. $4^{7/2}$
5. $125^{-1/3}$
6. $32^{-4/5}$
7. $a^{5/4}b^{-3/4}$
8. $x^{2/5}y^{-1/5}$

Convert to exponential notation and simplify.

9. $\sqrt[3]{20^2}$
10. $\sqrt[5]{17^3}$
11. $(\sqrt[4]{13})^5$
12. $(\sqrt[5]{12})^4$
13. $\sqrt[3]{\sqrt{11}}$
14. $\sqrt[3]{\sqrt[4]{7}}$
15. $\sqrt{5}\sqrt[3]{5}$
16. $\sqrt[3]{2}\sqrt{2}$
17. $\sqrt[5]{32^2}$
18. $\sqrt[3]{64^{-2}}$
19. $\sqrt[3]{8y^6}$
20. $\sqrt[5]{32c^{10}d^{15}}$
21. $\sqrt[3]{a^2+b^2}$
22. $\sqrt[4]{a^3-b^3}$
23. $\sqrt[3]{27a^3b^9}$
24. $\sqrt[4]{81x^8y^8}$
25. $\sqrt[6]{\dfrac{m^{12}n^{24}}{64}}$
26. $\sqrt[8]{\dfrac{m^{16}n^{24}}{2^8}}$

Simplify and then write radical notation, unless inappropriate.

27. $(2a^{3/2})(4a^{1/2})$
28. $(3a^{5/6})(8a^{2/3})$
29. $\left(\dfrac{x^6}{9b^{-4}}\right)^{-1/2}$
30. $\left(\dfrac{x^{2/3}}{4y^{-2}}\right)^{-1/2}$
31. $\dfrac{x^{2/3}y^{5/6}}{x^{-1/3}y^{1/2}}$
32. $\dfrac{a^{1/2}b^{5/8}}{a^{1/4}b^{3/8}}$

Write an expression containing a single radical and simplify.

33. $\sqrt[3]{6}\sqrt{2}$
34. $\sqrt{2}\sqrt[4]{8}$
35. $\sqrt[4]{xy}\sqrt[3]{x^2y}$
36. $\sqrt[3]{ab^2}\sqrt{ab}$
37. $\sqrt[3]{a^4}\sqrt{a^3}$
38. $\sqrt{a^3\sqrt[3]{a^2}}$
39. $\dfrac{\sqrt{(a+x)^3}\sqrt[3]{(a+x)^2}}{\sqrt[4]{a+x}}$
40. $\dfrac{\sqrt[4]{(x+y)^2}\sqrt[3]{(x+y)}}{\sqrt{(x+y)^3}}$

Simplify. (*Note:* Since $x^{1/4} = (x^{1/2})^{1/2}$ you can take a fourth root by taking a square root, and then the square root of the result. Round to three decimal places. Or, you can find decimal notation for the exponent, obtaining $x^{0.25}$, and use the power key x^y. Remember, answers can vary depending on the type and readout of your calculator.)

41. 🁢 $(\sqrt[4]{13})^5$
42. 🁢 $\sqrt[4]{17^3}$
43. 🁢 $12.3^{3/2}$
44. 🁢 $1.345^{5/3}$
45. 🁢 $105.6^{1.68}$
46. 🁢 $7.14^{-5.03}$

In a psychological study it was found that the length L, in feet, of the letters of a word printed on pavement which is most readable to the driver is given by

$$L = \frac{0.000169 d^{2.27}}{h},$$

where d is the distance of the car from the lettering and h is the height of the eye above the road. All units are in feet. According to the study and the formula, if a person h feet above the road is to read a message d feet away, that message will be the most readable if the length of the letters is L. Find L, given the values of d and h.

47. 🁢 $h = 4$ ft, $d = 180$ ft
48. 🁢 $h = 4$ ft, $d = 100$ ft
49. 🁢 $h = 4$ ft, $d = 200$ ft
50. 🁢 $h = 4$ ft, $d = 300$ ft

Simplify.

51. $\left(\sqrt{a^{\sqrt{a}}}\right)^{\sqrt{a}}$
52. $(2a^3 b^{5/4} c^{1/7})^4 \div (54 a^{-2} b^{2/3} c^{6/5})^{-1/3}$

1.11 HANDLING DIMENSION SYMBOLS

Speed

Speed is often measured by measuring a distance and a time and then dividing the distance by the time (this is *average* speed):

$$\text{Speed} = \frac{\text{Distance}}{\text{Time}}.$$

If a distance is measured in kilometers and the time required to travel that distance is measured in hours, the speed will be computed in *kilometers per hour* (km/h). For example, if a car travels 100 km in 2 h, the average speed is

$$\frac{100 \text{ km}}{2 \text{ h}}, \quad \text{or} \quad 50 \frac{\text{km}}{\text{h}}.$$

Dimension Symbols

The symbol 100 km/2 h makes it look as though we are dividing 100 km by 2 h. It may be argued that we cannot divide 100 km by 2 h (we can only divide 100 by 2). Nevertheless, it is convenient to treat dimension symbols such as *kilometers, hours, feet, seconds,* and *pounds* as if they were numerals or variables, for the reason that correct results can thus be obtained mechanically. Compare, for example,

$$\frac{100x}{2y} = \frac{100}{2} \cdot \frac{x}{y} = 50\frac{x}{y}$$

with

$$\frac{100 \text{ km}}{2 \text{ h}} = \frac{100}{2} \cdot \frac{\text{km}}{\text{h}} = 50 \frac{\text{km}}{\text{h}}.$$

The analogy holds in other situations, as shown in the following examples.

Example 1 Compare

$$3 \text{ ft} + 2 \text{ ft} = (3 + 2) \text{ ft} = 5 \text{ ft}$$

with

$$3x + 2x = (3 + 2)x = 5x.$$

This looks like a distributive law in use.

Example 2 Compare

$$4 \text{ m} \cdot 3 \text{ m} = 3 \cdot 4 \cdot \text{m} \cdot \text{m} = 12 \text{ m}^2 \text{ (sq m)}$$

with

$$4x \cdot 3x = 4 \cdot 3 \cdot x \cdot x = 12x^2.$$

Example 3 Compare

$$5 \text{ men} \cdot 8 \text{ hr} = 5 \cdot 8 \cdot \text{man-hr} = 40 \text{ man-hr}$$

with

$$5x \cdot 8y = 5 \cdot 8 \cdot x \cdot y = 40xy.$$

In each of the above examples, dimension symbols are treated as if they were variables or numerals, and as if a symbol such as "3 m" represents a product 3 times m. A symbol like km/h is treated as if it represents a division of km by h (*kilometers* by *hours*). Any two measures can be "multiplied" or "divided."

Changes of Unit

Changes of unit can be achieved by substitutions.

Example 4 Change to inches: 25 yd.

$$\begin{aligned} 25 \text{ yd} &= 25 \cdot 1 \text{ yd} \\ &= 25 \cdot 3 \text{ ft} \quad \text{Substituting 3 ft for 1 yd} \\ &= 25 \cdot 3 \cdot 1 \text{ ft} \\ &= 25 \cdot 3 \cdot 12 \text{ in.} \quad \text{Substituting 12 in. for 1 ft} \\ &= 900 \text{ in.} \end{aligned}$$

The notion of "multiplying by one" can also be used to change units.

Example 5 Change to yd: 7.2 in.

$$7.2 \text{ in.} = 7.2 \text{ in.} \cdot \underbrace{\frac{1 \text{ ft}}{12 \text{ in.}} \cdot \frac{1 \text{ yd}}{3 \text{ ft}}}_{\text{Both of these are equal to 1.}}$$

$$= \frac{7.2}{12 \cdot 3} \frac{\text{in.}}{\text{in.}} \frac{\text{ft}}{\text{ft}} \text{ yd}$$

$$= 0.2 \text{ yd}$$

Example 6 Change to $\frac{\text{m}}{\text{sec}}$: $60 \frac{\text{km}}{\text{h}}$.

$$60 \frac{\text{km}}{\text{h}} = 60 \frac{\text{km}}{\text{h}} \cdot \frac{1000 \text{ m}}{1 \text{ km}} \cdot \frac{1 \text{ h}}{60 \text{ min}} \cdot \frac{1 \text{ min}}{60 \text{ sec}}$$

$$= \frac{60 \cdot 1000}{60 \cdot 60} \cdot \frac{\text{km}}{\text{km}} \cdot \frac{\text{h}}{\text{h}} \cdot \frac{\text{min}}{\text{min}} \cdot \frac{\text{m}}{\text{sec}}$$

$$= 16.67 \frac{\text{m}}{\text{sec}} = 16.67 \text{ m/s.}*$$

EXERCISE SET 1.11

Perform these calculations and simplify if possible. Do not make any unit changes.

1. $36 \text{ ft} \cdot \frac{1 \text{ yd}}{3 \text{ ft}}$
2. $6 \text{ lb} \cdot \frac{16 \text{ oz}}{1 \text{ lb}}$
3. $6 \text{ kg} \cdot 8 \frac{\text{h}}{\text{kg}}$
4. $9 \frac{\text{km}}{\text{h}} \cdot 3 \text{ h}$
5. $3 \text{ cm} \cdot \frac{2 \text{ g}}{2 \text{ cm}}$
6. $\frac{9 \text{ km}}{3 \text{ days}} \cdot 6 \text{ days}$

---

*The standard abbreviation for meters per second is m/s and for kilometers per hour is km/h.

7. 6 m + 2 m

8. 10 tons + 6 tons

9. 5 ft³ + 7 ft³

10. 10 yd³ + 17 yd³

11. $\dfrac{3 \text{ kg}}{5 \text{ m}} \cdot \dfrac{7 \text{ kg}}{6 \text{ m}}$

12. 3 acres × 60 $\dfrac{1}{\text{acre}}$

13. $\dfrac{2000 \text{ lb} \cdot (6 \text{ mi/hr})^2}{100 \text{ ft}}$

14. $\dfrac{7 \text{ m} \cdot 8 \text{ kg/sec}}{4 \text{ sec}}$

15. $\dfrac{6 \text{ cm}^2 \cdot 5 \text{ cm/sec}}{2 \text{ sec}^2/\text{cm}^2 \cdot 2 \dfrac{1}{\text{kg}}}$

16. $\dfrac{320 \text{ lb} \cdot (5 \text{ ft/sec})^2}{2 \cdot 32 \dfrac{\text{ft}}{\text{sec}^2}}$

Perform the following changes of unit, using substitution or multiplying by one.

17. 72 in., change to ft

18. 17 hr, change to min

19. 2 days, change to sec

20. 360 sec, change to hr

21. 60 $\dfrac{\text{kg}}{\text{m}}$, change to g/cm

22. 44 $\dfrac{\text{ft}}{\text{sec}}$, change to mi/hr

23. 216 m², change to cm²

24. 60 $\dfrac{\text{lb}}{\text{ft}^3}$, change to ton/yd³

25. $\dfrac{\$36}{\text{day}}$, change to $\dfrac{¢}{\text{hr}}$

26. 1440 man-hr, change to man-days

27. 186,000 $\dfrac{\text{mi}}{\text{sec}}$ (speed of light), change to $\dfrac{\text{mi}}{\text{yr}}$ (Let 365 days = 1 yr.)

28. 1100 $\dfrac{\text{ft}}{\text{sec}}$ (speed of sound), change to $\dfrac{\text{mi}}{\text{yr}}$ (Let 365 days = 1 yr.)

Use Table 6 at the back of the book to do the following unit changes.

29. ▣ 89.2 $\dfrac{\text{ft}}{\text{sec}}$, change to m/min

30. ▣ 1013 yd³, change to m³

31. ▣ 640 mi², change to km²

32. ▣ 312.2 $\dfrac{\text{kg}}{\text{m}}$, change to lb/ft

CHAPTER 1 REVIEW

Given the numbers

$$-43.89, \ 12, \ -3, \ -\dfrac{1}{5}, \ \sqrt{7}, \ \sqrt[3]{10}, \ -1, \ -\dfrac{4}{3}, \ 7\dfrac{2}{3}, \ -19, \ 31, \ 0,$$

which are:

1. integers?
2. natural numbers?
3. rational numbers?
4. real numbers?
5. irrational numbers?
6. whole numbers?

Compute.

7. $15 + (-19)$
8. $-12 + (-4)$
9. $-2.5 + (-2.5)$
10. $22 - (-8)$
11. $\dfrac{18}{-3}$
12. $(-17)(-9)$
13. $-10(20)(-5)(-3)$
14. $-\dfrac{15}{16} \div \dfrac{3}{4}$
15. $\dfrac{5}{12} - \left(-\dfrac{7}{8}\right)$

What property is illustrated by each sentence?

16. $t + (-t) = 0$
17. $8(a + b) = 8a + 8b$
18. $-3(ab) = (-3a)b$
19. $tx = xt$

Convert to decimal notation.

20. 3.261×10^6
21. 4.1×10^{-4}

Convert to scientific notation.

22. 0.01432
23. $43{,}210$

Simplify.

24. $(7a^2b^4)(-2a^{-4}b^3)$
25. $\dfrac{54x^6y^{-4}z^2}{9x^{-3}y^2z^{-4}}$
26. $\sqrt[4]{81}$
27. $\sqrt[5]{-32}$
28. $\dfrac{b - a^{-1}}{a - b^{-1}}$
29. $\dfrac{\dfrac{x^2}{y} + \dfrac{y^2}{x}}{y^2 - xy + x^2}$
30. $(\sqrt{3} - \sqrt{7})(\sqrt{3} + \sqrt{7})$
31. $(5x^2 - \sqrt{2})^2$
32. $8\sqrt{5} + \dfrac{25}{\sqrt{5}}$
33. $(x + t)(x^2 - xt + t^2)$
34. $(5xy^4 - 7xy^2 + 4x^2 - 3) - (-3xy^4 + 2xy^2 - 2y + 4)$
35. $(2y^2 + 6x^2y)^2$

Factor.

36. $x^3 + 2x^2 - 3x - 6$
37. $12a^3 - 27ab^4$
38. $24x + 144 + x^2$
39. $9x^3 + 35x^2 - 4x$
40. $8x^3 - 1$
41. $27x^6 + 125y^6$

Write an expression containing a single radical.

42. $\sqrt{y^5} \cdot \sqrt[3]{y^2}$
43. $\dfrac{\sqrt{(a+b)^3} \cdot \sqrt[3]{(a+b)}}{\sqrt[6]{(a+b)^7}}$

44. Convert to radical notation: $b^{7/5}$.

45. Write rational exponents and simplify: $\sqrt[8]{\dfrac{m^{32}n^{16}}{3^8}}$.

Solve.

46. $y^2 - 3y = 18$

47. $3[x - 5(4 + 2x)] = 7x - 10(3x - 2)$

48. $(z^2 - 1) + z = 14 - z$

49. $(x - 2)(x + 3) + 4 = 0$

50. $14 - 4y < 22$

51. $(x - 5)(x + 5) \geq (x - 5)(x + 4)$

52. Divide and simplify.
$$\frac{3x^2 - 12}{x^2 + 4x + 4} \div \frac{x - 2}{x + 2}$$

53. Subtract and simplify.
$$\frac{x}{x^2 + 9x + 20} - \frac{4}{x^2 + 7x + 12}$$

54. Rationalize the denominator.
$$\frac{\sqrt{x} - \sqrt{y}}{\sqrt{x} + \sqrt{y}}$$

55. Change $10 \frac{\text{km}}{\text{h}}$ to $\frac{\text{m}}{\text{min}}$.

2
Equations

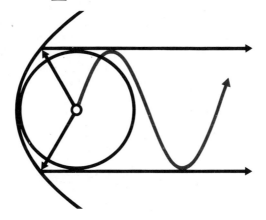

2.1 SOLVING EQUATIONS

Equivalent Equations

Equations that have the same solution set are called *equivalent equations*. The three equations in Example 1 are equivalent.

Example 1

$3x = 6$ $\qquad$ $3x + 5 = 11$ $\qquad$ $-12x = -24$
Solution set $\{2\}$. $\quad$ Solution set $\{2\}$. $\quad$ Solution set $\{2\}$.

The following pairs of equations are *not* equivalent.

Example 2

$3x = 4x$ $\qquad\qquad\qquad\qquad$ $\dfrac{3}{x} = \dfrac{4}{x}$

Solution set $\{0\}$. $\quad$ The solution set is $\emptyset$, the empty set (no solution, since division by 0 is not defined).

Example 3

$x = 1$ $\qquad\qquad$ $x^2 = x$
Solution set $\{1\}$. $\quad$ Solution set $\{0, 1\}$.

Equation-Solving Principles

In Chapter 1 we reviewed some equation-solving principles. Let us consider them in more detail.

The addition principle. **For any real numbers a, b, and c, if an equation $a = b$ is true, then $a + c = b + c$ is true.**

Let us look at a simple example illustrating this principle.

$$x + 5 = 9$$
$$x + 5 + (-5) = 9 + (-5) \qquad \text{Adding } -5 \text{ on both sides}$$
$$x = 4$$

In this example the original equation and the last equation had exactly the same solutions. They were equivalent. Whenever the steps in an argument are reversible, this will be the case. When we use the addition principle, this will be the case, unless an expression is added that has nonsensible replacements. To see this, note that we start with an equation $a = b$ and obtain $a + c = b + c$. By adding $-c$ we can always obtain $a = b$ again, so the steps are reversible.

The use of the addition principle produces an equation equivalent to the original, unless an expression is added that has nonsensible replacements. Checking by substituting is therefore not necessary except to detect errors in solving.

Now let us consider the multiplication principle.

The multiplication principle. **For any real numbers a, b, and c, if an equation $a = b$ is true, then $a \cdot c = b \cdot c$ is true.**

Does the multiplication principle yield equivalent equations? In other words, are the steps reversible? If the number c by which we multiply is not 0, then we can reverse the step by multiplying by $1/c$, but if we multiply by 0, then $1/c$ does not exist and the step is not reversible.

The use of the multiplication principle produces an equation equivalent to the original, only if we multiply by a nonzero number.

Now let us consider multiplying by expressions with variables, as in the following example:

$$\frac{3}{x} = \frac{4}{x}$$

$$\frac{3}{x} \cdot x^2 = \frac{4}{x} \cdot x^2 \quad \text{Here we multiplied by } x^2.$$

$$3x = 4x$$

$$0 = x$$

The number 0 is a solution of the last equation but is not a solution of the first. Hence we did not obtain equivalent equations. What can we do in practice?

When we use the multiplication principle and multiply by an expression with a variable, we may not obtain equivalent equations. We must check possible solutions by substituting in the original equation.

Example 4 Solve: $\dfrac{x-3}{x-7} = \dfrac{4}{x-7}$.

$$(x - 7) \cdot \frac{x-3}{x-7} = (x - 7) \cdot \frac{4}{x-7} \quad \begin{cases} \text{Caution! Here we multiplied} \\ \text{by an expression with a} \\ \text{variable. Thus we must check.} \end{cases}$$

$$x - 3 = 4$$

$$x = 7$$

The possible solution is 7. We check:

$$\frac{x-3}{x-7} = \frac{4}{x-7}$$

$\frac{7-3}{7-7}$	$\frac{4}{7-7}$
$\frac{4}{0}$	$\frac{4}{0}$

Division by 0 is undefined; 7 is not a solution. The equation has no solutions. The solution set is ∅.

Now let us consider the principle of zero products.

The principle of zero products. **For any real numbers a and b, if $ab = 0$, then $a = 0$ or $b = 0$; and if $a = 0$ or $b = 0$, then $ab = 0$.**

According to this principle, if we start with an equation $ab = 0$, we obtain $a = 0$ or $b = 0$. Also, if we start with $a = 0$ or $b = 0$, we can obtain $ab = 0$. Thus when we use this principle, that step is reversible and we have equivalent statements.

The use of the principle of zero products yields the solutions of the original equation. Checking by substituting is not necessary except to detect errors in solving.

Example 5 Solve: $2x^3 - x^2 = 3x$.

$2x^3 - x^2 - 3x = 0$ Addition principle
$x(2x^2 - x - 3) = 0$ Factoring
$x(2x - 3)(x + 1) = 0$
$x = 0$ or $2x - 3 = 0$ or $x + 1 = 0$ Principle of zero products
$x = 0$ or $x = \frac{3}{2}$ or $x = -1$

The solution set is $\left\{0, \frac{3}{2}, -1\right\}$.

In the following example we use the multiplication principle, multiplying by an expression with a variable, before we use the principle of zero products. Thus we must check possible solutions.

Example 6 Solve: $\dfrac{x^2}{x-3} = \dfrac{9}{x-3}$.

$(x-3) \cdot \dfrac{x^2}{x-3} = (x-3) \cdot \dfrac{9}{x-3}$ {Caution! Here we multiplied by an expression with a variable. Thus we must check.

$$x^2 = 9$$
$$x^2 - 9 = 0$$
$$(x+3)(x-3) = 0$$
$$x + 3 = 0 \quad \text{or} \quad x - 3 = 0$$
$$x = -3 \quad \text{or} \quad x = 3$$

The possible solutions are 3 and -3. We must check since we have multiplied by an expression with a variable.

For 3:
$$\dfrac{x^2}{x-3} = \dfrac{9}{x-3}$$
$$\dfrac{3^2}{3-3} \;\Big|\; \dfrac{9}{3-3}$$
$$\dfrac{9}{0} \;\Big|\; \dfrac{9}{0}$$

3 does not check.

For -3:
$$\dfrac{x^2}{x-3} = \dfrac{9}{x-3}$$
$$\dfrac{(-3)^2}{-3-3} \;\Big|\; \dfrac{9}{-3-3}$$
$$-\dfrac{9}{6} \;\Big|\; -\dfrac{9}{6}$$

-3 checks.

Thus, the solution set is $\{-3\}$.

In Examples 4 and 6 we solved *fractional equations*. These are equations that contain fractional expressions. A procedure for solving fractional equations involves multiplying by the LCM of all the denominators. It is called *clearing of fractions*. Let us solve another fractional equation.

Example 7 Solve: $\dfrac{14}{x+2} - \dfrac{1}{x-4} = 1$.

We multiply by the LCM of all the denominators: $(x+2)(x-4)$.

$$(x+2)(x-4) \cdot \dfrac{14}{x+2} - (x+2)(x-4) \cdot \dfrac{1}{x-4} = (x+2)(x-4) \cdot 1$$

$$14(x-4) - (x+2) = (x+2)(x-4)$$
$$14x - 56 - x - 2 = x^2 - 2x - 8$$
$$0 = x^2 - 15x + 50$$
$$0 = (x-10)(x-5)$$
$$x = 10 \quad \text{or} \quad x = 5$$

The possible solutions are 10 and 5. These check, so the solution set is $\{10, 5\}$.

EXERCISE SET 2.1

Determine which pairs of equations are equivalent.

1. $3x + 5 = 12$
   $3x = 7$

2. $x^2 = -7x$
   $x = -7$

3. $x = 3$
   $x^2 = 9$

4. $2y + 1 = -3$
   $8y + 4 = -12$

5. $\dfrac{(x - 2)(x + 8)}{(x - 2)} = x + 8$
   $x + 8 = x + 8$

6. $x^2 + x - 20 = 0$
   $x^2 - 25 = 0$

Solve.

7. $2x^2 - 6x = 0$
8. $9x^2 + 18x = 0$
9. $3y^3 - 5y^2 - 2y = 0$
10. $3t^3 - 5t^2 + 2t = 0$
11. $(2x - 3)(3x + 2)(x - 1) = 0$
12. $(y - 4)(4y + 12)(2y + 1) = 0$
13. $(2 - 4y)(y^2 + 3y) = 0$
14. $(y^2 - 9)(y^2 - 36) = 0$
15. $\dfrac{x + 2}{2} + \dfrac{3x + 1}{5} = \dfrac{x - 2}{4}$
16. $\dfrac{2x - 1}{3} - \dfrac{x - 2}{5} = \dfrac{x}{2}$
17. $\dfrac{1}{2} + \dfrac{2}{x} = \dfrac{1}{3} + \dfrac{3}{x}$
18. $\dfrac{1}{t} + \dfrac{1}{2t} + \dfrac{1}{3t} = 5$
19. $\dfrac{4}{x^2 - 1} - \dfrac{2}{x - 1} = \dfrac{3}{x + 1}$
20. $\dfrac{3y + 5}{y^2 + 5y} + \dfrac{y + 4}{y + 5} = \dfrac{y + 1}{y}$
21. $\dfrac{1}{2t} - \dfrac{2}{5t} = \dfrac{1}{10t} - 3$
22. $\dfrac{3}{m + 2} + \dfrac{2}{m - 2} = \dfrac{4m - 4}{m^2 - 4}$
23. $1 - \dfrac{3}{x} = \dfrac{40}{x^2}$
24. $1 - \dfrac{15}{y^2} = \dfrac{2}{y}$
25. $\dfrac{11 - t^2}{3t^2 - 5t + 2} = \dfrac{2t + 3}{3t - 2} - \dfrac{t - 3}{t - 1}$
26. $\dfrac{1}{3y^2 - 10y + 3} = \dfrac{6y}{9y^2 - 1} + \dfrac{2}{1 - 3y}$
27. ▦ $3.12x^2 - 6.715x = 0$
28. ▦ $9.25x^2 + 18.03x = 0$
29. ▦ $\dfrac{2.315}{y} - \dfrac{12.6}{17.4} = \dfrac{6.71}{7} + 0.763$
30. ▦ $\dfrac{6.034}{x} - 43.17 = \dfrac{0.793}{x} + 18.15$

☆

Solve.

31. $5x^3 + x^2 - 5x - 1 = 0$
    [Hint: $x^2(5x + 1) - 1(5x + 1) = 0.$]
32. $3x^3 + x^2 - 12x - 4 = 0$
33. $y^3 + 2y^2 - y - 2 = 0$
34. $t^3 + t^2 - 25t - 25 = 0$
35. Determine whether each equation is equivalent to the one that follows.

    a) $x^2 - x - 20 = x^2 - 25$
    b) $(x - 5)(x + 4) = (x - 5)(x + 5)$
    c) $\quad\quad x + 4 = x + 5$
    d) $\quad\quad\quad\quad 4 = 5$

36. Solve: $x^2 - x - 20 = x^2 - 25$.

Equations that are true for all sensible replacements of the variables are called *identities*. Determine which equations are identities.

37. $\dfrac{x^2 + 6x - 16}{x - 2} = x + 8$

38. $x + 4 = 4 + x$

39. $(x - 1)(x^2 + x + 1) = x^3 - 1$

40. $\dfrac{x^3 + 8}{x^2 - 4} = \dfrac{x^2 - 2x + 4}{x - 2}$

41. $(x + 7)^2 = x^2 + 49$

42. $\sqrt{x^2 - 16} = x - 4$

★ ──

43. Solve: $\dfrac{x + 3}{x + 2} - \dfrac{x + 4}{x + 3} = \dfrac{x + 5}{x + 4} - \dfrac{x + 6}{x + 5}$.

2.2 FORMULAS AND APPLIED PROBLEMS

Formulas

A formula is a recipe for doing a calculation. An example is $A = \pi rs + \pi r^2$, which gives the area A of a cone (see Fig. 1) in terms of the slant height s and radius of the base r.

Suppose we wanted to find the slant height s when the area A and radius r are known. Our knowledge of equations allows us to get s alone on one side, or as we say, "solve the formula for s."

Figure 1

Example 1 Solve $A = \pi rs + \pi r^2$, for s.

$$A - \pi r^2 = \pi rs$$

$$\dfrac{A - \pi r^2}{\pi r} = s \qquad \dfrac{A - \pi r^2}{\pi r} \text{ could also be expressed as } \dfrac{A}{\pi r} - r.$$

Example 2 Solve $\dfrac{1}{R} = \dfrac{1}{r_1} + \dfrac{1}{r_2}$, for R. (This is a formula from electricity.)

We first multiply by the LCM, which is Rr_1r_2:

$$Rr_1r_2 \cdot \dfrac{1}{R} = Rr_1r_2 \cdot \left(\dfrac{1}{r_1} + \dfrac{1}{r_2}\right)$$

$$Rr_1r_2 \cdot \dfrac{1}{R} = Rr_1r_2 \cdot \dfrac{1}{r_1} + Rr_1r_2 \cdot \dfrac{1}{r_2}$$

$$r_1r_2 = Rr_2 + Rr_1$$

$$r_1r_2 = (r_2 + r_1)R$$

$$\dfrac{r_1r_2}{r_2 + r_1} = R.$$

Applied Problems

By an *applied problem* we mean a problem in which mathematical techniques are used to answer some question. Problems like this may be posed orally. They may come about in the course of a conversation, or they can be hatched within the mind of one person. Thus to call them "word problems" or "story problems" is misleading.

There is no rule that will enable us to solve applied problems, because they are of many different kinds. We can, however, describe an overall, or general, strategy.

The idea is to translate the problem situation to mathematical language and then calculate to find a solution.

General Strategy for Solving Applied Problems

1. **Become familiar with the problem situation. If the problem is presented to you in written words, then of course this means to read carefully.**
    a) **Make a drawing, if it makes sense to do so. It is difficult to overemphasize the importance of this!**
    b) **Make a written list of the known facts and a list of what you wish to find out.**
2. **Translate the problem situation to mathematical language or symbolism. For most of the problems you will encounter in algebra this means to write one or more equations. You will, of course, also assign certain variables to represent unknown quantities.**
3. **Use your mathematical knowledge to find a possible solution. In algebra this usually means to solve an equation or system of equations.**
4. **Check to see if your possible solution actually fits the problem situation, and is thus really a solution of the problem.**

Problems stated in textbooks are of necessity somewhat contrived. Problems that you encounter in nonclassroom situations will almost invariably contain insufficient information to obtain a firm, or exact, answer. They also usually contain a good deal of extraneous, useless information. Here is an example that illustrates this point.

Example 3 There are 145 persons in the junior class at Notown College. The prom committee wishes to know how much they will have to charge each person who attends the spring prom, to be held May 19 in the College Gymnasium at 8:00 P.M.

The above problem is typical of problems encountered in real life. There is insufficient information to obtain an answer. To get an answer we need to know the following.

1. The cost of holding the prom. This would include such things as cost of music, refreshments, and hall rent.
2. The number of persons who will attend.

The committee will need to find out information about costs, and in the course of finding it, they may revise their formulation of the problem. For example, if a band costs too much, they will have to settle for using records. In any event, they may find that they can determine costs only approximately; in other words, they must estimate.

The number 145 may help in estimating the number who will attend, but at best this number will be an estimate. Thus, on the basis of the best information and/or estimates available, the committee will do its arithmetic.

Note that the date, time, and place, although they are interesting information, do not contribute to the solution of the problem. Thus this is extraneous information.

In the following examples we illustrate how a problem situation can be translated to mathematical language. In certain simple situations, the translation is easy because certain words translate directly to mathematical symbols. Note how the word "is" translates to an equals sign, the word "what" translates to a variable, and the word "of" translates to a multiplication sign.

Example 4 What percent of 84 is 11.76?

Translate: $x\% \cdot 84 = 11.76$

Solve: $x \cdot 0.01 \cdot 84 = 11.76$

$$x \cdot 0.84 = 11.76$$

$$x = \frac{11.76}{0.84} = 14$$

Check: $14\% \cdot 84 = 0.14 \cdot 84 = 11.76$

Example 5 14% of what is 11.76?

Translate: $14\% \cdot y = 11.76$

Solve: $0.14 \cdot y = 11.76$

$$y = \frac{11.76}{0.14} = 84$$

Check: $14\% \cdot 84 = 0.14 \cdot 84 = 11.76$

In the remainder of this section we consider applied problems of various types. Although there is no rule for solving applied problems because they can be so different, it does help somewhat to consider a few different types of problems. *The best way to learn to solve applied problems is to solve a lot of them.*

Compound Interest Problems

Example 6 An investment is made at 8%, compounded annually. It grows to $783 at the end of one year. How much was originally invested?

There is more than one way to translate the problem situation to mathematical language. The following is one method.

We first restate the situation as follows:

The invested amount *plus* the interest is $783.

Now the interest is 8% of the invested amount, so we have the following, which translates directly.

$$\underbrace{\text{Invested amount}}_{x} \text{ plus } \underbrace{8\%}_{8\%} \text{ of } \underbrace{\text{Invested amount}}_{x} \text{ is } \underbrace{\$783}_{=783}$$

Now we solve the equation:

$$x + 8\%x = 783$$
$$x + 0.08x = 783$$
$$(1 + 0.08)x = 783$$
$$1.08x = 783$$
$$x = \frac{783}{1.08} = 725.$$

The number 725 checks in the problem situation, so the answer is $725.

Let us now consider an investment over a period longer than one year. If we invest P dollars at an interest rate i, compounded annually, we will have an amount in the account at the end of a year that we will call A_1. Now $A_1 = P + Pi$, or

$$A_1 = P(1 + i),$$

or

$$A_1 = Pr,$$

where we have let $r = 1 + i$. Going into the second year we have Pr dollars. By the end of the second year we will have A_2 dollars, given by

$$A_2 = A_1 \cdot r.$$

But $A_1 = Pr$, so $A_2 = (Pr)r$, or

$$A_2 = Pr^2.$$

Similarly, the amount A_3 in the account at the end of three years is given by

$$A_3 = Pr^3,$$

and so on. In general, the following applies.

THEOREM 1

> If principal P is invested at an interest rate of i, compounded annually, in t years it will grow to an amount A given by
>
> $$A = P(1 + i)^t.$$

Example 7 Suppose $1000 is invested at 12%, compounded annually. What amount will be in the account at the end of ten years?

We use the equation $A = P(1 + i)^t$. We get

$$A = 1000(1 + 0.12)^{10} = 1000(1.12)^{10} \quad \text{Use the } x^y \text{ key.}$$
$$\approx 3105.85.$$

The answer is $3105.85.

Interest may be compounded more often than once a year. Suppose it is compounded four times a year, or *quarterly*. The formula derived above can be altered to apply. We consider one-fourth of a year to be an *interest period*. The rate of interest for such a period is then $i/4$. The number of periods will be four times the number of years. This is shown in Fig. 2.

Now suppose the number of interest periods per year is something other than 4, say n. Using the reasoning of the example above, we obtain a general formula.

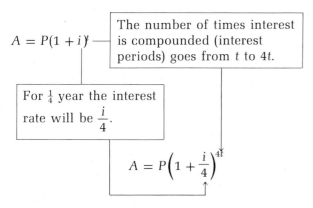

Figure 2

THEOREM 2

If principal P is invested at an interest rate i, compounded n times per year, in t years it will grow to an amount A given by

$$A = P\left(1 + \frac{i}{n}\right)^{nt}.$$

When problems involving compound interest are translated to mathematical language, the above formula is almost always used.

Example 8 Suppose $1000 is invested at 12%, compounded quarterly. How much will be in the account at the end of ten years?

In this case, $n = 4$ and $t = 10$. We substitute into the formula:

$$A = P\left(1 + \frac{i}{n}\right)^{nt} = 1000\left(1 + \frac{0.12}{4}\right)^{4 \cdot 10}$$

$$= 1000(1.03)^{40} \qquad \text{Use the } x^y \text{ key.}$$

$$\approx 3262.04.$$

The answer is $3262.04.

Area Problems

Example 9 The radius of a circular swimming pool is 10 ft. A sidewalk of uniform width is constructed around the outside and has an area of 44π ft². How wide is the sidewalk?

Figure 3

First make a drawing as in Fig. 3. Let x represent the width of the walk. Then, recalling that a formula for the area of a circle is $A = \pi r^2$, we have

Area of pool $= \pi \cdot 10^2 = 100\pi$;
Area of sidewalk plus pool $= \pi \cdot (10 + x)^2$
$= \pi(100 + 20x + x^2).$

Thus

(Area of sidewalk plus pool) − (Area of pool) = Area of sidewalk

$$\pi(100 + 20x + x^2) - 100\pi = 44\pi$$

$$100 + 20x + x^2 - 100 = 44 \qquad \text{Multiplying by } \frac{1}{\pi}$$

$$x^2 + 20x = 44$$

$$x^2 + 20x - 44 = 0$$

$$(x + 22)(x - 2) = 0$$

$$x = -22 \quad \text{or} \quad x = 2$$

We see that -22 ft is not a solution of the original problem since width has to be positive. The number 2 checks; that is, when the sidewalk is 2 ft wide, the area of the pool is 100π ft^2, and the area of the pool plus sidewalk is $\pi \cdot 12^2$, or 144π ft^2, so the area of the sidewalk is 44π ft^2.

Motion Problems

For problems that deal with distance, time, and speed, we almost always need to recall the definition of speed, or something equivalent to it.

Speed = distance/time or $r = d/t$, where r = speed, d = distance, and t = time.

If you memorize $r = d/t$, you can easily obtain either of the two equivalent equations $d = rt$ or $t = d/r$, as needed.

When translating these problems to mathematical language it is often helpful to look for some quantity that is constant in the problem. For example, two cars may travel for the same length of time, or two boats may travel the same distance. Such facts often provide the basis for setting up an equation.

Example 10 A boat travels 246 km downstream in the same time that it takes to travel 180 km upstream. The speed of the current in the stream is 5.5 km/h. Find the speed of the boat in still water.

We first make a drawing and lay out the known facts and any other information that is pertinent.

$\longrightarrow$ Downstream

$d_1 = 246$ $\quad t_1$ unknown $\quad r_1$ unknown, but equals speed of boat plus speed of current

$\longleftarrow$ Upstream

$d_2 = 180$ $\quad t_2$ unknown, but same as t_1 $\quad r_2$ unknown, but equals speed of boat minus speed of current

In this problem the times are the same, so we have an equation

$$t_1 = t_2.$$

Now, remembering that $t = d/r$, we substitute to obtain

$$\frac{d_1}{r_1} = \frac{d_2}{r_2}.$$

Now d_1, the distance downstream, is 246, and d_2 is 180.

The speed downstream is the speed of the boat *plus* that of the current. Let's call the boat's speed in still water r. Then r_1, the speed downstream, is $r + 5.5$, and r_2, the speed upstream, is $r - 5.5$. This gives us the equation

$$\frac{246}{r + 5.5} = \frac{180}{r - 5.5}.$$

Solving for r we get 35.5 km/h. This checks. Thus the speed of the boat in still water is 35.5 km/h.

In the following example, although the distance is the same in two cases, an additional piece of information is given that is the key to the translation.

Example 11 The speed of a boat in still water is 10 mph. It travels 24 miles upstream and 24 miles downstream in a total time of 5 hr. What is the speed of the current?

We first make a drawing and write out pertinent information.

```
─────────────────────────────────→ Downstream
 d₁ = 24 mi    t₁ unknown    r₁ unknown, but is
                              10 mph plus
                              speed of current

←─────────────────────────────────  Upstream
 d₂ = 24 mi    t₂ unknown, but   r₂ unknown, but is
               t₁ + t₂ = 5 hr    10 mph minus
                                 speed of current
```

This time the basis of our translation is the equation involving times:

$$t_1 + t_2 = 5.$$

This then leads to

$$\frac{d_1}{r_1} + \frac{d_2}{r_2} = 5,$$

and then

$$\frac{24}{10 + r} + \frac{24}{10 - r} = 5,$$

where r is the speed of the current. Solving for r we get $r = -2$ or $r = 2$. Since speed cannot be negative in this problem, -2 cannot be a solution. But 2 checks, so the speed of the current is 2 mph.

Work Problems

Suppose an employee can do a certain job in 4 hr. Then the employee can do $\frac{1}{4}$ of it in 1 hr and $\frac{3}{4}$ of it in 3 hr. The basic principle for translating work problems follows.

If a job can be done in time t, then $1/t$ of it can be done in 1 unit of time.

Example 12 Typist A can do a certain job in 3 hr. Typist B can do the same job in 5 hr. How long would it take both, working together, to do the same job?

a) Make a guess at a solution. Does 4 hr seem reasonable? Sometimes our intuition can fool us. Clearly the answer is less than 3 hr because one can do the job alone in 3 hr.

b) A can do the job in 3 hr. Thus in 1 hr $\frac{1}{3}$ of it can be done. Similarly, B can do $\frac{1}{5}$ of the job in 1 hr. Thus together they can do $\frac{1}{3} + \frac{1}{5}$ of it in 1 hr. Let t represent the amount of time it takes to do the job if they work together. Then together they can do $1/t$ of the job in 1 hr. Thus

$$\frac{1}{3} + \frac{1}{5} = \frac{1}{t}$$

$5t + 3t = 5 \cdot 3$ Multiplying by the LCM, $3 \cdot 5 \cdot t$

$8t = 15$

$t = \frac{15}{8}, \quad \text{or} \quad 1\frac{7}{8} \text{ hr.}$

And $t = 1\frac{7}{8}$ hr checks. Thus it takes them $1\frac{7}{8}$ hr to do the job if they work together.

Example 13 It takes Red 9 hr longer to build a wall than it does Mort. If they work together they can build the wall in 20 hr. How long would it take each, working alone, to build the wall?

Let $t =$ the amount of time it takes Mort working alone. Then $t + 9 =$ the amount of time it takes Red working alone. Then Mort can do $1/t$ of the work in 1 hr and Red can do $1/(t + 9)$ of it in 1 hr. Together they can do $1/t + 1/(t + 9)$ of the work in 1 hr. We also know that they can do $\frac{1}{20}$ of the work in 1 hr. Thus,

$$\frac{1}{t} + \frac{1}{t+9} = \frac{1}{20}.$$

Solving the equation, we get $t = -5$ or $t = 36$. Since negative time has no meaning in the problem, -5 is not a solution of the original problem. The number 36 checks in the original problem. Thus it would take Mort 36 hr and Red 45 hr.

EXERCISE SET 2.2

1. Solve $P = 2l + 2w$, for w.
2. Solve $F = ma$, for a.
3. Solve $E = IR$, for I.
4. Solve $F = \dfrac{km_1m_2}{d^2}$, for m_2.
5. Solve $\dfrac{P_1V_1}{T_1} = \dfrac{P_2V_2}{T_2}$, for T_1.
6. Solve $\dfrac{P_1V_1}{T_1} = \dfrac{P_2V_2}{T_2}$, for V_2.
7. Solve $S = \dfrac{H}{m(v_1 - v_2)}$, for v_1.
8. Solve $S = \dfrac{H}{m(v_1 - v_2)}$, for v_2.
9. Solve $\dfrac{1}{F} = \dfrac{1}{m} + \dfrac{1}{p}$, for p.
10. Solve $\dfrac{1}{F} = \dfrac{1}{m} + \dfrac{1}{p}$, for F.

Solve for x.

11. $(x + a)(x - b) = x^2 + 5$
12. $(c + d)x + (c - d)x = c^2$
13. $10(a + x) = 8(a - x)$
14. $4(a + b + x) + 3(a + b - x) = 8a$

Applied Problems

15. 79.2 is what percent of 180?
16. 6% of what number is 480?
17. What percent of 28 is 1.68?
18. What is 7% of 45.3?
19. A person gets an 11% raise, which is $1595. What was the old salary? the new salary?
20. A person gets a 12% raise, which is $2520. What was the old salary? the new salary?
21. An investment is made at 13%, compounded annually. It grows to $734.50 at the end of one year. How much was originally invested?
22. An investment is made at 14%, compounded annually. It grows to $912 at the end of one year. How much was originally invested?
23. In triangle ABC, angle B is five times as large as angle A. The measure of angle C is 2° less than that of angle A. Find the measures of the angles. (*Hint:* The sum of the angle measures is 180°.)
24. In triangle ABC, angle B is twice as large as angle A. Angle C measures 20° more than angle A. Find the measures of the angles.
25. The perimeter of a rectangle is 322 m. The length is 25 m more than the width. Find the dimensions.
26. The length of a rectangle is twice the width. The perimeter is 39 m. Find the dimensions.
27. A student's scores on three tests are 87%, 64%, and 78%. What must the student score on the fourth test so that the average will be 80%?
28. A student's scores on three tests are 74%, 55%, and 68%. What must the student score on the fourth test so that the average will be 70%?
29. An open box is made from a 10-cm by 20-cm piece of tin by cutting a square from each corner and folding up the edges. The area of the resulting base is 96 cm². What is the length of the sides of the squares?
30. The frame of a picture is 28 cm by 32 cm outside and is of uniform width. What is the width of the frame if 192 cm² of the picture shows?
31. After a 2% increase, the population of a city is 826,200. What was the former population?
32. After a 3% increase, the population of a city is 741,600. What was the former population?
33. An automobile was driven 144 mi. If it had gone 4 mph faster it could have made the trip in $\frac{1}{2}$ hr less time. What was the speed of the car?
34. An automobile was driven 280 mi. If it had gone 5 mph faster it could have made the trip in 1 hr less time. What was the speed of the car?

35. A boat goes 50 km downstream in the same time that it takes to go 30 km upstream. The speed of the stream is 3 km/h. Find the speed of the boat in still water.

36. A boat goes 50 km downstream in the same time that it takes to go 30 km upstream. The speed of the boat in still water is 16 km/h. Find the speed of the stream.

37. Person A can do a certain job in 3 hr, Person B can do the same job in 5 hr, and Person C can do the same job in 7 hr. How long would the job take with all working together?

38. Pipe A can fill a tank in 4 hr, pipe B can fill it in 10 hr, and pipe C can fill it in 12 hr. The pipes are connected to the same tank. How long does it take to fill the tank if all three are running together?

39. Person A can do a certain job, working alone, in 3.15 hr. Person A working with Person B can do the same job in 2.09 hr. How long will it take Person B, working alone, to do the job?

40. At a factory, smokestack A pollutes the air 2.13 times as fast as smokestack B. When both stacks operate together they yield a certain amount of pollution in 16.3 hr. Find the amount of time it would take each to yield the same amount of pollution if it operated alone.

41. Suppose $1000 is invested at $13\frac{3}{4}\%$. How much is in the account at the end of one year if interest is compounded (a) annually? (b) semiannually? (c) quarterly? (d) daily (use 365 days per yr)? (e) hourly?

42. Suppose $1000 is invested at 15.5%. How much is in the account at the end of five years, if interest is compounded (a) annually? (b) semiannually? (c) quarterly? (d) daily (use 365 days per yr)? (e) hourly?

★

43. A commuter drives to work at 45 mph and arrives one minute early. At 40 mph, the commuter would arrive one minute late. How far is it to work?

44. If b is 20% more than a, c is 25% more than b, and d is $k\%$ less than c, find k such that $a = d$.

2.3 QUADRATIC EQUATIONS

A *quadratic* equation, also known as an equation of *second degree*, is any equation equivalent to the following:

$$ax^2 + bx + c = 0, \quad \text{where } a \neq 0.$$

You have solved such equations by factoring. We now consider other methods. Let us first consider an equation $x^2 = p$, where p is not negative. We shall solve this equation in two ways.

1.
$$x^2 = p$$
$$x^2 - p = 0$$
$$(x - \sqrt{p})(x + \sqrt{p}) = 0 \quad \text{Factoring}$$
$$x - \sqrt{p} = 0 \quad \text{or} \quad x + \sqrt{p} = 0 \quad \text{Principle of zero products}$$
$$x = \sqrt{p} \quad \text{or} \quad x = -\sqrt{p}$$

2.
$$x^2 = p$$
$$\sqrt{x^2} = \sqrt{p} \quad \text{Taking principal square root}$$
$$|x| = \sqrt{p}$$
$$x = \sqrt{p} \quad \text{or} \quad -x = \sqrt{p} \quad \text{Definition of absolute value}$$
$$x = \sqrt{p} \quad \text{or} \quad x = -\sqrt{p}$$

With either method we obtain $\sqrt{p}$ and $-\sqrt{p}$, abbreviated $\pm\sqrt{p}$. If $p > 0$, there are two solutions. If $p = 0$, there is just one solution. If $p < 0$, there are no real-number solutions. We now use this method in another example.

Example 1 Solve: $(x + 5)^2 = 3$.

Taking the principal square root, we have $|x + 5| = \sqrt{3}$. Then,
$$x + 5 = \pm\sqrt{3}$$
$$x = -5 \pm \sqrt{3}.$$

The solution set is $\{-5 - \sqrt{3}, -5 + \sqrt{3}\}$.

Completing the Square

Now let us consider the equation $x^2 - 6x - 12 = 0$:
$$x^2 - 6x - 12 = 0$$
$$x^2 - 6x = 12.$$

We know that $x^2 - 6x + 9$ is a perfect square $(x - 3)^2$. So if we add 9 we can solve as follows:

$$x^2 - 6x + 9 = 12 + 9 \quad \text{Adding 9 makes}$$
$$(x - 3)^2 = 21 \quad \text{the left side a perfect square.}$$
$$x - 3 = \pm\sqrt{21}$$
$$x = 3 \pm \sqrt{21}$$

The solution set is $\{3 + \sqrt{21}, 3 - \sqrt{21}\}$, which we often abbreviate as $3 \pm \sqrt{21}$. Note that we can determine the number to add to make a perfect square as follows.

To complete the square on $x^2 + bx$, we take half the coefficient of x and square it. Then we add that number, $(b/2)^2$.

Example 2 Solve by completing the square.

$$x^2 + 3x - 5 = 0$$
$$x^2 + 3x = 5$$
$$x^2 + 3x + \frac{9}{4} = 5 + \frac{9}{4} \quad \boxed{\text{Take half of 3: } \tfrac{1}{2} \cdot 3 = \tfrac{3}{2}; \text{ square it: } (\tfrac{3}{2})^2 = \tfrac{9}{4}; \text{ and add.}}$$
$$\left(x + \frac{3}{2}\right)^2 = \frac{20}{4} + \frac{9}{4} = \frac{29}{4}$$
$$x + \frac{3}{2} = \pm\sqrt{\frac{29}{4}} = \pm\frac{\sqrt{29}}{2}$$
$$x = -\frac{3}{2} \pm \frac{\sqrt{29}}{2} = \frac{-3 \pm \sqrt{29}}{2}$$

The solution set is $\left\{\dfrac{-3 + \sqrt{29}}{2}, \dfrac{-3 - \sqrt{29}}{2}\right\}$, or $\dfrac{-3 \pm \sqrt{29}}{2}$.

In the previous examples we were solving $ax^2 + bx + c = 0$ with $a = 1$. For cases in which $a \neq 1$, we first multiply by $1/a$ to get an equation whose x^2-coefficient is 1.

Example 3 Solve by completing the square.

$$2x^2 - 3x - 1 = 0$$

$$x^2 - \frac{3}{2}x - \frac{1}{2} = 0 \qquad \text{Multiplying by } \frac{1}{2}$$

$$x^2 - \frac{3}{2}x = \frac{1}{2}$$

$$x^2 - \frac{3}{2}x + \frac{9}{16} = \frac{1}{2} + \frac{9}{16}$$

$$\left(x - \frac{3}{4}\right)^2 = \frac{8}{16} + \frac{9}{16} = \frac{17}{16}$$

$$x - \frac{3}{4} = \pm\sqrt{\frac{17}{16}} = \pm\frac{\sqrt{17}}{4}$$

$$x = \frac{3}{4} \pm \frac{\sqrt{17}}{4} = \frac{3 \pm \sqrt{17}}{4}$$

The solution set is $\left\{\dfrac{3 + \sqrt{17}}{4}, \dfrac{3 - \sqrt{17}}{4}\right\}$, or $\dfrac{3 \pm \sqrt{17}}{4}$.

The Quadratic Formula

To facilitate the solving of quadratic equations, we shall derive a formula. To do this, we will consider the standard form of the quadratic equation, with unspecified coefficients. Solving that equation gives us the formula we seek. We begin with

$$ax^2 + bx + c = 0.$$

We multiply on both sides by $4a$ to obtain

$$4a^2x^2 + 4abx + 4ac = 0.$$

Next we add $b^2 - b^2$ on the left and rearrange, as follows:

$$(4a^2x^2 + 4abx + b^2) - (b^2 - 4ac) = 0,$$

or

$$(2ax + b)^2 - (b^2 - 4ac) = 0.$$

The left side can be factored as a difference of squares, after which we

use the principle of zero products:
$$[(2ax + b) - \sqrt{b^2 - 4ac}][(2ax + b) + \sqrt{b^2 - 4ac}] = 0,$$
$$2ax + b - \sqrt{b^2 - 4ac} = 0 \quad \text{or} \quad 2ax + b + \sqrt{b^2 - 4ac} = 0,$$
$$x = \frac{-b + \sqrt{b^2 - 4ac}}{2a} \quad \text{or} \quad x = \frac{-b - \sqrt{b^2 - 4ac}}{2a}.$$

We can abbreviate the last line using the sign $\pm$. This gives us the quadratic formula.

THEOREM 3

The quadratic formula. **If a quadratic equation has solutions, they are given by**
$$x = \frac{-b \pm \sqrt{b^2 - 4ac}}{2a}.$$

When using the quadratic formula it is helpful to first find the standard form so that the coefficients a, b, and c can be determined.

Example 4 Solve: $3x^2 + 2x = 7$.

First find the standard form and determine a, b, and c:
$$3x^2 + 2x - 7 = 0,$$
$$a = 3, \quad b = 2, \quad c = -7.$$

Then use the quadratic formula:
$$x = \frac{-b \pm \sqrt{b^2 - 4ac}}{2a} = \frac{-2 \pm \sqrt{2^2 - 4 \cdot 3 \cdot (-7)}}{2 \cdot 3}$$

$$x = \frac{-2 \pm \sqrt{4 + 84}}{6} \qquad \text{To prevent careless mistakes it helps to write out } \textit{all} \text{ the steps.}$$

$$x = \frac{-2 \pm \sqrt{88}}{6} = \frac{-2 \pm \sqrt{4 \cdot 22}}{6}$$

$$= \frac{-2 \pm 2\sqrt{22}}{6} = \frac{2(-1 \pm \sqrt{22})}{2 \cdot 3} = \frac{-1 \pm \sqrt{22}}{3}.$$

Should such solutions arise in an applied problem, we can find approximations using Table 1 at the back of the book or a calculator:

$$\frac{-1 + \sqrt{22}}{3} \approx \frac{-1 + 4.69}{3}$$

$$\approx \frac{3.69}{3} \approx 1.2; \qquad \text{Rounded to the nearest tenth}$$

and

$$\frac{-1 - \sqrt{22}}{2} \approx \frac{-1 - 4.69}{3}$$

$$\approx \frac{-5.69}{3} \approx -1.9. \quad \text{Rounded to the nearest tenth}$$

The expression $b^2 - 4ac$, called the *discriminant,* determines the nature of the solutions. If x_1 and x_2 represent the solutions, then

$$x_1 = \frac{-b + \sqrt{b^2 - 4ac}}{2a}, \quad x_2 = \frac{-b - \sqrt{b^2 - 4ac}}{2a}.$$

Note that if $b^2 - 4ac = 0$, then x_1 and x_2 are the same, $-b/2a$. If $b^2 - 4ac > 0$, then the expression $\sqrt{b^2 - 4ac}$ represents a positive real number that is to be added to or subtracted from $-b$. In this case the solutions are different. If $b^2 - 4ac < 0$, the expression $\sqrt{b^2 - 4ac}$ does not represent a real number. Thus there are no real-number solutions. In a later chapter we will consider a number system in which such solutions exist.

THEOREM 4

If $b^2 - 4ac = 0$, $ax^2 + bx + c = 0$ has just one real-number solution.
If $b^2 - 4ac > 0$, $ax^2 + bx + c = 0$ has two real-number solutions.
If $b^2 - 4ac < 0$, $ax^2 + bx + c = 0$ has no real-number solution.

Writing Equations from Solutions

We can use the principle of zero products to write a quadratic equation whose solutions are known.

Example 5 Write a quadratic equation whose solutions are 3 and $-\frac{2}{5}$.

$$x = 3 \quad \text{or} \quad x = -\frac{2}{5}$$

$$x - 3 = 0 \quad \text{or} \quad x + \frac{2}{5} = 0$$

$$(x - 3)\left(x + \frac{2}{5}\right) = 0 \quad \text{Multiplying}$$

$$x^2 - \frac{13}{5}x - \frac{6}{5} = 0 \quad \text{or} \quad 5x^2 - 13x - 6 = 0$$

EXERCISE SET 2.3

Solve for x.
1. $2x^2 = 14$
2. $6x^2 = 36$
3. $9x^2 = 5$
4. $4x^2 = 3$
5. $ax^2 = b$
6. $\pi x^2 = k$
7. $(x - 7)^2 = 5$
8. $(x + 3)^2 = 2$
9. $(x - h)^2 = a$
10. $y = a(x - h)^2 + k$

Solve by completing the square. (It is important to practice this method since we will need it later.)
11. $x^2 + 6x + 4 = 0$
12. $x^2 - 6x - 4 = 0$
13. $y^2 + 7y - 30 = 0$
14. $y^2 - 7y - 30 = 0$
15. $5x^2 - 4x - 2 = 0$
16. $12y^2 - 14y + 3 = 0$
17. $2x^2 + 7x - 15 = 0$
18. $9x^2 - 30x = -25$

Solve, using the quadratic formula.
19. $x^2 + 4x - 5 = 0$
20. $x^2 - 2x = 15$
21. $2y^2 - 3y - 2 = 0$
22. $5m^2 + 3m - 2 = 0$
23. $3t^2 + 8t + 5 = 0$
24. $3u^2 = 18u - 6$
25. $u^2 = 12u - 3$
26. $5p^2 + 30p + 40 = 0$

Determine the nature of the solutions of each equation. Do not solve.
27. $x^2 - 2x + 10 = 0$
28. $4x^2 - 4\sqrt{3}x + 3 = 0$
29. $9x^2 - 6x = 0$
30. $2x^2 - 12x + 1 = 0$

Write quadratic equations whose solutions are as follows.
31. $-11, 9$
32. $-4, 4$
33. 7, only solution
34. $-\frac{2}{3}$, only solution
35. $-\frac{2}{5}, \frac{6}{5}$
36. $-\frac{1}{4}, -\frac{1}{2}$
37. $\frac{c}{2}, \frac{d}{2}$
38. $\frac{k}{3}, \frac{m}{4}$
39. $\sqrt{2}, 3\sqrt{2}$
40. $-\sqrt{3}, 2\sqrt{3}$

Solve.
41. ▦ $x^2 - 0.75x - 0.5 = 0$
42. ▦ $5.33x^2 - 8.23x - 3.24 = 0$

Solve, using any method. (In general, try factoring first. Then use the quadratic formula if factoring is not possible.)
43. $2x^2 - x - 6 = 0$
44. $3x^2 + 5x + 2 = 0$
45. $x^2 + 3x = 8$
46. $x^2 - 6x = 4$
47. $x + \frac{1}{x} = \frac{13}{6}$
48. $\frac{3}{x} + \frac{x}{3} = \frac{5}{2}$
49. $t^2 + 0.2t - 0.3 = 0$
50. $p^2 + 0.3p - 0.2 = 0$
51. $x^2 + x - \sqrt{2} = 0$
52. $x^2 - x - \sqrt{3} = 0$
53. $x^2 + \sqrt{5}x - \sqrt{3} = 0$
54. $2x^2 + \sqrt{3}x - \pi = 0$
55. $\sqrt{2}x^2 - \sqrt{3}x - \sqrt{5} = 0$
56. $\sqrt{2}x^2 + 5x + \sqrt{2} = 0$
57. $(2t - 3)^2 + 17t = 15$
58. $2y^2 - (y + 2)(y - 3) = 12$
59. $(x + 3)(x - 2) = 2(x + 11)$
60. $9t(t + 2) - 3t(t - 2) = 2(t + 4)(t + 6)$
61. $2x^2 + (x - 4)^2 = 5x(x - 4) + 24$
62. $(c + 2)^2 + (c - 2)(c + 2) = 44 + (c - 2)^2$

★

63. Solve for x in terms of y: $3x^2 + xy + 4y^2 - 9 = 0$
64. Derive the quadratic formula by completing the square.

65. One solution of $kx^2 + 3x - k = 0$ is -2. Find the other.

66. Find k so that the following equation has (a) two real-number solutions; (b) one solution; and (c) no real-number solutions.
$$3x^2 + 4x = k - 5$$

67. Prove that the solutions of $ax^2 + bx + c = 0$ are the reciprocals of the solutions of the equation $cx^2 + bx + a = 0$, $c \neq 0$.

68. Prove that the solutions of $ax^2 + bx + c = 0$ are the additive inverses of the solutions of the equation $ax^2 - bx + c = 0$.

2.4 FORMULAS AND APPLIED PROBLEMS

Formulas

To solve a formula for a certain variable, we use the principles of equation solving we have developed until we have an equation with the variable alone on one side. In most formulas the variables represent nonnegative numbers, so we usually do not need to use absolute values when taking principal square roots.

Example 1 Solve $V = \pi r^2 h$, for r.

$$\frac{V}{\pi h} = r^2$$

$$\sqrt{\frac{V}{\pi h}} = r$$

Example 2 Solve $A = \pi rs + \pi r^2$, for r.

$$\pi r^2 + \pi rs - A = 0$$

Then $a = \pi$, $b = \pi s$, $c = -A$, and we use the quadratic formula:

$$r = \frac{-b \pm \sqrt{b^2 - 4ac}}{2a} = \frac{-\pi s \pm \sqrt{(\pi s)^2 - 4 \cdot \pi \cdot (-A)}}{2\pi}$$

$$= \frac{-\pi s \pm \sqrt{\pi^2 s^2 + 4\pi A}}{2\pi}$$

or just

$$\frac{-\pi s + \sqrt{\pi^2 s^2 + 4\pi A}}{2\pi},$$

since the negative square root would result in a negative solution.

Applied Problems

Example 3 (*An interest problem*). $2560 is invested at interest rate r, compounded annually. In two years it grows to $3240. What is the interest rate?

We substitute 2560 for P, 3240 for A, and 2 for t in the formula $A = P(1 + i)^t$, and solve for i.

$$A = P(1 + i)^t$$

$$3240 = 2560(1 + i)^2$$

$$\frac{3240}{2560} = (1 + i)^2$$

$$\sqrt{\frac{324}{256}} = |1 + i| \quad \text{Taking the principal square root}$$

$$\pm \frac{18}{16} = 1 + i$$

$$-1 + \frac{18}{16} = i \quad \text{or} \quad -1 - \frac{18}{16} = i$$

$$\frac{2}{16} = i \quad \text{or} \quad -\frac{34}{16} = i$$

Since the interest rate cannot be negative, $i = \frac{2}{16} = \frac{1}{8} = 0.125 = 12.5\%$.

Example 4 A ladder 10 ft long leans against a wall. The bottom of the ladder is 6 ft from the wall. The bottom of the ladder is then pulled out 3 ft farther. How much does the top end move down the wall?

The first thing to do is to make a drawing and label it, with the known and unknown data. See Fig. 4. The dotted line shows the ladder in its original position, with the lower end 6 ft from the wall.

In the figure there are right triangles. This is a clue that we may wish to use the Pythagorean theorem. We use that theorem and begin to write equations. From the taller triangle, we get

$$h^2 + 6^2 = 10^2.$$

We solve this for h and find that $h = 8$ ft.

From the other triangle, we get

$$9^2 + (h - d)^2 = 10^2$$

but we know that $h = 8$, so we have

$$9^2 + (8 - d)^2 = 10^2$$

or

$$d^2 - 16d + 45 = 0.$$

Using the quadratic formula, we get

$$d = \frac{16 \pm \sqrt{76}}{2} = \frac{16 \pm 2\sqrt{19}}{2} = 8 \pm \sqrt{19}.$$

Figure 4

The length $8 + \sqrt{19}$ is not a solution since it exceeds the original length. Thus the solution is $8 - \sqrt{19} \approx 8 - 4.359 = 3.641$. Therefore the top moves down 3.641 ft when the bottom is moved out 3 ft.

When an object is dropped or thrown downward, the distance, in meters, that it falls in t seconds is given by the following formula:

$$s = 4.9t^2 + v_0 t.$$

In this formula v_0 is the initial velocity.

Example 5

a) An object is dropped from the top of the Gateway Arch in St. Louis, which is 195 meters high. How long does it take the object to reach the ground?

Since the object was *dropped* its initial velocity was 0. So we substitute 0 for v_0 and 195 for s and then solve for t:

$$195 = 4.9t^2 + 0 \cdot t$$
$$195 = 4.9t^2$$
$$t^2 = 39.8$$
$$t = \sqrt{39.8} \approx 6.31. \quad \text{Use the square root key.}$$

Thus it takes about 6.31 seconds to reach the ground.

b) An object is thrown downward from the arch at an initial velocity of 16 m/s. How long does it take to reach the ground?

We substitute 195 for s and 16 for v_0 and solve for t:

$$195 = 4.9t^2 + 16t$$
$$0 = 4.9t^2 + 16t - 195.$$

By the quadratic formula and a calculator we obtain

$$t = -8.15 \quad \text{or} \quad t = 4.88.$$

The negative answer is meaningless in this problem, so the answer is about 4.88 sec.

c) How far will an object fall in 3 seconds if it is thrown downard from the arch at an initial velocity of 16 m/s?

We substitute 16 for v_0 and 3 for t and solve for s:

$$s = 4.9t^2 + v_0 t$$
$$= 4.9(3)^2 + 16 \cdot 3 = 92.1.$$

Thus the object falls 92.1 meters in 3 seconds.

EXERCISE SET 2.4

Solve each formula for the given variable. Assume that all letters represent positive numbers.

1. $F = \dfrac{kM_1M_2}{d^2}$, for d
2. $E = mc^2$, for c
3. $S = \dfrac{1}{2}at^2$, for t
4. $V = 4\pi r^2$, for r
5. $s = -16t^2 + v_0 t$, for t
6. $A = 2\pi r^2 + 3\pi rh$, for r
7. $d = \dfrac{n^2 - 3n}{2}$, for n
8. $\sqrt{2}t^2 + 3k = \pi t$, for t
9. $A = P(1 + i)^2$, for i
10. $A = P\left(1 + \dfrac{i}{2}\right)^2$, for i

Applied problems

What is the interest rate if interest is compounded annually?

11. $2560 grows to $3610 in two years
12. $1000 grows to $1210 in two years
13. ▦ $8000 grows to $9856.80 in two years
14. ▦ $1000 grows to $1271.26 in two years

The number of diagonals, d, of a polygon of n sides is given by

$$d = \dfrac{n^2 - 3n}{2}.$$

15. A polygon has 27 diagonals. How many sides does it have?

16. A polygon has 44 diagonals. How many sides does it have?

17. A ladder 10 ft long leans against a wall. The bottom of the ladder is 6 ft from the wall. How much would the lower end of the ladder have to be pulled away so that the top end would be pulled down the same amount?

18. A ladder 13 ft long leans against a wall. The bottom of the ladder is 5 ft from the wall. How much would the lower end of the ladder have to be pulled away so that the top end would be pulled down the same amount?

19. The area of a triangle is 18 cm². The base is 3 cm longer than the height. Find the height.

20. A baseball diamond is a square 90 ft on a side. How far is it directly from second base to home?

21. Trains A and B leave the same city at right angles at the same time. Train B travels 5 mph faster than train A. After 2 hr they are 50 mi apart. Find the speed of each train.

22. Trains A and B leave the same city at right angles at the same time. Train A travels 14 km/h faster than train B. After 5 hr they are 130 km apart. Find the speed of each train.

For Exercises 23 and 24 use the formula $s = 4.9t^2 + v_0 t$.

23. a) An object is dropped 75 m from an airplane. How long does it take to reach the ground?
    b) An object is thrown downward from the plane at an initial velocity of 30 m/s. How long does it take to reach the ground?
    c) How far will an object fall in 2 sec, thrown downward at an initial velocity of 30 m/s?

24. a) An object is dropped 500 m from an airplane. How long does it take to reach the ground?
    b) An object is thrown downward from the plane at an initial velocity of 30 m/s. How long does it take to reach the ground?
    c) How far will an object fall in 5 sec, thrown downward at an initial velocity of 30 m/s?

The following formula will be helpful in Exercises 25 through 28:

$$T = c \cdot N.$$

Total cost = (Cost per item) · (Number of items)

Total cost = (Cost per person) · (Number of persons)

25. A group of students share equally in the $140 cost of a boat. At the last minute three students drop out and this raises the share of each remaining student $15. How many students were in the group at the outset?

26. An investor bought a group of lots for $8400. All but four of them were sold for the same price. The selling price for each lot was $350 greater than the cost. How many lots were bought?

27. An investor buys some stock for $720. If each share had cost $15 less four more shares could have been bought for the same $720. How many shares of stock were bought?

28. A sorority is going to spend $112 for a party. When 14 new pledges join the sorority, this reduces each student's cost by $4. How much did it cost each student before?

29. The diagonal of a square is 1.341 cm longer than a side. Find the length of the side.

30. The hypotenuse of a right triangle is 8.312 cm long. The sum of the lengths of the legs is 10.23 cm. Find the lengths of the legs.

Solve for x.   QUADRADIC EQUATION,

31. $kx^2 + (3 - 2k)x - 6 = 0$

32. $x^2 - 2x + kx + 1 = kx^2$

33. $(m + n)^2 x^2 + (m + n)x = 2$

34. Solve $x^2 - 3xy - 4y^2 = 0$
    (a) for x; (b) for y.

35. A rectangle of 12 cm² area is inscribed in the right triangle ABC as shown in the drawing below. What are its dimensions?

36. For interest compounded annually, what is the interest rate when $9826 grows to $13,704 in three years?

37. The world record for free-fall by a woman is 175 ft and is held by Kitty O'Neill. Approximately how long did the fall take?

38. The world record for free-fall by a man is 311 ft and is held by Dar Robinson. Approximately how long did the fall take?

2.5 RADICAL EQUATIONS

A *radical equation* is an equation in which variables occur in one or more radicands. For example, $\sqrt[3]{x} + \sqrt[3]{4x-7} = 2$. To solve such equations we need a new principle.

THEOREM 5

The principle of powers. For any number n, if an equation $a = b$ is true, then $a^n = b^n$ is true.

This principle does *not* always yield equivalent equations. For example, the solution of $x = 3$ is 3. If we square both sides, we get $x^2 = 9$, which has solutions -3 and 3. As another example, consider $\sqrt{x} = -3$. At the outset we should note that this equation has no solution because, by definition, $\sqrt{x}$ must be a *nonnegative* number. Suppose, though, that we should try to solve by squaring both sides. We would get $(\sqrt{x})^2 = (-3)^2$, or $x = 9$. The number 9 does not check.

It is imperative when using the principle of powers to check possible solutions in the original equation.

Example 1 Solve: $x - 5 = \sqrt{x + 7}$.

$(x - 5)^2 = (\sqrt{x + 7})^2$ Principle of powers; squaring both sides

$x^2 - 10x + 25 = x + 7$
$x^2 - 11x + 18 = 0$
$(x - 9)(x - 2) = 0$
$x = 9 \quad \text{or} \quad x = 2$

The possible solutions are 9 and 2. We check:

For 9:
$x - 5 = \sqrt{x + 7}$	
$9 - 5$	$\sqrt{9 + 7}$
4	4

For 2:
$x - 5 = \sqrt{x + 7}$	
$2 - 5$	$\sqrt{2 + 7}$
-3	3

Since 9 checks but 2 does not, the only solution is 9.

Example 2 Solve: $\sqrt[3]{4x^2 + 1} = 5$.

$(\sqrt[3]{4x^2 + 1})^3 = 5^3$ Principle of powers; cubing both sides

$4x^2 + 1 = 125$
$4x^2 = 124$
$x^2 = 31$
$x = \pm\sqrt{31}$

Both $\sqrt{31}$ and $-\sqrt{31}$ check. These are the solutions.

Sometimes we have to use the principle of powers more than once.

Example 3 Solve: $\sqrt{2x-5} = 1 + \sqrt{x-3}$.

$$(\sqrt{2x-5})^2 = (1 + \sqrt{x-3})^2$$
$$2x - 5 = 1 + 2\sqrt{x-3} + (x-3)$$
$$x - 3 = 2\sqrt{x-3}$$
$$(x-3)^2 = (2\sqrt{x-3})^2 \quad \text{Get the radical alone, then square.}$$
$$x^2 - 6x + 9 = 4(x-3)$$
$$x^2 - 6x + 9 = 4x - 12$$
$$x^2 - 10x + 21 = 0$$
$$(x-7)(x-3) = 0$$
$$x = 7 \quad \text{or} \quad x = 3$$

The numbers 7 and 3 check. Thus the solutions are 7 and 3.

Example 4 Solve $A = \sqrt{1 + \dfrac{a^2}{b^2}}$, for a. Assume the variables represent nonnegative numbers.

$$A^2 = 1 + \frac{a^2}{b^2}$$
$$b^2 A^2 = b^2 + a^2$$
$$b^2 A^2 - b^2 = a^2$$
$$\sqrt{b^2 A^2 - b^2} = a$$
$$\sqrt{b^2(A^2 - 1)} = a$$
$$b\sqrt{A^2 - 1} = a$$

EXERCISE SET 2.5

Solve.

1. $\sqrt{3x - 4} = 1$
2. $\sqrt[3]{2x + 1} = -5$
3. $\sqrt[4]{x^2 - 1} = 1$
4. $\sqrt{m + 1} - 5 = 8$
5. $\sqrt{y - 1} + 4 = 0$
6. $5 + \sqrt{3x^2 + \pi} = 0$
7. $\sqrt{x - 3} + \sqrt{x + 5} = 4$
8. $\sqrt{x} - \sqrt{x - 5} = 1$
9. $\sqrt{3x - 5} + \sqrt{2x + 3} + 1 = 0$
10. $\sqrt{2m - 3} = \sqrt{m + 7} - 2$
11. $\sqrt[3]{6x + 9} + 8 = 5$
12. $\sqrt[5]{3x + 4} = 2$
13. $\sqrt{6x + 7} = x + 2$
14. $\sqrt{6x + 7} - \sqrt{3x + 3} = 1$
15. $\sqrt{20 - x} = \sqrt{9 - x} + 3$
16. $\sqrt{n + 2} + \sqrt{3n + 4} = 2$
17. ▦ $\sqrt{7.35x + 8.051} = 0.345x + 0.067$
18. ▦ $\sqrt{1.213x + 9.333} = 5.343x + 2.312$

For Exercises 19 and 20, assume that the variables represent nonnegative numbers.

19. Solve $T = 2\pi\sqrt{\dfrac{L}{g}}$, for L; for g.

20. Solve $H = \sqrt{c^2 + d^2}$, for c.

The formula $V = 1.2\sqrt{h}$ can be used to approximate the distance V, in miles, that a person can see to the horizon from a height h, in feet.

21. How far can you see to the horizon through an airplane window at a height of 30,000 feet?

22. How far can a sailor see to the horizon from the top of a 72-ft mast?

23. A person can see 144 miles to the horizon from an airplane window. How high is the airplane?

24. A sailor can see 11 miles to the horizon from the top of a mast. How high is the mast?

Solve.

25. $(x - 5)^{2/3} = 2$

26. $(x - 3)^{2/3} = 2$

27. $\dfrac{x + \sqrt{x + 1}}{x - \sqrt{x + 1}} = \dfrac{5}{11}$

28. $\sqrt{\sqrt{x + 25} - \sqrt{x}} = 5$

29. $\sqrt{x + 2} - \sqrt{x - 2} = \sqrt{2x}$

30. $2\sqrt{x + 3} = \sqrt{x} + \sqrt{x + 8}$

31. $\sqrt[4]{x + 2} = \sqrt{3x + 1}$

32. $\sqrt[3]{2x - 1} = \sqrt[6]{x + 1}$

2.6 EQUATIONS REDUCIBLE TO QUADRATIC

Look for a pattern.

a) $x + 3\sqrt{x} - 10 = 0$; let $u = \sqrt{x}$. Then $u^2 + 3u - 10 = 0$.

b) $x^4 - 6x^2 + 7 = 0$; let $u = x^2$. Then $u^2 - 6u + 7 = 0$.

c) $(x^2 - x)^2 - 14(x^2 - x) + 24 = 0$; let $u = x^2 - x$. Then $u^2 - 14u + 24 = 0$.

The original equations are not quadratic, but after an appropriate substitution of an expression in another variable, we get a quadratic equation. Equations like the original ones are said to be *reducible to quadratic*.

To solve equations reducible to quadratic we first make a substitution, solve for the new variable, then solve for the original variable.

Example 1 Solve: $x + 3\sqrt{x} - 10 = 0$.

Let $u = \sqrt{x}$. Then we solve the equation resulting from substituting u for $\sqrt{x}$:

$$u^2 + 3u - 10 = 0$$
$$(u + 5)(u - 2) = 0$$
$$u = -5 \quad \text{or} \quad u = 2$$

Now we substitute $\sqrt{x}$ for u and solve these equations:

$$\sqrt{x} = -5 \quad \text{or} \quad \sqrt{x} = 2$$
$$\text{(no solution)} \quad \text{or} \quad x = 4.$$

The solution is 4.

Example 2 Solve: $x^4 - 6x^2 + 7 = 0$.

Let $u = x^2$. Then we solve the equation resulting from substituting u for x^2:

$$u^2 - 6u + 7 = 0$$
$$a = 1, \quad b = -6, \quad c = 7$$
$$u = \frac{-b \pm \sqrt{b^2 - 4ac}}{2a} = \frac{-(-6) \pm \sqrt{(-6)^2 - 4 \cdot 1 \cdot 7}}{2 \cdot 1}$$
$$u = \frac{6 \pm \sqrt{8}}{2} = \frac{2 \cdot 3 \pm 2\sqrt{2}}{2 \cdot 1} = 3 \pm \sqrt{2}.$$

Now we substitute x^2 for u and solve for x:

$$x^2 = 3 + \sqrt{2} \quad \text{or} \quad x^2 = 3 - \sqrt{2}$$
$$x = \pm\sqrt{3 + \sqrt{2}} \quad \text{or} \quad x = \pm\sqrt{3 - \sqrt{2}}.$$

Thus we have four solutions: $\sqrt{3 + \sqrt{2}}$, $-\sqrt{3 + \sqrt{2}}$, $\sqrt{3 - \sqrt{2}}$, and $-\sqrt{3 - \sqrt{2}}$.

Example 3 Solve: $(x^2 - x)^2 - 14(x^2 - x) + 24 = 0$.

Let $u = x^2 - x$. Then we solve the equation resulting from substituting u for $x^2 - x$:

$$u^2 - 14u + 24 = 0$$
$$(u - 12)(u - 2) = 0$$
$$u = 12 \quad \text{or} \quad u = 2.$$

Now we substitute $x^2 - x$ for u and solve:

$$x^2 - x = 12 \quad \text{or} \quad x^2 - x = 2$$
$$x^2 - x - 12 = 0 \quad \text{or} \quad x^2 - x - 2 = 0$$
$$(x - 4)(x + 3) = 0 \quad \text{or} \quad (x - 2)(x + 1) = 0$$
$$x = 4 \quad \text{or} \quad x = -3 \quad \text{or} \quad x = 2 \quad \text{or} \quad x = -1.$$

The solutions are 4, -3, 2, -1.

Example 4 Solve: $t^{2/5} - t^{1/5} - 2 = 0$.

Let $u = t^{1/5}$. Then we solve the equation resulting from substituting u

for $t^{1/5}$:
$$u^2 - u - 2 = 0$$
$$(u - 2)(u + 1) = 0$$
$$u = 2 \quad \text{or} \quad u = -1.$$

Now we substitute $t^{1/5}$ for u and solve:

$$t^{1/5} = 2 \quad \text{or} \quad t^{1/5} = -1$$
$$t = 32 \quad \text{or} \quad t = -1 \qquad \text{Principle of powers; raising to the 5th power}$$

The solutions are 32 and -1.

An Applied Problem

Example 5 (*A well problem*). An object is dropped into a well. Two seconds later the sound of the splash is heard at the top. The speed of sound is 1100 ft/sec. How deep is the well?

a) Let $s =$ the depth of the well. The formula for falling objects (p. 79) when the distance is in feet becomes

$$s = 16t^2 + v_0 t.$$

Since the object is dropped, $v_0 = 0$. The time t_1 that it takes for the object to reach the bottom of the well can be found as follows:

$$s = 16t_1^2, \quad \text{or} \quad t_1 = \frac{\sqrt{s}}{4}.$$

b) Now how long does it take the sound of the splash to reach the top of the well? We can use the formula $d = rt$, and let $d = s$, $r = 1100$, and $t = t_2$, the time it takes the sound to get to the top. Then

$$s = 1100 \cdot t_2, \quad \text{or} \quad t_2 = \frac{s}{1100}.$$

c) The total time for the object to fall and the sound to get back to the top is given by

$$t_1 + t_2 = 2, \quad \text{or} \quad \frac{\sqrt{s}}{4} + \frac{s}{1100} = 2. \qquad \text{See Fig. 5.}$$

Then multiplying by 1100, we get

$$275\sqrt{s} + s = 2200, \quad \text{or} \quad s + 275\sqrt{s} - 2200 = 0.$$

This equation is reducible to quadratic with $u = \sqrt{s}$. Substituting, we get

$$u^2 + 275u - 2200 = 0.$$

Figure 5

Using the quadratic formula we can solve for u:

$$u = \frac{-275 + \sqrt{275^2 + 8800}}{2} \quad \text{We want the positive solution.}$$

$$= \frac{-275 + \sqrt{84{,}425}}{2} \approx \frac{-275 + \sqrt{840 \cdot 100}}{2}$$

$$u = \frac{-275 + 10\sqrt{840}}{2}$$

$$\approx \frac{-275 + 10 \cdot 29}{2}$$

$$= \frac{15}{2} = 7.5.$$

Thus $u = 7.5 = \sqrt{s}$, so $s = 56.25$; that is, the well is about 56.25 ft deep.

EXERCISE SET 2.6

Solve.

1. $x - 10\sqrt{x} + 9 = 0$
2. $2x - 9\sqrt{x} + 4 = 0$
3. $x^4 - 10x^2 + 25 = 0$
4. $x^4 - 3x^2 + 2 = 0$
5. $t^{2/3} + t^{1/3} - 6 = 0$
6. $w^{2/3} - 2w^{1/3} - 8 = 0$
7. $z^{1/2} = z^{1/4} + 2$
8. $6 = m^{1/3} - m^{1/6}$
9. $(x^2 - 6x)^2 - 2(x^2 - 6x) - 35 = 0$
10. $(1 + \sqrt{x})^2 + (1 + \sqrt{x}) - 6 = 0$
11. $(y^2 - 5y)^2 + (y^2 - 5y) - 12 = 0$
12. $(2t^2 + t)^2 - 4(2t^2 + t) + 3 = 0$
13. $w^4 - 4w^2 - 2 = 0$
14. $t^4 - 5t^2 + 5 = 0$
15. $x^{-2} - x^{-1} - 6 = 0$
16. $4x^{-2} - x^{-1} - 5 = 0$
17. $2x^{-2} + x^{-1} = 1$
18. $10 - 9m^{-1} = m^{-2}$
19. $\left(\frac{x^2-2}{x}\right)^2 - 7\left(\frac{x^2-2}{x}\right) - 18 = 0$
20. $\left(\frac{x^2+1}{x}\right)^2 - 8\left(\frac{x^2+1}{x}\right) + 15 = 0$
21. $\frac{x}{x-1} - 6\sqrt{\frac{x}{x-1}} - 40 = 0$
22. $\frac{2x+1}{x} + 30 = 7\sqrt{\frac{2x+1}{x}}$
23. $5\left(\frac{x+2}{x-2}\right)^2 = 3\left(\frac{x+2}{x-2}\right) + 2$
24. $\left(\frac{x+1}{x+3}\right)^2 + \left(\frac{x+1}{x+3}\right) - 6 = 0$

25. A stone is dropped from a cliff. In 3 sec the sound of the stone striking the ground reaches the top of the cliff. Assuming the speed of sound is 1100 ft/sec, how high is the cliff?

26. A stone is dropped from a cliff. In 4 sec the sound of the stone striking the ground reaches the top of the cliff. Assuming the speed of sound is 1100 ft/sec, how high is the cliff?

Solve. Check possible solutions by substituting into the original equation.

27. ▦ $6.75x = \sqrt{35x} + 5.36$
28. ▦ $\pi x^4 - \sqrt{99.3} = \pi^2 x^2$

☆

Solve.

29. $9x^{3/2} - 8 = x^3$

30. $\sqrt[3]{2x + 3} = \sqrt[6]{2x + 3}$

★

Solve.

31. $\dfrac{2x + 1}{x} = 3 + 7\sqrt{\dfrac{2x + 1}{x}}$

2.7 VARIATION

Direct Variation

There are many situations that yield linear equations like $y = kx$, where k is some positive number. Note that as x increases, y increases. In such a situation we say we have *direct variation,* and k is called the *variation constant.* Usually only positive values of x and y are considered.

DEFINITION

If two variables x and y are related as in the equation $y = kx$, where k is a positive constant, we say that "y varies directly as x," or that "y is directly proportional to x."

For example, the circumference C of a circle varies directly as its diameter D: $C = \pi D$. A car moves at a constant speed of 65 mph. The distance d it travels varies directly as the time t: $d = 65t$.

Example 1 Find an equation of variation where y varies directly as x, and $y = 5.6$ when $x = 8$.

We know $y = kx$, so $5.6 = k \cdot 8$ and $0.7 = k$. Thus $y = 0.7x$.

Suppose y varies directly as x. Then we have $y = kx$. When (x_1, y_1) and (x_2, y_2) are solutions of the equation, we have $y_1 = kx_1$ and $y_2 = kx_2$, and

$$\frac{y_2}{y_1} = \frac{kx_2}{kx_1} = \frac{k}{k} \cdot \frac{x_2}{x_1} = \frac{x_2}{x_1}.$$

The equation $y_2/y_1 = x_2/x_1$ is called a *proportion.* Such equations are helpful in solving applied problems.

Example 2 (*A spring problem*). *Hooke's law* states that the distance d an elastic object such as a spring is stretched by placing a certain weight on it varies directly as the weight w of the object, as shown in Fig. 6. If the distance is 40 cm when the weight is 3 kg, what is the distance stretched when a 2-kg weight is attached?

Method 1. First find k using the first set of data. Then solve for d_2 using the second set of data.

$$d_1 = kw_1 \qquad d_2 = \frac{40}{3} \cdot w_2$$

$$40 = k \cdot 3 \qquad d_2 = \frac{40}{3} \cdot 2$$

$$\frac{40}{3} = k \qquad d_2 = \frac{80}{3}, \quad \text{or} \quad 26.7 \text{ cm}$$

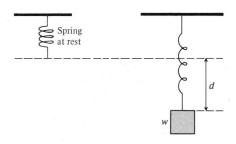

Figure 6

Method 2. Use a proportion to solve for d_2 without first finding the variation constant.

$$\frac{d_2}{d_1} = \frac{w_2}{w_1}$$

$$\frac{d_2}{40} = \frac{2}{3}$$

$$3 \cdot d_2 = 40 \cdot 2$$

$$d_2 = 26.7 \text{ cm}$$

Inverse Variation

There are also situations that yield equations of the type $y = k/x$, where k is a positive constant. Note that as positive values of x increase, y decreases. In such a situation we say we have *inverse variation,* and k is called the *variation constant.* In applications we are usually only considering positive values of x and y.

DEFINITION

If two variables x and y are related according to the equation $y = k/x$, where k is a positive constant, we say that "y varies inversely as x," or that "y is inversely proportional to x."

Suppose y varies inversely as x. Then we have $y = k/x$. When (x_1, y_1) and (x_2, y_2) are solutions of the equation, we have $y_1 = k/x_1$ and

$y_2 = k/x_2$. Then

$$\frac{y_2}{y_1} = \frac{\frac{k}{x_2}}{\frac{k}{x_1}} = \frac{k}{x_2} \cdot \frac{x_1}{k} = \frac{k}{k} \cdot \frac{x_1}{x_2} = \frac{x_1}{x_2}.$$

The equation $y_2/y_1 = x_1/x_2$ is a proportion that also can be helpful in solving applied problems. Compare it to the proportion for direct variation. Note how the variables are interchanged.

Example 3 (*Stocks and gold*). Certain economists theorize that stock prices are inversely proportional to the price of gold. That is, when the price of gold goes up, the prices of stock go down; and when the price of gold goes down, the prices of stock go up. Let us assume that the Dow-Jones Industrial Average, D, an index of the overall price of stock, is inversely proportional to the price of gold, G, in dollars per ounce. One day the Dow-Jones average was 818 and the price of gold was $673 per ounce. What will the Dow-Jones average be if the price of gold drops to $660 per ounce?

Method 1. Find k using the first set of data. Then solve for G_2.

$$D_1 = \frac{k}{G_1}$$

$$818 = \frac{k}{673}$$

$$k = 818(673)$$

$$= 550{,}514$$

$$D_2 = \frac{k}{G_2} = \frac{550{,}514}{660}$$

$$D_2 = 834$$

Method 2. Use a proportion to solve for D_2 without first finding the variation constant.

$$\frac{D_2}{D_1} = \frac{G_1}{G_2}$$

$$\frac{D_2}{818} = \frac{673}{660}$$

$$660 \cdot D_2 = 818(673)$$

$$D_2 = 834$$

Warning! Do not put too much "stock" in the equation of Example 3. It is meant to give us an idea of economic relationships. An equation to accurately predict the stock market has not been found!

Other Kinds of Variation

There are many other types of variation.

DEFINITION

y **varies directly as the square of** *x* **if there is some positive number** *k* **such that** $y = kx^2$.

For example, consider the equation for the area of a circle: $A = \pi r^2$.

DEFINITION

y **varies inversely as the square of** *x* **if there is some positive number** *k* **such that** $y = k/x^2$.

For example, the weight *W* of a body varies inversely as the square of the distance *d* from the center of the earth: $W = k/d^2$.

DEFINITION

y **varies *jointly* as** *x* **and** *z* **if there is some positive number** *k* **such that** $y = kxz$.

For example, consider the equation for the area of a triangle: $A = \frac{1}{2}bh$. The area varies jointly as *b* and *h*. The variation constant is $\frac{1}{2}$.

Several kinds of variation can occur together. For example, if $y = k \cdot xz^3/w^2$, then y varies jointly as x and the cube of z and inversely as the square of w.

Applied Problems

Example 4 (*The volume of a tree*). The volume of wood *V* in a tree varies jointly as the height *h* and the square of the girth *g* (girth is distance around). If the volume is 7750 ft^3 when the height is 100 ft and the girth is 5 ft, what is the height when the volume is 37,975 ft^3 and the girth is 7 ft?

Method 1. First find k using the first set of data. Then solve for h_2 using the second set of data.

$$V_1 = k \cdot h_1 \cdot g_1^2$$
$$7750 = k \cdot 100 \cdot 5^2$$
$$3.1 = k$$

$$37{,}975 = 3.1 \cdot h_2 \cdot 7^2$$
$$250 = h_2$$
$$h_2 = 250 \text{ ft}$$

Method 2. Use a proportion to solve for h_2 without first finding the variation constant.

$$\frac{V_2}{V_1} = \frac{h_2 \cdot g_2^2}{h_1 \cdot g_1^2}$$

$$\frac{37{,}975}{7750} = \frac{h_2 \cdot 7^2}{100 \cdot 5^2}$$

$$h_2 = 250 \text{ ft}$$

Example 5 (*The weight of an astronaut*). The weight W of an object varies inversely as the square of the distance d from the object to the center of the earth. At sea level (4000 mi from the center of the earth) an astronaut weighs 200 lb. Find his weight when he is 100 mi above the surface of the earth, and the spacecraft is not in motion.

We use the following proportion:

$$\frac{W_2}{W_1} = \frac{d_1^2}{d_2^2}$$

$$\frac{W_2}{200} = \frac{4000^2}{4100^2}$$

$$16{,}810{,}000 \cdot W_2 = 200(16{,}000{,}000)$$

$$W_2 = 190 \text{ lb (to the nearest pound)}.$$

EXERCISE SET 2.7

Find an equation of variation where:

1. y varies directly as x, and $y = 0.6$ when $x = 0.4$.

2. y varies inversely as x, and $y = 0.4$ when $x = 0.8$.

3. y varies inversely as the square of x, and $y = 0.15$ when $x = 0.1$.

4. y varies jointly as x and z, and $y = 56$ when $x = 7$ and $z = 10$.

5. y varies jointly as x and z and inversely as w, and $y = \frac{3}{2}$ when $x = 2$, $z = 3$, and $w = 4$.

6. y varies jointly as x and the square of z, and $y = 105$ when $x = 14$ and $z = 5$.

7. y varies jointly as x and z and inversely as the square of w, and $y = \frac{12}{5}$ when $x = 16$, $z = 3$, and $w = 5$.

8. y varies jointly as x and z and inversely as the product of w and p, and $y = \frac{3}{28}$ when $x = 3$, $z = 10$, $w = 7$, and $p = 8$.

9. Suppose y varies directly as x and x is doubled. What is the effect on y?

10. Suppose y varies inversely as x and x is tripled. What is the effect on y?

11. Suppose y varies inversely as the square of x and x is multiplied by n. What is the effect on y?

12. Suppose y varies directly as the square of x and x is multiplied by n. What is the effect on y?

13. The amount of pollution A entering the atmosphere varies directly as the amount of people N living in an area. If 60,000 people cause 42,600 tons of pollutants, how many tons entered the atmosphere in a city with a population of 750,000?

14. The volume V of a given mass of gas varies directly as the temperature T and inversely as the pressure P. If $V = 231$ in^3 when $T = 420°$ and $P = 20$ lb/sq in., what is the volume when $T = 300°$ and $P = 15$ lb/sq in.?

15. (*The strength of a beam*). The safe load (amount it supports without breaking) L of a beam varies jointly as its width w and the square of its height h and inversely as its length l. If the width and height are doubled at the same time the length is halved, what is the effect on L?

16. For a chord $\overline{PQ}$ through a fixed point A in a circle, the length of $\overline{PA}$ is inversely proportional to the length of $\overline{AQ}$. If the length of $\overline{PA} = 64$ when the length $\overline{AQ} = 16$, what is the length of $\overline{AQ}$ when the length of $\overline{PA} = 4$?

17. (*A sighting problem*). The distance d that one can see to the horizon varies directly as the square root of the height above sea level. If a person 19.5 m above sea level can see 28.97 km, how high above sea level must one be to see 54.32 km?

18. (*Electrical resistance*). At a fixed temperature and chemical composition, the resistance of a wire varies directly as the length l and inversely as the square of the diameter d. If the resistance of a certain kind of wire is 0.112 ohm when the diameter is 0.254 cm and the length is 15.24 m, what is the resistance of a wire whose length is 608.7 m and whose diameter is 0.478 cm?

19. Show that if p varies directly as q, then q varies directly as p.

20. Show that if u varies inversely as v, then v varies inversely as u, and 1/u varies directly as v.

21. The area of a circle varies directly as the square of the length of a diameter. What is the variation constant?

22. If P varies directly as the square of t, how does t vary in relation to P?

CHAPTER 2 REVIEW

Solve.

1. $\frac{3x + 2}{7} - \frac{5x}{3} = \frac{32}{21}$

2. $(p - 3)(3p + 2)(p + 2) = 0$

3. $3x^2 + 2x - 8 = 0$

4. $r^2 + 3r - 1 = 0$

5. $3t^2 = 4 + 3t$

6. $\frac{5}{2x + 3} + \frac{1}{x - 6} = 0$

7. $y^4 - 3y^2 + 1 = 0$

8. $x = 2x^{1/2} - 1$

9. $(x^2 - 1)^2 - (x^2 - 1) - 2 = 0$

10. $t^{2/3} - 10 = 3t^{1/3}$

11. $\sqrt{x - 1} - \sqrt{x - 4} = 1$

12. $\sqrt{5x + 1} - 1 = \sqrt{3x}$

13. Solve by completing the square: $y^2 - 6y = 16$.

Determine the nature of the solutions of each equation.

14. $4y^2 + 5y + 10 = 0$

15. $3z^2 + 2z - 1 = 0$

16. Write a quadratic equation whose solutions are -3 and $\frac{1}{2}$.

17. Solve $v = \sqrt{2gh}$, for h.

18. Solve $\frac{1}{a} + \frac{1}{b} = \frac{1}{t}$, for t.

19. A student scores 73% and 79% on two tests. If the third test counts as if it were two tests, what score must the student make on the third test so the average will be 85%?

20. A can mow a lawn in 4 hr. B can mow it in 2 hr. How long will it take if they work together?

21. A can mow a lawn in 3 hr. Working together, A and B can mow the lawn in 1 hr. How long will it take B, working alone, to mow the lawn?

22. 18 is 30% of what?

23. What is 9 percent of 50?

24. Two trains leave the same city at the same time at right angles. The first train travels 60 km/h. In one hour the trains are 100 km apart. How fast is the second train traveling?

25. In a right triangle the perimeter is 40 and the sum of the squares of the sides is 578. Find the lengths of the sides.

26. A boat travels 2 km upstream and 2 km downstream. The total time for both parts of the trip is 1 hr. The speed of the stream is 2 km/h. What is the speed of the boat in still water? Round to the nearest tenth.

27. Find an equation of variation where y varies inversely as the square of x, and $y = 0.005$ when $x = 10$.

28. Find an equation of variation where T varies directly as the square of x and inversely as p, and $T = 0.01$ when $x = 6$ and $p = 20$.

29. *Ohm's law* states that the voltage V, in an electric circuit, varies directly as the number of amperes I of electric current in the circuit. If the voltage is 10 volts when the current is 3 amperes, what is the voltage when the current is 15 amperes?

30. For a body falling freely from rest, the distance (s ft) that the body falls varies directly as the square of the time (t sec). Given that $s = 64$ when $t = 2$, find a formula for s in terms of t. How long will it take the body to fall 900 ft?

31. Suppose a, b, c, and d are nonzero real numbers such that c and d are solutions of

$$x^2 + ax + b = 0$$

and a and b are solutions of

$$x^2 + cx + d = 0.$$

Find $a + b + c + d$.

3

Relations, Functions, and Transformations

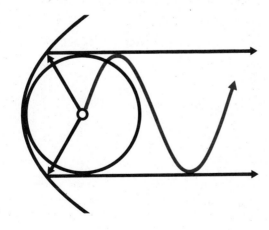

3.1 RELATIONS AND ORDERED PAIRS

Cartesian Products

Consider the sets $A = \{1, 2, 3\}$ and $B = \{a, b\}$. From these sets we can pick numbers and form ordered pairs; for example,

$$(1, a), \quad (2, a), \quad \text{and} \quad (3, b).$$

DEFINITION

The Cartesian product of two sets A and B, symbolized $A \times B$, is defined as follows. It is the set of *all* ordered pairs having the first member from set A and the second member from set B.

Example 1 Find the Cartesian product $A \times B$, where $A = \{1, 2, 3\}$ and $B = \{a, b\}$.

The Cartesian product, $A \times B$, is as follows:

$$(1, a), \quad (2, a), \quad (3, a),$$
$$(1, b), \quad (2, b), \quad (3, b).$$

Caution! Be sure in forming Cartesian products that in each pair of $A \times B$ the first member is taken from A and the second member is taken from B.

The sets A and B may be the same.

Example 2 Consider the set $\{2, 3, 4, 5\}$. Find the Cartesian product of this set by itself.

The Cartesian product of this set by itself is called a *Cartesian square* and is shown below.

```
5 | (2, 5)   (3, 5)   (4, 5)   (5, 5)
4 | (2, 4)   (3, 4)   (4, 4)   (5, 4)
3 | (2, 3)   (3, 3)   (4, 3)   (5, 3)
2 | (2, 2)   (3, 2)   (4, 2)   (5, 2)
   ‾‾‾‾‾‾‾‾‾‾‾‾‾‾‾‾‾‾‾‾‾‾‾‾‾‾‾‾‾‾‾‾
      2        3        4        5
```

The headings at the bottom and at the left are for reference only. The Cartesian square consists only of the ordered pairs.

Relations

In a Cartesian product we can pick out ordered pairs that make up common relations, such as $=$ or $<$ in the next examples.

Example 3 In the Cartesian product in Example 2 indicate all ordered pairs for which the first member is equal to the second. This set of ordered pairs is the relation $=$.

(2, 5)	(3, 5)	(4, 5)	(5, 5)
(2, 4)	(3, 4)	(4, 4)	(5, 4)
(2, 3)	(3, 3)	(4, 3)	(5, 3)
(2, 2)	(3, 2)	(4, 2)	(5, 2)

Example 4 In the Cartesian product in Example 2 indicate all ordered pairs for which the first member is less than the second. This set of ordered pairs is the relation $<$.

(2, 5)	(3, 5)	(4, 5)	(5, 5)
(2, 4)	(3, 4)	(4, 4)	(5, 4)
(2, 3)	(3, 3)	(4, 3)	(5, 3)
(2, 2)	(3, 2)	(4, 2)	(5, 2)

There are also many relations that do not have common names and relations with which we are not already familiar. Any time we select a set of ordered pairs from a Cartesian product, we have selected some relation. This is true even if we make the selection at random.

Example 5 The following set of ordered pairs is a relation, but is not a familiar one. It has no common name.

(2, 5)	(3, 5)	(4, 5)	(5, 5)
(2, 4)	(3, 4)	(4, 4)	(5, 4)
(2, 3)	(3, 3)	(4, 3)	(5, 3)
(2, 2)	(3, 2)	(4, 2)	(5, 2)

We shall now make our definition of relation.

DEFINITION

A *relation* from a set A to a set B is defined to be any set of ordered pairs in $A \times B$.

Domain and Range

DEFINITION

The set of all first members in a relation is called its *domain*. The set of all second members in a relation is called its *range*.

Example 6 Find the domain and range of the relation = in Example 3.

$$\text{Domain} = \{2, 3, 4, 5\}; \quad \text{range} = \{2, 3, 4, 5\}.$$

Example 7 Find the domain and range of the relation < in Example 4.

$$\text{Domain} = \{2, 3, 4\}; \quad \text{range} = \{3, 4, 5\}.$$

Example 8 Find the domain and range of the relation in Example 5.

$$\text{Domain} = \{2, 4, 5\}; \quad \text{range} = \{2, 3, 5\}.$$

EXERCISE SET 3.1

1. List all ordered pairs in the Cartesian product $A \times B$, where $A = \{0, 2, 4, 5\}$ and $B = \{a, b, c\}$. Remember that first coordinates come from A and second coordinates from B.

2. List all ordered pairs in the Cartesian product $A \times B$, where $A = \{1, 3, 5, 9\}$ and $B = \{d, e, f\}$. Remember that first coordinates come from A and second coordinates from B.

For Exercises 3–8, consider the set $\{-1, 0, 1, 2\}$.

3. Find the set of ordered pairs in the relation < (is less than).

4. Find the set of ordered pairs in the relation > (is greater than).

5. Find the set of ordered pairs in the relation ≤ (is less than or equal to).

6. Find the set of ordered pairs in the relation ≥ (is greater than or equal to).

7. Find the set of ordered pairs in the relation =.

8. Find the set of ordered pairs in the relation ≠.

9. a) List all the ordered pairs in the Cartesian square $D \times D$, where $D = \{-1, 0, 1, 2\}$.
   b) Graph the relation whose ordered pairs are $(0, 0)$, $(1, 1)$, $(0, 1)$, and $(1, 2)$.
   c) List the domain and the range of this relation.

10. a) List all the ordered pairs in the Cartesian square $E \times E$, where $E = \{-1, 1, 3, 5\}$.
    b) Graph the relation whose ordered pairs are $(-1, 1)$, $(1, 1)$, $(-1, 3)$, and $(1, 3)$.
    c) List the domain and the range of this relation.

3.2 GRAPHS OF EQUATIONS

Relations in Real Numbers

We shall be most interested in relations involving $R \times R$, where R is the set of real numbers. The set R is infinite. Thus relations involving R may be infinite, and therefore cannot be indicated by listing the ordered pairs one at a time. We usually indicate such relations with some sort of picture in $R \times R$. This kind of picture is called a graph.

Points in the Plane and Ordered Pairs

On a number line each point corresponds to a number. On a plane each point corresponds to a number pair from $R \times R$. To represent

$R \times R$ we draw an x-axis and a y-axis. The point at which they cross is called the *origin* and is labeled 0. The arrows show the positive directions. (See Fig. 1.) This is called a Cartesian coordinate system. The first member of an ordered pair is called the *first coordinate*. The second member is called the *second coordinate*.* Together these are called the *coordinates of a point*. The axes divide the plane into four *quadrants*, indicated by the Roman numerals.

Solutions of Equations

If an equation has two variables, its solutions are ordered pairs of numbers. A *solution* is an ordered pair which when substituted alphabetically for the variables produces a true equation.

Example 1 Determine whether the following ordered pairs are solutions of the equation $y = 3x - 1$: $(-1, -4)$, $(7, 5)$.

$$\begin{array}{c|c} y = 3x - 1 \\ \hline -4 & 3(-1) - 1 \\ -4 & -3 - 1 \\ & -4 \end{array}$$

We substitute -1 for x and -4 for y (alphabetical order of variables).

The equation becomes true: $(-1, -4)$ is a solution.

$$\begin{array}{c|c} y = 3x - 1 \\ \hline 5 & 3 \cdot 7 - 1 \\ 5 & 21 - 1 \\ & 20 \end{array}$$

We substitute.

The equation becomes false: $(7, 5)$ is not a solution.

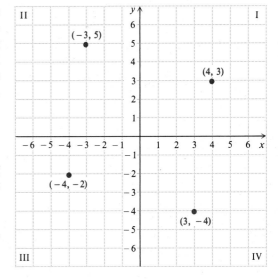

Figure 1

Graphs of Equations

The solutions of an equation are ordered pairs and thus constitute a relation. To *graph* an equation, or a relation, means to make a drawing of its solutions. Some general suggestions for graphing are as follows.

Graphing suggestions
a) **Use graph paper.**
b) **Label axes with symbols for the variables.**
c) **Use arrows to indicate positive directions.**
d) **Scale the axes; that is, mark numbers on the axes.**

---

*The first coordinate is sometimes called the *abscissa* and the second coordinate the *ordinate*.

e) **Plot solutions and complete the graph. When finished write down the equation or relation being graphed.**

Example 2 Graph: $y = 3x - 1$.

We find some ordered pairs that are solutions. To find an ordered pair, we choose *any* number that is a sensible replacement for x and then determine y. For example, if we choose 2 for x, then $y = 3(2) - 1$, or 5. We have found the solution (2, 5). We continue making choices for x and finding y. We make some negative, as well as positive, choices for x. We keep track of the solutions in a table.

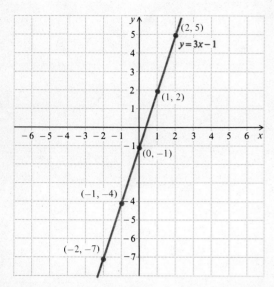

Figure 2

x	y
0	−1
1	2
2	5
−1	−4
−2	−7

The table gives us the ordered pairs (0, −1), (1, 2), (2, 5), and so on. Next, we plot these points. If we had enough of them, they would make a solid line. We can draw the line with a ruler, and label it $y = 3x - 1$. (See Fig. 2.)

Note that the equation $y = 3x - 1$ has an infinite (unending) set of solutions. The *graph of the equation* is a drawing of the relation that is made up of its solutions. Thus the relation consists of all pairs (x, y) such that $y = 3x - 1$ is true. That is, $\{(x, y) | y = 3x - 1\}$.

Example 3 Graph: $y = x^2 - 5$.

We select numbers for x and find the corresponding values for y. The table gives us the ordered pairs (0, −5), (−1, −4), and so on.

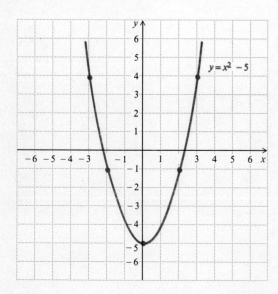

Figure 3

x	y
0	−5
−1	−4
1	−4
−2	−1
2	−1
−3	4
3	4

Next we plot these points. We note that as the absolute value of x increases, also $x^2 - 5$ increases. Thus the graph is a curve that rises gradually on either side of the y-axis, as shown in Fig. 3.

This graph shows the relation $\{(x, y) | y = x^2 - 5\}$.

Example 4 Graph the equation $x = y^2 + 1$.

Here we shall select numbers for y and then find the corresponding values for x. This time we arrange them in a horizontal table.

x	2	2	1	5	5
y	−1	1	0	−2	2

We must remember that x is the first coordinate and y is the second coordinate. Thus the table gives us the ordered pairs $(2, -1), (2, 1), (1, 0)$, and so on. We note that as the absolute value of y gradually increases, the value of x also gradually increases. Thus the graph is a curve that stretches farther and farther to the right as it gets farther from the x-axis. (See Fig. 4.)

This graph shows the relation $\{(x, y) \mid x = y^2 + 1\}$.

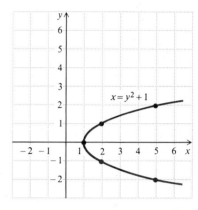
Figure 4

Example 5 Graph the equation $xy = 12$.

We find numbers that satisfy the equation.

x	1	−1	2	−2	3	−3	4	−4	6	−6	12	−12
y	12	−12	6	−6	4	−4	3	−3	2	−2	1	−1

We plot these points and connect them. To see how to do this, note that neither x nor y can be 0. Thus the graph does not cross either axis. As the absolute value of x gets small, the absolute value of y must get large, and conversely. Thus the graph consists of two curves, as shown in Fig. 5.

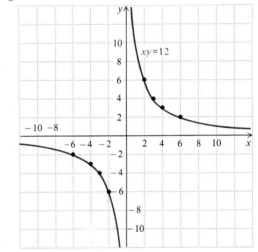
Figure 5

Example 6 Graph the equation $y = |x|$.

We find numbers that satisfy the equation.

x	0	1	−1	2	−2	3	−3	4	−4
y	0	1	1	2	2	3	3	4	4

We plot these points and connect them. To see how to do this, note that as we get farther from the origin, to the left or right, the absolute value of x increases. Thus the graph is a curve that rises to the left and right of the y-axis. It actually consists of parts of two straight lines, as shown in Fig. 6.

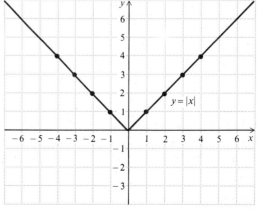
Figure 6

Domains and Ranges

In each example below we see a relation in real numbers, together with its domain and range.

Example 7

Example 8

Example 9

EXERCISE SET 3.2

1. Determine whether $(-1, -3)$ is a solution of $y = 5x + 2$.
2. Determine whether $(-2, 8)$ is a solution of $y = 3x - 7$.
3. Determine whether $(-2, 7)$ is a solution of $4x + 3y = 12$.
4. Determine whether $(0, -2)$ is a solution of $2x - 3y = 6$.
5. Determine whether $(3, 0)$ is a solution of $x^2 - 2y = 6$.
6. Determine whether $(-2, -1)$ is a solution of $x^2 - 2y = 6$.

Graph.

7. $y = x + 3$
8. $y = x - 2$
9. $y = 3x - 2$
10. $y = -4x + 1$
11. $y = x^2$
12. $y = -x^2$
13. $y = x^2 + 2$
14. $y = x^2 - 2$
15. $x = y^2 + 2$
16. $x = y^2 - 2$
17. $y = |x + 1|$
18. $y = |x - 1|$
19. $x = |y + 1|$
20. $x = |y - 1|$
21. $xy = 10$
22. $xy = -18$

Graph and compare.

23. $y = x^2 + 1$ and $y = (-x)^2 + 1$
24. $y = x^2 - 2$ and $y = 2 - x^2$

25. Graph a relation as follows.
    a) Draw a circle with radius of length 2, centered at (4, 3). Shade the circle and its interior.
    b) Shade (on the x-axis) the domain. Describe the domain.
    c) Shade (on the y-axis) the range. Describe the range.

26. Graph a relation as follows.
    a) Draw a triangle with vertices at (1, 1), (4, 2), and (3, 6). Shade the triangle and its interior.
    b) Shade (on the x-axis) the domain. Describe the domain.
    c) Shade (on the y-axis) the range. Describe the range.

In each of the following, all relations to be considered are in $R \times R$, where R is the set of real numbers.

27. Graph the relation in which the second coordinate is always 2 and the first coordinate may be any real number.

28. Graph the relation in which the first coordinate is always -3, and the second coordinate may be any real number.

29. Graph the relation in which the second coordinate is always 1 more than the first coordinate, and the first coordinate may be any real number.

30. Graph the relation in which the second coordinate is always 1 less than the first coordinate, and the first coordinate may be any real number.

31. Graph the relation in which the second coordinate is always twice the first coordinate, and the first coordinate may be any real number.

32. Graph the relation in which the second coordinate is always half the first coordinate, and the first coordinate may be any real number.

33. Graph the relation in which the second coordinate is always the square of the first coordinate, and the first coordinate may be any real number.

34. Graph the relation in which the first coordinate is always the square of the second coordinate, and the second coordinate may be any real number.

Graph the following equations. The definition of absolute value in Section 1.8 may be helpful for some of these exercises.

35. $y = |x| + x$
36. $y = x|x|$
37. $y = |x^2 - 4|$
38. $y = x^3$
39. $y = \sqrt{x}$
40. $y = |x^3|$
41. $|y| = |x|$
42. $|x| + |y| = 0$
43. $|xy| = 1$
44. Graph the relation $\{(x, y) | 1 < x < 4 \text{ and } -3 < y < -1\}$.

45. Graph $|x| + |y| = 1$.
46. Graph the relation $\{(x, y) | |x| \leq 1 \text{ and } |y| \leq 2\}$.

104 RELATIONS, FUNCTIONS, AND TRANSFORMATIONS

3.3 FUNCTIONS

Recognizing Graphs of Functions

A *function* is a special kind of relation. It is defined as follows.

DEFINITION

A *function* is a relation in which no two ordered pairs have the same first coordinate and different second coordinates.

In a function, given a member of the domain (a first coordinate), there is one and only one member of the range that goes with it (the second coordinate). Thus each member of the domain *determines* exactly one member of the range. It is easy to recognize the graph of a relation that is a function. If there are two or more points of the graph on the same vertical line, then the relation is not a function. Otherwise it is. Figure 7 shows some graphs of functions. In graph (c), the solid dot indicates that $(-1, 1)$ belongs to the graph. The open dot indicates that $(-1, -2)$ does not belong to the graph. Thus no vertical line crosses the graph more than once.

Figure 7

The graphs shown in Fig. 8 are not graphs of functions, because they fail the so-called *vertical line test*. That is, we can find a vertical line that meets the graph in more than one point.

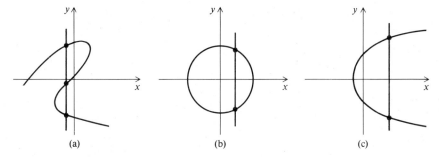

Figure 8

The vertical line test. **If it is possible for a vertical line to meet a graph more than once, the graph is not the graph of a function.**

Notation for Functions

Functions are often named by letters, such as f or g. A function f is thus a set of ordered pairs. If we represent the first coordinate of a pair by x, then we may represent the second coordinate by $f(x)$. The symbol $f(x)$ is read "f of x" or "f at x." The number represented by $f(x)$ is called the "value" of the function at x. [*Note*: "$f(x)$" does *not* mean "f times x."]

Example 1 Below let us call the function in color g.

$$\begin{array}{llll}(1, 4) & (2, 4) & (3, 4) & (4, 4)\\(1, 3) & (2, 3) & (3, 3) & (4, 3)\\(1, 2) & (2, 2) & (3, 2) & (4, 2)\\(1, 1) & (2, 1) & (3, 1) & (4, 1)\end{array}$$

Here $g(1) = 4$, $g(2) = 3$, $g(3) = 2$, and $g(4) = 4$.

Example 2 Let us call the function in Fig. 9 f. To find function values, we locate x on the x-axis, and then find $f(x)$ on the y-axis. From the graph in Fig. 9 we can see that $f(-2) = 0$, and that $f(0)$ is about 2.7, and $f(3)$ is about 0.6.

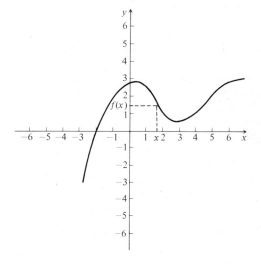

Figure 9

Some functions in real numbers can be defined by formulas, or equations. Here are some examples:

$$g(s) = 3, \quad p(t) = \frac{1}{t}, \quad f(x) = 3x^2 + 4x - 5, \quad u(y) = |y| + 3.$$

A function such as g above is called a *constant function* because all its function values are the same. The range contains only one number, 3.

Example 3 Let $f(x) = 2x^2 - 3$. We can find function values as follows:

a) $\quad f(0) = 2 \cdot 0^2 - 3 = -3;$

b) $\quad f(-1) = 2(-1)^2 - 3 = 2 \cdot 1 - 3 = -1;$

c) $\quad f(5a) = 2(5a)^2 - 3 = 2 \cdot 25a^2 - 3 = 50a^2 - 3;$

d) $\quad f(a - 4) = 2(a - 4)^2 - 3 = 2(a^2 - 8a + 16) - 3$
$\qquad\qquad = 2a^2 - 16a + 29;$

e) $\quad \dfrac{f(a + h) - f(a)}{h} = \dfrac{[2(a + h)^2 - 3] - [2a^2 - 3]}{h}$

$\qquad\qquad = \dfrac{2a^2 + 4ah + 2h^2 - 3 - 2a^2 + 3}{h}$

$\qquad\qquad = \dfrac{4ah + 2h^2}{h} = 4a + 2h.$

If you have trouble finding function values when a formula is given, think of the formula, in the case of Example 3, as

$$f(\ ) = 2(\ )^2 - 3.$$

Then whatever goes in the blank on the left between parentheses goes in the blank on the right between parentheses.

Finding the Domain of a Function

When a function in $R \times R$ is given by a formula, the domain is understood to be the set of all real numbers that are sensible replacements.

Example 4 Find the domain of $g(x) = \dfrac{x}{x^2 + 2x - 3}.$

The formula makes sense so long as a replacement for x does not make the denominator 0. To find those replacements that do make the denominator 0 we solve $x^2 + 2x - 3 = 0$:

$$x^2 + 2x - 3 = 0$$
$$(x - 1)(x + 3) = 0$$
$$x - 1 = 0 \quad \text{or} \quad x + 3 = 0$$
$$x = 1 \quad \text{or} \quad x = -3.$$

Thus the domain consists of the set of all real numbers except 1 and -3. We can name this set $\{x \mid x \neq 1 \text{ and } x \neq -3\}$.

Example 5 Find the domain of $f(x) = \sqrt{5x - 3}$.

The formula makes sense so long as a replacement makes the radicand nonnegative (no negative number has a square root). Thus to find the domain we solve the inequality $5x - 3 \geq 0$:

$$5x - 3 \geq 0$$
$$5x \geq 3$$
$$x \geq \frac{3}{5}.$$

The domain is $\{x \mid x \geq \frac{3}{5}\}$.

Example 6 Find the domain of $t(x) = x^3 + |x|$.

There are no restrictions on the numbers we can substitute into this formula. We can cube any real number, we can take the absolute value of any real number, and we can add the results. Thus the domain is the entire set of real numbers.

Figure 10

Functions, Mappings, and Machines

Functions can be thought of as mappings. A function f *maps* the set of first coordinates (the domain) to the set of second coordinates (the range). As shown in Fig. 10, each x in the domain corresponds to (or is mapped onto) just one y in the range. That y is the second coordinate of the ordered pair (x, y).

Example 7 Consider the function f for which $f(x) = 2x + 3$.

Since $f(0)$ is 3, this function maps 0 to 3. Since $f(3) = 9$, this function maps 3 to 9.

The concept of function is illuminated somewhat by considering a so-called *function machine*. In Fig. 11 we see a function machine de-

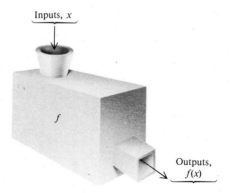

Figure 11

signed, or programmed, to do the mapping (function) f. The inputs acceptable to the machine are the members of the domain of f. The outputs are, of course, members of the range of f.

Composition of Functions

Functions can be combined in a way called *composition* of functions. Consider, for example,

$$f(x) = x^2 \qquad \text{This function squares each input.}$$

and

$$g(x) = x + 1. \qquad \text{This function adds 1 to each input.}$$

We define a new function that first does what g does (adds 1) and then does what f does (squares). The new function is called the *composition* of f and g, and is symbolized f ∘ g. Let us think of hooking two function machines f and g together to get the resultant function machine f ∘ g (see Fig. 12).

A composition machine for
f ∘ g

Figure 12

DEFINITION

The composed function f ∘ g is defined as follows: $f \circ g(x) = f(g(x))$.

Now let us see how to find a formula for $f \circ g$, and also the reverse composition $g \circ f$.

Example 8 Given that $f(x) = x^2$ and $g(x) = x + 1$, find formulas for $f \circ g(x)$ and $g \circ f(x)$.

By definition of $\circ$,
$$f \circ g(x) = f(g(x)).$$
Now
$$\begin{aligned} f(g(x)) &= f(x + 1) &&\text{Substituting } x + 1 \text{ for } g(x) \\ &= (x + 1)^2 &&\text{Substituting } x + 1 \text{ for } x \text{ in} \\ & &&\text{the formula for } f(x) \\ &= x^2 + 2x + 1. \end{aligned}$$

Similarly,
$$\begin{aligned} g \circ f(x) &= g(x^2) &&\text{Substituting } x^2 \text{ for } f(x) \\ &= x^2 + 1. &&\text{Substituting } x^2 \text{ for } x \text{ in the} \\ & &&\text{formula for } g(x) \end{aligned}$$

Note that the function $f \circ g$ is not the same as the function $g \circ f$.

In order for functions to be composable, such as f and g above, the outputs of g must be acceptable as inputs for f. In other words, if any number $g(x)$ is not in the domain of f, then x is not in the domain of $f \circ g$. Note also the order of happenings in $f \circ g$. The function $f \circ g$ does *first* what g does, *then* what f does.

Example 9 Let $f(x) = 2x$ and $g(x) = x^2 + 1$. Find $f \circ g(x)$ and $g \circ f(x)$.

By definition of $\circ$,
$$\begin{aligned} f \circ g(x) &= f(g(x)) \\ &= f(x^2 + 1) &&\text{Substituting } x^2 + 1 \text{ for } g(x) \\ &= 2(x^2 + 1) &&\text{Substituting } x^2 + 1 \text{ for } x \\ & &&\text{in the formula for } f(x) \\ &= 2x^2 + 2. \end{aligned}$$

Similarly,
$$\begin{aligned} g \circ f(x) &= g(f(x)) \\ &= g(2x) &&\text{Substituting } 2x \text{ for } f(x) \\ &= (2x)^2 + 1 &&\text{Substituting } 2x \text{ for } x \text{ in} \\ & &&\text{the formula for } g(x) \\ &= 4x^2 + 1. \end{aligned}$$

EXERCISE SET 3.3

1. Which of the following are graphs of functions?

a)   b)   c)   d)

  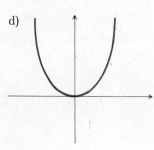

2. Which of the following are graphs of functions? An open circle means that point does not belong to the graph.

a)   b)   c)   d)

3. $f(x) = 5x^2 + 4x$. Find:
  a) $f(0)$;
  b) $f(-1)$;
  c) $f(3)$;
  d) $f(t)$;
  e) $f(t-1)$;
  f) $\dfrac{f(a+h) - f(a)}{h}$.

4. $g(x) = 3x^2 - 2x + 1$. Find:
  a) $g(0)$;
  b) $g(-1)$;
  c) $g(3)$;
  d) $g(t)$;
  e) $g(a+h)$;
  f) $\dfrac{g(a+h) - g(a)}{h}$.

5. $f(x) = 2|x| + 3x$. Find:
  a) $f(1)$;
  b) $f(-2)$;
  c) $f(-4)$;
  d) $f(2y)$;
  e) $f(a+h)$;
  f) $\dfrac{f(a+h) - f(a)}{h}$.

6. $g(x) = x^3 - 2x$. Find:
  a) $g(1)$;
  b) $g(-2)$;
  c) $g(-4)$;
  d) $g(3y)$;
  e) $g(2+h)$;
  f) $\dfrac{g(2+h) - g(2)}{h}$.

7. ▦ $f(x) = 4.3x^2 - 1.4x$. Find:
  a) $f(1.034)$;
  b) $f(-3.441)$;
  c) $f(27.35)$;
  d) $f(-16.31)$.

8. ▦ $g(x) = \sqrt{2.2|x| + 3.5}$. Find:
  a) $g(17.3)$;
  b) $g(-64.2)$;
  c) $g(0.095)$;
  d) $g(-6.33)$.

9. $f(x) = \dfrac{x^2 - x - 2}{2x^2 - 5x - 3}$. Find:
  a) $f(0)$;
  b) $f(4)$;
  c) $f(-1)$;
  d) $f(3)$.

10. $s(x) = \sqrt{\dfrac{3x - 4}{2x + 5}}$. Find:
  a) $s(10)$;
  b) $s(2)$;
  c) $s(1)$;
  d) $s(-1)$.

In Exercises 11–20, find the domain of the function.

11. $f(x) = 7x + 4$

12. $f(x) = |3x - 2|$

13. $f(x) = 4 - \dfrac{2}{x}$

14. $f(x) = \sqrt{x - 3}$

15. $f(x) = \sqrt{7x + 4}$

16. $f(x) = \dfrac{1}{9 - x^2}$

17. $f(x) = \dfrac{1}{x^2 - 4}$

18. $f(x) = \dfrac{2x + 6}{x^3 - 4x}$

19. $f(x) = \dfrac{4x^3 + 4}{4x^2 - 5x - 6}$

20. $f(x) = \dfrac{x^3 + 8}{x^2 - 4}$

In Exercises 21–26, find $f \circ g(x)$ and $g \circ f(x)$.

21. $f(x) = 3x^2 + 2$, $g(x) = 2x - 1$

22. $f(x) = 4x + 3$, $g(x) = 2x^2 - 5$

23. $f(x) = 4x^2 - 1$, $g(x) = \dfrac{2}{x}$

24. $f(x) = \dfrac{3}{x}$, $g(x) = 2x^2 + 3$

25. $f(x) = x^2 + 1$, $g(x) = x^2 - 1$

26. $f(x) = \dfrac{1}{x^2}$, $g(x) = x + 2$

☆ ───

27. From this graph, find approximately $g(-2)$, $g(-3)$, $g(0)$, and $g(2)$.

28. From this graph, find approximately $h(-2)$, $h(0)$, $h(3)$, and $h(-3)$.

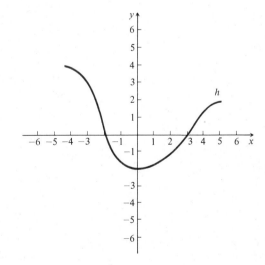

For each function find $\dfrac{f(x + h) - f(x)}{h}$.

29. $f(x) = \dfrac{1}{x}$

30. $f(x) = \dfrac{1}{x^2}$

31. $f(x) = \sqrt{x}$ (Rationalize the numerator.)

Find the domain of each function.

32. $f(x) = \dfrac{\sqrt{x}}{2x^2 - 3x - 5}$

33. $f(x) = \dfrac{\sqrt{x+3}}{x^2 - x - 2}$

34. $f(x) = \dfrac{\sqrt{x+1}}{x + |x|}$

35. $f(x) = \sqrt{x^2 + 1}$

36. For $f(x) = \dfrac{1}{1-x}$, find $f \circ f(x)$ and $f \circ [f \circ f(x)]$.

37. In Exercise 23, find the domain of $f \circ g$ and $g \circ f$.

38. Determine whether the relation $\{(x,y) \mid xy = 0\}$ is a function.

39. The *greatest integer function* $f(x) = [x]$ is defined as follows: $[x]$ is the greatest integer that is less than or equal to x. For example, if $x = 3.74$, then $[x] = 3$; and if $x = -0.98$, then $[x] = -1$. Graph the greatest integer function for values of x such that $-5 \leq x \leq 5$.

40. Graph the equation $[y] = [x]$. See Exercise 39. Is this the graph of a function?

3.4 SYMMETRY

Symmetry with Respect to a Line

In Fig. 13 points P and P_1 are said to be *symmetric* with respect to line l. They are the same distance from l.

Figure 13

DEFINITION

Two points P and P_1 are *symmetric with respect to a line l* if and only if l is the perpendicular bisector of the segment $\overline{PP_1}$. The line l is known as the *line of symmetry*.

We also say that the two points P and P_1 are *reflections* of each other across the line. The line is therefore also known as a *line of reflection*.

Now consider a set of points (geometric figure) as in the colored curve in Fig. 14. This figure is said to be symmetric with respect to the line l. This is because if you pick any point Q in the set, you can find another point Q_1 in the set such that Q and Q_1 are symmetric with respect to l.

Figure 14

3.4 SYMMETRY

DEFINITION

A figure, or set of points, is *symmetric with respect to a line l* if and only if for each point Q in the set there exists another point Q_1 in the set for which Q and Q_1 are symmetric with respect to line l.

Imagine picking up the curve in Fig. 14 and flipping it over. Points P and P_1 would be interchanged. Points Q and Q_1 would be interchanged. These are, then, pairs of symmetric points. The entire figure would look exactly like it did before flipping. This means that each point of the figure is symmetric with *some* point of the figure. Thus the figure is symmetric with respect to the line. A point and its reflection are known as *images* of each other. Thus P_1 is the image of P, for example. The line l is known as an *axis* of symmetry.

Symmetry with Respect to the Axes

There are special and interesting kinds of symmetry in which a coordinate axis is a line of symmetry. The following example shows graphs that are symmetric with respect to an axis and a graph that is not.

Example 1 In graph (a) in Fig. 15, flipping the figure about the y-axis would not change the figure. In graph (b), flipping the graph about the x-axis would not change the figure. In graph (c), flipping about either axis would change the figure.

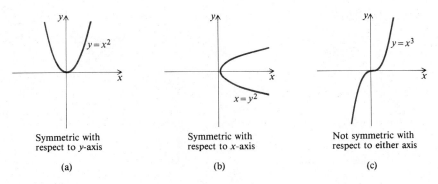

Figure 15

Let us consider a figure like graph (a) symmetric with respect to the y-axis. For every point of the figure there is another point the same

Figure 16

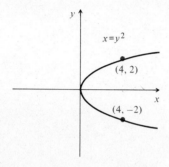

Figure 17

distance across the y-axis. The first coordinates of such a pair of points are additive inverses of each other.

Example 2 In the relation $y = x^2$ there are points $(2, 4)$ and $(-2, 4)$. (See Fig. 16.) The first coordinates, 2 and -2, are additive inverses of each other, whereas the second coordinates are the same. For every point of the figure (x, y), there is another point $(-x, y)$.

Let us consider a figure symmetric with respect to the x-axis. For every point of such a figure there is another point the same distance across the x-axis. The second coordinates of such a pair of points are additive inverses of each other.

Example 3 In the relation $x = y^2$ there are points $(4, 2)$ and $(4, -2)$. (See Fig. 17.) The second coordinates, 2 and -2, are additive inverses of each other, whereas the first coordinates are the same. For every point of the figure (x, y), there is another point $(x, -y)$.

Let us consider a figure symmetric with respect to the y-axis, as in Example 2. Suppose it is defined by an equation. If in this equation we replace x by $-x$, we obtain a new equation, but we get the same figure. This is true because any number x gives us the same y-value as its additive inverse, $-x$.

Let us consider a figure symmetric with respect to the x-axis, as in Example 3. Suppose it is defined by an equation. If we replace y by $-y$ in this equation, we obtain a new equation, but we get the same figure. This is true because any number y gives us the same x-value as $-y$. We thus have a means of testing a relation for symmetry, when it is defined by an equation.

When a relation is defined by an equation:

1. **If replacing *x* by *-x* produces an equivalent equation, then the graph is symmetric with respect to the *y*-axis.**
2. **If replacing *y* by *-y* produces an equivalent equation, then the graph is symmetric with respect to the *x*-axis.**

Example 4 Test $y = x^2 + 2$ for symmetry with respect to the y-axis.

a) Replace x by $-x$.

$$y = x^2 + 2 \quad (1)$$
$$y = (-x)^2 + 2$$

b) Simplify, if possible.

$$y = (-x)^2 + 2 = x^2 + 2 \quad (2)$$

c) Is the resulting equation (2) equivalent to the original (1)? Since the answer is yes, the graph is symmetric with respect to the y-axis.

Example 5 Test $y = x^2 + 2$ for symmetry with respect to the x-axis.

a) Replace y by $-y$.

$$y = x^2 + 2 \qquad (1)$$
$$\downarrow$$
$$-y = x^2 + 2$$

b) Simplify, if possible. The equation is simplified for the most part, although we could multiply on both sides by -1, obtaining

$$y = -(x^2 + 2). \qquad (2)$$

c) Is the resulting equation, equation (2), equivalent to the original (1)? This answer may be obvious to you and is no. To be sure one might use some trial and error reasoning as follows. Suppose we substitute 0 for x in equation (1). Then

$$y = 0^2 + 2 = 2,$$

so $(0, 2)$ is a solution of equation (1). In order for equations (1) and (2) to be equivalent, $(0, 2)$ must also be a solution of equation (2). We substitute to find out:

$$\begin{array}{c|c} y = -(x^2 + 2) \\ \hline 2 & -(0^2 + 2) \\ & -2 \end{array}$$

Thus $(0, 2)$ is not a solution of equation (2), so the equations are not equivalent and the graph is not symmetric with respect to the x-axis.

Example 6 Test $x^2 + y^4 + 5 = 0$ for symmetry with respect to the x-axis.

a) Replace y by $-y$.

$$x^2 + y^4 + 5 = 0 \qquad (1)$$
$$\downarrow$$
$$x^2 + (-y)^4 + 5 = 0$$

b) Simplify, if possible.

$$x^2 + (-y)^4 + 5 = x^2 + y^4 + 5 = 0 \qquad (2)$$

c) Is the resulting equation equivalent to the first? Since the answer is yes, the graph is symmetric with respect to the x-axis.

Symmetry with Respect to a Point

Two points are symmetric with respect to a point when they are situated as shown in Fig. 18. That is, the points are the same distance from that point, and all three points are on a line.

Figure 18

DEFINITION

Two points P and P_1 are *symmetric with respect to a point Q* if and only if Q (the point of symmetry) is the midpoint of segment $\overline{PP_1}$.

A *set* of points is symmetric with respect to a point when each point in the set is symmetric with some point in the set. This is illustrated in Fig. 19. Imagine sticking a pin in this figure at O and then rotating the figure 180°. Points P and P_1 would be interchanged. Points Q and Q_1 would be interchanged. These are pairs of symmetric points. The entire figure would look exactly as it did before rotating. This means that each point of the figure is symmetric with *some* point of the figure. Thus the figure is symmetric with respect to the point O.

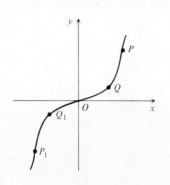

Figure 19

DEFINITION

A set of points is *symmetric with respect to a point B* if and only if for every point P in the set there exists another point P_1 in the set for which P and P_1 are symmetric with respect to B.

Symmetry with Respect to the Origin

A special kind of symmetry with respect to a point is symmetry with respect to the origin.

Example 7 In Fig. 20(a) and (b), rotating the graph about the origin 180° would not change the figure. In Fig. 20(c), such a rotation would change the figure.

Let us consider graphs like those in Fig. 20(a) and (b), symmetric with respect to the origin. For every point of the figure there is another point the same distance across the origin. The first coordinates of such a pair are additive inverses of each other, and the second coordinates are additive inverses of each other.

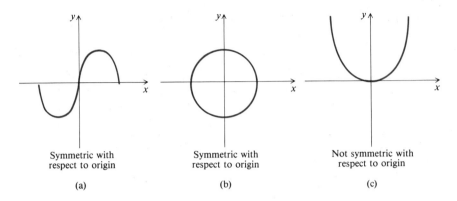

Symmetric with respect to origin
(a)

Symmetric with respect to origin
(b)

Not symmetric with respect to origin
(c)

Figure 20

Example 8 The relation $y = x^3$ is symmetric with respect to the origin. In this relation are the points (2, 8) and (−2, −8). (See Fig. 21.) The first coordinates are additive inverses of each other. The second coordinates are additive inverses of each other. For every point of the figure (x, y), there is another point (−x, −y).

In a relation symmetric with respect to the origin, as in Example 8, if we replace x by −x and y by −y, we obtain a new equation, but we will get the same figure. This is true because whenever a point (x, y) is in the relation, the point (−x, −y) is also in the relation. This gives us a means for testing a relation for symmetry when it is defined by an equation.

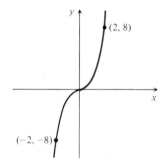

Figure 21

When a relation is defined by an equation, if replacing *x* by −*x* and replacing *y* by −*y* produces an equivalent equation, then the graph is symmetric with respect to the origin.

Example 9 Test $x^2 = y^2 + 2$ for symmetry with respect to the origin.

a) Replace x by −x and y by −y.

$$x^2 = y^2 + 2 \tag{1}$$
$$(-x)^2 = (-y)^2 + 2$$

b) Simplify, if possible. Since $(-x)^2 = x^2$ and $(-y)^2 = y^2$, $(-x)^2 = (-y)^2 + 2$ simplifies to

$$x^2 = y^2 + 2. \tag{2}$$

c) Is the resulting equation equivalent to the original? Since the answer is yes, the graph is symmetric with respect to the origin.

Example 10 Test $2x + 3y = 8$ for symmetry with respect to the origin.

a) Replace x by $-x$ and y by $-y$.

$$2x + 3y = 8$$
$$2(-x) + 3(-y) = 8$$

b) Simplify, if possible.

$$2(-x) + 3(-y) = -2x - 3y,$$

so $-(2x + 3y) = 8$ and $2x + 3y = -8$.

c) Is the resulting equation equivalent to the original? The equation $2x + 3y = -8$ is *not* equivalent to $2x + 3y = 8$, so the graph is not symmetric with respect to the origin.

EXERCISE SET 3.4

In Exercises 1–12, test for symmetry with respect to the coordinate axes and the origin.

1. $3y = x^2 + 4$
2. $5y = 2x^2 - 3$
3. $y^3 = 2x^2$
4. $3y^3 = 4x^2$
5. $2x^4 + 3 = y^2$
6. $3y^2 = 2x^4 - 5$
7. $2y^2 = 5x^2 + 12$
8. $3x^2 - 2y^2 = 7$
9. $2x - 5 = 3y$
10. $5y = 4x + 5$
11. $3b^3 = 4a^3 + 2$
12. $p^3 - 4q^3 = 12$

In Exercises 13–24, test for symmetry with respect to the origin.

13. $3x^2 - 2y^2 = 3$
14. $5y^2 = -7x^2 + 4$
15. $5x - 5y = 0$
16. $3x = 3y$
17. $3x + 3y = 0$
18. $7x = -7y$
19. $3x = \dfrac{5}{y}$
20. $3y = \dfrac{7}{x}$
21. $y = |2x|$
22. $3x = |y|$
23. $3a^2 + 4a = 2b$
24. $5v = 7u^2 - 2u$

Consider the figure at the right for Exercises 25–28.

25. Graph the reflection across the x-axis.
26. Graph the reflection across the y-axis.
27. Graph the reflection across the line $y = x$.
28. Graph the figure formed by reflecting each point through the origin.

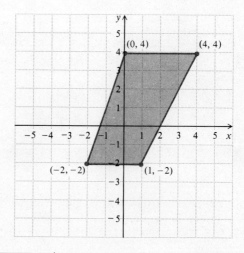

3.5 TRANSFORMATIONS

Given a relation, we can find various ways of altering it to obtain another relation. Such an alteration is called a *transformation*. If such an alteration consists merely of moving the graph without changing its shape or orientation, the transformation is called a *translation*.

Vertical Translations

Consider the relations $y = x^2$ and $y = 1 + x^2$ whose graphs are shown in Fig. 22. The graph of $y = 1 + x^2$ has the same shape as that of $y = x^2$, but is moved upward a distance of one unit. Consider any equation $y = f(x)$. Adding a constant a to produce $y = a + f(x)$ changes each function value by the same amount, a, hence produces no change in the shape of the graph, but merely translates it upward if the constant is positive. If a is negative, the graph will be moved downward.

Note that $y = 1 + x^2$ is equivalent to $y - 1 = x^2$. Thus the transformation above amounts to replacing y by $y - 1$ in the original equation.

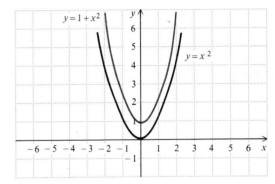

Figure 22

THEOREM 1

In an equation of a relation, replacing y by $y - a$, where a is a constant, translates the graph vertically a distance of $|a|$. If a is positive, the translation is upward. If a is negative, the translation is downward.*

If in an equation we replace y by $y + 3$, this is the same as replacing it by $y - (-3)$. In this case the constant a is -3 and the translation is downward. If we replace y by $y - 5$, the constant a is 5 and the translation is upward.

Example 1 Sketch a graph of $y = |x|$ and then $y = -2 + |x|$.

The graph of $y = |x|$ is shown in Fig. 23. Now consider $y = -2 + |x|$. Note that $y = -2 + |x|$ is equivalent to $y + 2 = |x|$ or $y - (-2) = |x|$. This shows that the new equation can be obtained by replacing y in

---

*In practice, when working with functions, we are more inclined to write $y = a + f(x)$ rather than $y - a = f(x)$, but for relations in general we usually consider replacing y by $y - a$.

$y = |x|$ by $y - (-2)$, so the translation is downward, two units, as shown in Fig. 23.

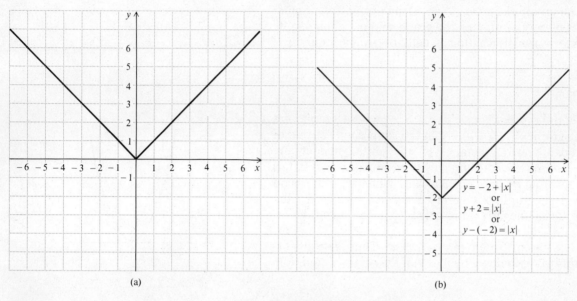

Figure 23

Horizontal Translations

Replacing y by $y - a$ in an equation translates vertically a distance of $|a|$. The translation is in the positive direction (upward) if a is positive. A similar thing happens in the horizontal direction. If we replace x by $x - b$ everywhere x occurs in the equation, we translate a distance of $|b|$ horizontally. If b is positive, we translate in the positive direction (to the right). If b is negative, we translate in the negative direction (to the left).

THEOREM 2

In an equation of a relation, replacing x by $x - b$, where b is a constant, translates the graph horizontally a distance of $|b|$. If b is positive, the translation is to the right. If b is negative, the translation is to the left.

Example 2 Given a graph of $y = |x|$, sketch a graph of $y = |x + 2|$.

Here we note that x is replaced by $x + 2$, or $x - (-2)$. Thus $b = -2$, and the translated graph will be moved two units in the negative direction (to the left). See Fig. 24.

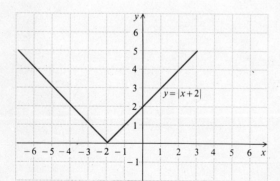

Figure 24

$y = |x|$ by $y - (-2)$, so the translation is downward, two units, as shown in Fig. 23.

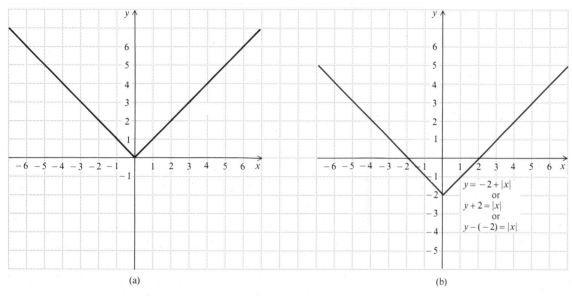

Figure 23

Horizontal Translations

Replacing y by $y - a$ in an equation translates vertically a distance of $|a|$. The translation is in the positive direction (upward) if a is positive. A similar thing happens in the horizontal direction. If we replace x by $x - b$ everywhere x occurs in the equation, we translate a distance of $|b|$ horizontally. If b is positive, we translate in the positive direction (to the right). If b is negative, we translate in the negative direction (to the left).

THEOREM 2

In an equation of a relation, replacing x by $x - b$, where b is a constant, translates the graph horizontally a distance of $|b|$. If b is positive, the translation is to the right. If b is negative, the translation is to the left.

Example 2 Given a graph of $y = |x|$, sketch a graph of $y = |x + 2|$.

Here we note that x is replaced by $x + 2$, or $x - (-2)$. Thus $b = -2$, and the translated graph will be moved two units in the negative direction (to the left). See Fig. 24.

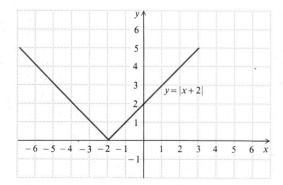

Figure 24

3.5 TRANSFORMATIONS

Given a relation, we can find various ways of altering it to obtain another relation. Such an alteration is called a *transformation*. If such an alteration consists merely of moving the graph without changing its shape or orientation, the transformation is called a *translation*.

Vertical Translations

Consider the relations $y = x^2$ and $y = 1 + x^2$ whose graphs are shown in Fig. 22. The graph of $y = 1 + x^2$ has the same shape as that of $y = x^2$, but is moved upward a distance of one unit. Consider any equation $y = f(x)$. Adding a constant a to produce $y = a + f(x)$ changes each function value by the same amount, a, hence produces no change in the shape of the graph, but merely translates it upward if the constant is positive. If a is negative, the graph will be moved downward.

Note that $y = 1 + x^2$ is equivalent to $y - 1 = x^2$. Thus the transformation above amounts to replacing y by $y - 1$ in the original equation.

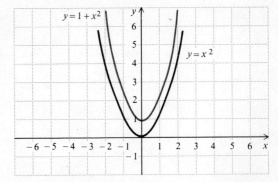

Figure 22

THEOREM 1

In an equation of a relation, replacing y by $y - a$, where a is a constant, translates the graph vertically a distance of $|a|$. If a is positive, the translation is upward. If a is negative, the translation is downward.*

If in an equation we replace y by $y + 3$, this is the same as replacing it by $y - (-3)$. In this case the constant a is -3 and the translation is downward. If we replace y by $y - 5$, the constant a is 5 and the translation is upward.

Example 1 Sketch a graph of $y = |x|$ and then $y = -2 + |x|$.

The graph of $y = |x|$ is shown in Fig. 23. Now consider $y = -2 + |x|$. Note that $y = -2 + |x|$ is equivalent to $y + 2 = |x|$ or $y - (-2) = |x|$. This shows that the new equation can be obtained by replacing y in

---

*In practice, when working with functions, we are more inclined to write $y = a + f(x)$ rather than $y - a = f(x)$, but for relations in general we usually consider replacing y by $y - a$.

3.5 TRANSFORMATIONS

Example 3 A circle centered at the origin with radius of length 1 has an equation $x^2 + y^2 = 1$. If we replace x by $x - 1$ and y by $y + 2$, we translate the circle so that the center is at the point $(1, -2)$. See Fig. 25.

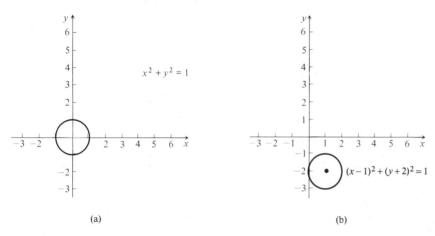

Figure 25

Vertical Stretchings and Shrinkings

Consider the function $y = |x|$. We will compare its graph with that of $y = 2|x|$ and $y = \frac{1}{2}|x|$. See Fig. 26. The graph of $y = 2|x|$ looks like that of $y = |x|$ but every output is doubled, so the graph is stretched in a vertical direction. The graph of $y = \frac{1}{2}|x|$ is flattened or shrunk in a vertical direction since every output is cut in half.

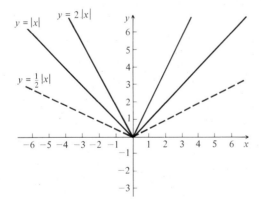

Figure 26

Consider any equation such as $y = f(x)$. Multiplying on the right by the constant 2 will double every function value, thus stretching the

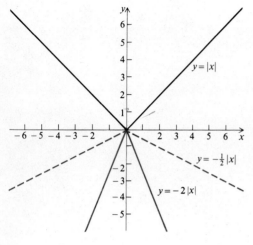

Figure 27

graph both ways from the horizontal axis. A similar thing is true for any constant greater than 1. If the constant is between 0 and 1, then the graph will be flattened or shrunk vertically.

When we multiply by a negative constant, the graph is reflected across the x-axis as well as being stretched or shrunk.

Example 4 Compare the graphs of $y = |x|$, $y = -2|x|$, and $y = -\frac{1}{2}|x|$ shown in Fig. 27.

Multiplying by c on the right is, of course, equivalent to dividing by c on the left.

THEOREM 3

In an equation of a relation, dividing y by a constant does the following to the graph.
1. If $|c| > 1$, the graph is stretched vertically.
2. If $|c| < 1$, the graph is shrunk vertically.
3. If c is negative, the graph is also reflected across the x-axis.*

Note that if $c = -1$, this has the effect of replacing y by $-y$ and we obtain a reflection without stretching or shrinking.

Example 5 Here is a graph of $y = f(x)$. Sketch a graph of $y/2 = f(x)$ or $y = 2f(x)$.

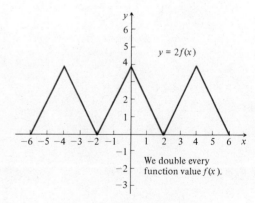

We double every function value $f(x)$.

---

*Again, with functions, we are more inclined to write $y = cf(x)$ rather than $y/c = f(x)$, but for relations in general we usually consider replacing y by y/c.

Example 6 Here is a graph of $y = g(x)$. Sketch a graph of $y/-\frac{1}{2} = g(x)$ or $y = -\frac{1}{2}g(x)$.

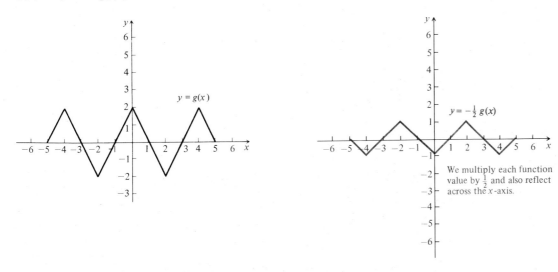

We multiply each function value by $\frac{1}{2}$ and also reflect across the x-axis.

Horizontal Stretchings and Shrinkings

For vertical stretchings and shrinkings we divided y by a constant c. Similarly, if we divide x by a constant wherever it occurs, we will get a horizontal stretching or shrinking.

THEOREM 4

In an equation of a relation, dividing x wherever it occurs by d does the following to the graph.
1. **If $|d| < 1$, the graph is shrunk horizontally.**
2. **If $|d| > 1$, the graph is stretched horizontally.**
3. **If d is negative, the graph is also reflected across the y-axis.***

Note that if $d = -1$, this has the effect of replacing x by $-x$ and we obtain a reflection without stretching or shrinking.

Example 7 Here is a graph of $y = f(x)$. Sketch a graph of $y = f(x/\frac{1}{2})$ or $y = f(2x)$, a graph of $y = f(x/2)$ or $y = f(\frac{1}{2}x)$, and a graph of $y =$

---

*Again, with functions, we are more inclined to write $y = f(kx)$ rather than $y = f(x/d)$, but for relations in general we usually consider replacing x by x/d.

$f(x/-2)$ or $y = f(-\frac{1}{2}x)$.

A shrinking. Cut each x-value in half.

A stretching and a reflection across the y-axis.

A stretching. Double each x-value.

EXERCISE SET 3.5

In Exercises 1–14 sketch graphs by transforming the graph of $y = |x|$.

1. $y + 3 = |x|$
2. $y = 2 + |x|$
3. $y = |x - 1|$
4. $y = |x + 2|$
5. $y = -4|x|$
6. $\frac{y}{3} = |x|$
7. $y = \frac{1}{3}|x|$
8. $y = -\frac{1}{4}|x|$
9. $y = |2x|$
10. $y = \left|\frac{x}{3}\right|$
11. $y = |x - 2| + 3$
12. $y = 2|x + 1| - 3$
13. $y = -3|x - 2|$
14. $y = \frac{1}{3}|x + 2| + 1$

Here is a graph of y = f(x). No formula will be given for this function. In Exercises 15–33 sketch graphs by transforming this one.

15. $y = 2 + f(x)$
16. $y + 1 = f(x)$
17. $y = f(x - 1)$
18. $y = f(x + 2)$
19. $\dfrac{y}{-2} = f(x)$
20. $y = 3f(x)$
21. $y = \dfrac{1}{3}f(x)$
22. $y = -\dfrac{1}{2}f(x)$
23. $y = f(2x)$
24. $y = f(3x)$
25. $y = f(-2x)$
26. $y = f(-3x)$
27. $y = f\left(\dfrac{x}{-2}\right)$
28. $y = f\left(\dfrac{1}{3}x\right)$
29. $y = f(x - 2) + 3$
30. $y = -3f(x - 2)$
31. $y = 2 \cdot f(x + 1) - 2$
32. $y = \dfrac{1}{2}f(x + 2) - 1$
33. $y = -\dfrac{1}{2}f(x - 3) + 2$

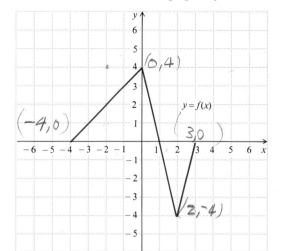

Here is a graph of y = f(x). No formula will be given for this function. In Exercises 34–38 sketch graphs by transforming this one.

34. $y = -2f(x + 1) - 1$
35. $y = 3f(x + 2) + 1$
36. $y = \dfrac{5}{2}f(x - 3) - 2$

37. $y = -\sqrt{2}f(x + 1.8)$
38. $y = \dfrac{\sqrt{3}}{2} \cdot f(x - 2.5) - 5.3$

★

39. A *linear transformation* of a coordinatized line is one that takes any point x to the point x', where x' = ax + b (a and b constants). For example, the transformation x' = 3x + 5 takes the point 2 to the point 3·2 + 5, or 11. Suppose for a particular linear transformation the point 2 goes to 5 and the point 3 goes to 7.

a) Describe the transformation as x' = ax + b.
b) This transformation leaves one point fixed. What point is it?
c) The set $\{x \mid -2 \leq x \leq 1\}$ is mapped to what set under this transformation?

40. How does the linear transformation $x' = x - 3$ transform each of the following sets?
a) $\{x \mid 0 < x < 1\}$
b) $\{x \mid -4 \leq x < -1\}$
c) $\{x \mid 8 \leq x \leq 20\}$

41. Show that any linear transformation $x' = ax + b, a \neq 1$, leaves one point fixed.

42. A professor gives a test with 80 possible points. She decides that a score of 55 should be passing. She performs a linear transformation on the scores so that a perfect paper gets 100 and 70 is passing.
a) Describe the linear transformation that the professor used.
b) What grade remains unchanged under the transformation?

43. Suppose the average score on the test of Exercise 42 was 60. What will be the average of the transformed scores?

Given the graph of $y = f(x)$ for Exercises 34–36, graph each of the following.

44. $\dfrac{y}{3} = f\left(2x + \dfrac{1}{2}\right)$

45. $y = -4 \cdot f(5x + 10)$

3.6 SOME SPECIAL CLASSES OF FUNCTIONS

Even and Odd Functions

If the graph of a function is symmetric with respect to the y-axis, it is an *even* function. Recall (Section 3.4) that a function will be symmetric to the y-axis if in its equation we can replace x by $-x$ and obtain an equivalent equation. Thus if we have a function given by $y = f(x)$, then $y = f(-x)$ will give the same function if the function is even. In other words, an even function is one for which $f(x) = f(-x)$ for all x in its domain. This is the definition of even function.

DEFINITION

A function f is an *even* function in case $f(x) = f(-x)$ for all x in the domain of f.

Example 1 Determine whether the function $f(x) = x^2 + 1$ is even.

a) Find $f(-x)$ and simplify.

$$f(-x) = (-x)^2 + 1 = x^2 + 1$$

b) Compare $f(x)$ and $f(-x)$. Since $f(x) = f(-x)$ for all x in the domain, f is an even function.

Note that the graph is symmetric with respect to the y-axis (Fig. 28).

Example 2 Determine whether the function $f(x) = x^2 + 8x^3$ is even.

a) Find $f(-x)$ and simplify.
$$f(-x) = (-x)^2 + 8(-x)^3 = x^2 - 8x^3$$

b) Compare $f(x)$ and $f(-x)$. Since $f(x)$ and $f(-x)$ are *not* the same for all x in the domain, f is *not* an even function.

If the graph of a function is symmetric with respect to the origin, it is an *odd* function. Recall that a function will be symmetric with respect to the origin if in its equation we can replace x by $-x$ and y by $-y$ and obtain an equivalent equation. Thus if we have a function given by $y = f(x)$, then $-y = f(-x)$ will be equivalent if f is an odd function. In other words, an odd function is one for which $f(-x) = -f(x)$ for all x in the domain. Let us make this our definition.

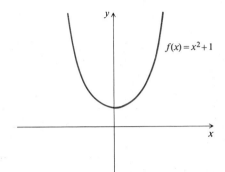

Figure 28

DEFINITION

A function f is an *odd* function when $f(-x) = -f(x)$ for all x in the domain of f.

Example 3 Determine whether $f(x) = x^3$ is an odd function.

a) Find $f(-x)$ and $-f(x)$ and simplify.
$$f(-x) = (-x)^3 = -x^3,$$
$$-f(x) = -x^3.$$

b) Compare $f(-x)$ and $-f(x)$. Since $f(-x) = -f(x)$ for all x in the domain, f is odd. Note that the graph is symmetric with respect to the origin (Fig. 29).

Example 4 Determine whether $f(x) = x^2 - 4x^3$ is even, odd, or neither even nor odd.

a) Find $f(-x)$ and $-f(x)$ and simplify.
$$f(-x) = (-x)^2 - 4(-x)^3 = x^2 + 4x^3,$$
$$-f(x) = -x^2 + 4x^3.$$

b) Compare $f(x)$ and $f(-x)$ to determine whether f is even. Since $f(x)$ and $f(-x)$ are *not* the same for all x in the domain, f is *not* even.

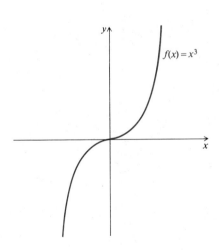

Figure 29

c) Compare $f(-x)$ and $-f(x)$ to determine whether f is odd. Since $f(-x)$ and $-f(x)$ are *not* the same for all x in the domain, f is *not* odd. Thus f is *neither* even nor odd.

Periodic Functions

Certain functions with a repeating pattern are called *periodic*. Some examples are shown in Fig. 30.

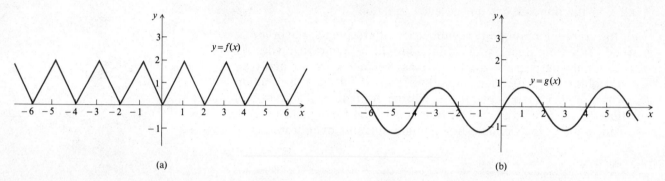

Figure 30

The function values of the function f repeat themselves every two units as we move from left to right. In other words, for any x, we have $f(x) = f(x + 2)$. To see this another way, think of the part of the graph between 0 and 2 and note that the rest of the graph consists of copies of it. In terms of translations, if we translate the graph two units to the left or right, the original graph will be obtained.

In the function g, the function values repeat themselves every four units. Hence $g(x) = g(x + 4)$ for any x, and if the graph is translated four units to the left, or right, it will coincide with itself. Or, think of the part of the graph between 0 and 4. The rest of the graph consists of copies of it.

We say that f has a *period* of 2, and that g has a period of 4.

DEFINITION

If a function f has the property that $f(x + p) = f(x)$ whenever x and $x + p$ are in the domain, where p is a positive constant, then f is said to be *periodic*. The smallest positive number p (if there is one) for which $f(x + p) = f(x)$ for all x is called the *period* of the function.

Interval Notation

For a and b real numbers such that $a < b$, we define the *open interval* (a, b) as follows:

(a, b) is the set of all numbers x such that $a < x < b$,

or
$$\{x \mid a < x < b\}.$$

Its graph is as follows:

Note that the endpoints are not included. Be careful not to confuse this notation with that of an ordered pair. The context of the writing should make the meaning clear. If not, we might say "the interval $(-2, 3)$." When we mean an ordered pair, we might say "the pair $(-2, 3)$."

The *closed interval* $[a, b]$ is defined as follows:

$[a, b]$ is the set of all x such that $a \leq x \leq b$,

or
$$\{x \mid a \leq x \leq b\}.$$

Its graph is as follows:

Note that endpoints are included. For example, the graph of $[-2, 3]$ is as follows:

There are two kinds of *half-open intervals* defined as follows:

a) $$(a, b] = \{x \mid a < x \leq b\}.$$

This is open on the left. Its graph is as follows:

b) $$[a, b) = \{x \mid a \leq x < b\}.$$

This is open on the right. Its graph is as follows:

Continuous Functions

Some functions have graphs that are continuous curves, without breaks or holes in them. Such functions are called *continuous functions*.

Example 5 The function f in Fig. 31 is continuous because it has no breaks, jumps, or holes in it. The function g has *discontinuities* where $x = -2$ and $x = 5$. The function g is continuous on the interval $[-2, 5]$ or on any interval contained therein. It is also continuous on other intervals, but it is not continuous on an interval such as $[-3, 1]$ or $(3, 7)$.

Figure 31

Increasing and Decreasing Functions

If the graph of a function rises from left to right it is said to be an *increasing* function. If the graph of a function drops from left to right, it is said to be a *decreasing* function. (See Fig. 32.) This can be stated more formally.

Figure 32

DEFINITION

1. A function f is an *increasing* function when for all a and b in the domain of f, if $a < b$, then $f(a) < f(b)$.

2. A function f is a *decreasing* function when for all a and b in the domain of f, if $a < b$, then $f(a) > f(b)$.

Examples 6-9 Determine whether each function is increasing, decreasing, or neither.

6.

f is increasing

7.

g is decreasing

8.

h is neither increasing nor decreasing

9.

m is neither increasing nor decreasing

Note that while the function m is neither increasing nor decreasing, it is increasing on the interval $[0, 2]$ and decreasing on the interval $[-2, 0]$.

Functions Defined Piecewise

Sometimes functions are defined piecewise. That is, we have different output formulas for different parts of the domain.

Example 10 Graph the function defined as follows.

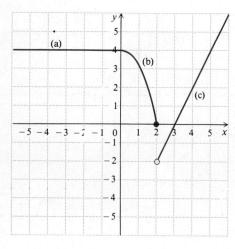

$$f(x) = \begin{cases} 4 & \text{for } x \leq 0; \\ & \text{(This means that for any input x less than or equal to 0, the output is 4.)} \\ 4 - x^2 & \text{for } 0 < x \leq 2; \\ & \text{(This means that for any input x greater than 0 and less than or equal to 2, the output is } 4 - x^2.) \\ 2x - 6 & \text{for } x > 2. \\ & \text{(This means that for any input x greater than 2, the output is } 2x - 6.) \end{cases}$$

See the graph in Fig. 33.
a) We graph $f(x) = 4$ for inputs less than or equal to 0 (that is, $x \leq 0$).
b) We graph $f(x) = 4 - x^2$ for inputs greater than 0 and less than or equal to 2 (that is, $0 < x \leq 2$).
c) We graph $f(x) = 2x - 6$ for inputs greater than 2 (that is, $x > 2$).

Figure 33

EXERCISE SET 3.6

1. Determine whether each function is even, odd, or neither even nor odd.

a)

b)

c)

d)

2. Determine whether each function is even, odd, or neither even nor odd.

a)

b)

c)

d)

Determine whether each function is even, odd, or neither even nor odd.

3. $f(x) = 2x^2 + 4x$
4. $f(x) = -3x^3 + 2x$
5. $f(x) = 3x^4 - 4x^2$
6. $f(x) = 5x^2 + 2x^4 - 1$
7. $f(x) = 7x^3 + 4x - 2$
8. $f(x) = 4x$
9. $f(x) = |3x|$
10. $f(x) = x^{24}$
11. $f(x) = x^{17}$
12. $f(x) = x + \dfrac{1}{x}$
13. $f(x) = x - |x|$
14. $f(x) = \sqrt{x}$
15. $f(x) = \sqrt[3]{x}$
16. $f(x) = 7$
17. $f(x) = 0$
18. $f(x) = \sqrt[3]{x - 2}$
19. $f(x) = \sqrt{x^2 + 1}$
20. $f(x) = \dfrac{x^2 + 1}{x^3 - x}$

21. Determine whether each function is periodic.

a)
b)
c)
d)

22. Determine whether each function is periodic.

a)
b)
c)
d)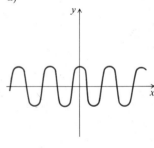

23. What is the period of this function?

24. What is the period of this function?

25. Write interval notation for each set pictured.

a)
b)

c)

26. Write interval notation for each set pictured.

a) ![number line -3 to 2]

d)

b) ![number line c to d]

c) ![number line 0 to 5]

d) ![number line -2 to 2]

27. Write interval notation for each set.
a) $\{x \mid -2 < x < 4\}$
b) $\{x \mid -\frac{1}{4} < x \leq \frac{1}{4}\}$
c) $\{x \mid 7 \leq x < 10\pi\}$
d) $\{x \mid -9 \leq x \leq -6\}$

28. Write interval notation for each set.
a) $\{x \mid -5 < x < 0\}$
b) $\{x \mid -\sqrt{2} \leq x < \sqrt{2}\}$
c) $\{x \mid -\frac{\pi}{2} < x \leq \frac{\pi}{2}\}$
d) $\{x \mid -12 \leq x \leq -\frac{1}{2}\}$

29. Is this function continuous

a) in the interval [0, 2]?
b) in the interval (−2, 0)?
c) in the interval [0, 3]?
d) in the interval [−3, −1]?
e) in the interval (−3, −1)?

30. Is this function continuous

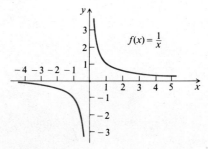

a) in the interval [−3, −1]?
b) in the interval (1, 4)?
c) in the interval [−1, 1]?
d) in the interval [−2, 4]?
e) in the interval (0, 1]?

31. Where are the discontinuities of the function in Exercise 29?

32. Where are the discontinuities of the function in Exercise 30?

33. Determine whether each function is increasing, decreasing, or neither increasing nor decreasing.

34. Determine whether each function is increasing, decreasing, or neither increasing nor decreasing.

a)   b)   c)   d)

Graph.

35. $f(x) = \begin{cases} 1, & \text{for } x < 0, \\ -1, & \text{for } x \geq 0 \end{cases}$

36. $f(x) = \begin{cases} 2, & \text{for } x \text{ an integer,} \\ -2, & \text{for } x \text{ not an integer} \end{cases}$

37. $f(x) = \begin{cases} 3, & \text{for } x \leq -3, \\ |x|, & \text{for } -3 < x \leq 3, \\ -3, & \text{for } x > 3 \end{cases}$

38. $f(x) = \begin{cases} -2x - 6, & \text{for } x \leq -2, \\ 2 - x^2, & \text{for } -2 < x < 2, \\ 2x - 6, & \text{for } x \geq 2 \end{cases}$

39. *The postage function.* Postage rates are as follows: 15¢ for the first ounce plus 13¢ for each additional ounce or fraction thereof. Thus, if x is the weight of a letter in ounces, then $p(x)$ is the cost of mailing the letter, where

$$p(x) = \begin{cases} 15\text{¢}, & \text{if } 0 < x \leq 1, \\ 28\text{¢}, & \text{if } 1 < x \leq 2, \\ 41\text{¢}, & \text{if } 2 < x \leq 3, \end{cases}$$

and so on, up to 12 ounces, after which postal cost also depends on distance. Graph this function for x such that $0 < x \leq 12$.

40. Graph.

$f(x) = \begin{cases} 3 + x, & \text{for } x \leq 0, \\ \sqrt{x}, & \text{for } 0 < x < 4, \\ x^2 - 4x - 1, & \text{for } x \geq 4 \end{cases}$

☆ ──────────────────────────────

41. For the function in Exercise 33(d), find an interval on which the function is (a) increasing; (b) decreasing.

42. For the function in Exercise 34(d), find an interval on which the function is (a) increasing; (b) decreasing.

43. Graph each function. Then determine whether it is increasing, decreasing, or neither increasing nor decreasing.

a) $f(x) = 3x + 4$ b) $f(x) = -3x + 4$
c) $f(x) = x^2 + 1$ d) $f(x) = -2$
e) $f(x) = x^3 + 1$ f) $f(x) = |x|$

44. Graph each function. Then determine whether it is increasing, decreasing, or neither increasing nor decreasing.

a) $f(x) = -2x - 3$ b) $f(x) = 2x - 3$
c) $f(x) = 3x^2$ d) $f(x) = \sqrt{3}$
e) $f(x) = |x| + 2$ f) $f(x) = x^3 - 2$

45. To what interval is each of the following mapped under the linear transformation $x' = x + 2$?

a) [0, 1]   b) (−2, 7]   c) (−8, −1)

46. To what interval is each of the following mapped under the linear transformation $x' = 0.4x - 3$?

a) [0, 1]   b) (−4, −20)   c) [6, 11]

3.7 INVERSES

Inverses of Relations

If in a relation we interchange first and second members in each ordered pair, we obtain a new relation. The new relation is called the *inverse* of the original relation. A relation is shown here in color; its inverse is shaded.

(1, 4)	(2, 4)	(3, 4)	(4, 4)
(1, 3)	(2, 3)	(3, 3)	(4, 3)
(1, 2)	(2, 2)	(3, 2)	(4, 2)
(1, 1)	(2, 1)	(3, 1)	(4, 1)

When a relation is defined by an equation, interchanging x and y produces an equation of the inverse relation.

Example 1 Find an equation of the inverse of $y = x^2 - 5$.

We interchange x and y, and obtain $x = y^2 - 5$. This is an equation of the inverse relation.

Graphs of Inverse Relations

Interchanging first and second coordinates in each ordered pair of a relation has the effect of interchanging the x-axis and the y-axis.

Interchanging the x-axis and the y-axis has the effect of reflecting across the diagonal line whose equation is $y = x$, as shown in Fig. 34. Thus the graphs of a relation and its inverse are always reflections of each other across the line $y = x$. (This assumes, of course, that the same scale is used on both axes.)

Figure 34

Examples 2–4 In each case a relation is shown in black. The graph of its inverse is shown in color.

2.

3.

4.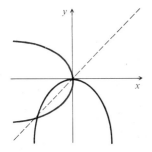

Symmetry with Respect to the Line $y = x$

It can happen that a relation is its own inverse; that is, when x and y are interchanged, or the relation is reflected across the line $y = x$, there is no change. Such a relation is symmetric with respect to the line $y = x$. Figures 35 and 36 are two examples of relations that are symmetric with respect to the line $y = x$.

Each of the relations shown in black in Examples 2-4 is *not* symmetric with respect to the line $y = x$.

For a relation defined by an equation:

If interchanging x and y produces an equivalent equation, the relation is its own inverse, and the graph of the equation is symmetric with respect to the line $y = x$.

Example 5 Test $3x + 3y = 5$ for symmetry with respect to the line $y = x$.

a) Interchange x and y. This amounts to replacing each occurrence of x by a y and each y by an x.

$$3x + 3y = 5 \quad (1)$$
$$3y + 3x = 5 \quad (2)$$

b) Is the resulting equation equivalent to the original? The commutative law of addition guarantees that the resulting equation is equivalent to the original. Thus the graph is symmetric with respect to the line $y = x$.

Example 6 Test $y = x^2$ for symmetry with respect to the line $y = x$.

a) Interchange x and y.

$$y = x^2$$
$$x = y^2$$

Figure 35

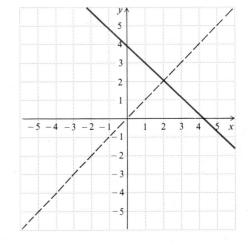

Figure 36

b) Is the resulting equation equivalent to the original? Note that (2, 4) is a solution of the original equation $y = x^2$, but it is not a solution of the resulting equation $x = y^2$.

$y = x^2$		$x = y^2$	
4	2^2	2	4^2
	4		16

Thus the equations are not equivalent, so the graph of $y = x^2$ is *not* symmetric with respect to the line $y = x$.

Inverses of Functions

Every function has an inverse, but that inverse may not be a function, as the following examples show.

Example 7 The relation g given by

$$g = \{(5, 3), (3, 1), (-7, 2), (2, 3)\}$$

is a function. Find the inverse of g and determine whether it is a function.

The inverse of g is found by interchanging first and second members in each ordered pair, and is

$$\{(3, 5), (1, 3), (2, -7), (3, 2)\}.$$

It is *not* a function because the pairs (3, 5) and (3, 2) have the same first coordinates but different second coordinates.

Example 8 Graph the relation $y = x^2$ and determine whether it is a function. Then graph its inverse and decide whether it is a function. See Figs. 37 and 38.

Finding Formulas for Inverses of Functions

All functions have inverses, but in only some cases is the inverse also a function. If the inverse of a function f is also a function, we denote it by f^{-1} (read "f inverse"). [*Caution:* This is *not* exponential notation!] Recall that we obtain the inverse of any relation by reversing the coordinates in each ordered pair. In terms of mapping, let us see what this means.

A function f *maps* the set of first coordinates (the domain) D to the set of second coordinates (the range) R. (See Fig. 39.) The inverse mapping f^{-1} maps the range of f onto the domain of f. Each y in R is mapped

A function

Figure 37

Its inverse: not a function

Figure 38

Figure 39

onto just one x in D, provided f^{-1} is a function. Note that the domain of f is the range of f^{-1} and the range of f is the domain of f^{-1}.

Let us consider inverses of functions in terms of function machines. Suppose that the function f programmed into a machine has an inverse that is also a function. Suppose then that the function machine has a reverse switch. When the switch is thrown the machine is then programmed to do the inverse mapping f^{-1}. Inputs then enter at the opposite end and the entire process is reversed. See Fig. 40.

When a function is defined by a formula, we can sometimes find a formula for its inverse by thinking of interchanging x and y.

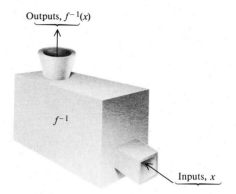

Figure 40

Example 9 Given $f(x) = x + 1$, find a formula for $f^{-1}(x)$.

a) Let us think of this as $y = x + 1$.
b) To find the inverse we interchange x and y: $x = y + 1$.
c) Now we solve for y: $y = x - 1$.
d) Thus $f^{-1}(x) = x - 1$.

Note in Example 9 that f maps any x onto $x + 1$ (this function adds 1 to each number of the domain). Its inverse, f^{-1}, maps any number x onto $x - 1$ (this inverse function subtracts 1 from each member of its domain). Thus the function and its inverse do opposite things.

Example 10 Let $g(x) = 2x + 3$. Find a formula for $g^{-1}(x)$.

a) Let us think of this as $y = 2x + 3$.
b) To find the inverse we interchange x and y: $x = 2y + 3$.
c) Now we solve for y: $y = (x - 3)/2$.
d) Thus $g^{-1}(x) = (x - 3)/2$.

Note in Example 10 that g maps any x onto $2x + 3$ (this function doubles each input and then adds 3). Its inverse, g^{-1}, maps any input onto $(x - 3)/2$ (this inverse function subtracts 3 from each input and then divides it by 2). Thus the function and its inverse do opposite things.

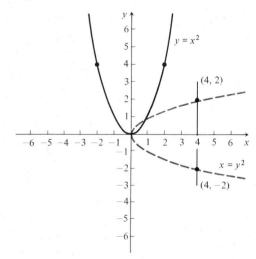

Figure 41

Example 11 Consider $f(x) = x^2$. Let us think of this as $y = x^2$.

To find the inverse we interchange x and y: $x = y^2$. Now we solve for y: $y = \pm\sqrt{x}$. Note that for each positive x we get two y's. For example, the pairs (4, 2) and (4, −2) belong to the relation. Look at the graphs in Fig. 41. Note also that the inverse fails the vertical line test.

If we restrict the domain of $f(x) = x^2$ to nonnegative numbers, then its inverse is a function, $f^{-1}(x) = \sqrt{x}$. See Fig. 42.

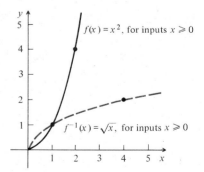

Figure 42

Finding $f^{-1}(f(x))$ and $f(f^{-1}(x))$

Suppose the inverse of a function f is also a function. Let us suppose that we do the mapping f and then do the inverse mapping f^{-1}. We will be back where we started. In other words, if we find $f(x)$ for some x and then find f^{-1} for this number, we will be back at x. In function notation the statement looks like this:

$$f^{-1}(f(x)) = x.$$

This is read "f inverse of f of x equals x." It means, working from the inside out, to take x, then find $f(x)$, and then find f^{-1} for that number. When we do, we will be back where we started, at x. For similar reasons, the following is also true:

$$f(f^{-1}(x)) = x.$$

For the above statements to be true, x must of course be in the domain of the function being considered.

Let us consider function machines. To find $f^{-1}(f(x))$ we use x as an input, when the machine is set on f. We then use the output for an input, running the machine backward. The last output is the same as the first input. The first output must of course be an acceptable input when the machine is set on f^{-1}. Following is a precise statement of the property.

THEOREM 5

For any function f whose inverse is a function, $f^{-1}(f(a)) = a$ for any a in the domain of f. Also, $f(f^{-1}(a)) = a$ for any a in the domain of f^{-1}.

Proof of Theorem 5. Suppose a is in the domain of f. Then

$$f(a) = b, \text{ for some } b \text{ in the range of } f.$$

Then the ordered pair (a, b) is in f, and by definition of f^{-1}, (b, a) is in f^{-1}. It follows that $f^{-1}(b) = a$. Then substituting $f(a)$ for b, we get $f^{-1}(f(a)) = a$. A similar proof shows that $f(f^{-1}(a)) = a$, for some a in the domain of f^{-1}. □

Example 12 For the function g in Example 10, find $g(4)$. Then find $g^{-1}(g(4))$.

$$g(x) = 2x + 3, \quad \text{so} \quad g(4) = 11$$

Now
$$g^{-1}(x) = \frac{x - 3}{2},$$

so
$$g^{-1}(11) = \frac{11-3}{2} = 4.$$

Thus
$$g^{-1}(g(4)) = 4.$$

Example 13 For the function g in Example 12 find $g^{-1}(g(283))$. Find also $g(g^{-1}(-12{,}045))$.

We note that every real number is in the domain of both g and g^{-1}. Thus we may immediately write the answers, without calculating.

$$g^{-1}(g(283)) = 283,$$
$$g(g^{-1}(-12{,}045)) = -12{,}045.$$

EXERCISE SET 3.7

Write an equation of the inverse relation.

1. $y = 4x - 5$
2. $y = 3x + 5$
3. $x^2 - 3y^2 = 3$
4. $2x^2 + 5y^2 = 4$
5. $y = 3x^2 + 2$
6. $y = 5x^2 - 4$
7. $xy = 7$
8. $xy = -5$
9. Graph $y = x^2 + 1$. Then by reflection across the line $y = x$, graph its inverse.
10. Graph $y = x^2 - 3$. Then by reflection across the line $y = x$, graph its inverse.
11. Graph $y = |x|$. Then by reflection across the line $y = x$, graph its inverse.
12. Graph $x = |y|$. Then by reflection across the line $y = x$, graph its inverse.

Test for symmetry with respect to the line $y = x$.

13. $3x + 2y = 4$
14. $5x - 2y = 7$
15. $4x + 4y = 3$
16. $5x + 5y = -1$
17. $xy = 10$
18. $xy = 12$
19. $3x = \dfrac{4}{y}$
20. $4y = \dfrac{5}{x}$
21. $y = |2x|$
22. $3x = |2y|$
23. $4x^2 + 4y^2 = 3$
24. $3x^2 + 3y^2 = 5$

25. Which of the following have inverses that are functions?

a)
b)
c)
d)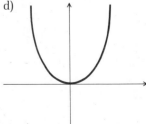

26. Which of the following have inverses that are functions?

a)   b)   c)   d)

27. $f(x) = 2x + 5$. Find a formula for $f^{-1}(x)$.

28. $g(x) = 3x - 1$. Find a formula for $g^{-1}(x)$.

29. $f(x) = \sqrt{x + 1}$. Find a formula for $f^{-1}(x)$.

30. $g(x) = \sqrt{x - 1}$. Find a formula for $g^{-1}(x)$.

31. $f(x) = 35x - 173$. Find $f^{-1}(f(3))$.
Find $f(f^{-1}(-125))$.

32. $g(x) = \dfrac{-173x + 15}{3}$. Find $g^{-1}(g(5))$.
Find $g(g^{-1}(-12))$.

33. $f(x) = x^3 + 2$. Find $f^{-1}(f(12{,}053))$.
Find $f(f^{-1}(-17{,}243))$.

34. $g(x) = x^3 - 486$. Find $g^{-1}(g(489))$.
Find $g(g^{-1}(-17{,}422))$.

35. Carefully graph $y = x^2$, using a large scale. Then use the graph to approximate $\sqrt{3.1}$.

36. Carefully graph $y = x^3$, using a large scale. Then use the graph to approximate $\sqrt[3]{-5.2}$.

37. Graph this equation and its inverse. Then test for symmetry with respect to the x-axis, the y-axis, the origin, and the line $y = x$.

$$y = \frac{1}{x^2}$$

38. Ice melts at 0° Celsius or 32° Fahrenheit. Water boils at 100° Celsius or 212° Fahrenheit.

 a) What linear transformation converts Celsius temperature to Fahrenheit?

 b) Find the inverse of your answer to (a). Is it a function? What kind of conversion does it accomplish?

 c) At what temperature are the Celsius and Fahrenheit scales the same?

Graph each equation and its inverse. Then test for symmetry with respect to the x-axis, the y-axis, the origin, and the line $y = x$.

39. $|x| - |y| = 1$

40. $y = \dfrac{|x|}{x}$

CHAPTER 3 REVIEW

1. List all the ordered pairs in the Cartesian square $G \times G$, where $G = \{1, 3, 5, 7\}$.

2. List the domain and the range of the relation whose ordered pairs are (3, 1), (5, 3), (7, 7), and (3, 5).

Graph.

3. $x = |y|$

4. $y = (x + 1)^2$

5. $y = |x| - 2$

6. $f(x) = \sqrt{x}$

7. $f(x) = \sqrt{x - 2}$

8. $f(x) = 2\sqrt{x + 3}$

9. $f(x) = \frac{1}{4}\sqrt{x - 1} + 2$

10. $|x - y| = 1$

Consider the following relations for Exercises 11–14.

a) $y = 7$
b) $x^2 + y^2 = 4$
c) $x^3 = y^3 - y$
d) $y^2 = x^2 + 3$
e) $x + y = 3$
f) $x = 3$
g) $y = x^2$
h) $y = x^3$

11. Which are symmetric with respect to the x-axis?

12. Which are symmetric with respect to the y-axis?

13. Which are symmetric with respect to the origin?

14. Which are symmetric with respect to the line $y = x$?

Write an equation of the inverse.

15. $y = 3x^2 + 2x - 1$

16. $y = \sqrt{x + 2}$

17. Which of the following are graphs of functions?

a)

b)

c)

d)
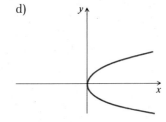

18. Which of the relations in Exercise 17 have inverses that are functions?

Use $f(x) = x^2 - x - 3$ to answer Exercises 19–21. Find:

19. $f(0)$.

20. $f(-3)$.

21. $f(a + h)$.

Use $g(x) = 2\sqrt{x-1}$ to answer Exercises 22-24. Find:

22. $g(1)$.   **23.** $g(5)$.   **24.** $g(a+2)$.

25. $f(x) = \dfrac{\sqrt{x}}{2} + 2$. Find a formula for $f^{-1}(x)$.

26. $f(x) = x^2 + 2$. Find a formula for $f^{-1}(x)$.

Find the domain of each function.

27. $f(x) = \sqrt{7 - 3x}$ **28.** $f(x) = \dfrac{1}{x^2 - 6x + 5}$

29. $f(x) = (x - 9x^{-1})^{-1}$ **30.** $f(x) = \dfrac{\sqrt{1-x}}{x - |x|}$

Find $f \circ g(x)$ and $g \circ f(x)$ in Exercises 31 and 32.

31. $f(x) = \dfrac{4}{x^2}$; $g(x) = 3 - 2x$ **32.** $f(x) = 3x^2 + 4x$; $g(x) = 2x - 1$

33. $f(x) = x^3 + 2$. Find $f(f^{-1}(a))$.

34. $h(x) = x^{17} + x^{65}$. Find $h^{-1}(h(t))$.

35. Here is a graph of $y = f(x)$.

Sketch graphs of the following.

a) $y = 1 + f(x)$ b) $y = \dfrac{1}{2}f(x)$ c) $y = f(x+1)$

Use the following to answer Exercises 36-38.

a)

b)

c) $f(x) = 3x^2 - 2$ d) $f(x) = x + 3$

e) $f(x) = 3x^3$ f) $f(x) = x^5 - x^3$

36. Which of the previous are even?

37. Which of the previous are odd?

38. Which of the previous are neither even nor odd?

39. Which of the following functions are periodic?

a)

b)

c)

d)

40. What is the period of this function?

41. Is this function continuous

a) in the interval $[-3, -1]$?

b) in the interval $[-1, 1]$?

Write interval notation for:

42. $\{x \mid -\pi \leq x \leq 2\pi\}$.

43. $\{x \mid 0 < x \leq 1\}$.

146 RELATIONS, FUNCTIONS, AND TRANSFORMATIONS

Use the following for Exercises 44–46.

a)

b)

c)

d)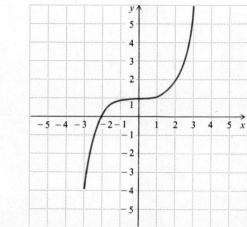

44. Which of the above are increasing?

45. Which of the above are decreasing?

46. Which of the above are neither increasing nor decreasing?

47. Graph.

$$f(x) = \begin{cases} x^2 + 2, & \text{for } x < 0, \\ x^3, & \text{for } 0 \leq x < 2, \\ -4x + 5, & \text{for } x \geq 2 \end{cases}$$

48. Graph several functions of the type $y = |f(x)|$. Describe a procedure, involving transformations, for graphing such functions.

Linear and Quadratic Functions and Inequalities

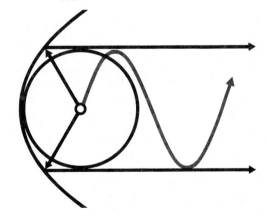

LINEAR AND QUADRATIC FUNCTIONS AND INEQUALITIES

Figure 1

Figure 2

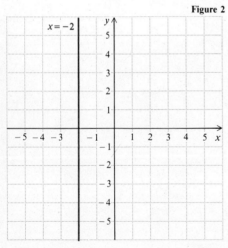

Figure 3

4.1 LINES AND LINEAR FUNCTIONS

Graphs of Linear Equations

An equation of the type $Ax + By = C$ is called a *linear* equation because its graph is a straight line. In the above, A, B, and C are constants, but A and B cannot both be 0. Any equation that is equivalent to one of this form has a straight-line graph.

Example 1

a) The equation $5x + 8y - 3 = 0$ is linear because it is equivalent to $5x + 8y = 3$. Here, $A = 5$, $B = 8$, and $C = 3$.
b) The equation $3x^2 - 4y + 5 = 0$ is not linear because x is squared.
c) The equation $4x + 3xy = 25$ is not linear because the product xy occurs.

Since two points determine a line, we can graph a linear equation by finding two of its points. Then we draw a line through those points.

A third point should always be used as a check. The easiest points to find are often the intercepts (the points where the line crosses the axes).

Example 2 Graph $4x + 5y = 20$.

We set $x = 0$ and find that $y = 4$. Thus $(0, 4)$ is a point of the graph (the y-intercept).

We set $y = 0$ and find that $x = 5$. Thus $(5, 0)$ is a point of the graph (the x-intercept). The graph is shown in Fig. 1. The point $(-2, 5\frac{3}{5})$ was used as a check.

If a graph, such as $y = 7x$, goes through the origin, it has only one intercept, and other points will have to be used for graphing.

If an equation has a missing variable ($A = 0$ or $B = 0$), then its graph is parallel to one of the axes.

Example 3.

a) Graph $y = 3$.

The graph is shown in Fig. 2.

b) Graph $x = -2$.

The graph is shown in Fig. 3.

Slope

The graph of a linear equation may slant upward or downward in various ways. Let us see how this relates to equations.

4.1 LINES AND LINEAR FUNCTIONS

Suppose points P_1 and P_2 with coordinates (x_1, y_1) and (x_2, y_2) are two different points on a line not parallel to an axis. Consider a right triangle as shown with legs parallel to the axes. The point P_3 with coordinates (x_2, y_1) is the third vertex of the triangle. As we move from P_1 to P_2, y changes from y_1 to y_2. The change in y is $y_2 - y_1$. Similarly, the change in x is $x_2 - x_1$. The ratio of these changes is called the *slope*. See Fig. 4.

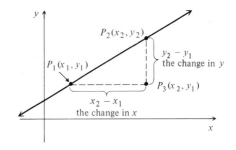

Figure 4

DEFINITION

The *slope* of a line containing two points (x_1, y_1) and (x_2, y_2) is defined to be

$$\frac{y_2 - y_1}{x_2 - x_1} \left(\frac{\text{change in } y}{\text{change in } x} \right).$$

Example 4 Graph the line through the points (1, 2) and (3, 6) and find its slope.

Let us call the slope m. Then

$$m = \frac{\text{change in } y}{\text{change in } x} = \frac{6 - 2}{3 - 1} = \frac{4}{2} = 2.$$

Note that we can also use the points in the opposite order, so long as we are consistent. We get the same slope:

$$m = \frac{\text{change in } y}{\text{change in } x} = \frac{2 - 6}{1 - 3} = \frac{-4}{-2} = 2.$$

See Fig. 5 for the graph.

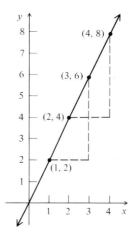

Figure 5

From Example 4 we see that it does not matter in which order we choose the points, so long as we take differences in the same order. From Example 4 we can also see that it does not matter which two points of a line we choose to determine the slope. No matter what points we choose we get the same number for the slope. For example, if we choose (2, 4) and (4, 8) we get for the slope

$$\frac{8 - 4}{4 - 2} = \frac{4}{2} = 2.$$

If a line slants upward from left to right, it has a positive slope, as in Example 4. If a line slants downward from left to right, the change in x and the change in y have opposite signs, so the line has a negative slope.

If a line is horizontal, the change in y for any two points is 0. Thus a horizontal line has zero slope.

If a line is vertical, the change in x for any two points is 0. Thus the slope is not defined, because we cannot divide by zero. A vertical line does not have a slope.

Point-Slope Equations of Lines

Suppose we have a nonvertical line and that the coordinates of one point P_1 are (x_1, y_1). We think of point P_1 as fixed. Suppose, also, that we have a movable point P on the line with coordinates (x, y). Thus the slope would be given by

$$\frac{(y - y_1)}{(x - x_1)} = m. \tag{1}$$

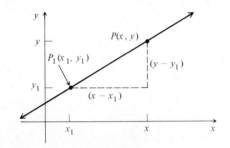

Figure 6

(See Fig. 6.) Note that this is true only when (x, y) is a point different from (x_1, y_1). If we use the multiplication principle, we get*

$$(y - y_1) = m(x - x_1). \quad \textit{Point-slope equation} \tag{2}$$

Equation (2) will be true even if $(x, y) = (x_1, y_1)$. Equation (2) is called the *point-slope equation* of a line. Thus if we know the slope of a line and the coordinates of a point on the line, we can find an equation of the line.

Example 5 Find an equation of the line containing the point $(\frac{1}{2}, -1)$ with slope 5.

If we substitute in $(y - y_1) = m(x - x_1)$, we get $y - (-1) = 5(x - \frac{1}{2})$, which simplifies to

$$y + 1 = 5\left(x - \frac{1}{2}\right)$$

or

$$y = 5x - \frac{5}{2} - 1$$

or

$$y = 5x - \frac{7}{2}.$$

Two-Point Equations of Lines

Suppose a nonvertical line contains the points $P_1(x_1, y_1)$ and $P_2(x_2, y_2)$. The slope of the line is

$$\frac{y_2 - y_1}{x_2 - x_1}.$$

---

*Hint: It may be easier to remember Eq. (1), since it relates to the formula. We can easily get Eq. (2) from it.

If we substitute $(y_2 - y_1)/(x_2 - x_1)$ for m in the point-slope equation,
$$y - y_1 = m(x - x_1),$$
we have*
$$y - y_1 = \frac{y_2 - y_1}{x_2 - x_1}(x - x_1). \quad \textit{Two-point equation}$$

This is known as the *two-point equation* of a line. Note that either of the two given points can be called P_1 or P_2 and (x, y) is any point on the line.

Example 6 Find an equation of the line containing the points $(2, 3)$ and $(1, -4)$.

If we take $(2, 3)$ as P_1 and use the two-point equation, we get
$$y - 3 = \frac{-4 - 3}{1 - 2}(x - 2),$$
which simplifies to $y = 7x - 11$.

Slope-Intercept Equations of Lines

Suppose a nonvertical line has slope m and y-intercept $(0, b)$. We sometimes, for brevity, refer to the number b as the y-intercept. Let us substitute m and $(0, b)$ in the point-slope equation. We get
$$y - y_1 = m(x - x_1)$$
$$y - b = m(x - 0).$$
This simplifies to
$$y = mx + b. \quad \textit{Slope-intercept equation}$$

This is called the *slope-intercept equation* of a line. The advantage of such an equation is that we can read the slope m and the y-intercept b from the equation.

Example 7 Find the slope and the y-intercept of $y = 5x - \frac{1}{4}$.

$$y = 5x - \frac{1}{4}$$
Slope: 5   y-intercept: $-\frac{1}{4}$

---

*Hint: It may be easier to remember
$$\frac{y - y_1}{x - x_1} = \frac{y_2 - y_1}{x_2 - x_1},$$
interpreting each side as slope.

Example 8 Find the slope and y-intercept of $y = 8$.

We can rewrite this equation as $y = 0x + 8$. We then see that the slope is 0 and the y-intercept is 8.

To find the slope-intercept equation of a line, when given another equation, we solve for y.

Example 9 Find the slope and y-intercept of the line having the equation $3x - 6y - 7 = 0$.

We solve for y, obtaining $y = \frac{1}{2}x - \frac{7}{6}$. Thus the slope is $\frac{1}{2}$ and the y-intercept is $-\frac{7}{6}$.

If a line is vertical it has no slope. Thus it has no slope-intercept equation. Such a line does have a simple equation, however. All vertical lines have equations $x = c$, where c is some constant.

Linear Functions

Any nonvertical straight line is the graph of a function. Such a function is called a *linear function*. Any nonvertical line has an equation $y = mx + b$. Thus a function f is a linear function if and only if it has an equation $f(x) = mx + b$. If the slope m is zero, the equation simplifies to $f(x) = b$. A function like this is called a *constant function*.

EXERCISE SET 4.1

1. Which of the following are linear equations?
   a) $3y = 2x - 5$
   b) $5x + 3 = 4y$
   c) $3y = x^2 + 2$
   d) $y = 3$
   e) $xy = 5$
   f) $3x^2 + 2y = 4$
   g) $3x + \frac{1}{y} = 4$
   h) $5x - 2 = 4y$

2. Which of the following are linear equations?
   a) $5y = 3x - 4$
   b) $3x + 5 = 7y$
   c) $4y = 3x^2 - 4$
   d) $4x - \frac{2}{y} = 3$
   e) $2xy = 4$
   f) $5x^2 + 3y = -4$
   g) $x = -4$
   h) $6x - 7 = 3y$

Graph the following equations.

3. $8x - 3y = 24$
4. $5x - 10y = 50$
5. $3x + 12 = 4y$
6. $4x - 20 = 5y$
7. $y = -2$
8. $2y - 3 = 9$
9. $5x + 2 = 17$
10. $19 = 5 - 2x$

Find the slopes of the lines containing these points.

11. $(6, 2)$ and $(-2, 1)$
12. $(-2, 1)$ and $(-4, -2)$
13. $(2, -4)$ and $(4, -3)$
14. $(5, -3)$ and $(-5, 8)$

15. $(2\pi, 5)$ and $(\pi, 4)$
16. $(\sqrt{2}, -4)$ and $(\pi, -4)$

Find equations of the following lines.

17. Through $(3, 2)$ with $m = 4$
18. Through $(4, 7)$ with $m = -2$
19. With y-intercept -5 and $m = 2$
20. With y-intercept π and $m = \dfrac{1}{4}$
21. Containing $(1, 4)$ and $(5, 6)$
22. Containing $(-2, 0)$ and $(2, 3)$

Find the slope and y-intercept of each line.

23. $y = 2x + 3$
24. $y = 6 - x$
25. $2y = -6x + 10$
26. $-3y = -12x + 9$
27. $3x - 4y = 12$
28. $5x + 2y = -7$
29. $3y + 10 = 0$
30. $y = 7$

Find equations of the following lines.

31. ▨ Through $(3.014, -2.563)$ with slope 3.516.
32. ▨ Through $(-173.4, -17.6)$ with slope -0.00014.
33. ▨ Through the points $(1.103, 2.443)$ and $(8.114, 11.012)$.
34. ▨ Through the points $(473.78, 910.2)$ and $(993.55, 171.43)$.

☆ ─────────────────────────────

Suppose f is a linear function. Find a formula for $f(x)$ given that

35. $f(3x) = 3f(x)$.
36. $f(kx) = kf(x)$, for some number k.
37. $f(x + 2) = f(x) + 2$.
38. $[f(x)]^2$ is linear.

Assume f is a linear function. Are the following true or false?

39. $f(a + b) = f(a) + f(b)$
40. $f(ab) = f(a)f(b)$
41. $f(kx) = kf(x)$
42. $f(a - b) = f(a) - f(b)$

43. Determine whether these three points are on a line. [Hint: Compare the slopes of $\overline{AB}$ and $\overline{BC}$. ($\overline{AB}$ refers to the segment from A to B.)]

$$A(9, 4), \quad B(-1, 2), \quad C(4, 3)$$

44. Determine whether these three points are on a line. (See the hint for Exercise 43.)

$$A(-1, -1), \quad B(2, 2), \quad C(-3, -4)$$

45. Use graph paper. Plot the points $A(0, 0)$, $B(8, 2)$, $C(11, 6)$, and $D(3, 4)$. Draw $\overline{AB}$, $\overline{BC}$, $\overline{CD}$, and $\overline{DA}$. Find the slopes of these four segments. Compare the slopes of $\overline{AB}$ and $\overline{CD}$. Compare the slopes of $\overline{BC}$ and $\overline{DA}$. (Figure $ABCD$ is a parallelogram and its opposite sides are parallel.)

46. Use graph paper. Plot the points $E(-2, -5)$, $F(2, -2)$, $G(7, -2)$, and $H(3, -5)$. Draw $\overline{EF}$, $\overline{FG}$, $\overline{GH}$, $\overline{HE}$, $\overline{EG}$, and $\overline{FH}$. Compare the slopes of $\overline{EG}$ and $\overline{FH}$. (Figure $EFGH$ is a rhombus and its diagonals are perpendicular.)

47. (Fahrenheit temperature as a function of Celsius temperature). Fahrenheit temperature F is a linear function of Celsius (or Centigrade) temperature C. When C is 0, F is 32. When C is 100, F is 212. Use these data to express F as a linear function of C.

48. (Celsius temperature as a function of Fahrenheit temperature). Celsius (Centigrade) temperature C is a linear function of Fahrenheit temperature F. When F is 32, C is 0. When F is 212, C is 100. Use these data to express C as a linear function of F.

49. Suppose P is a nonconstant linear function of Q. Show that Q is a linear function of P.

50. Suppose y is directly proportional to x. Show that y is a linear function of x.

4.2 PARALLEL AND PERPENDICULAR LINES; THE DISTANCE FORMULA

Parallel and Perpendicular Lines

If two lines are vertical, they are parallel. Thus equations such as $x = c_1$ and $x = c_2$ (where c_1 and c_2 are constants) have graphs that are parallel lines. Now we consider nonvertical lines. For such lines to be parallel, they must have the same slope, but different y-intercepts. Thus equations such as $y = mx + b_1$ and $y = mx + b_2$, $b_1 \neq b_2$, have graphs that are parallel lines.

THEOREM 1

Vertical lines are parallel. Nonvertical lines are parallel if and only if they have the same slope and different y-intercepts.

If two equations are equivalent, they represent the same line. Thus if two lines have the same slope and the same y-intercept, they are not really two different lines. They are the same line. In such a case we sometimes speak of *coincident lines*.

If one line is vertical and the other is horizontal, such as $x = c_1$ and $y = c_2$, they are perpendicular. Otherwise, how can we tell whether two lines are perpendicular? Consider a line $\overleftrightarrow{AB}$ as shown in Fig. 7, with slope a/b. Then think of rotating the figure $90°$ to get a line perpendicular to $\overleftrightarrow{AB}$. For the new line the change in x and the change in y are interchanged, but the change in x is now negative. Thus the slope of the new line is $-b/a$. Let us multiply the slopes:

$$\frac{a}{b}\left(-\frac{b}{a}\right) = -1.$$

This is the condition under which lines will be perpendicular.

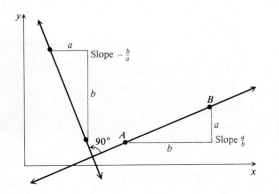

Figure 7

THEOREM 2

Two lines are perpendicular if the product of their slopes is -1. (If one line has slope m, the slope of a line perpendicular to it is $-1/m$. That is, we take the reciprocal and change the sign.)

Example 1 Determine whether the following lines are parallel, perpendicular, or neither.

a) $y + 2 = 5x$, $\quad 5y + x = -15$.

We solve for y:
$$y = 5x - 2, \qquad y = -\frac{1}{5}x - 3.$$

The slopes are 5 and $-\frac{1}{5}$. Their product is -1, so the lines are perpendicular.

b) $2y + 4x = 8$, $5 + 2x = -y$.

By solving for y we determine that $m_1 = -2$ and $m_2 = -2$, so the lines are parallel.

c) $2x + 1 = y$, $y + 3x = 4$.

By solving for y we determine that $m_1 = 2$ and $m_2 = -3$, so the lines are neither parallel nor perpendicular.

Example 2 Write equations of the lines perpendicular and parallel to the line $4y - x = 20$ and containing the point $(2, -3)$.

We first solve for y: $y = \frac{1}{4}x + 5$, so the slope is $\frac{1}{4}$. The slope of the perpendicular line is -4.

Now we use the point-slope equation to write an equation with slope -4 and containing the point $(2, -3)$:
$$y - y_1 = m(x - x_1)$$
$$y - (-3) = -4(x - 2).$$

This simplifies to $y = -4x + 5$.

The line parallel to the given line will have slope $\frac{1}{4}$. The equation is
$$y - (-3) = \frac{1}{4}(x - 2).$$

This simplifies to $y = \frac{1}{4}x - \frac{7}{2}$.

Vertical Lines

If a line is vertical, it has no slope. Every vertical line has an equation $x = c$, where c is a constant. Any line parallel to a vertical line must also be vertical, hence must also have an equation $x = k$, where k is a constant.

A line will be perpendicular to a vertical line if and only if that line is horizontal. Every horizontal line has zero slope and has an equation $y = c$, where c is a constant. Thus if a line is vertical, it is simple to determine whether another line is parallel to it or perpendicular to it.

Example 3 Find equations of the lines parallel and perpendicular to the line $x = 4$ and containing the point $(-2, 3)$.

The line x = 4 is vertical, so any line parallel to it must be vertical. The line we seek has one x-coordinate, which is −2, so all x-coordinates on the line must be −2. The equation is x = −2.

The perpendicular line has a y-coordinate, which is 3, so all y-coordinates must be 3. The equation is y = 3.

The Distance Formula

We will develop a formula for finding the distance between two points whose coordinates are known. Suppose the points are on a horizontal line, thus having the same second coordinates. We can find the distance between them by subtracting their first coordinates. This difference may be negative, depending on the order in which we subtract. So to make sure we get a positive number, we take the absolute value of this difference. The distance between two points on a horizontal line (x_1, y) and (x_2, y) is thus $|x_1 - x_2|$. Similarly, the distance between two points on a vertical line (x, y_1) and (x, y_2) is $|y_1 - y_2|$.

Now consider in Fig. 8 any two points not on a horizontal or vertical line (x_1, y_1) and (x_2, y_2). These points are vertices of a right triangle, as shown. The other vertex is (x_2, y_1). The legs of this triangle have the lengths $|x_1 - x_2|$ and $|y_1 - y_2|$. Now by the Pythagorean theorem we obtain a relation between the length of the hypotenuse d, and the lengths of the legs:

$$d^2 = |x_1 - x_2|^2 + |y_1 - y_2|^2.$$

We may now dispense with the absolute value signs because squares of numbers are never negative. Thus we have

$$d^2 = (x_1 - x_2)^2 + (y_1 - y_2)^2.$$

By taking the square root we obtain the distance between the two points.

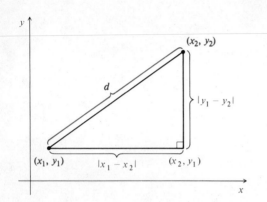

Figure 8

THEOREM 3

The distance formula. **The distance between any two points (x_1, y_1) and (x_2, y_2) is given by**

$$d = \sqrt{(x_1 - x_2)^2 + (y_1 - y_2)^2}.$$

Although we derived the distance formula by considering two points not on a horizontal or vertical line, the formula holds for *any* two points.

4.2 PARALLEL AND PERPENDICULAR LINES; THE DISTANCE FORMULA

Example 4 Find the distance between the points $(8, 7)$ and $(3, -5)$.

Using the distance formula, we have
$$d = \sqrt{(8-3)^2 + [7-(-5)]^2} = \sqrt{5^2 + 12^2}$$
$$= \sqrt{169} = 13.$$

Example 5 Decide whether the points $(1, 4)$, $(2, 1)$, and $(3, 2)$ are vertices of a right triangle.

First we find the squares of the distances between the points:
$$d_1^2 = (1-2)^2 + (4-1)^2 = (-1)^2 + 3^2 \quad = 1 + 9 = 10,$$
$$d_2^2 = (1-3)^2 + (4-2)^2 = (-2)^2 + 2^2 \quad = 4 + 4 = 8,$$
$$d_3^2 = (2-3)^2 + (1-2)^2 = (-1)^2 + (-1)^2 = 1 + 1 = 2.$$

Since $d_2^2 + d_3^2 = d_1^2$, it follows by the converse of the Pythagorean theorem that the points are the vertices of a right triangle.

Midpoints of Segments

The distance formula can be used to verify or derive a formula for finding the coordinates of the midpoint of a segment when the coordinates of its endpoints are known. We shall not derive this formula but simply state it.

THEOREM 4

The midpoint formula. **If the endpoints of a segment are (x_1, y_1) and (x_2, y_2), then the coordinates of the midpoint are**
$$\left(\frac{x_1 + x_2}{2}, \frac{y_1 + y_2}{2}\right).$$

Note that we obtain the coordinates of the midpoint by averaging the coordinates of the endpoints. This is an easy way to remember this formula.

Example 6 Find the midpoint of the segment with endpoints $(-3, 5)$ and $(4, -7)$.

Using the midpoint formula, we obtain
$$\left(\frac{-3+4}{2}, \frac{5+(-7)}{2}\right) \quad \text{or} \quad \left(\frac{1}{2}, -1\right).$$

EXERCISE SET 4.2

Determine whether the lines are parallel, perpendicular, or neither.

1. $2x - 5y = -3$,
   $2x + 5y = 4$
2. $x + 2y = 5$,
   $2x + 4y = 8$
3. $y = 4x - 5$,
   $4y = 8 - x$
4. $y = -x + 7$,
   $y = x + 3$

Find an equation of the line containing the given point and parallel to the given line.

5. $(0, 3)$, $3x - y = 7$
6. $(-4, -5)$, $2x + y = -4$
7. $(3, 8)$, $x = 2$
8. $(3, -3)$, $x = -1$
9. $(-2, -3)$, $y = 4$
10. $(-3, 2)$, $y = -3$

Find an equation of the line containing the given point and perpendicular to the given line.

11. $(-3, -5)$, $5x - 2y = 4$
12. $(3, -2)$, $3x + 4y = 5$
13. $(0, 3)$, $x = 1$
14. $(-2, -2)$, $x = 3$
15. $(-3, -7)$, $y = 2$
16. $(4, -5)$, $y = -1$

17. ▦ Find an equation of the line parallel to the one given, and containing the given point.

$$4.323x - 7.071y = 16.61, \quad (-2.603, 1.818)$$

18. ▦ Find an equation of the line containing the given point and perpendicular to the given line.

$$6.232x + 4.001y = 4.881, \quad (3.149, -2.908)$$

Find the distance between the points of each pair.

19. $(-3, -2)$ and $(1, 1)$
20. $(5, 9)$ and $(-1, 6)$
21. $(0, -7)$ and $(3, -4)$
22. $(2, 2)$ and $(-2, -2)$
23. $(a, -3)$ and $(2a, 5)$
24. $(5, 2k)$ and $(-3, k)$
25. $(0, 0)$ and (a, b)
26. $(\sqrt{2}, \sqrt{3})$ and $(0, 0)$
27. $(\sqrt{a}, \sqrt{b})$ and $(-\sqrt{a}, \sqrt{b})$
28. $(c - d, c + d)$ and $(c + d, d - c)$

Decide whether the points are vertices of a right triangle.

29. $(9, 6)$, $(-1, 2)$, and $(1, -3)$
30. $(-5, -8)$, $(1, 6)$, and $(5, -4)$

Find the midpoints of the segments having the following endpoints.

31. $(-4, 7)$ and $(3, -9)$
32. $(4, 5)$ and $(6, -7)$
33. (a, b) and $(a, -b)$
34. $(-c, d)$ and (c, d)

Find the distance between the points of each pair.

35. ▦ $(7.3482, -3.0991)$ and $(18.9431, -17.9054)$
36. ▦ $(-25.414, 175.31)$ and $(275.34, -95.144)$

Find the midpoints of the segments having the following endpoints.

37. ▦ $(-3.895, 8.1212)$ and $(2.998, -8.6677)$
38. ▦ $(4.1112, 6.9898)$ and $(5.1928, -6.9143)$

☆

39. Find an equation of the line containing the point $(4, -2)$ and parallel to the line containing $(-1, 4)$ and $(2, -3)$.
40. Find an equation of the line containing the point $(-1, 3)$ and parallel to the line containing $(3, -5)$ and $(-2, -7)$.
41. Find the point on the x-axis that is equidistant from the points $(1, 3)$ and $(8, 4)$.
42. Find the point on the y-axis that is equidistant from the points $(-2, 0)$ and $(4, 6)$.

43. Consider any right triangle with base b and height h, situated as shown. Show that the midpoint of the hypotenuse P is equidistant from the three vertices of the triangle.

45. Prove that the distance formula holds when two points are on either a vertical line or a horizontal line.

44. Consider any quadrilateral situated as shown. Show that the segments joining the midpoints of the sides, in order as shown, form a parallelogram.

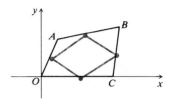

46. The distance formula can be interpreted as a function with two inputs, $d(P, Q)$.
  a) What is the domain of the function?
  b) What is the range of the function?
  c) Under what conditions is an output 0?

4.3 QUADRATIC FUNCTIONS

If a function can be described by a second-degree polynomial, it is called *quadratic*. The following is a more precise definition.

DEFINITION

A *quadratic* function is one that can be described as follows:
$$f(x) = ax^2 + bx + c, \quad \text{where } a \neq 0.$$

In this definition we insist that $a \neq 0$; otherwise the polynomial would not be of degree two. One or both of the constants b and c can be 0.

Consider $f(x) = x^2$. This is an even function, so the y-axis is a line of symmetry. The graph opens upward, as shown in Fig. 9. If we multiply by a constant a to get $f(x) = ax^2$, we obtain a vertical stretching or shrinking, and a reflection if a is negative. Examples of this are shown in Fig. 9. Bear in mind that if a is negative the graph opens downward.

Graphs of quadratic functions are called *parabolas*. In each parabola (Fig. 9), the point $(0, 0)$ is the *vertex*.

Let us consider $f(x) = a(x - h)^2$. We have replaced x by $x - h$ in ax^2, and will therefore obtain a horizontal translation. The translation

LINEAR AND QUADRATIC FUNCTIONS AND INEQUALITIES

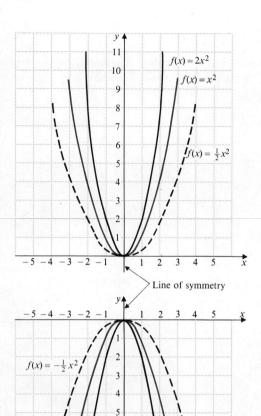

Graphs of quadratic functions. Parabolas.

Figure 9

will be to the right if h is positive. The graphs in Fig. 10 illustrate translations, and a summary is given below.

The graph of $f(x) = a(x - h)^2$

a) opens upward if $a > 0$, downward if $a < 0$;
b) has $(h, 0)$ as a vertex;
c) has $x = h$ as a line of symmetry;
d) has a minimum 0 if $a > 0$, a maximum 0 if $a < 0$.

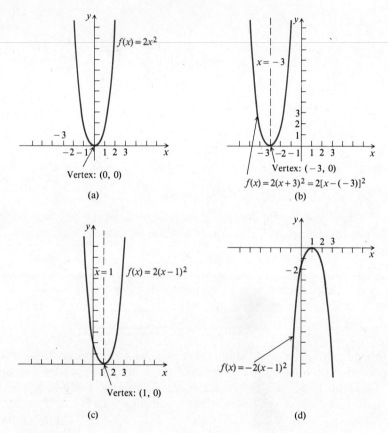

Figure 10

Now consider $f(x) = a(x - h)^2 + k$, or $f(x) - k = a(x - h)^2$. We have replaced $f(x)$ by $f(x) - k$ in the equation $f(x) = a(x - h)^2$. Thus we have a translation. If k is positive, the translation is upward. If k is negative, the translation is downward. Consider the examples in Fig. 11. Note that the vertex has been moved off the x-axis.

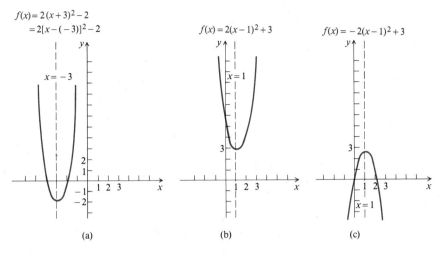

Figure 11

The graph of $f(x) = a(x - h)^2 + k$

a) opens upward if $a > 0$, downward if $a < 0$;
b) has (h, k) as a vertex;
c) has $x = h$ as the line of symmetry;
d) has k as a minimum value (output) if $a > 0$, has k as a maximum value (output) if $a < 0$.

Thus without graphing, we can determine a lot of information about a function described by $f(x) = a(x - h)^2 + k$. The following table is an example.

FUNCTION	$f(x) = 3(x - \frac{1}{4})^2 - 2$	$g(x) = -3(x + 5)^2 + 7$ $= -3[x - (-5)]^2 + 7$
a) What is the vertex?	$(\frac{1}{4}, -2)$	$(-5, 7)$
b) What is the line of symmetry?	$x = \frac{1}{4}$	$x = -5$
c) Is there a maximum? What is it?	No; graph extends upward; $3 > 0$	Yes; 7; graph extends downward; $-3 < 0$
d) Is there a minimum? What is it?	Yes; -2; graph extends upward; $3 > 0$	No; graph extends downward; $-3 < 0$

Now let us consider a quadratic function described by $f(x) = ax^2 + bx + c$. Note that it is not in the form $f(x) = a(x - h)^2 + k$. We

can get it in this form by completing the square. For example, consider
$$f(x) = x^2 - 6x + 4$$
$$= (x^2 - 6x) + 4.$$

We complete the square using the two terms in parentheses, but we do it in a different way than before. We take half the x-coefficient: $\frac{1}{2} \cdot (-6) = 3$; square it: $3^2 = 9$; then add and subtract 9 inside the parentheses and proceed as follows:

$$f(x) = (x^2 - 6x + 9 - 9) + 4$$
$$= (x^2 - 6x + 9) + (-9 + 4)$$
$$= (x - 3)^2 - 5$$
$$= 1 \cdot (x - 3)^2 - 5.$$

Vertex: $(3, -5)$,
Line of symmetry: $x = 3$,
Minimum: -5, since the coefficient 1 is positive.

Example 1 For the following function, (a) find the vertex, (b) find the line of symmetry, and (c) determine whether the second coordinate of the vertex is a maximum or a minimum, and find the maximum or minimum.

$$f(x) = -2x^2 + 10x - 7$$
$$= -2(x^2 - 5x) - 7$$
$$= -2\left(x^2 - 5x + \frac{25}{4} - \frac{25}{4}\right) - 7 \qquad \left[\frac{1}{2} \cdot (-5)\right]^2 = \frac{25}{4}$$
$$= -2\left(x^2 - 5x + \frac{25}{4}\right) - 2\left(-\frac{25}{4}\right) - 7$$
$$= -2\left(x - \frac{5}{2}\right)^2 + \frac{25}{2} - \frac{14}{2}$$
$$= -2\left(x - \frac{5}{2}\right)^2 + \frac{11}{2}$$

Vertex: $\left(\frac{5}{2}, \frac{11}{2}\right)$,

Line of symmetry: $x = \frac{5}{2}$,

Maximum: $\frac{11}{2}$, since the coefficient -2 is negative.

Example 2 For the following function, (a) find the vertex, (b) find the line of symmetry, and (c) determine whether the second coordinate of the vertex is a maximum or a minimum, and find the maximum or minimum output.

$$f(x) = \frac{3}{4}x^2 + 6x$$

$$= \frac{3}{4}(x^2 + 8x)$$

$$= \frac{3}{4}(x^2 + 8x + 16 - 16) \qquad \left[\frac{1}{2} \cdot 8\right]^2 = 16$$

$$= \frac{3}{4}(x^2 + 8x + 16) + \frac{3}{4}(-16) = \frac{3}{4}(x + 4)^2 - 12$$

$$= \frac{3}{4}[x - (-4)]^2 - 12$$

Vertex: $(-4, -12)$,
Line of symmetry: $x = -4$,

Minimum: -12, since the coefficient $\frac{3}{4}$ is positive.

The preceding examples illustrate the following.

THEOREM 5

For any quadratic function, we can complete the square and obtain an equivalent equation as follows:

$$f(x) = a(x - h)^2 + k.$$

The function f has

a) vertex (h, k);
b) line of symmetry $x = h$;
c) k as a minimum if $a > 0$;
d) k as a maximum if $a < 0$.

x-Intercepts

The x-intercepts of the graph of $y = f(x) = ax^2 + bx + c$ occur where the graph crosses the x-axis; that is, when y, the output, is 0.

LINEAR AND QUADRATIC FUNCTIONS AND INEQUALITIES

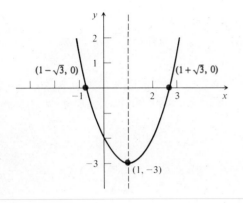

Figure 12

Thus to find the x-coordinates of such points, we solve the equation $0 = ax^2 + bx + c$.

Example 3 Find the x-intercepts of the graph of $y = f(x) = x^2 - 2x - 2$. Use these and the vertex to graph the function.

We solve $0 = x^2 - 2x - 2$. Using the quadratic formula, we find the solutions are $1 \pm \sqrt{3}$. Thus the x-intercepts are $(1 + \sqrt{3}, 0)$ and $(1 - \sqrt{3}, 0)$. When we complete the square, we get $f(x) = (x - 1)^2 - 3$, so the vertex is $(1, -3)$. The graph opens upward so we can complete the graph as shown in Fig. 12.

The discriminant of $ax^2 + bx + c$, which is $b^2 - 4ac$, indicates how many x-intercepts there are. Compare the graphs in Fig. 13.

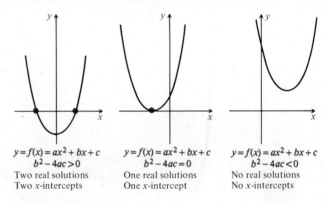

Figure 13

EXERCISE SET 4.3

For each of the following functions,

a) find the vertex;
b) find the line of symmetry; and
c) determine whether the second coordinate of the vertex is a maximum or a minimum, and find the maximum or minimum.

1. $f(x) = x^2$
2. $f(x) = -5x^2$
3. $f(x) = 2\left(x - \dfrac{1}{4}\right)^2$
4. $f(x) = 5(x - 7)^2$
5. $f(x) = -2(x - 9)^2$
6. $f(x) = -3\left(x - \dfrac{1}{2}\right)^2$
7. $f(x) = -2\left(x + \dfrac{1}{2}\right)^2$
8. $f(x) = 3(x - 9)^2 + 5$
9. $f(x) = (x - 4)^2 + 3$ (Also graph.)
10. $f(x) = -(x + 4)^2 - 3$ (Also graph.)
11. $f(x) = -\dfrac{1}{2}(x + 3)^2 + 5$ (Also graph.)
12. $f(x) = 2(x - 1)^2 - 4$ (Also graph.)

For each of the following functions,

a) find an equation of the type $f(x) = a(x - h)^2 + k$;
b) find the vertex;
c) find the line of symmetry; and
d) determine whether the second coordinate of the vertex is a maximum or a minimum, and find the maximum or minimum.

13. $f(x) = -x^2 + 2x + 3$ **14.** $f(x) = -x^2 + 8x - 7$ **15.** $f(x) = x^2 + 3x$

16. $f(x) = x^2 - 9x$ **17.** $f(x) = -\dfrac{3}{4}x^2 + 6x$ **18.** $f(x) = \dfrac{3}{2}x^2 + 3x$

19. $f(x) = 3x^2 + x - 4$ **20.** $f(x) = -2x^2 + x - 1$

Find the x-intercepts. Use these and the vertex to graph each function.

21. $f(x) = -x^2 + 2x + 3$ **22.** $f(x) = x^2 - 3x - 4$

Find the x-intercepts.

23. $f(x) = x^2 - 8x + 5$ **24.** $f(x) = 2x^2 + x - 6$

25. $f(x) = -x^2 - 3x - 3$ **26.** $f(x) = 5x^2 - 6x + 5$

☆
___

Find an equation of the type $f(x) = a(x - h)^2 + k$.

27. $f(x) = 3x^2 + mx + m^2$ **28.** $f(x) = mx^2 - 2x + m^2$

Graph.

29. $f(x) = |x^2 - 1|$ **30.** $f(x) = |3 - 2x - x^2|$

Find the maximum or minimum value for each of the following functions.

31. ▦ $f(x) = 2.31x^2 - 3.135x - 5.89$ **32.** ▦ $f(x) = -18.8x^2 + 7.92x + 6.18$

33. ▦ What is the minimum product of two numbers whose difference is 4.932? What are the numbers?

34. ▦ What is the minimum product of two numbers whose sum is 21.355? What are the numbers?

★
___

35. Find the dimensions and area of the largest rectangle that can be inscribed as shown in a right triangle ABC whose sides have lengths 9 cm, 12 cm, and 15 cm.

36. A farmer wants to build a rectangular fence near a river. If 120 ft of fencing will be used, what is the area of the largest region that can be enclosed? Note that the side next to the river is not fenced.

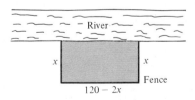

4.4 MATHEMATICAL MODELS

Mathematics is often constructed to fit certain situations. When this is done, we say that we have a *mathematical model*. One of the simplest examples is the natural numbers, i.e., the numbers 1, 2, 3, 4, and so on. These numbers and operations on them were invented so that they would apply to situations in which counting is involved, in some way or other. By working within a mathematical model, we hope to be able to predict what will happen in an actual situation.

Example 1 Use the mathematical model of the natural numbers to predict how much money a theater owner will collect if $2 is charged per admission and 147 people attend.

We translate the problem situation to the language of the mathematical model. In this case, the total number of dollars collected will be the product of the admission price and the number of admissions. We multiply, to get an answer of 294. On this basis we predict that when the owner counts the receipts, there will be $294.

In an example such as the preceding one, we expect our prediction to be exact. In most situations, mathematical models are not that good. Predictions will be only approximate.

Example 2 Use the mathematical model of the natural numbers to predict how much mixture will result from mixing one liter of water and one liter of alcohol.

We solve this problem by adding 1 and 1, to get 2. On this basis, the answer is 2 liters.

In Example 2, our result is not exact. The actual result will be something less than 2 liters, because for some reason there is shrinkage when water and alcohol are mixed.

If a mathematical model does not give precise answers, we ordinarily try revising it so that it does, or we may even look for an entirely new model.

Functions as Mathematical Models

As a result of experiment or experience, we often acquire data that indicate that a function of some sort would be a good mathematical model. In Fig. 14 data have been plotted on a graph. It looks as if the points lie more or less on a straight line, and therefore a linear function would be a good model.

Data approximated by a linear function

Figure 14

4.4 MATHEMATICAL MODELS

We can find a formula for such a function and use it to make predictions. Some sets of data are well approximated by linear functions, whereas others are well approximated by quadratic functions. Still others are approximated best by other kinds of functions.

Example 3 (*World records in the mile run*). It has been found experimentally that running records, within certain restrictions, can be approximated by linear functions. In 1900 the record for the mile run was 4:20 (4 minutes, 20 seconds). In 1975 it was 3:50. Let R represent the record in the mile run, in seconds, and t the number of years since 1900.

a) Find a linear function that fits the data.

b) ▦ In July of 1979, Sebastian Coe of England set a record of 3:49. What would the record have been at that time according to the function in (a)?

c) Predict the record in the year 2000.

a) To determine R as a function of t we use the two-point equation

$$R - R_1 = \frac{R_2 - R_1}{t_2 - t_1}(t - t_1)$$

and the data (ordered pairs) (0, 260) and (75, 230). We have converted the records to seconds. We substitute and simplify:

$$R - 260 = \frac{230 - 260}{75 - 0}(t - 0) = -\frac{30}{75}t = -0.4t$$

$$R = -0.4t + 260.$$

b) ▦ Using the function in (a) and $t = 79\frac{7}{12}$, we get

$$R = -0.4(79\tfrac{7}{12}) + 260 \approx 228.167 \text{ seconds} \approx 3:48.2.$$

Thus the prediction is fairly close.

c) To predict the world record in the year 2000, we substitute 100 for t:

$$R = -0.4(100) + 260 = 220 \text{ seconds} \approx 3:40.$$

Most track authorities think the world record will never fall below this level, although at one time no one thought the four-minute mile would be broken. One should keep in mind that models may be valid only within certain restrictions. For example, if we set $R = 0$ in (a) and solve for t, we find that the world record will be 0 minutes in the year 2550. It might also be pointed out in this context that any function "looks" linear if you restrict its domain sufficiently.

Example 4 (*Profit and loss analysis*). Boxowits, Inc., a computer manufacturing firm, is going to produce a new minicalculator. During the first year, the costs for setting up the new production line are $100,000. These are fixed costs such as rent, tools, etc. These costs must be incurred before any calculators are produced. The additional cost of producing a calculator is $20. This cost is directly related to production, such as material, wages, fuels, and so on. It is a variable cost, according to the number of calculators produced. If x calculators are produced, the variable cost is then 20x dollars. The *total* cost of producing x calculators is given by a function C:

$$\text{Total cost} = \text{Fixed cost plus Variable cost}$$
$$C(x) = 100{,}000 + 20x.$$

The firm determines that its revenue (money coming in) from the sales of the calculators is $45 per calculator, or 45x dollars for x calculators. Thus we have a *revenue* function R:

$$R(x) = 45x.$$

The stockholders are most interested in the *profit* function P:

$$\text{Profit} = \text{Revenue} - \text{Total cost}$$
$$P(x) = 45x - (100{,}000 + 20x)$$
$$P(x) = 25x - 100{,}000.$$

Find those values of x for which (a) the company will break even; (b) the company will make a profit; and (c) the company will suffer a loss.

In this example, we have a model consisting of three linear functions. The function that will provide the answer to the question is the profit function P. When the profit is 0 the company breaks even. When it is positive the company makes money. When it is negative the company loses money.

Let us first find the value(s) of x that make $P(x) = 0$:

$$0 = P(x) = 25x - 100{,}000$$
$$25x = 100{,}000$$
$$x = 4000.$$

Figure 15

The break-even point occurs when 4000 calculators are sold. When $x > 4000$ there is a profit. When $x < 4000$ there is a loss.

A graph of the profit function of Example 4 is shown in Fig. 15. Graphs are often of considerable help with mathematical models.

Example 5 (*A projectile problem*). When an object such as a bullet or a ball is shot or thrown upward with an initial velocity v_0, its height is given, approximately, by a quadratic function:

$$s = -4.9t^2 + v_0 t + h.$$

In this function, h is the starting height in meters, s is the actual height (also in meters), and t is the time from projection in seconds.

This model is constructed from theoretical principles, rather than experiment. It is based on the assumption that there is no air resistance, and that the force of gravity pulling the object earthward is constant. Neither of these conditions exists precisely, so this model (as is the case with most mathematical models) gives only approximate results.

A model rocket is fired upward. At the end of the burn it has an upward velocity of 49 m/s and is 155 m high. Find (a) its maximum height and when it is attained; (b) when it reaches the ground.

We will start counting time at the end of the burn. Thus $v_0 = 49$ and $h = 155$. We will graph the appropriate function, and we begin by completing the square:

$$\begin{aligned} s &= -4.9\left(t^2 - \frac{49}{4.9}t\right) + 155 \\ &= -4.9(t^2 - 10t + 25 - 25) + 155 \\ &= -4.9(t - 5)^2 + 4.9 \times 25 + 155 \\ &= -4.9(t - 5)^2 + 277.5. \end{aligned}$$

The vertex of the graph is the point (5, 277.5) and the graph is shown in Fig. 16. The maximum height reached is 277.5 m and it is attained 5 seconds after the end of the burn.

To find when the rocket reaches the ground, we set $s = 0$ in our equation and solve for t:

$$-4.9(t - 5)^2 + 277.5 = 0$$

$$(t - 5)^2 = \frac{277.5}{4.9}$$

$$t - 5 = \sqrt{\frac{277.5}{4.9}} \approx 7.525$$

$$t \approx 12.525.$$

The rocket will reach the ground about 12.525 seconds after the end of the burn.

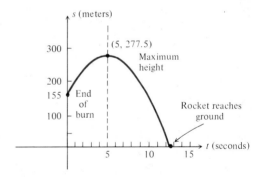

Figure 16

EXERCISE SET 4.4

1. (*Life expectancy of females in the United States*). In 1950 the life expectancy of females was 72 years. In 1970 it was 75 years. Let E represent the life expectancy and t the number of years since 1950. ($t = 0$ gives 1950 and $t = 10$ gives 1960.)

    a) Express E as a linear function of t. [*Hint:* Data points are (0, 72) and (20, 75).]

    b) Use the equation of part (a) to predict the life expectancy of females in 1980; in 1985.

2. (*Life expectancy of males in the United States*). In 1950 the life expectancy of males was 65 years. In 1970 it was 68 years. Let E represent life expectancy and t the number of years since 1950.

    a) Express E as a linear function of t.

    b) Use the equation of part (a) to predict the life expectancy of males in 1980; in 1985.

3. (*World record in the 100-meter dash*). In 1920 the world record for the 100-meter dash was 10.43 seconds. In 1970 it was 9.93 seconds. Let R represent the record in the 100-meter dash and t the number of years since 1920.

    a) Find a linear function to fit the data points.

    b) Use the function in part (a) to predict the record in 1984; in 1990.

    c) According to this model, in what year will the record be 9.0 seconds?

4. (*Natural gas demand*). In 1950 natural gas demand in the United States was 19 quadrillion BTU. In 1960 the demand was 21 quadrillion BTU. Let D represent the demand for natural gas t years after 1950.

    a) Express D as a linear function of t.

    b) Use the equation of part (a) to predict the natural gas demand in 1980; in 2000.

5. A rocket is fired upward. At the end of the burn it has an upward velocity of 147 m/sec and is 560 m high. Find (a) its maximum height and when it is attained; (b) when it reaches the ground.

6. A rocket is fired upward. At the end of the burn it has an upward velocity of 245 m/sec and is 1240 m high. Find (a) its maximum height and when it is attained; (b) when it reaches the ground.

7. The sum of the base and height of a triangle is 20. Find the dimensions for which the area is a maximum.

8. The sum of the base and height of a triangle is 14. Find the dimensions for which the area is a maximum.

9. (*Maximizing yield*). An orange grower finds that she gets an average yield of 40 bu per tree when she plants 20 trees on an acre of ground. Each time she adds a tree to an acre the yield per tree decreases by 1 bu, due to congestion. How many trees per acre should she plant for maximum yield?

10. (*Maximizing revenue*). When a theater owner charges $2 for admission he averages 100 people attending. For each 10¢ increase in admission price the average number attending decreases by 1. What should he charge to make the most money?

11. (*Maximizing revenue*). When a theater charges $4 for admission it averages 500 people in attendance. For each 20¢ decrease in admission price the average number of people increases by 30. What should it charge to make the most money?

12. (*Maximizing area*). A farmer wants to enclose two adjacent rectangular regions, as shown below, near a river, one for sheep and one for cattle. No fencing will be used next to the river. If 60 m of fencing will be used, what is the area of the largest region that can be enclosed?

13. (*Profit and loss in the ski business*). A ski manufacturer is planning a new line of skis. For the first year the fixed costs for setting up the new production line are $22,500. Variable costs for producing each pair of skis are estimated to be $40. The sales department projects that 3000 pairs of skis can be sold the first year. The revenue from each pair is to be $85.
   a) Formulate a function $C(x)$ for the total cost of producing x pairs of skis.
   b) Formulate a function $R(x)$ for the total revenue from the sale of x pairs of skis.
   c) Formulate a function $P(x)$ for the profit from the sale of x pairs of skis.
   d) What profit or loss will the company realize if 3000 pairs are actually sold?
   e) Find the break-even value of x.
   f) Find the values of x that will result in a profit.
   g) Find the values of x that will result in a loss.

4.5 SETS, SENTENCES, AND INEQUALITIES

The *intersection* of two sets consists of those elements common to the sets. Intersection is illustrated in Fig. 17. The intersection of two sets may be the empty set as shown in Fig. 18. The intersection of sets A and B is written as $A \cap B$.

Figure 17

Graphs are set diagrams. They often show pictorially the solution set of an equation or inequality. We can find intersections of solution sets using graphs. In the following example, we find the intersection of the set of all x greater than 3 and the set of all x less than or equal to 5. These sets are indicated, and the symbolism is read, as follows:

Figure 18

$\{x \mid x > 3\}$ "the set of all x such that x is greater than 3,"
$\{x \mid x \leq 5\}$ "the set of all x such that x is less than or equal to 5."

Example 1 Find and graph: $\{x \mid x > 3\} \cap \{x \mid x \leq 5\}$.

We graph the two solution sets separately and then find the intersection.

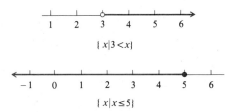

The open circle at 3 indicates that 3 is not in the solution set. The solid circle at 5 indicates that 5 is in the solution set. The intersection is shown below.

Example 2 Graph: $\{x \mid -3 \leq x\} \cap \{x \mid x \leq 0\}$.

Again, we graph the solution sets separately and then find the intersection.

Figure 19

The *union* of two sets consists of the members that are in one or both of the sets. Union is illustrated in Fig. 19. The union of sets A and B is written as $A \cup B$.

In the following examples, we find unions of solution sets. Note that we graph the individual sets separately and then combine them.

Example 3 Graph: $\{x \mid -1 \leq x\} \cup \{x \mid x < 2\}$.

Example 4 Graph: $\{x \mid x \leq -2\} \cup \{x \mid x > 1\}$.

Compound Sentences

When two sentences are joined by the word *and*, a compound sentence is formed. Such a sentence is called a *conjunction*. (The word conjunction used in this way is a logical term; the meaning is not the same as in ordinary grammar.) A conjunction of two sentences is true when both parts are true. Thus the solution set is the intersection of the solution sets of the parts. Consider, for example, the conjunction

$$-2 \leq x \quad \text{and} \quad x < 1.$$

Any number that makes this sentence true must make both parts true. We can graph the solution set by graphing the intersection of the two solution sets of the parts as shown in Fig. 20.

Figure 20

We often abbreviate certain conjunctions of inequalities. In this case, $-2 \leq x \quad \text{and} \quad x < 1$ is abbreviated $-2 \leq x < 1$.

The latter is read "-2 is less than or equal to x and x is less than 1," or "-2 is less than or equal to x is less than 1." Thus

$$\{x \mid -2 \leq x < 1\} = \{x \mid -2 \leq x \quad \text{and} \quad x < 1\}$$
$$= \{x \mid -2 \leq x\} \cap \{x \mid x < 1\}.$$

The word *and* corresponds to set *intersection*.

Example 5 Solve and graph: $-3 < 2x + 5 < 7$.

Method 1. $-3 < 2x + 5$ and $2x + 5 < 7$ Rewriting using *and*
  $-8 < 2x$ and $2x < 2$ Adding -5
  $-4 < x$ and $x < 1$ Multiplying by $\frac{1}{2}$.

Method 2. $-3 < 2x + 5 < 7$
  $-8 < 2x < 2$
  $-4 < x < 1$

Figure 21

The solution set is

$$\{x \mid -4 < x\} \cap \{x \mid x < 1\}, \text{ or } \{x \mid -4 < x < 1\}.$$

The graph is shown in Fig. 21.

Example 6 Solve and graph: $-4 < \dfrac{5 - 3x}{2} \leq 5$.

$-8 < 5 - 3x \leq 10$ Multiplying by 2
$-13 < -3x \leq 5$ Adding -5
$\dfrac{13}{3} > x \geq -\dfrac{5}{3}$ Multiplying by $-\dfrac{1}{3}$

$-\dfrac{5}{3} \leq x < \dfrac{13}{3}$ $x \geq -\dfrac{5}{3}$ means $-\dfrac{5}{3} \leq x$, and $\dfrac{13}{3} > x$ means $x < \dfrac{13}{3}$

The solution set is

$$\left\{x \mid -\dfrac{5}{3} \leq x < \dfrac{13}{3}\right\}.$$

Figure 22

The graph is shown in Fig. 22.

When two sentences are joined by the word *or*, a compound sentence is formed. Such a sentence is called a *disjunction*. A disjunction of two sentences is true when either part is true. It is also true when both parts are true. The solution set of a disjunction is thus the union of the solution sets of the parts. Consider the disjunction

$$x < -2 \quad \text{or} \quad x > \dfrac{1}{4}.$$

Any number that makes either or both of the parts true makes the disjunction true.

We can graph the solution set by graphing the union of the solution sets of the parts as shown in Fig. 23. The word *or* corresponds to set *union*.

Any number in this set makes the first part true.
(a)

Any number in this set makes the second part true.
(b)

Figure 23

Any number in this set makes one or both of the parts true.
(c)

Note that $3x \leq 15$ is an abbreviation for $3x < 15$ or $3x = 15$. There is no compact way to abbreviate disjunctions of inequalities, ordinarily. *Be careful about this!* For example, if you try to abbreviate "$-3 < x$ or $x < 4$" as $-3 < x < 4$ you will be *wrong*, because $-3 < x < 4$ is an abbreviation for the conjunction $-3 < x$ and $x < 4$.

Example 7 Solve the following. Then graph.

$$\begin{array}{lll} 2x - 5 < -7 & \text{or} & 2x - 5 > 7 \\ 2x < -2 & \text{or} & 2x > 12 \quad \text{Adding 5} \\ x < -1 & \text{or} & x > 6 \quad \text{Multiplying by } \tfrac{1}{2} \end{array}$$

The solution set is

$$\{x \mid x < -1 \text{ or } x > 6\},$$

which is the union

$$\{x \mid x < -1\} \cup \{x \mid x > 6\}.$$

The graph is shown in Fig. 24.

Figure 24

Example 8 Solve the following. Then graph.

$$\begin{array}{lll} \dfrac{4 - 3x}{2} < -1 & \text{or} & \dfrac{4 - 3x}{2} \geq 1 \\ 4 - 3x < -2 & \text{or} & 4 - 3x \geq 2 \quad \text{Multiplying by 2} \\ 4 < -2 + 3x & \text{or} & 4 \geq 2 + 3x \quad \text{Adding } 3x \\ 6 < 3x & \text{or} & 2 \geq 3x \\ 2 < x & \text{or} & \dfrac{2}{3} \geq x \end{array}$$

The solution set is

$$\left\{x \mid 2 < x \text{ or } \tfrac{2}{3} \geq x\right\},$$

which is the union

$$\{x \mid 2 < x\} \cup \left\{x \mid \tfrac{2}{3} \geq x\right\}.$$

The graph is shown in Fig. 25.

Figure 25

EXERCISE SET 4.5

Find these unions or intersections.

1. $\{3, 4, 5, 8, 10\} \cap \{1, 2, 3, 4, 5, 6, 7\}$
2. $\{3, 4, 5, 8, 10\} \cup \{1, 2, 3, 4, 5, 6, 7\}$
3. $\{0, 2, 4, 6, 8\} \cup \{4, 6, 9\}$
4. $\{0, 2, 4, 6, 8\} \cap \{4, 6, 9\}$
5. $\{a, b, c\} \cap \{c, d\}$
6. $\{a, b, c\} \cup \{c, d\}$

Graph.

7. $\{x \mid 7 \leq x\} \cup \{x \mid x < 9\}$
8. $\left\{x \mid -\frac{1}{2} \leq x\right\} \cup \left\{x \mid x < \frac{1}{2}\right\}$
9. $\left\{x \mid -\frac{1}{2} \leq x\right\} \cap \left\{x \mid x < \frac{1}{2}\right\}$
10. $\left\{x \mid x > \frac{1}{4}\right\} \cap \{x \mid 1 \geq x\}$
11. $\{x \mid x < -\pi\} \cup \{x \mid x > \pi\}$
12. $\{x \mid -\pi \leq x\} \cap \{x \mid x < \pi\}$
13. $\{x \mid x < -7\} \cup \{x \mid x = -7\}$
14. $\left\{x \mid x > \frac{1}{2}\right\} \cup \left\{x \mid x = \frac{1}{2}\right\}$
15. $\{x \mid x \geq 5\} \cap \{x \mid x \leq -3\}$
16. $\{x \mid x \geq -3\} \cup \{x \mid x \leq 0\}$

Solve. Then graph.

17. $-2 < x + 1$
18. $x - 5 > -7$

Solve.

19. $2x - 3 \geq -6$
20. $5x + 7 < -8$
21. $7 - 3x < x - 9$
22. $5 - 2x \geq 7x + 4$
23. $2 - 3x > 7 - 3x$
24. $3 + 5x < 5x + 7$

Solve. Then graph.

25. $-2 \leq x + 1 < 4$
26. $-3 < x + 2 \leq 5$
27. $5 \leq x - 3 \leq 7$
28. $-1 < x - 4 < 7$
29. $-3 \leq x + 4 \leq -3$
30. $-5 < x + 2 < -5$

Solve.

31. $-2 < 2x + 1 < 5$
32. $-3 \leq 5x + 1 \leq 3$
33. $-4 \leq 6 - 2x < 4$
34. $-3 < 1 - 2x \leq 3$
35. $-5 < \frac{1}{2}(3x + 1) \leq 7$
36. $\frac{2}{3} \leq -\frac{4}{5}(x - 3) < 1$

Solve. Then graph.

37. $3x \leq -6$ or $x - 1 > 0$
38. $2x < 8$ or $x + 3 \geq 1$
39. $2x + 3 \leq -4$ or $2x + 3 \geq 4$
40. $3x - 1 < -5$ or $3x - 1 > 5$

Solve.

41. $2x - 20 < -0.8$ or $2x - 20 > 0.8$
42. $5x + 11 \leq -4$ or $5x + 11 \geq 4$
43. $x + 14 \leq -\frac{1}{4}$ or $x + 14 \geq \frac{1}{4}$
44. $x - 9 < -\frac{1}{2}$ or $x - 9 > \frac{1}{2}$

☆

Solve.

45. $x \leq 3x - 2 \leq 2 - x$
46. $2x \leq 5 - 7x < 7 + x$

47. $(x + 1)^2 > x(x - 3)$

48. $(x + 4)(x - 5) > (x + 1)(x - 7)$

49. $(x + 1)^2 \le (x + 2)^2 \le (x + 3)^2$

50. $(x - 1)(x + 1) < (x + 1)^2 \le (x - 3)^2$

51. ▪ The length of a rectangle is 15.23 cm. What widths will give a perimeter greater than 40.23 cm and less than 137.8 cm?

52. The height of a triangle is 15 m. What lengths of the base will keep the area less than or equal to 305.4 m² (and, of course, positive)?

53. To get an A in a course, a student's average must be greater than or equal to 90%. It will, of course, be less than or equal to 100%. On the first three tests a student scored 83%, 87%, and 93%. What scores on the fourth test will produce an A? Is an A possible?

54. In Exercise 53, suppose the scores on the first three tests are 75%, 70%, and 83%. What scores on the fourth test will produce an A? Is an A possible?

Find the domain of each function.

55. $f(x) = \dfrac{\sqrt{x + 2}}{\sqrt{x - 2}}$

56. $f(x) = \dfrac{\sqrt{3 - x}}{\sqrt{x + 5}}$

4.6 EQUATIONS AND INEQUALITIES WITH ABSOLUTE VALUE

Absolute value was defined in Chapter 1 (p. 3). An informal way of thinking of absolute value is that it is the distance from 0 of a number on a number line. For example, $|4|$ is 4 because 4 is 4 units from 0; $|-5|$ is 5 because -5 is 5 units from 0. This idea is helpful in solving equations and inequalities with absolute value.

Example 1 Solve: $|x| = 3$.

To solve we look for all numbers x whose distance from 0 is 3. There are two of them, so there are two solutions, 3 and -3. The graph is shown in Fig. 26.

Figure 26

Example 2 Solve: $|x| < 3$.

This time we look for all numbers x whose distance from 0 is less than 3. These are the numbers between -3 and 3. The solution set and its graph are shown in Fig. 27.

$\{x \mid -3 < x < 3\}$

Figure 27

Example 3 Solve: $|x| \ge 3$.

This time we look for all numbers x whose distance from 0 is 3 or greater. The solution set and its graph are shown in Fig. 28.

$\{x \mid x \le -3 \text{ or } x \ge 3\}$

Figure 28

The results of the above examples can be generalized as follows.

For any $a > 0$,

i) $|x| = a$ **is equivalent to** $x = -a$ **or** $x = a$.

ii) $|x| < a$ **is equivalent to** $-a < x < a$.

iii) $|x| > a$ **is equivalent to** $x < -a$ **or** $x > a$.

Similar statements hold true for $|x| \leq a$ and $|x| \geq a$.

Example 4 Solve: $|x - 2| = 3$.

Note that this is a translation of $|x| = 3$, two units to the right. We first graph $|x| = 3$ in Fig. 29. This consists of the numbers that are a distance of 3 from 0. We now translate, as in Fig. 30. This solution set consists of the numbers that are a distance of 3 from 2 (note that 2 is where 0 went in the translation).

The solutions of $|x - 2| = 3$ are -1 and 5.

Figure 29

Figure 30

The results of this example can be generalized as follows.

For any a, $|x - a|$ is the distance from a to x.

Example 5

a) $|x - 5|$ is the distance from 5 to x.

b) $|x + 7|$ is the distance from -7 to x [because $x + 7 = x - (-7)$].

Here are some further examples of solving inequalities.

Example 6 Solve: $|x + 2| < 3$.

Method 1. We translate the graph of $|x| < 3$ to the left two units as shown in Fig. 31.

The solution set is $\{x \mid -5 < x < 1\}$.

Figure 31

Method 2. The solutions are those numbers x whose distance from -2 is less than 3. Thus to find the solutions graphically we locate -2 (see Fig. 32). Then we locate those numbers that are less than three units to the left and less than three units to the right. Thus the solution set is $\{x \mid -5 < x < 1\}$.

Figure 32

Method 3. We use Property (ii), replacing x by $x + 2$.

$$|x + 2| < 3$$
$$-3 < x + 2 < 3$$
$$-5 < x < 1$$

The solution set is $\{x \mid -5 < x < 1\}$.

The sentences in the following examples are more complicated. Although we could continue using translations and graphs, this actually gets more difficult than if we use Properties (i) to (iii).

Example 7 Solve: $|2x + 3| = 1$.

$$\begin{array}{lll} 2x + 3 = -1 & \text{or} \quad 2x + 3 = 1 & \text{Property (i)} \\ 2x = -4 & \text{or} \quad 2x = -2 & \text{Adding } -3 \\ x = -2 & \text{or} \quad x = -1 & \end{array}$$

The solution set is $\{-2, -1\}$.

Example 8 Solve: $|2x + 3| \leq 1$.

$$-1 \leq 2x + 3 \leq 1 \quad \text{Property (ii)}$$
$$-4 \leq 2x \leq -2 \quad \text{Adding } -3$$
$$-2 \leq x \leq -1$$

The solution set is $\{x \mid -2 \leq x \leq -1\}$.

Example 9 Solve: $|3 - 4x| > 2$.

$$3 - 4x < -2 \quad \text{or} \quad 3 - 4x > 2 \quad \text{Property (iii)}$$
$$-4x < -5 \quad \text{or} \quad -4x > -1 \quad \text{Adding } -3$$
$$x > \frac{5}{4} \quad \text{or} \quad x < \frac{1}{4}$$

The solution set is $\{x \mid x > \frac{5}{4} \text{ or } x < \frac{1}{4}\}$.

EXERCISE SET 4.6

Solve and graph.

1. $|x| = 7$
2. $|x| = \pi$
3. $|x| < 7$
4. $|x| \leq \pi$
5. $|x| \geq \pi$
6. $|x| > 7$

Solve. Use three methods.

7. $|x - 1| = 4$
8. $|x - 7| = 5$
9. $|x + 8| < 9$
10. $|x + 6| \leq 10$
11. $|x + 8| \geq 9$
12. $|x + 6| > 10$
13. $\left|x - \frac{1}{4}\right| < \frac{1}{2}$
14. $|x - 0.5| \leq 0.2$

Solve. Use any method.

15. $|3x| = 1$
16. $|5x| = 4$
17. $|3x + 2| = 1$
18. $|7x - 4| = 8$
19. $|3x| < 1$
20. $|5x| \leq 4$
21. $|2x + 3| \leq 9$
22. $|2x + 3| < 13$
23. $|x - 5| > 0.1$
24. $|x - 7| \geq 0.4$
25. $\left|x + \frac{2}{3}\right| \leq \frac{5}{3}$
26. $\left|x + \frac{3}{4}\right| < \frac{1}{4}$
27. $|6 - 4x| \leq 8$
28. $|5 - 2x| > 10$
29. $\left|\frac{2x + 1}{3}\right| > 5$
30. $\left|\frac{3x + 2}{4}\right| \leq 5$
31. $\left|\frac{13}{4} + 2x\right| > \frac{1}{4}$
32. $\left|\frac{5}{6} + 3x\right| < \frac{7}{6}$
33. $\left|\frac{3 - 4x}{2}\right| \leq \frac{3}{4}$
34. $\left|\frac{2x - 1}{3}\right| \geq \frac{5}{6}$
35. $|x| = -3$
36. $|x| < -3$
37. $|2x - 4| < -5$
38. $|3x + 5| < 0$
39. ▦ $|x - 2.0245| < 0.1011$
40. ▦ $|x + 17.217| > 5.0012$
41. ▦ $|3.0147x - 8.9912| \leq 6.0243$
42. ▦ $|-2.1437x + 7.8814| \geq 9.1132$

☆ ──

Solve. The definition of absolute value in Section 1.8 may be helpful for some of these exercises.

43. $|4x - 5| = x + 1$
44. $|2x + 3| = |x| + 8$
45. $||x| - 1| = 3$
46. $|x + 2| > x$
47. $|x + 2| \leq |x - 5|$
48. $|3x - 1| > 5x - 2$
49. $|x| + |x - 1| < 10$
50. $|x| - |x - 3| < 7$
51. $|x - 3| + |2x + 5| > 6$

★

52. $|x - 3| + |2x + 5| + |3x - 1| = 12$

Prove the following for any real numbers a and b.

53. $-|a| \leq a \leq |a|$

54. **The triangle inequality:** $|a + b| \leq |a| + |b|$

55. Show that if $|a| < \frac{e}{2}$ and $|b| < \frac{e}{2}$, then $|a + b| < e$.

56. a) Prove that $\left| x - \frac{a+b}{2} \right| < \frac{b-a}{2}$ is equivalent to $a < x < b$.

Use graphs or the result of part (a) to find an inequality with absolute value for each of the following:
b) $-5 < x < 5$; c) $-6 < x < 6$; d) $-1 < x < 7$; e) $-5 < x < 1$.

57. Use absolute value to prove that the number halfway between a and b is $\frac{a+b}{2}$.

4.7 QUADRATIC, POLYNOMIAL, AND RATIONAL INEQUALITIES

Quadratic Inequalities

Inequalities such as the following are called *quadratic* inequalities:
$$x^2 - 2x - 2 > 0, \qquad 3x^2 + 5x + 1 \leq 0.$$
In each case we have a polynomial of degree 2 on the left. One way of solving quadratic inequalities is by graphing.

Example 1 Solve: $x^2 + 2x - 3 > 0$.

We consider the function $f(x) = x^2 + 2x - 3$. The inputs that produce positive outputs are the solutions of the inequality. We graph the function (see Fig. 33) to see where the outputs (function values) are positive.
We set $f(x) = 0$ and factor to find the intercepts, and graph:
$$(x + 3)(x - 1) = 0 \quad \text{so} \quad x = -3 \quad \text{or} \quad x = 1.$$
The solution set of the inequality is $\{x \mid x > 1 \text{ or } x < -3\}$.

Example 2 Solve: $x^2 - 2x \leq 2$.

We first find standard form, with 0 on one side:
$$x^2 - 2x - 2 \leq 0.$$

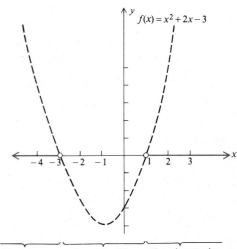

These inputs give positive outputs.   Inputs in this interval give negative or 0 outputs.   These inputs give positive outputs.

Figure 33

Figure 34

Now we graph the function $f(x) = x^2 - 2x - 2$ (see Fig. 34). The quadratic formula is needed to find the intercepts. They are $1 + \sqrt{3}$ and $1 - \sqrt{3}$. The solution set is $\{x \mid 1 - \sqrt{3} \leq x \leq 1 + \sqrt{3}\}$.

It should be pointed out that we need not actually draw graphs as in the preceding examples. Merely visualizing the graph will usually suffice.

Let us now consider another method of solving inequalities.

Example 3 Solve: $x^2 + 2x - 3 > 0$.

We factor, obtaining the inequality $(x + 3)(x - 1) > 0$. The solutions of $(x + 3)(x - 1) = 0$, -3 and 1, though not solutions of the inequality, divide the real-number line in a natural way, pictured as follows. The product $(x + 3)(x - 1)$ is positive or negative, for values other than -3 and 1, depending on the signs of the factors $x + 3$ and $x - 1$. We tabulate signs in these intervals.

INTERVAL	$x + 3$	$x - 1$	PRODUCT: $(x + 3)(x - 1)$
$x < -3$	−	−	+
$-3 < x < 1$	+	−	−
$1 < x$	+	+	+

In order for the product $(x + 3)(x - 1)$ to be positive, both of the factors must be positive or both must be negative. We see from the table that the solution set of the inequality is

$$\{x \mid x > 1\} \cup \{x \mid x < -3\}, \quad \text{or} \quad \{x \mid x > 1 \quad \text{or} \quad x < -3\}.$$

Polynomial Inequalities

We can extend the preceding method to apply to polynomials with more than two factors.

Example 4 Solve: $4x(x + 1)(x - 1) < 0$.

The solutions of $4x(x + 1)(x - 1) = 0$ are $-1, 0$, and 1. They divide the real-number line in a natural way, pictured as follows. The product $4x(x + 1)(x - 1)$ is positive or negative depending on the signs of the factors $4x$, $x + 1$, and $x - 1$. We tabulate signs in these intervals.

INTERVAL	4x	x + 1	x − 1	PRODUCT: 4x(x + 1)(x − 1)
x < −1	−	−	−	−
−1 < x < 0	−	+	−	+
0 < x < 1	+	+	−	−
1 < x	+	+	+	+

The product of three numbers is negative when it has an odd number of negative factors. We see from the table above that the solution set is

$\{x | x < -1\} \cup \{x | 0 < x < 1\}$, or $\{x | x < -1 \text{ or } 0 < x < 1\}$.

Rational Inequalities

We can also use the preceding method when an inequality involves functions that are quotients of polynomials. Such functions are called *rational*.

Example 5 Solve: $\dfrac{x+1}{x-2} \geq 3$.

We add -3, to get 0 on one side:

$$\frac{x+1}{x-2} - 3 \geq 0.$$

Next, we obtain a single fractional expression:

$$\frac{x+1}{x-2} - 3\frac{x-2}{x-2} = \frac{x+1-3x+6}{x-2} \geq 0$$

$$\frac{-2x+7}{x-2} \geq 0.$$

Let us consider the equality portion of $\geq$:

$$\frac{-2x+7}{x-2} = 0.$$

This has the solution $\frac{7}{2}$. Thus $\frac{7}{2}$ is in the solution set.
Next, consider the inequality portion of $\geq$:

$$\frac{-2x+7}{x-2} > 0.$$

The solutions of $-2x + 7 = 0$ and $x - 2 = 0$ are 2 and $\frac{7}{2}$. They divide the real-number line in a natural way, pictured as follows. The quotient is positive or negative depending on the signs of $-2x + 7$ and $x - 2$. We tabulate signs in these intervals.

INTERVAL	$-2x + 7$	$x - 2$	QUOTIENT: $\frac{-2x + 7}{x - 2}$
$x < 2$	+	−	−
$2 < x < \frac{7}{2}$	+	+	+
$\frac{7}{2} < x$	−	+	−

We see from the table that the solution set of $(-2x + 7)/(x - 2) > 0$ is $\{x \mid 2 < x < \frac{7}{2}\}$. Thus the solution set of the inequality in question is $\{x \mid 2 < x \leq \frac{7}{2}\}$.

EXERCISE SET 4.7

Solve by graphing.

1. $x^2 - x - 2 < 0$
2. $x^2 - 4 > 0$
3. $x^2 \geq 1$
4. $x^2 - 2x < 5$
5. $x^2 - 2x + 1 \geq 0$
6. $x^2 + 6x + 9 < 0$
7. $x^2 < 6x - 4$
8. $5x - 2 > x^2$

Solve.

9. $x^2 - 8x - 20 \leq 0$
10. $x^2 - 8x - 20 \geq 0$
11. $x^2 - 2x > 8$
12. $x^2 - 2x < 8$
13. $4x^2 + 7x \leq 15$
14. $4x^2 + 7x \geq 15$
15. $2x^2 + x > 5$
16. $2x^2 + x \leq 2$
17. $5x(x + 1)(x - 1) > 0$
18. $3x(x + 2)(x - 2) < 0$
19. $(x + 3)(x + 2)(x - 1) < 0$
20. $(x - 2)(x - 3)(x + 1) > 0$
21. $\frac{1}{4 - x} < 0$
22. $\frac{-4}{2x + 5} > 0$
23. $\frac{3x + 2}{x - 3} > 0$
24. $\frac{5 - 2x}{4x + 3} < 0$
25. $\frac{x + 1}{2x - 3} \geq 1$
26. $\frac{x - 1}{x - 2} \geq 3$
27. $\frac{x + 1}{x + 2} \leq 3$
28. $\frac{x + 1}{2x - 3} \leq 1$
29. $(x + 1)(x - 2) > (x + 3)^2$
30. $(x - 4)(x + 3) \leq (x - 1)^2$

31. $x^3 - x^2 > 0$
32. $x^3 - 4x > 0$
33. $x + \dfrac{4}{x} > 4$
34. $\dfrac{1}{x^2} \leq \dfrac{1}{x^3}$
35. $\dfrac{1}{x^3} \leq \dfrac{1}{x^2}$
36. $x + \dfrac{1}{x} > 2$
37. $\dfrac{2 + x - x^2}{x^2 + 5x + 6} < 0$
38. $\dfrac{4}{x^2} - 1 > 0$

☆

Solve.

39. $\left|\dfrac{x+3}{x-4}\right| < 2$
40. $|x^2 - 5| = 5 - x^2$
41. $(7 - x)^{-2} < 0$
42. $(1 - x)^3 > 0$
43. $\left|1 + \dfrac{1}{x}\right| < 3$
44. $(x + 5)^{-2} > 0$
45. $\left|2 - \dfrac{1}{x}\right| \leq 2 + \left|\dfrac{1}{x}\right|$
46. $\dfrac{(x-2)^2(x-3)^3(x+1)}{(x+2)(4-x)} \geq 0$
47. $|x^2 + 3x - 1| < 3$
48. $|1 + 5x - x^2| \geq 5$

49. The base of a triangle is 4 cm greater than the height. Find the possible heights h such that the area of the triangle will be greater than 10 cm².

50. The length of a rectangle is 3 m greater than the width. Find the possible widths w such that the area of the rectangle will be greater than 15 m².

51. A company has the following total cost and total revenue functions to use in producing and selling x units of a certain product:

$$R(x) = 50x - x^2, \quad C(x) = 5x + 350.$$

(For reference, see p. 168.)
  a) Find the break-even values.
  b) Find the values of x that produce a profit.
  c) Find the values of x that result in a loss.

52. A company has the following total cost and total revenue functions to use in producing and selling x units of a certain product:

$$R(x) = 80x - x^2, \quad C(x) = 10x + 600.$$

  a) Find the break-even values.
  b) Find the values of x that produce a profit.
  c) Find the values of x that result in a loss.

53. Find the numbers k for which the quadratic equation $x^2 + kx + 1 = 0$ has (a) two real-number solutions; (b) no real-number solution.

54. Find the numbers k for which the quadratic equation $2x^2 - kx + 1 = 0$ has (a) two real-number solutions; (b) no real-number solution.

Find the domain of each function.

55. $f(x) = \sqrt{1 - x^2}$
56. $f(x) = \dfrac{1}{\sqrt{1 - x^2}}$
57. $g(x) = \sqrt{x^2 + 2x - 3}$

CHAPTER 4 REVIEW

1. Find the slope of the line $-2x - y = 7$.
2. Find the slope of the line through $(7, -2)$ and $(1, 4)$.
3. Find an equation of the line through $(-2, -1)$ with $m = 3$.
4. Find an equation of the line containing $(4, 1)$ and $(-2, -1)$.
5. Find the distance between $(3, 7)$ and $(-2, 4)$.
6. Find the midpoint of the segment with endpoints $(3, 7)$ and $(-2, 4)$.

Given the point $(1, -1)$ and the line $2x + 3y = 4$:

7. Find an equation of the line containing the given point and parallel to the given line.

8. Find an equation of the line containing the given point and perpendicular to the given line.

Determine whether the lines are parallel, perpendicular, or neither.

9. $3x - 2y = 8$,
$6x - 4y = 2$

10. $y - 2x = 4$,
$2y - 3x = -7$

11. $y = \frac{3}{2}x + 7$,
$y = -\frac{2}{3}x - 4$

For the functions in Questions 12 and 13:
  a) find an equation of the type $f(x) = a(x - h)^2 + k$;
  b) find the vertex;
  c) find the line of symmetry; and
  d) determine whether the second coordinate of the vertex is a maximum or a minimum, and find the maximum or minimum.

12. $f(x) = 3x^2 + 6x + 1$ (Also graph.)

13. $f(x) = -2x^2 - 3x + 6$

14. Find the x-intercepts: $y = f(x) = -x^2 - x - 1$.

Solve.

15. $-3x + 2 \leq -4$ *and* $x - 1 \leq 3$

16. $|x - 6| < 5$

17. $|-3x - 2| > 2$

18. $\left|\frac{1}{2}x + 4\right| \leq 6$

19. $|2x + 5| = 9$

20. $x^2 - 9 < 0$

21. $2x^2 - 3x - 2 > 0$

22. $\frac{x - 2}{x + 3} < 4$

23. $(1 - x)(x + 4)(x - 2) < 0$

24. $4x \leq 5x + 2 < 4 - x$

25. $\left|1 - \frac{1}{x^2}\right| < 3$

26. $(x - 2)^{-3} < 0$

27. A student decided to predict his final exam score S from his midterm test score d. To do this, he checked with students who had taken the course before. One student scored 70 on the midterm and 75 on the final. Another scored 85 on the midterm and 89 on the final. The student used a linear model to predict his own final score.
  a) Express S as a linear function of d.
  b) Use the equation obtained in part (a) to predict the student's final exam score, assuming that he scored 81 on his midterm.

28. Ball bearings cost 3¢ each. A wheel manufacturer knows that between 12,000 and 15,000 ball bearings will be needed. What is the range for the total cost?

29. Find two numbers such that their sum is 30 and their product is a maximum.

30. The sum of the length and width of a rectangle is 40. Find the dimensions for which the area is a maximum.

31. A company has the following total cost and total revenue functions to use in producing and selling x units of a product:

$$R(x) = 100 + x^2, \quad C(x) = 15x + 200.$$

a) Find the break-even values.

b) Find the values of x that produce a profit.

c) Find the values of x that result in a loss.

32. Under what conditions is the square of a linear function a quadratic function?

Find the domain of each function.

33. $f(x) = \sqrt{1 - |3x - 2|}$

34. $g(x) = \dfrac{1}{\sqrt{5 - |7x + 2|}}$

5

Systems of Linear Equations and Inequalities

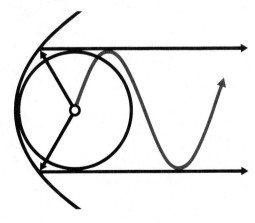

5.1 SYSTEMS OF EQUATIONS IN TWO VARIABLES

Recall that a *conjunction* of sentences is formed by joining sentences with the word *and*. Here is an example:

$$x + y = 11 \quad \text{and} \quad 3x - y = 5.$$

A solution of a sentence with two variables such as $x + y = 11$ is an ordered pair. Some pairs in the solution set of $x + y = 11$ are

$$(5, 6), \quad (12, -1), \quad (4, 7), \quad (8, 3).$$

Some pairs in the solution set of $3x - y = 5$ are

$$(0, -5), \quad (4, 7), \quad (-2, -11), \quad (9, 22).$$

The solution set of the sentence

$$x + y = 11 \quad \text{and} \quad 3x - y = 5$$

consists of all pairs that make *both* sentences true. That is, it is the *intersection* of the solution sets. Note that $(4, 7)$ is a solution of the conjunction; in fact, it is the only solution.

One way to find solutions is by trial and error. Another way is to graph the equations and look for points of intersection. For example, the graph in Fig. 1 shows the solution sets of $x + y = 11$ and $3x - y = 5$. Their intersection is the single ordered pair $(4, 7)$.

We often refer to a conjunction of equations as a *system* of equations. We usually omit the word *and*, and very often write one equation under the other.

We now consider two algebraic methods for solving systems of linear equations.

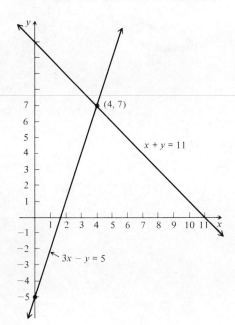

Figure 1

The Substitution Method

To use the substitution method we solve one equation for one of the variables. Then we substitute in the other equation and solve.

Example 1 Solve the following system:

$$x + y = 11, \quad (1)$$
$$3x - y = 5. \quad (2)$$

First we solve equation (1) for y*:

$$y = 11 - x.$$

---

*We could just as well solve for x.

Then we substitute $11 - x$ for y in equation (2). This gives an equation in one variable, which we know how to solve from earlier work.

$$3x - (11 - x) = 5$$
$$x = 4.$$

Now we substitute 4 for x in either equation (1) or (2) and solve for y. Let us use equation (1):

$$4 + y = 11$$
$$y = 7.$$

The solution is $(4, 7)$. Note the alphabetical listing: 4 for x, 7 for y. Check:

$$\begin{array}{c|c} x + y = 11 \\ \hline 4 + 7 & 11 \\ 11 & \end{array} \qquad \begin{array}{c|c} 3x - y = 5 \\ \hline 3 \cdot 4 - 7 & 5 \\ 5 & \end{array}$$

The Addition Method

The *addition method* makes use of the addition and multiplication principles for solving equations. Consider the system

$$3x - 4y = -1,$$
$$-3x + 2y = 0.$$

We add the left-hand sides and obtain $-2y$, and add the right-hand sides* and obtain -1. Thus we have a new equation: $-2y = -1$. We can now solve for y and then substitute into one of the original equations to find x:

$$-2y = -1 \qquad -3x + 2\left(\frac{1}{2}\right) = 0 \qquad \text{Substituting in the second equation.}$$
$$y = \frac{1}{2} \qquad -3x + 1 = 0$$
$$-3x = -1$$
$$x = \frac{1}{3}$$

We now know that the solution of the original system is $(\frac{1}{3}, \frac{1}{2})$, because the solutions of this last system are obvious and we know that this

---

*When we do this we often say that we "added the two equations."

system is equivalent to the original. We know that the last system is equivalent to the original, because the only computations we did consisted of using the addition principle, adding only expressions for which all replacements were sensible, and multiplying by a nonzero constant. Each produces results equivalent to the original equations. For this reason we do not need to check results, except to detect errors in computation.

Example 2 Solve the following system:

$$5x + 3y = 7,$$
$$3x - 5y = -23.$$

We first multiply the second equation by 5 to make the x-coefficient a multiple of 5:

$$5x + 3y = 7$$
$$15x - 25y = -115.$$

Now we multiply the first equation by -3 and add it to the second equation. This eliminates the x-term:

$$5x + 3y = 7$$
$$-34y = -136.$$

Next we solve the second equation for y. Then we substitute the result in the first equation to find x.

$$-34y = -136 \qquad 5x + 3(4) = 7$$
$$y = 4 \qquad 5x + 12 = 7$$
$$5x = -5$$
$$x = -1$$

The solution is $(-1, 4)$.

Shortcuts

There are some shortcuts when using the addition method. One is to first write the equation in the form $Ax + By = C$. Another shortcut is to interchange two equations before beginning. For example, if we have the system

$$5x + y = -2,$$
$$x + 7y = 3,$$

we may prefer to write the second equation first. This accomplishes two things. One is that the x-coefficient is 1 in the first equation. Having found y later, this will make solving for x easier. The other is that this makes the x-coefficient in the second equation a multiple of the first.

Another shortcut consists of multiplying one or more equations by a power of 10 before beginning in order to eliminate decimal points. For example, if we have the system

$$-0.3x + 0.5y = 0.03,$$
$$0.01x - 0.4y = 1.2,$$

we can multiply both equations by 100 to clear of decimal points, obtaining

$$-30x + 50y = 3,$$
$$x - 40y = 120.$$

Still other shortcuts are possible. However, it is recommended that no others be used at this point, because we are establishing an *algorithm*, which could be done mechanically by a computer.

Example 3 Solve:

$$5x + y = -2,$$
$$x + 7y = 3.$$

We first interchange the equations so that the x-coefficient of the second equation will be a multiple of the first:

$$x + 7y = 3,$$
$$5x + y = -2.$$

Next, we multiply the first equation by -5 and add the result to the second equation. This eliminates the x-term in the second equation:

$$x + 7y = 3,$$
$$-34y = -17.$$

Now we solve the second equation for y. Then we substitute the result in the first equation to find x.

$$-34y = -17 \qquad x + 7\left(\frac{1}{2}\right) = 3$$

$$y = \frac{1}{2} \qquad x + \frac{7}{2} = 3$$

$$x = 3 - \frac{7}{2}, \text{ or } -\frac{1}{2}$$

The solution is $(-\frac{1}{2}, \frac{1}{2})$.

EXERCISE SET 5.1

1. Determine whether $\left(\frac{1}{2}, 1\right)$ is a solution.

$3x + y = \frac{5}{2}$,
$2x - y = \frac{1}{4}$

2. Determine whether $\left(-2, \frac{1}{4}\right)$ is a solution.

$x + 4y = -1$,
$2x + 8y = -2$

Solve graphically.

3. $x + y = 2$,
$3x + y = 0$

4. $x + y = 1$,
$3x + y = 7$

Solve, using the substitution method.

5. $x - 5y = 4$,
$2x + y = 7$

6. $3x - y = 5$,
$x + y = \frac{1}{2}$

Solve, using the addition method.

7. $x - 3y = 2$,
$6x + 5y = -34$

8. $x + 3y = 0$,
$20x - 15y = 75$

9. $0.3x + 0.2y = -0.9$,
$0.2x - 0.3y = -0.6$

10. $0.2x - 0.3y = 0.3$,
$0.4x + 0.6y = -0.2$

11. $\frac{1}{5}x + \frac{1}{2}y = 6$,
$\frac{3}{5}x - \frac{1}{2}y = 2$

12. $\frac{2}{3}x + \frac{3}{5}y = -17$,
$\frac{1}{2}x - \frac{1}{3}y = -1$

13. $2a = 5 - 3b$,
$4a = 11 - 7b$

14. $7(a - b) = 14$,
$2a = b + 5$

☆

15. $\frac{x+y}{4} - \frac{x-y}{3} = 1$

$\frac{x-y}{2} + \frac{x+y}{4} = -9$

16. $\frac{x+y}{2} - \frac{y-x}{3} = 0$

$\frac{x+y}{3} - \frac{x+y}{4} = 0$

Solve. Check by substituting.

17. ▦ $2.35x - 3.18y = 4.82$,
$1.92x + 6.77y = -3.87$

18. ▦ $0.0375x + 0.912y = -1.003$,
$463x - 801y = 946$

Each of the following is a system of equations that is *not* linear. But each is *linear in form,* in that an appropriate substitution, say u for 1/x and v for 1/y, yields a linear system. Solve for the new variable and then solve for the original variable.

19. $\frac{1}{x} - \frac{3}{y} = 2$,

$\frac{6}{x} + \frac{5}{y} = -34$

20. $2\sqrt[3]{x} + \sqrt{y} = 0$,
$5\sqrt[3]{x} + 2\sqrt{y} = -5$

21. $3|x| + 5|y| = 30$,
$5|x| + 3|y| = 34$

22. $15x^2 + 2y^3 = 6$
$25x^2 - 2y^3 = -6$

5.2 INCONSISTENT AND DEPENDENT EQUATIONS; APPLIED PROBLEMS

Inconsistent Equations

Some systems of equations have no solution. Such systems are called *inconsistent*. For a system of two linear equations, this means that their graphs are parallel lines and do not intersect.

Example 1 Solve:
$$x - 3y = 1,$$
$$-2x + 6y = 5.$$

We multiply the first equation by 2 and add. This gives us
$$x - 3y = 1$$
$$0 = 7.$$

The second equation says that $0 \cdot x + 0 \cdot y = 7$. There are no numbers x and y for which this is true.

Whenever we obtain a statement such as $0 = 7$, which is obviously false, we will know that the system we are trying to solve has no solutions. It is inconsistent. The solution set is ∅.

The graphs of the above equations are shown in Fig. 2.

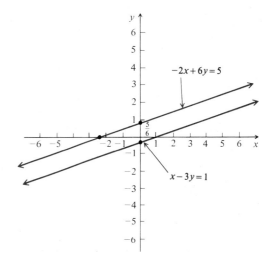

Figure 2

Dependent Equations

Consider the following system:
$$2x + 3y = 1,$$
$$4x + 6y = 2.$$

Note that if we multiply the first equation by 2 we get the second equation. Thus the two equations are equivalent. Or, to put it another way, the conjunction of the *two* equations is equivalent to *one* of the equations. Such equations are called *dependent*. If a system of *three* equations is equivalent to *one* of them or to the conjunction of *two* of them, then it is a dependent system.

DEFINITION

If a system of *n* linear equations is equivalent to a system of fewer than *n* of them, then the system is *dependent*.

Suppose we try to solve a dependent system of two equations:

$$2x + 3y = 1,$$
$$4x + 6y = 2.$$

We multiply the first equation by -2 and add. This gives us

$$2x + 3y = 1$$
$$0 + 0 = 0.$$

When, in solving, we get an obviously true statement, such as $0 + 0 = 0$, we know that the system is dependent.

A dependent system of equations may have an infinite number of solutions. We usually describe the solutions by expressing one variable in terms of the other.

Example 2 Describe the solutions of the system

$$2x + 3y = 1,$$
$$4x + 6y = 2.$$

In attempting to solve this system, above, we obtained

$$2x + 3y = 1$$
$$0 + 0 = 0.$$

The last equation contributes nothing, so we consider only the first one. Let us solve it for x. We obtain

$$x = \frac{1}{2} - \frac{3}{2}y, \text{ or } \frac{1 - 3y}{2}.$$

We can now describe the ordered pairs in the solution set, in terms of y only, as follows:

$$\left(\frac{1 - 3y}{2}, y \right).$$

Any value we choose for y then gives us a value for x, and thus an ordered pair in the solution set. Some of these solutions are

$$(-4, 3), \quad \left(-\frac{5}{2}, 2\right), \quad \left(\frac{7}{2}, -2\right).$$

Applied Problems

Translating problems to equations becomes easier in many cases if we translate to more than one equation in more than one variable.

Example 3 An airplane flies the 3000-mi distance from Los Angeles to New York, with a tail wind, in 5 hr. On the return trip, against the wind, it makes the trip in 6 hr. Find the speed of the plane and the speed of the wind.

Figure 3

Translate: Let p = speed of the plane, and w = speed of the wind. We make a drawing as shown in Fig. 3. Then from $d_1 = r_1 \cdot t_1$, we have

$$3000 = (p + w)5 = 5p + 5w.$$

From $d_2 = r_2 \cdot t_2$, we have

$$3000 = (p - w)6 = 6p - 6w.$$

Then we solve the system:

$$5p + 5w = 3000,$$
$$6p - 6w = 3000.$$

The solution is (550, 50); that is, the speed of the plane is 550 mph and the speed of the wind is 50 mph. This checks in the original problem.

Example 4 Wine A is 5% alcohol and wine B is 15% alcohol. How many liters of each should be mixed to get a 10-liter mixture that is 12% alcohol?

a) First consider the amount of wine. Let x = the amount of wine A, and y = the amount of wine B. The amount of the mixture is to be 10 L. We get the equation

$$x + y = 10.$$

b) Second, consider the amount of alcohol. The amount of alcohol in wine A is 5% · x, and the amount in wine B is 15% · y. We know that the amount in the mixture is 12% · 10. We get the equation

$$5\%x + 15\%y = 12\% \cdot 10,$$

or

$$0.05x + 0.15y = 1.2,$$

or

$$5x + 15y = 120.$$

c) Now we solve the system:

$$x + y = 10,$$
$$5x + 15y = 120.$$

The solution is (3, 7). Thus 3 L of wine A should be mixed with 7 L of wine B.

EXERCISE SET 5.2

In Exercises 1-6, solve. If a system is dependent, describe the solutions by expressing one variable in terms of the other.

1. $2a = 4 - 3b$,
   $4a = 12 - 7b$

2. $7(a - b) = 21$,
   $2a = b + 10$

3. $9x - 3y = 15$,
   $6x - 2y = 10$

4. $2s - 3t = 9$,
   $4s - 6t = 9$

5. $5c + 2d = 24$,
   $30c + 12d = 10$

6. $3x + 2y = 18$,
   $9x + 6y = 5$

7. Classify the systems in Exercises 1-6 as consistent or inconsistent.

8. Classify the systems in Exercises 1-6 as dependent or independent.

Solve the following systems.

9. $3x + 2y = 5$,
   $4y = 10 - 6x$

10. $5x + 2 = 7y$,
    $-14y + 4 = -10x$

11. $12y - 8x = 6$,
    $4x + 3 = 6y$

12. $16x - 12y = 10$,
    $6y + 5 = 8x$

Solve the following applied problems.

13. Find two numbers whose sum is -10 and whose difference is 1.

14. Find two numbers whose sum is -1 and whose difference is 10.

15. A boat travels 46 km downstream in 2 hr. It travels 51 km upstream in 3 hr. Find the speed of the boat and the speed of the stream.

16. An airplane travels 3000 km with a tail wind in 3 hr. It travels 3000 km with a head wind in 4 hr. Find the speed of the plane and the speed of the wind.

17. Antifreeze A is 18% alcohol. Antifreeze B is 10% alcohol. How many liters of each should be mixed to get 20 L of a mixture that is 15% alcohol?

18. Beer A is 6% alcohol and beer B is 2% alcohol. How many liters of each should be mixed to get 50 L of a mixture that is 3.2% alcohol?

Solve.

19. 🖩 $4.026x - 1.448y = 18.32$,
    $0.724y = -9.16 + 2.013x$

20. 🖩 $0.0284y = 1.052 - 8.114x$,
    $0.0142y + 4.057x = 0.526$

Determine the constant k such that each system is dependent.

21. $6x - 9y = -3$,
    $-4x + 6y = k$

22. $8x - 16y = 20$,
    $10x - 20y = k$

5.3 SYSTEMS OF EQUATIONS IN THREE OR MORE VARIABLES

A solution of an equation in three variables is an ordered triple of numbers. A solution of a system of equations in three variables is a triple that makes all of them true. Graphical methods of solving linear equations in three variables are unsatisfactory, because a three-dimensional coordinate system is required. The substitution method becomes cumbersome for most systems of more than two equations. Therefore,

5.3 SYSTEMS OF EQUATIONS IN THREE OR MORE VARIABLES

we will use the addition method. It is essentially the same as for systems of two equations.

Example 1 Solve:

$2x - 4y + 6z = 22,$ ①
$4x + 2y - 3z = 4,$ ② These numbers indicate the equations in the first, second, and third positions, respectively.
$3x + 3y - z = 4.$ ③

We begin by multiplying (3) by 2, to make each x-coefficient a multiple of the first.* Then we have the following:

$$2x - 4y + 6z = 22,$$
$$4x + 2y - 3z = 4,$$
$$6x + 6y - 2z = 8.$$

Next, we multiply (1) by -2 and add it to (2). We also multiply (1) by -3 and add it to (3). Then we have the following:

$$2x - 4y + 6z = 22,$$
$$10y - 15z = -40,$$
$$18y - 20z = -58.$$

Now we multiply (3) by -5 to make the y-coefficient a multiple of the y-coefficient in (2):

$$2x - 4y + 6z = 22,$$
$$10y - 15z = -40,$$
$$-90y + 100z = 290.$$

Next, we multiply (2) by 9 and add it to (3):

$$2x - 4y + 6z = 22,$$
$$10y - 15z = -40,$$
$$-35z = -70.$$

Now we solve (3) for z:

$$2x - 4y + 6z = 22,$$
$$10y - 15z = -40,$$
$$z = 2.$$

Next we substitute 2 for z in (2), and solve for y:

$$10y - 15(2) = -40$$
$$10y - 30 = -40$$
$$10y = -10$$
$$y = -1.$$

---
*By proceeding in this manner, we avoid fractions. The method will work when fractions are allowed, but it is more difficult that way.

Finally, we substitute -1 for y and 2 for z in (1):

$$2x - 4(-1) + 6(2) = 22$$
$$2x + 4 + 12 = 22$$
$$2x + 16 = 22$$
$$2x = 6$$
$$x = 3.$$

The solution is $(3, -1, 2)$. To be sure computational errors have not been made, one can check by substituting 3 for x, -1 for y, and 2 for z in all three equations. If all are true, then the triple is a solution.

Dependent and Inconsistent Equations

If a system of equations is inconsistent, we will obtain, at some stage of our solving, a false equation, such as $1 = 5$. If a system of three equations is dependent, this means that it is equivalent to a system of two of them, or to one of them alone. When we try to solve such a system, we may find that at some stage two of the equations are identical. We may also obtain an obviously true statement, such as $0 = 0$. When a system of three equations is dependent, its solutions may sometimes be described by expressing two of the variables in terms of the third.

Example 2 Solve:

$$x + 3y - 7z = 1, \quad \text{①}$$
$$2x - 4y + z = 3, \quad \text{②}$$
$$-2x + 4y - z = 10. \quad \text{③}$$

Each x-coefficient is a multiple of the first. Thus we can begin by multiplying (1) by -2 and adding it to (2). We also multiply (1) by 2 and add it to (3):

$$x + 3y - 7z = 1,$$
$$-10y + 15z = 1,$$
$$10y - 15z = 12.$$

Since the y-coefficient of (2) is a multiple of the y-coefficient of (3), we proceed by simply adding (2) to (3):

$$x + 3y - 7z = 1$$
$$-10y + 15z = 1$$
$$0 = 13.$$

Since in (3) we obtain the false equation $0 = 13$, the system is inconsistent; it has no solution. We need work no further.

Example 3 Solve:

$$x + 2y + 3z = 4, \qquad (1)$$
$$2x - y + z = 3, \qquad (2)$$
$$3x + y + 4z = 7. \qquad (3)$$

Each x-coefficient is a multiple of the first. Thus we can begin by multiplying (1) by -2 and adding it to (2). We also multiply (1) by -3 and add it to (3):

$$x + 2y + 3z = 4,$$
$$-5y - 5z = -5,$$
$$-5y - 5z = -5.$$

Now (2) and (3) are identical. We no longer have a system of three equations, but a system of two. Thus we know that the system is dependent. If we were to multiply (2) by -1 and add it to (3), we would obtain $0 = 0$. We proceed by multiplying (2) by $-\frac{1}{5}$ since the coefficients have the common factor -5:

$$x + 2y + 3z = 4,$$
$$y + z = 1.$$

We will express two of the variables in terms of the other one. Let us choose z. Then we solve (2) for y:

$$y = 1 - z.$$

We substitute this value of y in (1), obtaining

$$x + 2(1 - z) + 3z = 4.$$

Solving for x, we get

$$x = 2 - z.$$

The solutions, then, are all of the form

$$(2 - z, 1 - z, z).$$

We obtain the solutions by choosing various values of z. If we let $z = 0$, we obtain the triple $(2, 1, 0)$. If we let $z = 2$, we obtain $(0, -1, 2)$, and so on.

Homogeneous Equations

When all the terms of a polynomial have the same degree, we say that the polynomial is *homogeneous*. Here are some examples:

$$3x^2 + 5y^2, \qquad 4x + 5y - 2z, \qquad 17x^3 - 4y^3 + 57z^3.$$

An equation formed by a homogeneous polynomial set equal to 0 is called a *homogeneous equation*. Let us now consider a system of homogeneous equations.

Example 4 Solve:
$$4x - 3y + z = 0, \quad \text{①}$$
$$2x \qquad - 3z = 0, \quad \text{②}$$
$$-8x + 6y - 2z = 0. \quad \text{③}$$

Any homogeneous system like this always has a solution (can never be inconsistent), because $(0, 0, 0)$ is a solution. This is called the *trivial* solution. There may or may not be other solutions. To find out, we proceed as in the case of nonhomogeneous equations.

First we interchange the first two equations so all the x-coefficients are multiples of the first:
$$2x \qquad - 3z = 0,$$
$$4x - 3y + z = 0,$$
$$-8x + 6y - 2z = 0.$$

Now we multiply (1) by -2 and add it to (2). We also multiply (1) by 4 and add it to (3):
$$2x \qquad - 3z = 0,$$
$$-3y + 7z = 0,$$
$$6y - 14z = 0.$$

Next we multiply (2) by 2 and add it to (3):
$$2x \qquad - 3z = 0,$$
$$-3y + 7z = 0,$$
$$0 = 0.$$

Now we know that the system is dependent, hence it has an infinite set of solutions. Now we solve (2) for y:
$$y = \frac{7}{3}z.$$

Typically, we would now substitute $\frac{7}{3}z$ for y in (1), but since the y-term is missing, we need only solve for x:
$$x = \frac{3}{2}z.$$

We can now describe the members of the solution set as follows:
$$\left(\frac{3}{2}z, \frac{7}{3}z, z\right).$$

Some of the ordered triples in the solution set are

$$\left(\frac{3}{2}, \frac{7}{3}, 1\right), \quad \left(3, \frac{14}{3}, 2\right), \quad \left(-\frac{3}{2}, -\frac{7}{3}, -1\right).$$

Mathematical Models and Applied Problems

In a situation in which a quadratic function will serve as a mathematical model, we may wish to find an equation, or formula, for the function. Recall that for a linear model, we can find an equation if we know two data points. For a quadratic function we need three data points.

Example 5 In a certain situation, it is believed that a quadratic function will be a good model. Find an equation of the function, given the data points $(1, -4)$, $(-1, -6)$, and $(2, -9)$.

We wish to find a quadratic function

$$f(x) = ax^2 + bx + c$$

containing the three given data points, i.e., a function for which the equation will be true when we substitute any of the ordered pairs of numbers into it. When we substitute, we get:

for $(1, -4)$, $-4 = a \cdot 1^2 + b \cdot 1 + c$,
for $(-1, -6)$, $-6 = a(-1)^2 + b(-1) + c$,
for $(2, -9)$, $-9 = a \cdot 2^2 + b \cdot 2 + c$.

We now have a system of equations in the three unknowns a, b, and c:

$$\begin{aligned} a + b + c &= -4, \\ a - b + c &= -6, \\ 4a + 2b + c &= -9. \end{aligned}$$

We solve this system of equations, obtaining $(-2, 1, -3)$. Thus the function we are looking for is

$$f(x) = -2x^2 + x - 3.$$

Example 6 (*The cost of operating an automobile at various speeds*). It is found that the cost of operating an automobile as a function of speed is approximated by a quadratic function. Use the data given on the following page to find an equation of the function. Then use the equation to determine the cost of operating the automobile at 60 mph and at 80 mph.

SPEED IN MPH	OPERATING COST PER MILE, IN CENTS
10	22
20	20
50	20

We use the three data points to obtain a, b, and c in $f(x) = ax^2 + bx + c$:

$$22 = 100a + 10b + c,$$
$$20 = 400a + 20b + c, \quad \text{Substituting}$$
$$20 = 2500a + 50b + c.$$

We solve this system of equations, obtaining $(0.005, -0.35, 25)$. Thus

$$f(x) = 0.005x^2 - 0.35x + 25.$$

To find the cost of operating at 60 mph, we find $f(60)$:

$$f(60) = 0.005(60)^2 - 0.35(60) + 25 = 22¢.$$

We also find $f(80)$:

$$f(80) = 0.005(80)^2 - 0.35(80) + 25 = 29¢.$$

A graph of the cost function is shown in Fig. 4. It should be noted that this cost function can give approximate results only within a certain interval. For example, $f(0) = 25$, meaning that it costs 25¢ per mile to stand still, and this, of course, is absurd.

Figure 4

EXERCISE SET 5.3

Consider the system

$$2x + 3y - 5z = 1,$$
$$6x - 6y + 10z = 3,$$
$$4x - 9y + 5z = 0.$$

1. Determine whether $\left(\frac{1}{2}, \frac{1}{3}, \frac{1}{5}\right)$ is a solution of the system.

2. Determine whether $(-1, 1, 0)$ is a solution of the system.

Solve. If a system has more than one solution, list three of them.

3. $x + y + z = 2,$
$6x - 4y + 5z = 31,$
$5x + 2y + 2z = 13$

4. $x + 6y + 3z = 4,$
$2x + y + 2z = 3,$
$3x - 2y + z = 0$

5. $x - y + 2z = -3,$
$x + 2y + 3z = 4,$
$2x + y + z = -3$

6. $x + y + z = 6,$
$2x - y - z = -3,$
$x - 2y + 3z = 6$

5.3 SYSTEMS OF EQUATIONS IN THREE OR MORE VARIABLES

7. $4a + 9b = 8,$
   $8a + 6c = -1,$
   $6b + 6c = -1$

8. $3p + 2r = 11,$
   $q - 7r = 4,$
   $p - 6q = 1$

9. $x + 2y - z = -8,$
   $2x - y + z = 4,$
   $8x + y + z = 2$

10. $x + 2y - z = 4,$
    $4x - 3y + z = 8,$
    $5x - y = 12$

11. $2x + y - 3z = 1,$
    $x - 4y + z = 6,$
    $4x - 16y + 4z = 24$

12. $4x + 12y + 16z = 4,$
    $3x + 4y + 5z = 3,$
    $x + 8y + 11z = 1$

13. $2x + y - 3z = 0,$
    $x - 4y + z = 0,$
    $4x - 16y + 4z = 0$

14. $4x + 12y + 16z = 0,$
    $3x + 4y + 5z = 0,$
    $x + 8y + 11z = 0$

15. $x + y - z = -3,$
    $x + 2y + 2z = -1$

16. $x + y + 13z = 0,$
    $x - y - 6z = 0$

17. $2x + y + z = 0,$
    $x + y - z = 0,$
    $x + 2y + 2z = 0$

18. $5x + 4y + z = 0,$
    $10x + 8y - z = 0,$
    $x - y - z = 0$

19. (*Curve fitting*). Find numbers a, b, and c, such that a quadratic function $ax^2 + bx + c$ fits the data points (1, 4), (−1, −2), and (2, 13).

20. (*Curve fitting*). Find numbers a, b, and c, such that a quadratic function $ax^2 + bx + c$ fits the data points (1, 4), (−1, 6), and (−2, 16).

21. (*Predicting earnings*). A business earns $38 in the first week, $66 in the second week, and $86 in the third week. The manager graphs the points (1, 38), (2, 66), and (3, 86), and finds that a quadratic function might fit the data.

    a) Find a quadratic function that fits the data.
    b) Using your model, predict the earnings for the fourth week.

22. (*Predicting earnings*). A business earns $1000 in its first month, $2000 in the second month, and $8000 in the third month. The manager plots the points (1, 1000), (2, 2000), and (3, 8000), and finds that a quadratic function might fit the data.

    a) Find a quadratic function that fits the data.
    b) Using your model, predict the earnings for the fourth month.

☆ ─────────────────────────────────────

Hint for Exercises 23 and 24: First solve for $\frac{1}{x}$, $\frac{1}{y}$, and $\frac{1}{z}$.

23. $\frac{2}{x} - \frac{1}{y} - \frac{3}{z} = -1,$
    $\frac{2}{x} - \frac{1}{y} + \frac{1}{z} = -9,$
    $\frac{1}{x} + \frac{2}{y} - \frac{4}{z} = 17$

24. $\frac{2}{x} + \frac{2}{y} - \frac{3}{z} = 3,$
    $\frac{1}{x} - \frac{2}{y} - \frac{3}{z} = 9,$
    $\frac{7}{x} - \frac{2}{y} + \frac{9}{z} = -39$

25. When A, B, and C are working together, they can do a job in 2 hr. When B and C work together, they can do the job in 4 hr. When A and B work together, they can do the job in $\frac{12}{5}$ hr. How long would it take each, working alone, to do the job?

26. Pipes A, B, and C are connected to the same tank. When all three pipes are running, they can fill the tank in 3 hr. When pipes A and C are running, they can fill the tank in 4 hr. When pipes A and B are running, they can fill the tank in 8 hr. How long would it take each, running alone, to fill the tank?

Solve.

27. $x + y + z + w = 2,$
    $2x + 2y + 4z + w = 1,$
    $x - y - z - w = -6,$
    $3x + y - z - w = -2$

28. $x - y + z + w = 0,$
    $2x + 2y + z - w = 5,$
    $3x + y - z - w = -4,$
    $x + y - 3z - 2w = -7$

29. Solve.

$3.12x + 2.14y - 0.988z = 3.79,$
$3.84x - 3.53y + 1.96z = 7.80,$
$4.63x - 1.08y + 0.011z = 6.34$

30. Solve.

$1.01x - 0.905y + 2.12z = -2.54,$
$1.32x + 2.05y + 2.97z = 3.97,$
$2.21x + 1.35y + 0.001z = -3.15$

5.4 MATRICES AND SYSTEMS OF EQUATIONS

In solving systems of equations, we perform computations with the constants. The variables play no important role in the process. We can simplify writing by omitting the variables. For example, the system

$$3x + 4y = 5,$$
$$x - 2y = 1$$

simplifies to

$$\begin{array}{ccc} 3 & 4 & 5 \\ 1 & -2 & 1 \end{array}$$

if we leave off the variables and omit the operation and equals signs.

In the example above we have written a rectangular array of numbers. Such an array is called a *matrix* (plural, *matrices*). We ordinarily write brackets around matrices, although this is not necessary when calculating with matrices. The following are matrices.

$$\begin{bmatrix} 4 & 1 & 3 & 5 \\ 1 & 0 & 1 & 2 \\ 6 & 3 & -2 & 0 \end{bmatrix} \quad \begin{bmatrix} 6 & 2 & 1 & 4 & 7 \\ 1 & 2 & 1 & 3 & 1 \\ 4 & 0 & -2 & 0 & -3 \end{bmatrix} \quad \begin{bmatrix} 1 & 2 \\ 145 & 0 \\ -7 & 9 \\ 8 & 1 \\ 0 & 0 \end{bmatrix}$$

The *rows* of a matrix are horizontal, and the *columns* are vertical.

$$\begin{bmatrix} 5 & -2 & 2 \\ 1 & 0 & 1 \\ 0 & 1 & 2 \end{bmatrix} \begin{array}{l} \leftarrow \text{Row 1} \\ \leftarrow \text{Row 2} \\ \leftarrow \text{Row 3} \end{array}$$

Column 1 Column 2 Column 3

Let us now use matrices to solve systems of linear equations.

5.4 MATRICES AND SYSTEMS OF EQUATIONS

Example 1 Solve:

$$2x - y + 4z = -3,$$
$$x \quad\quad - 4z = 5,$$
$$6x - y + 2z = 10.$$

We first write a matrix, using only the constants. Note that where there are missing terms we must write 0's:

$$\begin{bmatrix} 2 & -1 & 4 & -3 \\ 1 & 0 & -4 & 5 \\ 6 & -1 & 2 & 10 \end{bmatrix}.$$

We do exactly the same calculations using the matrix that we would do if we wrote the entire equations.

The first thing to do, if possible, is to interchange the rows so that each number below the first number in the first row is a multiple of that number. We do this by interchanging rows 1 and 2:

$$\begin{bmatrix} 1 & 0 & -4 & 5 \\ 2 & -1 & 4 & -3 \\ 6 & -1 & 2 & 10 \end{bmatrix}.$$ This corresponds to interchanging equation (1) with equation (2).

Next we multiply the first row by -2 and add it to the second row:

$$\begin{bmatrix} 1 & 0 & -4 & 5 \\ 0 & -1 & 12 & -13 \\ 6 & -1 & 2 & 10 \end{bmatrix}.$$ This corresponds to multiplying equation (1) by -2 and adding it to equation (2).

Now we multiply the first row by -6 and add it to the third row:

$$\begin{bmatrix} 1 & 0 & -4 & 5 \\ 0 & -1 & 12 & -13 \\ 0 & -1 & 26 & -20 \end{bmatrix}.$$ This corresponds to multiplying equation (1) by -6 and adding it to equation (3).

Next we multiply row 2 by -1 and add it to the third row:

$$\begin{bmatrix} 1 & 0 & -4 & 5 \\ 0 & -1 & 12 & -13 \\ 0 & 0 & 14 & -7 \end{bmatrix}.$$ This corresponds to multiplying equation (2) by -1 and adding it to equation (3).

If we now put the variables back in, we have

$$x \quad\quad - 4z = 5,$$
$$- y + 12z = -13,$$
$$14z = -7.$$

Now we proceed as before. We solve (3) for z and get $z = -\frac{1}{2}$. Next we substitute $-\frac{1}{2}$ for z in (2) and solve for y: $-y + 12(-\frac{1}{2}) = -13$, so $y = 7$. Since there is no y-term in (1) we need only substitute $-\frac{1}{2}$ for z in (1) and solve for x: $x - 4(-\frac{1}{2}) = 5$, so $x = 3$. The solution is $(3, 7, -\frac{1}{2})$.

Note in the preceding discussion that our goal was to get the matrix in the form

$$\begin{bmatrix} a & b & c & d \\ 0 & e & f & g \\ 0 & 0 & h & k \end{bmatrix},$$

where there are 0's below the *main diagonal*, formed by a, e, and h. Then we put the variables back in and complete the solution.

Each of the operations used in the preceding steps corresponds to operations with the equations and produces equivalent systems of equations. We call the matrices *row equivalent*, and the operations row equivalent.

THEOREM 1

Each of the following row-equivalent operations produces equivalent matrices:

a) **Interchanging any two rows of a matrix;**

b) **Multiplying each element of a row by the same nonzero constant;**

c) **Multiplying each element of a row by a nonzero number and adding the result to another row.**

The best overall method for solving systems of equations is the use of row-equivalent matrices; even computers are programmed to do so.

EXERCISE SET 5.4

Solve, using matrices.

1. $4x + 2y = 11$,
   $3x - y = 2$

2. $3x - 3y = 11$,
   $9x - 2y = 5$

3. $x + 2y - 3z = 9$,
   $2x - y + 2z = -8$,
   $3x - y - 4z = 3$

4. $x - y + 2z = 0$,
   $x - 2y + 3z = -1$,
   $2x - 2y + z = -3$

5. $5x - 3y = -2$,
   $4x + 2y = 5$

6. $3x + 4y = 7$,
   $-5x + 2y = 10$

7. $4x - y - 3z = 1$,
   $8x + y - z = 5$,
   $2x + y + 2z = 5$

8. $3x + 2y + 2z = 3$,
   $x + 2y - z = 5$,
   $2x - 4y + z = 0$

9. $p + q + r = 1$,
   $p + 2q + 3r = 4$,
   $4p + 5q + 6r = 7$

10. $m + n + t = 9,$
    $m - n - t = -15,$
    $m + n + t = 3$

11. $2x + 2y - 2z - 2w = -10,$
    $x + y + z + w = -5,$
    $x - y + 4z + 3w = -2,$
    $3x - 2y + 2z + w = -6$

12. $2x - 3y + z - w = -8,$
    $x + y - z - w = -4,$
    $x + y + z + w = 22,$
    $x - y - z - w = -14$

13. A collection of 34 coins consists of dimes and nickels. The total value is $1.90. How many dimes and how many nickels are there?

14. A collection of 43 coins consists of dimes and quarters. The total value is $7.60. How many dimes and how many quarters are there?

15. A collection of 22 coins consists of nickels, dimes, and quarters. The total value is $2.90. There are six more nickels than there are dimes. How many of each type of coin are there?

16. A collection of 18 coins consists of nickels, dimes, and quarters. The total value is $2.55. There are two more quarters than there are dimes. How many of each type of coin are there?

17. A tobacco dealer has two kinds of tobacco. One is worth $4.05 per lb and the other is worth $2.70 per lb. The dealer wants to blend the two tobaccos to get a 15-lb mixture worth $3.15 per lb. How much of each kind of tobacco should be used?

18. A grocer mixes candy worth $0.80 per lb with nuts worth $0.70 per lb to get a 20-lb mixture worth $0.77 per lb. How many pounds of candy and how many pounds of nuts should be used?

Recall the formula $I = prt$, for simple interest.

19. One year $8950 was received in interest from two investments. A certain amount was invested at $12\frac{1}{2}\%$ and $10,000 more than this invested at 13%. Find the amount of principal invested at each rate.

20. One year a person had some money invested at $13\frac{1}{2}\%$ and another amount invested at $13\frac{3}{4}\%$. The income from the investments was $3275. The income from the $13\frac{1}{2}\%$ investment was $575 less than that from the $13\frac{3}{4}\%$ investment. How much was invested at each rate?

Solve.

21. $4.83x + 9.06y = -39.42,$
    $-1.35x + 6.67y = -33.99$

22. $3.11x - 2.04y = -24.39,$
    $7.73x + 5.19y = -35.48$

23. $3.55x - 1.35y + 1.03z = 9.16,$
    $-2.14x + 4.12y + 3.61z = -4.50,$
    $5.48x - 2.44y - 5.86z = 0.813$

24. $4.12x - 1.35y - 18.2z = 601.3,$
    $-3.41x + 68.9y + 38.7z = 1777,$
    $0.955x - 0.813y - 6.53z = 160.2$

25. $\sqrt{2}x + \pi y = 3,$
    $\pi x - \sqrt{2}y = 1$

26. $ax + by = c,$
    $dx + ey = f$

5.5 DETERMINANTS AND CRAMER'S RULE

If a matrix has the same number of rows and columns, it is called a *square* matrix. With every square matrix is associated a number called its determinant, defined as follows for two-by-two matrices.

DEFINITION

The determinant of the matrix
$$\begin{bmatrix} a & c \\ b & d \end{bmatrix}$$
is denoted
$$\begin{vmatrix} a & c \\ b & d \end{vmatrix}$$
and is defined as follows:
$$\begin{vmatrix} a & c \\ b & d \end{vmatrix} = ad - bc.$$

Example 1 Evaluate: $\begin{vmatrix} \sqrt{2} & -3 \\ -4 & -\sqrt{2} \end{vmatrix}$.

$\begin{vmatrix} \sqrt{2} & -3 \\ -4 & -\sqrt{2} \end{vmatrix}$ The arrows indicate the products involved.

$= \sqrt{2}(-\sqrt{2}) - (-4)(-3) = -2 - 12 = -14$

Determinants have many uses. One of these is in solving systems of nonhomogeneous linear equations, where the number of variables is the same as the number of equations,. Let us consider a system of two equations:

$$a_1 x + b_1 y = c_1$$
$$a_2 x + b_2 y = c_2.$$

Using the methods of the preceding sections we can solve. We obtain

$$x = \frac{c_1 b_2 - c_2 b_1}{a_1 b_2 - a_2 b_1}, \qquad y = \frac{a_1 c_2 - a_2 c_1}{a_1 b_2 - a_2 b_1}.$$

The numerators and denominators of the expressions for x and y are determinants:

$$x = \frac{\begin{vmatrix} c_1 & b_1 \\ c_2 & b_2 \end{vmatrix}}{\begin{vmatrix} a_1 & b_1 \\ a_2 & b_2 \end{vmatrix}}, \qquad y = \frac{\begin{vmatrix} a_1 & c_1 \\ a_2 & c_2 \end{vmatrix}}{\begin{vmatrix} a_1 & b_1 \\ a_2 & b_2 \end{vmatrix}}.$$

The equations above make sense only if the denominator determinant is not 0. If the denominator *is* 0, then one of two things happens.

1. If the denominator is 0 and the other two determinants are also 0, then the system of equations is dependent.
2. If the denominator is 0 and at least one of the other determinants is not 0, then the system is inconsistent.

The equations with determinants above describe *Cramer's rule* for solving systems of equations. To use this rule, we find the three determinants and compute x and y as shown above. Note that the denominator in both cases contains the coefficients of x and y, in the same position as in the original equations. For x the numerator is obtained by replacing the x-coefficients (the a's) by the c's. For y, the numerator is obtained by replacing the y-coefficients (the b's) by the c's.

Example 2 Solve, using Cramer's rule:
$$2x + 5y = 7,$$
$$5x - 2y = -3.$$

$$x = \frac{\begin{vmatrix} 7 & 5 \\ -3 & -2 \end{vmatrix}}{\begin{vmatrix} 2 & 5 \\ 5 & -2 \end{vmatrix}} = \frac{7(-2) - (-3)5}{2(-2) - 5 \cdot 5} = -\frac{1}{29}$$

$$y = \frac{\begin{vmatrix} 2 & 7 \\ 5 & -3 \end{vmatrix}}{\begin{vmatrix} 2 & 5 \\ 5 & -2 \end{vmatrix}} = \frac{2(-3) - 5 \cdot 7}{-29} = \frac{41}{29}$$

The solution is $\left(-\frac{1}{29}, \frac{41}{29}\right)$.

Three-by-Three Determinants

DEFINITION

The *determinant* of a three-by-three matrix is defined as follows:
$$\begin{vmatrix} a_1 & b_1 & c_1 \\ a_2 & b_2 & c_2 \\ a_3 & b_3 & c_3 \end{vmatrix} = a_1 \cdot \begin{vmatrix} b_2 & c_2 \\ b_3 & c_3 \end{vmatrix} - a_2 \cdot \begin{vmatrix} b_1 & c_1 \\ b_3 & c_3 \end{vmatrix} + a_3 \cdot \begin{vmatrix} b_1 & c_1 \\ b_2 & c_2 \end{vmatrix}.$$

The two-by-two determinants on the right can be found by crossing out the row and column in which the a-coefficient occurs.

Example 3 Evaluate.

$$\begin{vmatrix} -1 & 0 & 1 \\ -5 & 1 & -1 \\ 4 & 8 & 1 \end{vmatrix} = -1 \cdot \begin{vmatrix} 1 & -1 \\ 8 & 1 \end{vmatrix} - (-5) \cdot \begin{vmatrix} 0 & 1 \\ 8 & 1 \end{vmatrix} + 4 \cdot \begin{vmatrix} 0 & 1 \\ 1 & -1 \end{vmatrix}$$

$$= -1(1 + 8) + 5(-8) + 4(-1)$$
$$= -9 - 40 - 4 = -53$$

Let us consider three equations in three variables. Consider

$$a_1x + b_1y + c_1z = d_1,$$
$$a_2x + b_2y + c_2z = d_2,$$
$$a_3x + b_3y + c_3z = d_3,$$

and the following determinants:

$$D = \begin{vmatrix} a_1 & b_1 & c_1 \\ a_2 & b_2 & c_2 \\ a_3 & b_3 & c_3 \end{vmatrix}, \qquad D_x = \begin{vmatrix} d_1 & b_1 & c_1 \\ d_2 & b_2 & c_2 \\ d_3 & b_3 & c_3 \end{vmatrix},$$

$$D_y = \begin{vmatrix} a_1 & d_1 & c_1 \\ a_2 & d_2 & c_2 \\ a_3 & d_3 & c_3 \end{vmatrix}, \qquad D_z = \begin{vmatrix} a_1 & b_1 & d_1 \\ a_2 & b_2 & d_2 \\ a_3 & b_3 & d_3 \end{vmatrix}.$$

If we solve the system of equations, we obtain the following:

$$x = \frac{D_x}{D}, \qquad y = \frac{D_y}{D}, \qquad z = \frac{D_z}{D}.$$

Note that we obtain the determinant D_x in the numerator for x from D by replacing the x-coefficients by $d_1, d_2,$ and d_3. A similar thing happens with D_y and D_z. We have thus extended *Cramer's rule* to solve systems of three equations in three variables. As before, when $D = 0$, Cramer's rule cannot be used. If $D = 0$ and D_x, D_y, and D_z are 0, the system is dependent. If $D = 0$ and one of D_x, D_y, or D_z is not zero, then the system is inconsistent.

Example 4 Solve, using Cramer's rule:

$$x - 3y + 7z = 13,$$
$$x + y + z = 1,$$
$$x - 2y + 3z = 4.$$

$$D = \begin{vmatrix} 1 & -3 & 7 \\ 1 & 1 & 1 \\ 1 & -2 & 3 \end{vmatrix} = -10, \qquad D_x = \begin{vmatrix} 13 & -3 & 7 \\ 1 & 1 & 1 \\ 4 & -2 & 3 \end{vmatrix} = 20,$$

$$D_y = \begin{vmatrix} 1 & 13 & 7 \\ 1 & 1 & 1 \\ 1 & 4 & 3 \end{vmatrix} = -6, \qquad D_z = \begin{vmatrix} 1 & -3 & 13 \\ 1 & 1 & 1 \\ 1 & -2 & 4 \end{vmatrix} = -24.$$

Then

$$x = \frac{D_x}{D} = \frac{20}{-10} = -2, \quad y = \frac{D_y}{D} = \frac{-6}{-10} = \frac{3}{5}, \quad z = \frac{D_z}{D} = \frac{-24}{-10} = \frac{12}{5}.$$

The solution is $(-2, \frac{3}{5}, \frac{12}{5})$. In practice, it is not necessary to evaluate D_z. When we have found values for x and y we can substitute them into one of the equations and find z.

EXERCISE SET 5.5

Evaluate.

1. $\begin{vmatrix} -2 & -\sqrt{5} \\ -\sqrt{5} & 3 \end{vmatrix}$

2. $\begin{vmatrix} \sqrt{5} & -3 \\ 4 & 2 \end{vmatrix}$

3. $\begin{vmatrix} x & 4 \\ x & x^2 \end{vmatrix}$

4. $\begin{vmatrix} y^2 & -2 \\ y & 3 \end{vmatrix}$

5. $\begin{vmatrix} 3 & 1 & 2 \\ -2 & 3 & 1 \\ 3 & 4 & -6 \end{vmatrix}$

6. $\begin{vmatrix} 3 & -2 & 1 \\ 2 & 4 & 3 \\ -1 & 5 & 1 \end{vmatrix}$

7. $\begin{vmatrix} x & 0 & -1 \\ 2 & x & x^2 \\ -3 & x & 1 \end{vmatrix}$

8. $\begin{vmatrix} x & 1 & -1 \\ x^2 & x & x \\ 0 & x & 1 \end{vmatrix}$

Solve, using Cramer's rule.

9. $-2x + 4y = 3,$
   $3x - 7y = 1$

10. $5x - 4y = -3,$
    $7x + 2y = 6$

11. $\sqrt{3}x + \pi y = -5,$
    $\pi x - \sqrt{3}y = 4$

12. $\pi x - \sqrt{5}y = 2,$
    $\sqrt{5}x + \pi y = -3$

13. $3x + 2y - z = 4,$
    $3x - 2y + z = 5,$
    $4x - 5y - z = -1$

14. $3x - y + 2z = 1,$
    $x - y + 2z = 3,$
    $-2x + 3y + z = 1$

15. $6y + 6z = -1,$
    $8x + 6z = -1,$
    $4x + 9y = 8$

16. $3x + 5y = 2,$
    $2x - 3z = 7,$
    $4y + 2z = -1$

☆

Solve.

17. $\begin{vmatrix} x & 5 \\ -4 & x \end{vmatrix} = 24$

18. $\begin{vmatrix} y & 2 \\ 3 & y \end{vmatrix} = y$

19. $\begin{vmatrix} x & -3 \\ -1 & x \end{vmatrix} \geq 0$

20. $\begin{vmatrix} y & -5 \\ -2 & y \end{vmatrix} < 0$

21. $\begin{vmatrix} x+3 & 4 \\ x-3 & 5 \end{vmatrix} = -7$

22. $\begin{vmatrix} m+2 & -3 \\ m+5 & -4 \end{vmatrix} = 3m - 5$

23. $\begin{vmatrix} 2 & x & 1 \\ 1 & 2 & -1 \\ 3 & 4 & -2 \end{vmatrix} = -6$

24. $\begin{vmatrix} x & 2 & x \\ 3 & -1 & 1 \\ 1 & -2 & 2 \end{vmatrix} = -10$

Rewrite each expression using determinants. Answers may vary.

25. $2L + 2W$

26. $\pi r + \pi h$

27. $a^2 + b^2$

28. $\frac{1}{2}h(a + b)$

29. $2\pi r^2 + 2\pi rh$

30. $x^2y^2 - Q^2$

Figure 5

Figure 6

Figure 7

5.6 SYSTEMS OF INEQUALITIES

A *solution* of an inequality in two variables is an ordered pair of numbers that makes it true.

Example 1 Determine whether $(-3, 2)$ is a solution of the inequality $5x - 4y \leq 13$.

We replace x by -3 and y by 2:

$$\begin{array}{c|c} 5x - 4y \leq 13 & \\ \hline 5(-3) - 4 \cdot 2 & 13 \\ -15 - 8 & \\ -23 & \end{array}$$

Since -23 is less than 13, this replacement makes the inequality true.

To graph inequalities, we use our knowledge and skill in graphing equations. In fact, the first step in graphing an inequality is usually the graphing of an equation.

Example 2 Graph: $y < x + 2$.

We first graph the equation $y = x + 2$, drawing the line dashed (see Fig. 5). For any point on the line, the value of y is $x + 2$. For any point above the line, y is greater than this; in other words, $y > x + 2$. For any point below the line, y is less than $x + 2$, or $y < x + 2$. Thus the graph is the half-plane *below* the line $y = x + 2$. We show this by shading the lower half-plane (see Fig. 6). The fact that the line is drawn dashed means that points on the line are not in the graph.

The graph of any linear inequality in two variables is either a half-plane or a half-plane together with the line along the edge.

Example 3 Graph: $3y - 2x \geq 1$.

Method 1. Solve for y: $y \geq (2x + 1)/3$.

Now graph the line $y = (2x + 1)/3$. For any point on the line the value of y is $(2x + 1)/3$. Such values make the inequalities true, hence are in the graph. Therefore we draw the line solid this time.

For any point above the line the value of y is greater than $(2x + 1)/3$. Hence any such point is in the graph. The graph of the inequality is the half-plane above the line, together with the line (see Fig. 7).

Method 2. We are to graph $3y - 2x \geq 1$.

We graph the line $3y - 2x = 1$, by any method. This time let us find the intercepts. They are $(0, \frac{1}{3})$ and $(-\frac{1}{2}, 0)$. We plot them and then use them to draw the line.

Next, we know that the rest of the graph is a half-plane, but we need to determine which one. All we need to do is choose any point on one of the half-planes, substitute it into the inequality, and see if we get a true sentence. The origin is an easy point to use, if the line doesn't contain the origin. We try $(0, 0)$:

$$3 \cdot 0 - 2 \cdot 0 \geq 1, \quad \text{or} \quad 0 \geq 1.$$

This gives us a false sentence, hence $(0, 0)$ is not a solution. Thus we shade the half-plane not containing $(0, 0)$.

Example 4 Graph: $x \leq 3$.

We first graph the equation $x = 3$, drawing it solid. Then we shade the appropriate half-plane (see Fig. 8).

Example 5 Graph: $-1 < y \leq 2$.

This is a conjunction of two inequalities,

$$-1 < y \quad \text{and} \quad y \leq 2.$$

It will be true for any y that is both greater than -1 and less than or equal to 2. The graph is shown in Fig. 9.

Since our inequality is a conjunction, the graph is the intersection of the graphs of the parts. This is shown in Fig. 10.

To get a picture of the solution set of a system or *conjunction* of inequalities, we will graph the inequalities separately and find their intersection. A system of linear inequalities may have a graph that is a polygon and its interior. In certain applications it is important to find the vertices.

Example 6 Graph this system of inequalities. Find the coordinates of any vertices formed.

$$2x + y \geq 2,$$
$$4x + 3y \leq 12,$$
$$\frac{1}{2} \leq x \leq 2,$$
$$y \geq 0.$$

Figure 8

Figure 9

Figure 10

Figure 11

Figure 12

The separate graphs are shown in Fig. 11, and the graph of the intersection, which is the graph of the system, is shown in Fig. 12.

We find the vertex $(\frac{1}{2}, 1)$ by solving the system

$$2x + y = 2,$$
$$x = \frac{1}{2}.$$

We find the vertex $(1, 0)$ by solving the system

$$2x + y = 2,$$
$$y = 0.$$

We find the vertex $(2, 0)$ from the system

$$x = 2,$$
$$y = 0.$$

The vertices $(2, \frac{4}{3})$ and $(\frac{1}{2}, \frac{10}{3})$ were found by solving, respectively, the systems

$$x = 2,$$
$$4x + 3y = 12,$$

and

$$x = \frac{1}{2},$$
$$4x + 3y = 12.$$

EXERCISE SET 5.6

Graph.

1. $y < x$
2. $y \geq x$
3. $y + x \geq 0$
4. $y + x < 0$
5. $3x - 2y < 6$
6. $2x - 5y > 10$
7. $2x + 3y \geq 6$
8. $x + 2y \leq 4$
9. $3x - 2 \leq 5x + y$
10. $2x - 6y \geq 8 + 4y$
11. $x < -4$
12. $y \geq 5$
13. $0 \leq x < 5\frac{1}{2}$
14. $-4 < y < -1$
15. $y \geq |x|$
16. $y < |x|$

Graph these systems. Find the coordinates of any vertices formed.

17. $x + y \leq 1$,
    $x - y \leq 2$
18. $x + y \leq 3$,
    $x - y \leq 4$
19. $y - 2x > 1$,
    $y - 2x < 3$
20. $y + x > 0$,
    $y + x < 2$
21. $2y - x \leq 2$,
    $y - 3x \geq -1$
22. $x + 3y \geq 9$,
    $3x - 2y \leq 5$
23. $y \leq 2x + 1$,
    $y \geq -2x + 1$,
    $x \leq 2$
24. $x - y \leq 2$,
    $x + 2y \geq 8$,
    $y \leq 4$
25. $x + 2y \leq 12$,
    $2x + y \leq 12$,
    $x \geq 0$,
    $y \geq 0$
26. $8x + 5y \leq 40$,
    $x + 2y \leq 8$,
    $x \geq 0$,
    $y \geq 0$
27. $3x + 4y \geq 12$,
    $5x + 6y \leq 30$,
    $1 \leq x \leq 3$
28. $y - 2x \geq 3$,
    $y - 2x \leq 5$,
    $6 \leq y \leq 8$

Graph.

29. $y \geq x^2 - 2$,
    $y \leq 2 - x^2$
30. $x \geq 2y^2 - 1$,
    $x < y^2$
31. $y < x + 1$,
    $y \geq x^2$
32. $y \leq -x^2 + 5$,
    $y > \frac{1}{2}x^2 - 1$

Graph.

33. $|x| + |y| \leq 1$
34. $|x + y| \leq 1$
35. $|x - y| > 0$
36. $|x| > |y|$

5.7 LINEAR PROGRAMMING

Suppose you are taking a test in which different items are worth different numbers of points. Presumably those with higher point values take more time. How many items of each kind should you do, in order to make the best score? A kind of mathematics called *linear programming* (developed during World War II) may provide an answer. Let us consider an example.

Example 1 You are taking a test in which items of type A are worth 10 points and items of type B are worth 15 points. It takes three minutes for each item of type A and six minutes for each item of type B. The total time allowed is 60 minutes and you are not allowed to answer

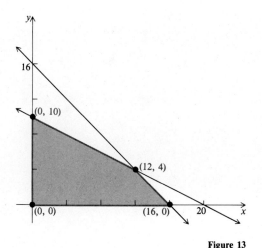

Figure 13

more than 16 questions. Assuming that all your answers are correct, how many items of each type should you answer to get the best score?

Let $x =$ the number of items of type A, and $y =$ the number of items of type B. The total score T is a function of the two variables x and y:

$$T = 10x + 15y.$$

This function has a domain that is a set of ordered pairs of numbers (x, y). This domain is determined by the following conditions, called *constraints*:

Total number of questions allowed, not more than 16	$x + y \leq 16$
Time, not more than 60 min	$3x + 6y \leq 60$
Number of items answered will not be negative	$x \geq 0,$ $y \geq 0.$

We now graph the domain of the function T (see Fig. 13). This is the graph of the system of inequalities above. We will determine the vertices, if any are formed. The domain consists of a *convex polygon** and its interior. Under this condition, our linear function does have a maximum value and a minimum value. Moreover, the maximum and minimum values occur at the vertices of the polygon. All we need do to find these values is substitute the coordinates of the vertices in $T = 10x + 15y$.

VERTICES (x, y)	SCORE $T = 10x + 15y$
(0, 0)	0
(16, 0)	160
(12, 4)	180
(0, 10)	150

From the table above we see that the minimum value is 0 and the maximum is 180. To get this maximum you would answer 12 items of type A and 4 items of type B.

We now state the main theorem needed to perform linear programming.

---

*A *convex polygon*, roughly speaking, is one that has no indentations, unlike this one:

THEOREM 2

If a linear function $F = ax + by + c$ is defined on a domain described by a system of linear inequalities (constraints), then any maximum or minimum value of F will occur at a vertex. If the domain consists of a convex polygon and its interior, then both maximum and minimum values of F actually exist.

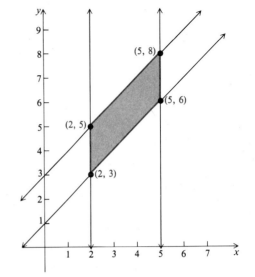

Figure 14

Example 2 Find the maximum and minimum values of $F = 9x + 40y$ subject to the constraints

$$y - x \geq 1,$$
$$y - x \leq 3,$$
$$2 \leq x \leq 5.$$

We graph the system of inequalities, determine the vertices, and find the function values for those ordered pairs (see Fig. 14). The maximum value of F is 365 when $x = 5$ and $y = 8$. The minimum value of F is 138 when $x = 2$ and $y = 3$.

VERTICES (x, y)	SCORE $F = 9x + 40y$
(2, 3)	138
(2, 5)	218
(5, 6)	285
(5, 8)	365

Example 3 A company manufactures motorcycles and bicycles. To stay in business it must produce at least 10 motorcycles each month, but it does not have facilities to produce more than 60 motorcycles. It also does not have facilities to produce more than 120 bicycles. The total production of motorcycles and bicycles cannot exceed 160. The profit on a motorcycle is $134 and on a bicycle is $20. Find the number of each that should be manufactured to maximize profit.

Let $x =$ the number of motorcycles to be produced, and $y =$ the number of bicycles to be produced. The profit P is given by

$$P = \$134x + \$20y,$$

subject to the constraints

$$10 \leq x \leq 60,$$
$$0 \leq y \leq 120,$$
$$x + y \leq 160.$$

We graph the system of inequalities, determine the vertices, and find the function values for those ordered pairs (see Fig. 15).

VERTICES (x, y)	PROFIT $P = \$134x + \$20y$
(10, 0)	$ 1,340
(60, 0)	$ 8,040
(60, 100)	$10,040
(40, 120)	$ 7,760
(10, 120)	$ 3,740

Thus the company will make a maximum profit of $10,040 by producing 60 motorcycles and 100 bicycles.

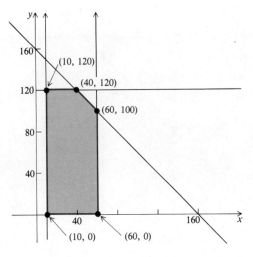

Figure 15

EXERCISE SET 5.7

In Exercises 1–4, maximize and also minimize (find maximum and minimum values of the function, and the values of x and y where they occur).

1. $F = 4x + 28y$,
subject to
$5x + 3y \leq 34$,
$3x + 5y \leq 30$,
$x \geq 0$,
$y \geq 0$.

2. $G = 14x + 16y$,
subject to
$3x + 2y \leq 12$,
$7x + 5y \leq 29$,
$x \geq 0$,
$y \geq 0$.

3. $P = 16x - 2y + 40$,
subject to
$6x + 8y \leq 48$,
$0 \leq y \leq 4$,
$0 \leq x \leq 7$.

4. $Q = 24x - 3y + 52$,
subject to
$5x + 4y \geq 20$,
$0 \leq y \leq 4$,
$0 \leq x \leq 3$.

5. You are about to take a test that contains questions of type A worth 4 points and questions of type B worth 7 points. You must do at least 5 questions of type A but time restricts doing more than 10. You must do at least 3 questions of type B but time restricts doing more than 10. In total, you can do no more than 18 questions. How many of each type of question must you do to maximize your score? What is this maximum score?

6. You are about to take a test that contains questions of type A worth 10 points and questions of type B worth 25 points. You must do at least 3 questions of type A but time restricts doing more than 12. You must do at least 4 questions of type B but time restricts doing more than 15. In total you can do no more than 20 questions. How many of each type of question must you do to maximize your score? What is this maximum score?

7. A man is planning to invest up to $22,000 in bank X or bank Y or both. He wants to invest at least $2000 but no more than $14,000 in bank X. Bank Y does not insure more than a $15,000 investment so he will invest no more than that in bank Y. The interest in bank X is 6% and in bank Y it is $6\frac{1}{2}$% and this will be simple interest for one year. How much should he invest in each bank to maximize his income? What is the maximum income?

8. A woman is planning to invest up to $40,000 in corporate or municipal bonds or both. The least she is allowed to invest in corporate bonds is $6000 and she does not want to invest more than $22,000 in corporate bonds. She also does not want to invest more than $30,000 in municipal bonds. The interest on corporate bonds is 8% and on municipal bonds it is $7\frac{1}{2}$%. This is simple interest for one year. How much should she invest in each type of bond to maximize her income? What is the maximum income?

9. It takes a tailoring firm 2 hr of cutting and 4 hr of sewing to make a knit suit. To make a worsted suit it takes 4 hr of cutting and 2 hr of sewing. At most, 20 hr per day are available for cutting and, at most, 16 hr per day are available for sewing. The profit on a knit suit is $34 and on a worsted suit is $31. How many of each kind of suit should be made to maximize profit? What is the maximum profit?

10. A pipe tobacco company has 3000 lb of English tobacco, 2000 lb of Virginia tobacco, and 500 lb of Latakia tobacco. To make one batch of Smello tobacco it takes 12 lb of English tobacco and 4 lb of Latakia. To make one batch of Roppo tobacco it takes 8 lb of English and 8 lb of Virginia tobacco. The profit is $10.56 per batch for Smello and $6.40 for Roppo. How many batches of each kind of tobacco should be made to yield maximum profit? What is the maximum profit?

11. An airline with two types of airplanes, P1 and P2, has contracted with a tour group to provide accommodations for a minimum of each of 2000 first-class, 1500 tourist, and 2400 economy-class passengers. Airplane P1 costs $12,000 per mile to operate and can accommodate 40 first-class, 40 tourist, and 120 economy-class passengers, whereas airplane P2 costs $10,000 per mile to operate and can accommodate 80 first-class, 30 tourist, and 40 economy-class passengers. How many of each type of airplane should be used to minimize the operating cost?

12. A new airplane P3 becomes available, having an operating cost of $15,000 per mile and accommodating 40 first-class, 80 tourist, and 80 economy-class passengers. If airplane P1 of Exercise 11 were replaced by airplane P3, how many of P2 and P3 would be needed to minimize the operating cost?

CHAPTER 5 REVIEW

Solve.

1. $5x - 3y = -4,$
$3x - y = -4$

2. $2x + 3y = 2,$
$5x - y = -29$

3. $\dfrac{2}{3x} + \dfrac{4}{5y} = 8,$
$\dfrac{5}{4x} - \dfrac{3}{2y} = -6$

4. $\dfrac{3}{x} - \dfrac{4}{y} + \dfrac{1}{z} = -2,$
$\dfrac{5}{x} + \dfrac{1}{y} - \dfrac{2}{z} = 1,$
$\dfrac{7}{x} + \dfrac{3}{y} + \dfrac{2}{z} = 19$

5. $x - y = 5,$
$y - z = 6,$
$z - w = 7,$
$w + x = 8$

6. The value of 75 coins, consisting of nickels and dimes, is $5.95. How many of each kind of coin are there?

7. A family invested $5000, part at 10% and the remainder at 10.5%. The annual income from both investments is $517. What is the amount invested at each rate?

Solve, using matrices. If there is more than one solution, list three of them.

8. $x + 2y = 5,$
$2x - 5y = -8$

9. $3x + 4y + 2z = 3,$
$5x - 2y - 13z = 3,$
$4x + 3y - 3z = 6$

10. $3x + 5y + z = 0$,
$2x - 4y - 3z = 0$,
$x + 3y + z = 0$

11. $x + y + z + w = -2$,
$-2x + 3y + 2z - 3w = 10$,
$3x + 2y - z + 2w = -12$,
$4x - y + z + 2w = 1$

Classify as consistent or inconsistent, dependent or independent.

12. $6x - 2y = 1$,
$-6x + 4y = -2$

13. $x + 5y = 12$,
$5x + 25y = 12$

14. $2x - 4y + 3z = -3$,
$-5x + 2y - z = 7$,
$3x + 2y - 2z = 4$

15. $x + 5y + 3z = 4$,
$3x - 2y + 4z = 21$,
$2x + 3y - z = -13$

16. Find numbers a, b, and c such that the function $y = ax^2 + bx + c$ fits the data points $(0, 3)$, $(1, 0)$, and $(-1, 4)$. Then write the equation.

17. One year a person invested a total of $40,000, part at 12%, part at 13%, and the rest at $14\frac{1}{2}$%. The total interest received on the investments was $5370. The interest received on the $14\frac{1}{2}$% investment was $1050 more than the interest received on the 13% investment. How much was invested at each rate?

Evaluate.

18. $\begin{vmatrix} 1 & 2 \\ 3 & 4 \end{vmatrix}$

19. $\begin{vmatrix} \sqrt{3} & -5 \\ -2 & -\sqrt{3} \end{vmatrix}$

20. $\begin{vmatrix} -2 & -3 \\ 4 & x \end{vmatrix}$

21. $\begin{vmatrix} 2 & -1 & 1 \\ 1 & 2 & -1 \\ 3 & 4 & -3 \end{vmatrix}$

22. $\begin{vmatrix} -5.8 & 7.5 & 4.6 \\ 0 & 2.2 & 8.9 \\ 0 & 0 & 1.3 \end{vmatrix}$

23. $\begin{vmatrix} 3a & 3b & 3c \\ 5a & 5b & 5c \\ d & e & f \end{vmatrix}$

Solve for (x, y), using Cramer's rule.

24. $5x - 2y = 19$.
$7x + 3y = 15$

25. $ax - by = a^2$,
$bx + ay = ab$

Solve using Cramer's rule.

26. $x - 3y - 7z = 6$,
$2x + 3y + z = 9$,
$4x + y = 7$

27. Graph this system. Find the coordinates of any vertices formed.
$2x + y \geq 9$,
$4x + 3y \geq 23$,
$x + 3y \geq 8$,
$x \geq 0$,
$y \geq 0$

28. Maximize and minimize $T = 6x + 10y$ subject to
$x + y \leq 10$,
$5x + 10y \leq 50$,
$x \geq 2$,
$y \geq 0$.

29. You are about to take a test that contains questions of type A worth 7 points and questions of type B worth 12 points. The total number of questions worked must be at least 8. If you know that type A questions take 10 min each and type B questions take 8 min each and that the maximum time for the test is 80 min, how many of each type of question must you do to maximize your score? What is this maximum score?

Graph.

30. $|xy| > 1$

31. $|x| - |y| \leq 1$

Matrices and Determinants

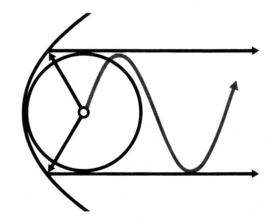

6.1 OPERATIONS ON MATRICES*

A matrix of m rows and n columns is called a matrix with *dimensions* $m \times n$ (read "m by n").

Example 1 Find the dimensions of each matrix:

$$\begin{bmatrix} 2 & -3 & 4 \\ -1 & \frac{1}{2} & \pi \end{bmatrix}, \quad \begin{bmatrix} -3 & 8 & 9 \\ \pi & -2 & 5 \\ -6 & 7 & 8 \end{bmatrix},$$

A 2×3 matrix   A 3×3 matrix

$$[-3 \quad 4], \quad \begin{bmatrix} 10 \\ -7 \end{bmatrix}.$$

A 1×2 matrix   A 2×1 matrix

A matrix with the same number of rows as columns is called a *square* matrix.

Matrix Addition

To add matrices, we add the corresponding members. In order to do this the matrices must have the same dimensions.

Example 2

$$\begin{bmatrix} -5 & 0 \\ 4 & \frac{1}{2} \end{bmatrix} + \begin{bmatrix} 6 & -3 \\ 2 & 3 \end{bmatrix} = \begin{bmatrix} -5+6 & 0-3 \\ 4+2 & \frac{1}{2}+3 \end{bmatrix}$$

$$= \begin{bmatrix} 1 & -3 \\ 6 & 3\frac{1}{2} \end{bmatrix}$$

Example 3

$$\begin{bmatrix} 1 & 3 & 2 \\ -1 & 5 & 4 \\ 6 & 0 & 1 \end{bmatrix} + \begin{bmatrix} -1 & -2 & 1 \\ 1 & -2 & 2 \\ -3 & 1 & 0 \end{bmatrix} = \begin{bmatrix} 0 & 1 & 3 \\ 0 & 3 & 6 \\ 3 & 1 & 1 \end{bmatrix}$$

Addition of matrices is both commutative and associative.

Zero Matrices

A matrix having zeros for all of its entries is called a *zero matrix* and is often denoted by O. When a zero matrix is added to another

---

*This chapter is optional.

matrix of the same dimensions, that same matrix is obtained. Thus a zero matrix is an *additive identity*.

Example 4

$$\begin{bmatrix} 2 & -1 & 3 \\ 1 & 0 & -1 \end{bmatrix} + \begin{bmatrix} 0 & 0 & 0 \\ 0 & 0 & 0 \end{bmatrix} = \begin{bmatrix} 2 & -1 & 3 \\ 1 & 0 & -1 \end{bmatrix}$$

Inverses and Subtraction

To subtract matrices, we subtract the corresponding members. Of course, the matrices must have the same dimensions in order to do this.

Example 5

$$\begin{bmatrix} 1 & 2 \\ -2 & 0 \\ -3 & -1 \end{bmatrix} - \begin{bmatrix} 1 & -1 \\ 1 & 3 \\ 2 & 3 \end{bmatrix} = \begin{bmatrix} 0 & 3 \\ -3 & -3 \\ -5 & -4 \end{bmatrix}$$

The additive inverse of a matrix can be obtained by replacing each member by its additive inverse. Of course, when two matrices that are inverses of each other are added, a zero matrix is obtained.

Example 6

$$\begin{bmatrix} 1 & 0 & 2 \\ 3 & -1 & 5 \end{bmatrix} + \begin{bmatrix} -1 & 0 & -2 \\ -3 & 1 & -5 \end{bmatrix} = \begin{bmatrix} 0 & 0 & 0 \\ 0 & 0 & 0 \end{bmatrix}.$$
$$\quad A \quad\quad + \quad\quad (-A) \quad\quad = \quad O.$$

With numbers, we can subtract by adding an inverse. This is also true of matrices. If we denote matrices by A and B and an additive inverse by $-B$, this fact can be stated as follows:

$$A - B = A + (-B).$$

Example 7

$$\begin{bmatrix} 3 & -1 \\ -2 & 4 \end{bmatrix} - \begin{bmatrix} 2 & 1 \\ 3 & -2 \end{bmatrix} = \begin{bmatrix} 1 & -2 \\ -5 & 6 \end{bmatrix}$$
$$\quad A \quad\quad - \quad\quad B$$

$$\begin{bmatrix} 3 & -1 \\ -2 & 4 \end{bmatrix} + \begin{bmatrix} -2 & -1 \\ -3 & 2 \end{bmatrix} = \begin{bmatrix} 1 & -2 \\ -5 & 6 \end{bmatrix}.$$
$$\quad A \quad\quad + \quad\quad (-B)$$

Multiplying Matrices and Numbers

We define a product of a matrix and a number, and call this product a *scalar* product.

DEFINITION

The (scalar) product of a number k and a matrix A is the matrix, denoted kA, obtained by multiplying each number in A by the number k.

Examples Let
$$A = \begin{bmatrix} -3 & 0 \\ 4 & 5 \end{bmatrix}.$$
Then

8. $$3A = 3\begin{bmatrix} -3 & 0 \\ 4 & 5 \end{bmatrix} = \begin{bmatrix} -9 & 0 \\ 12 & 15 \end{bmatrix},$$

9. $$(-1)A = -1\begin{bmatrix} -3 & 0 \\ 4 & 5 \end{bmatrix} = \begin{bmatrix} 3 & 0 \\ -4 & -5 \end{bmatrix}.$$

Products of Matrices

We do not multiply two matrices by multiplying their corresponding members. The motivation for defining matrix products comes from systems of equations.

Let us begin by considering one equation:
$$3x + 2y - 2z = 4.$$

We will write the coefficients on the left side in a 1×3 matrix (a *row* matrix) and the variables in a 3×1 matrix (a *column* matrix). The 4 on the right is written in a 1×1 matrix:

$$[3 \quad 2 \quad -2]\begin{bmatrix} x \\ y \\ z \end{bmatrix} = [4].$$

We can return to our original equation by multiplying the members of the row matrix by those of the column matrix, and adding:

$$[3 \quad 2 \quad -2]\begin{bmatrix} x \\ y \\ z \end{bmatrix} = [3x + 2y - 2z].$$

We define multiplication accordingly. In this special case, we have a row matrix A and a column matrix B. Their product AB is a 1×1 matrix, having the single member 4 (also called $3x + 2y - 2z$).

Example 10 Find the product of these matrices.

$$[3 \quad 2 \quad -1] \begin{bmatrix} 1 \\ -2 \\ 3 \end{bmatrix} = [3 \cdot 1 + 2(-2) + (-1) \cdot 3] = [-4]$$

Let us continue by considering a system of equations:

$$3x + 2y - 2z = 4,$$
$$2x - y + 5z = 3,$$
$$-x + y + 4z = 7.$$

Consider the following matrices:

$$\underbrace{\begin{bmatrix} 3 & 2 & -2 \\ 2 & -1 & 5 \\ -1 & 1 & 4 \end{bmatrix}}_{A} \underbrace{\begin{bmatrix} x \\ y \\ z \end{bmatrix}}_{X} \underbrace{\begin{bmatrix} 4 \\ 3 \\ 7 \end{bmatrix}}_{B}.$$

If we multiply the first row of A by the (only) column of X, as we did above, we get $3x + 2y - 2z$. If we multiply the second row of A by the column in X, in the same way we get the following:

$$\begin{bmatrix} 3 & 2 & -2 \\ 2 & -1 & 5 \\ -1 & 1 & 4 \end{bmatrix} \begin{bmatrix} x \\ y \\ z \end{bmatrix} = 2x - y + 5z.$$

Note that the first members are multiplied, the second members are multiplied, the third members are multiplied, and the results are added, to get the single number $2x - y + 5z$. What do we get when we multiply the third row of A by the column in X?

$$\begin{bmatrix} 3 & 2 & -2 \\ 2 & -1 & 5 \\ -1 & 1 & 4 \end{bmatrix} \begin{bmatrix} x \\ y \\ z \end{bmatrix} = -x + y + 4z$$

We define the product AX to be the column matrix

$$\begin{bmatrix} 3x + 2y - 2z \\ 2x - y + 5z \\ -x + y + 4z \end{bmatrix}.$$

Now consider this matrix equation:

$$\begin{bmatrix} 3x + 2y - 2z \\ 2x - y + 5z \\ -x + y + 4z \end{bmatrix} = \begin{bmatrix} 4 \\ 3 \\ 7 \end{bmatrix}.$$

Equality for matrices is the same as for numbers, that is, a sentence such as $a = b$ says that a and b are two names for the same thing. Thus if the above matrix equation is true, the "two" matrices are really the same one. This means that $3x + 2y - 2z$ is 4, $2x - y + 5z$ is 3, and $-x + y + 4z$ is 7, or that

$$3x + 2y - 2z = 4,$$
$$2x - y + 5z = 3,$$
and
$$-x + y + 4z = 7.$$

Thus the matrix equation $AX = B$ is equivalent to the original system of equations.

Example 11 Write a matrix equation equivalent to this system of equations:

$$4x + 2y - z = 3,$$
$$9x + z = 5,$$
$$4x + 5y - 2z = 1,$$
$$x + y + z = 0.$$

We write the coefficients on the left in a matrix. We multiply that matrix by the column matrix containing the variables, and set this equal to the column matrix containing the constants on the right:

$$\begin{bmatrix} 4 & 2 & -1 \\ 9 & 0 & 1 \\ 4 & 5 & -2 \\ 1 & 1 & 1 \end{bmatrix} \begin{bmatrix} x \\ y \\ z \end{bmatrix} = \begin{bmatrix} 3 \\ 5 \\ 1 \\ 0 \end{bmatrix}.$$

Example 12 Multiply.

$$\begin{bmatrix} 3 & 1 & -1 \\ 1 & 2 & 2 \\ -1 & 0 & 5 \\ 4 & 1 & 2 \end{bmatrix} \begin{bmatrix} 1 \\ 2 \\ 1 \end{bmatrix} = \begin{bmatrix} 3 \cdot 1 + 1 \cdot 2 - 1 \cdot 1 \\ 1 \cdot 1 + 2 \cdot 2 + 2 \cdot 1 \\ -1 \cdot 1 + 0 \cdot 2 + 5 \cdot 1 \\ 4 \cdot 1 + 1 \cdot 2 + 2 \cdot 1 \end{bmatrix} = \begin{bmatrix} 4 \\ 7 \\ 4 \\ 8 \end{bmatrix}.$$

In all the examples so far, the second matrix had only one column. If the second matrix has more than one column, we treat it in the same way when multiplying that we treated the single column. The product matrix will have as many columns as the second matrix.

Example 13 Multiply (compare with Example 12).

$$\begin{bmatrix} 3 & 1 & -1 \\ 1 & 2 & 2 \\ -1 & 0 & 5 \\ 4 & 1 & 2 \end{bmatrix} \begin{bmatrix} 1 & 0 \\ 2 & 1 \\ 1 & 3 \end{bmatrix} = \begin{bmatrix} 4 & 3\cdot 0 + 1\cdot 1 + (-1)3 \\ 7 & 1\cdot 0 + 2\cdot 1 + 2\cdot 3 \\ 4 & -1\cdot 0 + 0\cdot 1 + 5\cdot 3 \\ 8 & 4\cdot 0 + 1\cdot 1 + 2\cdot 3 \end{bmatrix} = \begin{bmatrix} 4 & -2 \\ 7 & 8 \\ 4 & 15 \\ 8 & 7 \end{bmatrix}$$

A B

Same as in Example 12

The rows of A multiplied by the second column of B

Example 14 Multiply.

$$\begin{bmatrix} 3 & 1 & -1 \\ 2 & 0 & 3 \end{bmatrix} \begin{bmatrix} 1 & 4 & 6 \\ 3 & -1 & 9 \\ 2 & 5 & 1 \end{bmatrix}$$

$$= \begin{bmatrix} 3\cdot 1 + 1\cdot 3 - 1\cdot 2 & 3\cdot 4 + 1(-1) - 1\cdot 5 & 3\cdot 6 + 1\cdot 9 - 1\cdot 1 \\ 2\cdot 1 + 0\cdot 3 + 3\cdot 2 & 2\cdot 4 + 0\cdot (-1) + 3\cdot 5 & 2\cdot 6 + 0\cdot 9 + 3\cdot 1 \end{bmatrix}$$

$$= \begin{bmatrix} 4 & 6 & 26 \\ 8 & 23 & 15 \end{bmatrix}$$

If matrix A has n columns and matrix B has n rows, then we can compute the product AB, regardless of the other dimensions. The product will have as many rows as A and as many columns as B.

EXERCISE SET 6.1

For Exercises 1–16, let

$$A = \begin{bmatrix} 1 & 2 \\ 4 & 3 \end{bmatrix}, \quad B = \begin{bmatrix} -3 & 5 \\ 2 & -1 \end{bmatrix}, \quad C = \begin{bmatrix} 1 & -1 \\ -1 & 1 \end{bmatrix}, \quad D = \begin{bmatrix} 1 & 1 \\ 1 & 1 \end{bmatrix},$$

$$E = \begin{bmatrix} 1 & 3 \\ 2 & 6 \end{bmatrix}, \quad F = \begin{bmatrix} 3 & 3 \\ -1 & -1 \end{bmatrix}, \quad O = \begin{bmatrix} 0 & 0 \\ 0 & 0 \end{bmatrix}, \quad \text{and} \quad I = \begin{bmatrix} 1 & 0 \\ 0 & 1 \end{bmatrix}.$$

Find.

1. $A + B$
2. $B + A$
3. $E + O$
4. $2A$
5. $3F$
6. $(-1)D$
7. $3F + 2A$
8. $A - B$
9. $B - A$
10. AB
11. BA
12. OF
13. CD
14. EF
15. AI
16. IA

In Exercises 17–20, let

$$A = \begin{bmatrix} 1 & 0 & -2 \\ 0 & -1 & 3 \\ 3 & 2 & 4 \end{bmatrix}, \quad B = \begin{bmatrix} -1 & -2 & 5 \\ 1 & 0 & -1 \\ 2 & -3 & 1 \end{bmatrix}, \quad C = \begin{bmatrix} -2 & 9 & 6 \\ -3 & 3 & 4 \\ 2 & -2 & 1 \end{bmatrix}, \text{ and } I = \begin{bmatrix} 1 & 0 & 0 \\ 0 & 1 & 0 \\ 0 & 0 & 1 \end{bmatrix}.$$

Find.

17. AB **18.** BA **19.** CI **20.** IC

Multiply.

21. $\begin{bmatrix} -3 & 2 \end{bmatrix} \begin{bmatrix} 4 \\ -2 \end{bmatrix}$ **22.** $\begin{bmatrix} -2 & 0 & 4 \end{bmatrix} \begin{bmatrix} 8 \\ -6 \\ \frac{1}{2} \end{bmatrix}$ **23.** $\begin{bmatrix} -5 & 1 & 2 \end{bmatrix} \begin{bmatrix} 1 & 3 \\ -1 & 0 \\ 4 & -2 \end{bmatrix}$ **24.** $\begin{bmatrix} -3 & 2 \\ 0 & 1 \\ -4 & 5 \end{bmatrix} \begin{bmatrix} 4 \\ 2 \end{bmatrix}$

Write a matrix equation equivalent to each of the following systems of equations.

25. $3x - 2y + 4z = 17,$
$2x + y - 5z = 13$

26. $3x + 2y + 5z = 9,$
$4x - 3y + 2z = 10$

27. $x - y + 2z - 4w = 12,$
$2x - y - z + w = 0,$
$x + 4y - 3z - w = 1,$
$3x + 5y - 7z + 2w = 9$

28. $2x + 4y - 5z + 12w = 2,$
$4x - y + 12z - w = 5,$
$-x + 4y + 2w = 13,$
$2x + 10y + z = 5$

Compute.

29. $\begin{bmatrix} 3.61 & -2.14 & 16.7 \\ -4.33 & 7.03 & 12.9 \\ 5.82 & -6.95 & 2.34 \end{bmatrix} \begin{bmatrix} 3.05 & 0.402 & -1.34 \\ 1.84 & -1.13 & 0.024 \\ -2.83 & 2.04 & 8.81 \end{bmatrix}$

30. $\begin{bmatrix} -1.23 & 4.51 & -17.4 \\ 61.2 & -8.81 & 0.123 \\ 14.14 & 6.92 & -14.4 \end{bmatrix} \begin{bmatrix} 4.24 & 16.1 & 41.3 \\ 0.146 & -6.06 & -18.9 \\ -8.43 & 1.12 & 0.0245 \end{bmatrix}$

☆

For Exercises 31 and 32, let

$$A = \begin{bmatrix} -1 & 0 \\ 2 & 1 \end{bmatrix}, \text{ and } B = \begin{bmatrix} 1 & -1 \\ 0 & 2 \end{bmatrix}.$$

31. Show that $(A + B)(A - B) \neq A^2 - B^2$, where $A^2 = AA$ and $B^2 = BB$.

32. Show that $(A + B)(A + B) \neq A^2 + 2AB + B^2$.

★

Let $A = \begin{bmatrix} a_{11} & a_{12} \\ a_{21} & a_{22} \end{bmatrix}, \quad B = \begin{bmatrix} b_{11} & b_{12} \\ b_{21} & b_{22} \end{bmatrix}, \quad C = \begin{bmatrix} c_{11} & c_{12} \\ c_{21} & c_{22} \end{bmatrix}, \quad I = \begin{bmatrix} 1 & 0 \\ 0 & 1 \end{bmatrix}.$

Prove the following.

33. $A + B = B + A$

34. $A + (B + C) = (A + B) + C$

35. $k(A + B) = kA + kB$

36. $(k + m)A = kA + mA$

37. $A(BC) = (AB)C$

38. $AI = IA = A$

6.2 DETERMINANTS OF HIGHER ORDER

Further Notation

We want to give a definition of the determinant of any square matrix. To do this we need some new notation. The members, or elements, of a matrix will now be denoted by lower-case letters with two subscripts:

$$A = \begin{bmatrix} a_{11} & a_{12} & a_{13} \\ a_{21} & a_{22} & a_{23} \\ a_{31} & a_{32} & a_{33} \end{bmatrix}.$$

The element in the ith row and jth column is denoted a_{ij}. We may also name the above matrix A as

$$[a_{ij}].$$

Example 1 Consider

$$[a_{ij}] = \begin{bmatrix} -8 & 0 & 6 \\ 4 & -6 & 7 \\ -1 & -3 & 5 \end{bmatrix}.$$

Find a_{12}, a_{23}, and a_{33}.

$a_{12} = 0$ (This is the intersection of the first row and the second column.)

$a_{23} = 7$ (This is the intersection of the second row and third column.)

$a_{33} = 5$ (This is the intersection of the third row and third column.)

Minors

We shall restrict our attention to square matrices.

DEFINITION

In a matrix $[a_{ij}]$, the *minor* M_{ij} of an element a_{ij} is the determinant of the matrix found by deleting the ith row and jth column.

Example 2 In the matrix in Example 1, find M_{11} and M_{23}.

To find M_{11} we delete the first row and the first column.

$$\begin{bmatrix} -8 & 0 & 6 \\ 4 & -6 & 7 \\ -1 & -3 & 5 \end{bmatrix}$$

We calculate the determinant of the matrix formed by the remaining elements:

$$M_{11} = \begin{vmatrix} -6 & 7 \\ -3 & 5 \end{vmatrix} = (-6) \cdot 5 - (-3) \cdot 7$$
$$= -30 - (-21) = -30 + 21 = -9.$$

To find M_{23} we delete the second row and the third column.

$$\begin{bmatrix} -8 & 0 & 6 \\ 4 & -6 & 7 \\ -1 & -3 & 5 \end{bmatrix}$$

We calculate the determinant of the matrix formed by the remaining elements:

$$M_{23} = \begin{vmatrix} -8 & 0 \\ -1 & -3 \end{vmatrix} = -8(-3) - (-1)0 = 24.$$

DEFINITION

In a matrix $[a_{ij}]$, the *cofactor* of an element a_{ij} is denoted A_{ij} and is given by

$$A_{ij} = (-1)^{i+j} M_{ij},$$

where M_{ij} is the minor of a_{ij}. In other words, to find the cofactor of an element, find its minor and multiply it by $(-1)^{i+j}$.

Note that $(-1)^{i+j}$ is 1 if $(i + j)$ is even and is -1 if $(i + j)$ is odd. Thus in calculating a cofactor, find the minor. Then add the number of the row and the number of the column, $i + j$. If this number is odd, change the sign of the minor. If this number is even, leave the minor as is.*

---

*The value $(-1)^{i+j}$ can also be found by counting through the matrix horizontally and/or vertically, starting with a_{11} and $(+)$, saying $+, -, +, -$, and so on, until you come to a_{ij}. Start here $(+)$.

$$\begin{bmatrix} + & - & + & - \\ a_{11} \rightarrow \rightarrow \rightarrow \downarrow & & & \\ \downarrow & & & + \\ \downarrow & & & - \\ & & & +\, a_{ij} \end{bmatrix}$$ The path does not matter.

Example 3 In the matrix in Example 1, find A_{11} and A_{23}.

In Example 2 we found that $M_{11} = -9$. In A_{11} the sum of the subscripts is even, so

$$A_{11} = -9.$$

In Example 2 we found that $M_{23} = 24$. In A_{23} the sum of the subscripts is odd, so

$$A_{23} = -24.$$

Evaluating Determinants Using Cofactors

Consider the matrix A given by

$$A = \begin{bmatrix} a_{11} & a_{12} & a_{13} \\ a_{21} & a_{22} & a_{23} \\ a_{31} & a_{32} & a_{33} \end{bmatrix}.$$

The determinant of the matrix, denoted $|A|$, can be found as follows:

$$|A| = a_{11}A_{11} + a_{21}A_{21} + a_{31}A_{31}.$$

That is, multiply each element of the first column by its cofactor and add. This can be expressed as:

$$|A| = a_{11} \cdot \begin{vmatrix} a_{22} & a_{23} \\ a_{32} & a_{33} \end{vmatrix} - a_{21} \cdot \begin{vmatrix} a_{12} & a_{13} \\ a_{32} & a_{33} \end{vmatrix} + a_{31} \cdot \begin{vmatrix} a_{12} & a_{13} \\ a_{22} & a_{23} \end{vmatrix}.$$

The last line is equivalent to the definition on p. 211. It can be shown that $|A|$ can be found by picking *any* row or column, multiplying each element by its cofactor, and adding. This is called *expanding* the determinant about a row or column. We just expanded $|A|$ about the first column. We now define the determinant of any square matrix.

DEFINITION

For any $n \times n$, $n > 1$, square matrix A we define the determinant of A, $|A|$, to be that number as follows. Pick any row or column. Multiply each element in that row or column by its cofactor and add the results. The determinant of a 1×1 matrix is simply the element of the matrix.

The value of a determinant will be the same no matter how it is evaluated.

Example 4 Consider the matrix A in Example 1. Find $|A|$ by expanding about the third row.

$$\begin{aligned}|A| &= (-1)A_{31} + (-3)A_{32} + 5A_{33}\\ &= (-1)(-1)^{3+1} \cdot \begin{vmatrix} 0 & 6 \\ -6 & 7 \end{vmatrix} + (-3)(-1)^{3+2} \cdot \begin{vmatrix} -8 & 6 \\ 4 & 7 \end{vmatrix}\\ &\quad + 5(-1)^{3+3} \cdot \begin{vmatrix} -8 & 0 \\ 4 & -6 \end{vmatrix}\\ &= (-1) \cdot 1 \cdot [0 \cdot 7 - (-6)6] + (-3)(-1)[-8 \cdot 7 - 4 \cdot 6]\\ &\quad + 5 \cdot 1 \cdot [-8(-6) - 4 \cdot 0]\\ &= -[36] + 3[-80] + 5[48]\\ &= -36 - 240 + 240\\ &= -36\end{aligned}$$

EXERCISE SET 6.2

Use the following matrix for Exercises 1-10.

$$A = \begin{bmatrix} 7 & -4 & -6 \\ 2 & 0 & -3 \\ 1 & 2 & -5 \end{bmatrix}$$

1. Find a_{11}, a_{32}, and a_{22}.
2. Find a_{13}, a_{31}, and a_{23}.
3. Find M_{11}, M_{32}, and M_{22}.
4. Find M_{13}, M_{31}, and M_{23}.
5. Find A_{11}, A_{32}, and A_{22}.
6. Find A_{13}, A_{31}, and A_{23}.
7. Expand $|A|$ about the second row.
8. Expand $|A|$ about the second column.
9. Expand $|A|$ about the third column.
10. Expand $|A|$ about the first row.

Use the following matrix for Exercises 11-16.

$$A = \begin{bmatrix} 1 & 0 & 0 & -2 \\ 4 & 1 & 0 & 0 \\ 5 & 6 & 7 & 8 \\ -2 & -3 & -1 & 0 \end{bmatrix}$$

11. Find M_{41} and M_{33}.
12. Find M_{12} and M_{44}.
13. Find A_{24} and A_{43}.
14. Find A_{22} and A_{34}.
15. Expand $|A|$ about the first row.
16. Expand $|A|$ about the third column.

Evaluate.

17. $\begin{vmatrix} 5 & -4 & 2 & -2 \\ 3 & -3 & -4 & 7 \\ -2 & 3 & 2 & 4 \\ -8 & 9 & 5 & -5 \end{vmatrix}$

18. $\begin{vmatrix} x & p & q & r \\ 0 & y & s & t \\ 0 & 0 & z & u \\ 0 & 0 & 0 & w \end{vmatrix}$

19. If a line contains the points (x_1, y_1) and (x_2, y_2), an equation of the line can be written as follows:

$$\begin{vmatrix} x & y & 1 \\ x_1 & y_1 & 1 \\ x_2 & y_2 & 1 \end{vmatrix} = 0.$$

Prove this.

20. Show that three points (x_1, y_1), (x_2, y_2), and (x_2, y_3) are collinear (on the same straight line) if and only if

$$\begin{vmatrix} x_1 & y_1 & 1 \\ x_2 & y_2 & 1 \\ x_3 & y_3 & 1 \end{vmatrix} = 0.$$

6.3 PROPERTIES OF DETERMINANTS

We can simplify the evaluation of determinants using the following properties.

THEOREM 1

If a row (or column) of a matrix A has all elements 0, then $|A| = 0$.

Proof. Just expand the determinant about the row (or column) that has all 0's. □

Examples Evaluate.

1. $\begin{vmatrix} 0 & 6 \\ 0 & 7 \end{vmatrix} = 0$

2. $\begin{vmatrix} 4 & 5 & 7 \\ 0 & 0 & 0 \\ -3 & 9 & 6 \end{vmatrix} = 0$

THEOREM 2

If two rows (or columns) of a matrix A are interchanged to obtain a new matrix B, then $|A| = -|B|$.

Proof. Pick one of the rows (or columns) to be interchanged and expand $|A|$ about that row. Use that same row to expand $|B|$. For that row each $(-1)^{i+j}$ has changed signs, so $|A| = -|B|$. □

Examples

3. $\begin{vmatrix} 6 & 7 & 8 \\ 4 & 1 & 2 \\ 2 & 9 & 0 \end{vmatrix} = -1 \cdot \begin{vmatrix} 6 & 8 & 7 \\ 4 & 2 & 1 \\ 2 & 0 & 9 \end{vmatrix}$

4. $\begin{vmatrix} -6 & 8 \\ 4 & -3 \end{vmatrix} = -1 \cdot \begin{vmatrix} 4 & -3 \\ -6 & 8 \end{vmatrix}$

THEOREM 3

If two rows (or columns) of a matrix A are the same, then $|A| = 0$.

Proof. Interchanging the rows (or columns) does not change A. Thus by Theorem 2, $|A| = -|A|$. The only way this can happen is when $|A| = 0$. That is, if $a = -a$, $a + a = 0$, $2a = 0$, so $a = 0$. $\square$

Examples Evaluate.

5. $\begin{vmatrix} 6 & 7 & 8 \\ -2 & 6 & 5 \\ -2 & 6 & 5 \end{vmatrix} = 0$

6. $\begin{vmatrix} -5 & 4 & -5 \\ 3 & 7 & 3 \\ 0 & 12 & 0 \end{vmatrix} = 0$

THEOREM 4

If all the elements of a row (or column) of a matrix A are multiplied by k, $|A|$ is multiplied by k. Or, if all the elements of a row (or column) of A have a common factor, we can factor it out of the determinant of A.

Examples

7. $\begin{vmatrix} 2 & 4 & 6 \\ -2 & 5 & 9 \\ 4 & -1 & -3 \end{vmatrix} = 3 \cdot \begin{vmatrix} 2 & 4 & 2 \\ -2 & 5 & 3 \\ 4 & -1 & -1 \end{vmatrix}$

8. $\begin{vmatrix} 10 & 25 \\ -4 & -7 \end{vmatrix} = 5 \cdot \begin{vmatrix} 2 & 5 \\ -4 & -7 \end{vmatrix}$

Proof of Theorem 4. Expand the determinants about the row (or column) in question. The cofactors are the same and the k can be factored out. Consider the case of a 3 by 3 matrix and the second column. Let

$$A = \begin{vmatrix} a_{11} & a_{12} & a_{13} \\ a_{21} & a_{22} & a_{23} \\ a_{31} & a_{32} & a_{33} \end{vmatrix} \text{ and } B = \begin{vmatrix} a_{11} & ka_{12} & a_{13} \\ a_{21} & ka_{22} & a_{23} \\ a_{31} & ka_{32} & a_{33} \end{vmatrix}.$$

Then
$$|B| = ka_{12}A_{12} + ka_{22}A_{22} + ka_{32}A_{32}$$
$$= k(a_{12}A_{12} + a_{22}A_{22} + a_{32}A_{32})$$
$$= k|A|. \quad \square$$

Example 9 Without expanding, find $|A|$.

$$A = \begin{bmatrix} -6 & 3 & 8 \\ 15 & -9 & -20 \\ -9 & -1 & 12 \end{bmatrix},$$

$$|A| = (-3) \cdot \begin{vmatrix} 2 & 3 & 8 \\ -5 & -9 & -20 \\ 3 & -1 & 12 \end{vmatrix} \qquad \text{Factoring } -3 \text{ out of the first column}$$

$$= (-3)(4) \cdot \begin{vmatrix} 2 & 3 & 2 \\ -5 & -9 & -5 \\ 3 & -1 & 3 \end{vmatrix} \qquad \text{Factoring 4 out of the third column}$$

$$= 0. \qquad \text{By Theorem 3}$$

THEOREM 5

If each element in a row (or column) is multiplied by a number k, and the products are added to the corresponding elements of another row (or column), we still get the same determinant.

Example 10 Find a determinant equal to the one on the left by adding three times the second column to the first column.

$$\begin{vmatrix} 0 & 1 & 2 \\ 4 & 5 & 6 \\ 7 & 8 & 9 \end{vmatrix} = \begin{vmatrix} 0 + 3(1) & 1 & 2 \\ 4 + 3(5) & 5 & 6 \\ 7 + 3(8) & 8 & 9 \end{vmatrix} = \begin{vmatrix} 3 & 1 & 2 \\ 19 & 5 & 6 \\ 31 & 8 & 9 \end{vmatrix}$$

Example 11 Find a determinant equal to the one on the left by adding two times the third row to the first row.

$$\begin{vmatrix} 0 & 1 & 2 \\ 4 & 5 & 6 \\ 7 & 8 & 9 \end{vmatrix} = \begin{vmatrix} 0 + 2(7) & 1 + 2(8) & 2 + 2(9) \\ 4 & 5 & 6 \\ 7 & 8 & 9 \end{vmatrix} = \begin{vmatrix} 14 & 17 & 20 \\ 4 & 5 & 6 \\ 7 & 8 & 9 \end{vmatrix}$$

Proof of Theorem 5. We prove the theorem for the case of a 3 by 3 matrix where k times the first column has been added to the third column. Let

$$A = \begin{bmatrix} a_{11} & a_{12} & a_{13} \\ a_{21} & a_{22} & a_{23} \\ a_{31} & a_{32} & a_{33} \end{bmatrix} \quad \text{and} \quad B = \begin{bmatrix} a_{11} & a_{12} & ka_{11} + a_{13} \\ a_{21} & a_{22} & ka_{21} + a_{23} \\ a_{31} & a_{32} & ka_{31} + a_{33} \end{bmatrix}.$$

To show that $|A| = |B|$, we expand $|B|$ about the third column:

$$\begin{aligned} |B| &= (ka_{11} + a_{13})A_{13} + (ka_{21} + a_{23})A_{23} + (ka_{31} + a_{33})A_{33} \\ &= k(a_{11}A_{13} + a_{21}A_{23} + a_{31}A_{33}) + (a_{13}A_{13} + a_{23}A_{23} + a_{33}A_{33}) \\ &= k(a_{11}A_{13} + a_{21}A_{23} + a_{31}A_{33}) + |A| \\ &= k \cdot \begin{vmatrix} a_{11} & a_{12} & a_{11} \\ a_{21} & a_{22} & a_{21} \\ a_{31} & a_{32} & a_{31} \end{vmatrix} + |A| = k(0) + |A| \quad \text{By Theorem 3} \\ &= |A|. \quad \square \end{aligned}$$

We can use the properties of determinants to simplify their evaluation. We try to find another determinant in which in a particular row or column one element is 1 and the rest are 0.

Example 12 Evaluate by first simplifying to a determinant in which in one row or column one element is 1 and the rest are 0.

$$\begin{vmatrix} 6 & 2 & 3 \\ 6 & -1 & 5 \\ -2 & 3 & 1 \end{vmatrix}$$

We will try to get two 0's and a 1 in the third row. It already has a 1; that is why we chose the third row. We first factor a 2 out of column 1:

$$2 \cdot \begin{vmatrix} 3 & 2 & 3 \\ 3 & -1 & 5 \\ -1 & 3 & 1 \end{vmatrix}. \quad \text{By Theorem 4}$$

Now we multiply each element of column 3 by -3 and add the corresponding elements to column 2:

$$2 \cdot \begin{vmatrix} 3 & -7 & 3 \\ 3 & -16 & 5 \\ -1 & 0 & 1 \end{vmatrix}. \quad \text{By Theorem 5}$$

Then we add the elements of column 3 to the corresponding elements of column 1 (Theorem 5):

$$2 \cdot \begin{vmatrix} 6 & -7 & 3 \\ 8 & -16 & 5 \\ 0 & 0 & 1 \end{vmatrix}.$$

Now we evaluate the determinant about the last row:

$$2 \cdot \left(0 - 0 + 1 \cdot \begin{vmatrix} 6 & -7 \\ 8 & -16 \end{vmatrix} \right) = 2 \cdot [6(-16) - 8(-7)] = -80.$$

Example 13. Factor: $\begin{vmatrix} 1 & x & x^2 \\ 1 & y & y^2 \\ 1 & z & z^2 \end{vmatrix}$.

$\begin{vmatrix} 1 & x & x^2 \\ 1 & y & y^2 \\ 1 & z & z^2 \end{vmatrix} = \begin{vmatrix} 0 & x-y & x^2-y^2 \\ 1 & y & y^2 \\ 0 & z-y & z^2-y^2 \end{vmatrix}$ By Theorem 5, adding -1 times the second row to the first row; and -1 times the second row to the third row

$= (x-y)(z-y) \cdot \begin{vmatrix} 0 & 1 & x+y \\ 1 & y & y^2 \\ 0 & 1 & z+y \end{vmatrix}$ By Theorem 4, factoring $x-y$ out of the first row and $z-y$ out of the third row

$= (x-y)(z-y) \cdot \begin{vmatrix} 0 & 0 & x-z \\ 1 & y & y^2 \\ 0 & 1 & z+y \end{vmatrix}$ By Theorem 5, adding -1 times the third row to the first row

$= (x-y)(z-y)(x-z) \cdot \begin{vmatrix} 0 & 0 & 1 \\ 1 & y & y^2 \\ 0 & 1 & z+y \end{vmatrix}$ Factoring $x-z$ out of the first row

$= (x-y)(z-y)(x-z)$ Expanding the determinant about the first row, we get 1.

EXERCISE SET 6.3

1. For the matrices below, find $|A|$ and $|B|$ and compare.

$$A = \begin{bmatrix} -2 & 3 \\ 4 & -1 \end{bmatrix}, \quad B = \begin{bmatrix} 3 & -2 \\ -1 & 4 \end{bmatrix}$$

2. For the matrices below, find $|A|$ and $|B|$ and compare.

$$A = \begin{bmatrix} -1 & -2 \\ -6 & 5 \end{bmatrix}, \quad B = \begin{bmatrix} 2 & 4 \\ 12 & -10 \end{bmatrix}$$

3. For the matrices below, find $|A|$ and $|B|$ and compare.

$$A = \begin{bmatrix} 3 & -2 & 3 \\ -2 & 2 & -1 \\ -3 & 1 & -2 \end{bmatrix}, \quad B = \begin{bmatrix} -2 & 3 & 3 \\ 2 & -2 & -1 \\ 1 & -3 & -2 \end{bmatrix}$$

4. For the matrices below, find $|C|$ and $|D|$ and compare.

$$C = \begin{bmatrix} 2 & -2 & 1 \\ -3 & 6 & 3 \\ -3 & 1 & 2 \end{bmatrix}, \quad D = \begin{bmatrix} 2 & -2 & 1 \\ 1 & -2 & -1 \\ -3 & 1 & 2 \end{bmatrix}$$

Consider

$$|A| = \begin{vmatrix} 3 & -2 & 3 \\ 1 & 2 & -3 \\ 4 & -3 & 1 \end{vmatrix}$$

for Exercises 5 and 6.

5. Find a matrix whose determinant is $|A|$ by adding three times the first column to the second column.

6. Find a matrix whose determinant is $|A|$ by adding three times the first row to the third row.

Evaluate by first simplifying to a determinant in which in one row or column one element is 1 and the rest are 0.

7. $\begin{vmatrix} -4 & 5 \\ 6 & 10 \end{vmatrix}$

8. $\begin{vmatrix} 3 & -9 \\ -2 & 4 \end{vmatrix}$

9. $\begin{vmatrix} 2 & 1 & 1 \\ 2 & -3 & -1 \\ -4 & 5 & 2 \end{vmatrix}$

10. $\begin{vmatrix} 1 & 2 & 4 \\ 2 & 3 & 5 \\ 3 & 1 & 6 \end{vmatrix}$

11. $\begin{vmatrix} 11 & -15 & 20 \\ 16 & 24 & -8 \\ 6 & 9 & 15 \end{vmatrix}$

12. $\begin{vmatrix} 4 & -24 & 15 \\ -3 & 18 & -6 \\ 5 & -4 & 3 \end{vmatrix}$

13. $\begin{vmatrix} -3 & 0 & 2 & 6 \\ 2 & 4 & 0 & -1 \\ -1 & 0 & -5 & 2 \\ 0 & -1 & -2 & -3 \end{vmatrix}$

14. $\begin{vmatrix} -2 & 1 & 0 & 5 \\ 3 & 0 & -4 & -2 \\ 4 & -6 & -8 & -1 \\ 8 & 0 & -2 & -3 \end{vmatrix}$

Find each determinant without expanding.

15. $\begin{vmatrix} x & y & z \\ 0 & 0 & 0 \\ p & q & r \end{vmatrix}$

16. $\begin{vmatrix} 5 & 5 & 5 \\ 3 & 3 & 3 \\ 2 & -7 & 8 \end{vmatrix}$

17. $\begin{vmatrix} 2a & t & -7a \\ 2b & u & -7b \\ 2c & v & -7c \end{vmatrix}$

18. $\begin{vmatrix} a & -1 & 4a \\ b & 2 & 4b \\ x & -3 & 4x \end{vmatrix}$

Factor.

19. $\begin{vmatrix} x^2 & x & 1 \\ y^2 & y & 1 \\ z^2 & z & 1 \end{vmatrix}$

20. $\begin{vmatrix} 1 & 1 & 1 \\ a & b & c \\ a^2 & b^2 & c^2 \end{vmatrix}$

21. $\begin{vmatrix} x & x^2 & x^3 \\ y & y^2 & y^3 \\ z & z^2 & z^3 \end{vmatrix}$

22. $\begin{vmatrix} 1 & 1 & 1 \\ a & b & c \\ a^3 & b^3 & c^3 \end{vmatrix}$

☆ Solve for x.

23. $\begin{vmatrix} -16 & 32 \\ -5 & -3 \end{vmatrix} = x \cdot \begin{vmatrix} 4 & -8 \\ -5 & -3 \end{vmatrix}$

24. $\begin{vmatrix} -3 & 12 & 2 \\ 5 & -6 & 3 \\ 0 & 18 & 5 \end{vmatrix} = x \cdot \begin{vmatrix} -3 & 2 & 2 \\ 5 & -1 & 3 \\ 0 & 3 & 5 \end{vmatrix}$

25. Consider a triangle with vertices (x_1, y_1), (x_2, y_2), and (x_3, y_3). The area of this triangle is the absolute value of

$$\frac{1}{2} \cdot \begin{vmatrix} x_1 & y_1 & 1 \\ x_2 & y_2 & 1 \\ x_3 & y_3 & 1 \end{vmatrix}.$$

Prove this. (*Hint:* Look at this drawing. The area of triangle ABC is the area of trapezoid $ABDE$ plus the area of trapezoid $AEFC$ minus the area of trapezoid $BDFC$.)

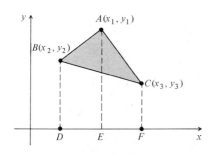

26. Prove that the lines $a_1x + b_1y = c_1$ and $a_2x + b_2y = c_2$ are parallel when

$$\begin{vmatrix} a_1 & b_1 \\ a_2 & b_2 \end{vmatrix} = 0$$

and either

$$\begin{vmatrix} c_1 & b_1 \\ c_2 & b_2 \end{vmatrix} \neq 0 \quad \text{or} \quad \begin{vmatrix} a_1 & c_1 \\ a_2 & c_2 \end{vmatrix} \neq 0.$$

6.4 INVERSES OF MATRICES

We use the symbol I to represent matrices of the type

$$\begin{bmatrix} 1 & 0 \\ 0 & 1 \end{bmatrix}, \quad \begin{bmatrix} 1 & 0 & 0 \\ 0 & 1 & 0 \\ 0 & 0 & 1 \end{bmatrix}.$$

Note that these are square matrices with 1's extending from the upper left to the lower right along the main diagonal, and 0's elsewhere.

The following property can be shown easily for the 2×2 and 3×3 cases (see Exercise 33).

THEOREM 6

For any $n \times n$ matrices A and I, $AI = IA = A$ (I is a multiplicative identity).

We can use matrices in a different way than before, to solve certain kinds of systems. Before giving a more formal explanation, let us consider an example. Consider the system

$$3x_1 + 5x_2 = -1,$$
$$x_1 - 2x_2 = 4.$$

We write a matrix equation equivalent to this system:

$$\begin{bmatrix} 3 & 5 \\ 1 & -2 \end{bmatrix} \cdot \begin{bmatrix} x_1 \\ x_2 \end{bmatrix} = \begin{bmatrix} -1 \\ 4 \end{bmatrix}.$$

Now we let

$$\begin{bmatrix} 3 & 5 \\ 1 & -2 \end{bmatrix} = A, \quad \begin{bmatrix} x_1 \\ x_2 \end{bmatrix} = X, \quad \text{and} \quad \begin{bmatrix} -1 \\ 4 \end{bmatrix} = B.$$

Then we have

$$A \cdot X = B.$$

To solve this equation, we want to find a matrix A^{-1}, called the *inverse* of A, such that $A^{-1} \cdot A = I$. In fact,

$$A^{-1} = -\frac{1}{11} \begin{bmatrix} -2 & -5 \\ -1 & 3 \end{bmatrix};$$

we will explain later how we found it.

We solve the matrix equation $A \cdot X = B$:

$$\begin{array}{ll} A^{-1} \cdot A \cdot X = A^{-1} \cdot B & \text{Multiplying by } A^{-1} \\ I \cdot X = A^{-1} \cdot B & \text{Since } A^{-1} \cdot A = I \\ X = A^{-1} \cdot B & \text{Since } I \text{ is an identity} \end{array}$$

Now we have, substituting,

$$X = A^{-1} \cdot B$$

$$\begin{bmatrix} x_1 \\ x_2 \end{bmatrix} = -\frac{1}{11} \begin{bmatrix} -2 & -5 \\ -1 & 3 \end{bmatrix} \cdot \begin{bmatrix} -1 \\ 4 \end{bmatrix}$$

$$= -\frac{1}{11} \begin{bmatrix} -18 \\ 13 \end{bmatrix}$$

$$= \begin{bmatrix} \frac{18}{11} \\ -\frac{13}{11} \end{bmatrix}.$$

The solution of the system of equations is therefore $x_1 = \frac{18}{11}$ and $x_2 = -\frac{13}{11}$.

Calculating Matrix Inverses

In this section we consider two ways of calculating the inverse of a square matrix, if it exists. We shall see later that such inverses exist only when the determinant of the matrix is nonzero.

The Cofactor Method

Before stating this method we need some additional notation.

DEFINITION

The *transpose* of a matrix A, denoted A^t, is found by interchanging the rows and columns of A.

Example 1 Find A^t, B^t, C^t, and D^t.

$$A = \begin{bmatrix} 2 & 4 & 6 \\ 9 & 8 & -2 \\ 0 & -1 & 4 \end{bmatrix}, \quad B = \begin{bmatrix} -3 & 0 & 4 \\ 7 & 1 & 6 \end{bmatrix},$$

$$C = \begin{bmatrix} -4 \\ 3 \\ 2 \end{bmatrix}, \quad D = \begin{bmatrix} -1 & 2 & 3 & 0 \end{bmatrix}.$$

The transposes are as follows:

$$A^t = \begin{bmatrix} 2 & 9 & 0 \\ 4 & 8 & -1 \\ 6 & -2 & 4 \end{bmatrix}$$ The transpose of a square matrix can be found by "reflecting" across the main diagonal.

$$B^t = \begin{bmatrix} -3 & 7 \\ 0 & 1 \\ 4 & 6 \end{bmatrix}, \quad C^t = \begin{bmatrix} -4 & 3 & 2 \end{bmatrix}, \quad D^t = \begin{bmatrix} -1 \\ 2 \\ 3 \\ 0 \end{bmatrix}.$$

The following is a procedure for calculating the inverse of a square matrix.

The cofactor method. To find the inverse A^{-1} of a square matrix A:

1. Find the cofactor of each element;
2. Replace each element by its cofactor;
3. Find the transpose of the matrix found in (2);
4. Multiply the matrix in (3) by $1/|A|$. This is A^{-1}.

Example 2 Find A^{-1}.

$$A = \begin{bmatrix} 3 & 5 \\ 1 & -2 \end{bmatrix}$$

a) Find the cofactor of each element:

$A_{11} = (-1)^{1+1}(-2)$
$\quad = -2,$
$A_{12} = (-1)^{1+2}(1) = -1,$
$A_{21} = (-1)^{2+1}(5) = -5,$
$A_{22} = (-1)^{2+2}(3) = 3.$

The determinant of a 1 by 1 matrix is just the number. For example, $|[-2]| = -2$. Don't confuse determinant notation with absolute value notation.

b) Replace each element by its cofactor:

$$A = \begin{bmatrix} 3 & 5 \\ 1 & -2 \end{bmatrix} \quad \begin{bmatrix} -2 & -1 \\ -5 & 3 \end{bmatrix}.$$

c) Find the transpose of the matrix found in (b):

The transpose of $\begin{bmatrix} -2 & -1 \\ -5 & 3 \end{bmatrix}$ is $\begin{bmatrix} -2 & -5 \\ -1 & 3 \end{bmatrix}$.

d) Multiply by $\dfrac{1}{|A|}$:

$$|A| = 3(-2) - 1(5) = -11; \quad A^{-1} = \frac{1}{-11} \cdot \begin{bmatrix} -2 & -5 \\ -1 & 3 \end{bmatrix} = \begin{bmatrix} \frac{2}{11} & \frac{5}{11} \\ \frac{1}{11} & -\frac{3}{11} \end{bmatrix}.$$

Example 3 Find A^{-1} if

$$A = \begin{bmatrix} 2 & -1 & 1 \\ 1 & -2 & 3 \\ 4 & 1 & 2 \end{bmatrix}.$$

a) Find the cofactor of each element:

$A_{11} = (-1)^{1+1} \cdot \begin{vmatrix} -2 & 3 \\ 1 & 2 \end{vmatrix} = -7, \quad A_{12} = (-1)^{1+2} \cdot \begin{vmatrix} 1 & 3 \\ 4 & 2 \end{vmatrix} = 10,$

$A_{13} = (-1)^{1+3} \cdot \begin{vmatrix} 1 & -2 \\ 4 & 1 \end{vmatrix} = 9, \quad A_{21} = (-1)^{2+1} \cdot \begin{vmatrix} -1 & 1 \\ 1 & 2 \end{vmatrix} = 3,$

$A_{22} = (-1)^{2+2} \cdot \begin{vmatrix} 2 & 1 \\ 4 & 2 \end{vmatrix} = 0, \quad A_{23} = (-1)^{2+3} \cdot \begin{vmatrix} 2 & -1 \\ 4 & 1 \end{vmatrix} = -6,$

$A_{31} = (-1)^{3+1} \cdot \begin{vmatrix} -1 & 1 \\ -2 & 3 \end{vmatrix} = -1, \quad A_{32} = (-1)^{3+2} \cdot \begin{vmatrix} 2 & 1 \\ 1 & 3 \end{vmatrix} = -5,$

$A_{33} = (-1)^{3+3} \cdot \begin{vmatrix} 2 & -1 \\ 1 & -2 \end{vmatrix} = -3.$

b) Replace each element by its cofactor:

$$A = \begin{bmatrix} 2 & -1 & 1 \\ 1 & -2 & 3 \\ 4 & 1 & 2 \end{bmatrix} \longrightarrow \begin{bmatrix} -7 & 10 & 9 \\ 3 & 0 & -6 \\ -1 & -5 & -3 \end{bmatrix}.$$

c) Find the transpose of the matrix found in (b).

The transpose of $\begin{bmatrix} -7 & 10 & 9 \\ 3 & 0 & -6 \\ -1 & -5 & -3 \end{bmatrix}$ is $\begin{bmatrix} -7 & 3 & -1 \\ 10 & 0 & -5 \\ 9 & -6 & -3 \end{bmatrix}.$

d) Multiply by $\dfrac{1}{|A|}$:

$$|A| = a_{11}A_{11} + a_{21}A_{21} + a_{31}A_{31} \quad \text{Expanding } |A| \text{ about the first column}$$
$$= 2(-7) + 1(3) + 4(-1)$$
$$= -15;$$

$$A^{-1} = \frac{1}{-15}\begin{bmatrix} -7 & 3 & -1 \\ 10 & 0 & -5 \\ 9 & -6 & -3 \end{bmatrix} = \begin{bmatrix} \tfrac{7}{15} & -\tfrac{1}{5} & \tfrac{1}{15} \\ -\tfrac{2}{3} & 0 & \tfrac{1}{3} \\ -\tfrac{3}{5} & \tfrac{2}{5} & \tfrac{1}{5} \end{bmatrix}.$$

If $|A|$ is 0, then $1/|A|$ is not defined and A^{-1} does not exist.

The Gauss–Jordan Reduction Method

We now consider a different method. Suppose we want to find the inverse of the matrix in Example 3:

$$A = \begin{bmatrix} 2 & -1 & 1 \\ 1 & -2 & 3 \\ 4 & 1 & 2 \end{bmatrix}.$$

First we form a new matrix consisting, on the left, of the matrix A and, on the right, of the corresponding identity matrix I.

$$\left[\begin{array}{ccc|ccc} 2 & -1 & 1 & 1 & 0 & 0 \\ 1 & -2 & 3 & 0 & 1 & 0 \\ 4 & 1 & 2 & 0 & 0 & 1 \end{array}\right]$$

The matrix A ⤴   ⤴ The identity matrix I

If we were solving a system of equations, A would be the coefficient matrix. An *augmented* matrix is formed by adjoining two matrices. The above is an example.

We now proceed in a manner very much like that described on pp. 206–208. We attempt to transform A to an identity matrix, but whatever operations we perform, we do on the entire augmented matrix. When we finish we will get a matrix like the following:

$$\begin{bmatrix} 1 & 0 & 0 & a & b & c \\ 0 & 1 & 0 & d & e & f \\ 0 & 0 & 1 & g & h & i \end{bmatrix}.$$

The matrix on the right

$$\begin{bmatrix} a & b & c \\ d & e & f \\ g & h & i \end{bmatrix}$$

will be A^{-1}.

Example 4 Find A^{-1} if

$$A = \begin{bmatrix} 2 & -1 & 1 \\ 1 & -2 & 3 \\ 4 & 1 & 2 \end{bmatrix}.$$

a) Find the augmented matrix consisting of A and I:

$$\begin{bmatrix} 2 & -1 & 1 & 1 & 0 & 0 \\ 1 & -2 & 3 & 0 & 1 & 0 \\ 4 & 1 & 2 & 0 & 0 & 1 \end{bmatrix}.$$

b) We interchange the first and second rows so that the elements of the first column are multiples of the top number on the main diagonal:

$$\begin{bmatrix} 1 & -2 & 3 & 0 & 1 & 0 \\ 2 & -1 & 1 & 1 & 0 & 0 \\ 4 & 1 & 2 & 0 & 0 & 1 \end{bmatrix}.$$

c) Next we obtain 0's in the rest of the first column. We multiply the first row by -2 and add to the second row. Then we multiply the first row by -4 and add to the third row:

$$\begin{bmatrix} 1 & -2 & 3 & 0 & 1 & 0 \\ 0 & 3 & -5 & 1 & -2 & 0 \\ 0 & 9 & -10 & 0 & -4 & 1 \end{bmatrix}.$$

d) Next, we move down the main diagonal to the number 3. We note that the number below it, 9, is a multiple of 3. We multiply the second row by -3 and add it to the third:

$$\begin{bmatrix} 1 & -2 & 3 & 0 & 1 & 0 \\ 0 & 3 & -5 & 1 & -2 & 0 \\ 0 & 0 & 5 & -3 & 2 & 1 \end{bmatrix}.$$

e) Next, we move down the main diagonal to the number 5. We check to see if each number above 5 in the third column is a multiple of 5. Since this is not the case, we multiply the first row by -5:

$$\begin{bmatrix} -5 & 10 & -15 & 0 & -5 & 0 \\ 0 & 3 & -5 & 1 & -2 & 0 \\ 0 & 0 & 5 & -3 & 2 & 1 \end{bmatrix}.$$

f) Now we work back up. We add the third row to the second. We also multiply the third row by 3 and add it to the first:

$$\begin{bmatrix} -5 & 10 & 0 & -9 & 1 & 3 \\ 0 & 3 & 0 & -2 & 0 & 1 \\ 0 & 0 & 5 & -3 & 2 & 1 \end{bmatrix}.$$

g) We move back to the number 3 on the main diagonal. We multiply the first row by -3, so the element on the top of the second column is a multiple of 3:

$$\begin{bmatrix} 15 & -30 & 0 & 27 & -3 & -9 \\ 0 & 3 & 0 & -2 & 0 & 1 \\ 0 & 0 & 5 & -3 & 2 & 1 \end{bmatrix}.$$

h) We multiply the second row by 10 and add it to the first.

$$\begin{bmatrix} 15 & 0 & 0 & 7 & -3 & 1 \\ 0 & 3 & 0 & -2 & 0 & 1 \\ 0 & 0 & 5 & -3 & 2 & 1 \end{bmatrix}.$$

i) Finally, we get all 1's on the main diagonal. We multiply the first row by $\frac{1}{15}$, the second by $\frac{1}{3}$, and the third by $\frac{1}{5}$:

$$\begin{bmatrix} 1 & 0 & 0 & \frac{7}{15} & -\frac{1}{5} & \frac{1}{15} \\ 0 & 1 & 0 & -\frac{2}{3} & 0 & \frac{1}{3} \\ 0 & 0 & 1 & -\frac{3}{5} & \frac{2}{5} & \frac{1}{5} \end{bmatrix}.$$

We now have the matrix I on the left. Thus

$$A^{-1} = \begin{bmatrix} \frac{7}{15} & -\frac{1}{5} & \frac{1}{15} \\ -\frac{2}{3} & 0 & \frac{1}{3} \\ -\frac{3}{5} & \frac{2}{5} & \frac{1}{5} \end{bmatrix}.$$

EXERCISE SET 6.4

Find A^{-1}, if it exists. Use the cofactor method.

1. $A = \begin{bmatrix} 3 & 2 \\ 5 & 3 \end{bmatrix}$
2. $A = \begin{bmatrix} 3 & 5 \\ 1 & 2 \end{bmatrix}$
3. $A = \begin{bmatrix} 11 & 3 \\ 7 & 2 \end{bmatrix}$
4. $A = \begin{bmatrix} 8 & 5 \\ 5 & 3 \end{bmatrix}$
5. $A = \begin{bmatrix} 4 & -3 \\ 1 & 2 \end{bmatrix}$
6. $A = \begin{bmatrix} 0 & -1 \\ 1 & 0 \end{bmatrix}$
7. $A = \begin{bmatrix} 3 & 1 & 0 \\ 1 & 1 & 1 \\ 1 & -1 & 2 \end{bmatrix}$
8. $A = \begin{bmatrix} 1 & 0 & 1 \\ 2 & 1 & 0 \\ 1 & -1 & 1 \end{bmatrix}$
9. $A = \begin{bmatrix} 1 & -1 & 2 \\ 0 & 1 & 3 \\ 2 & 1 & -2 \end{bmatrix}$
10. $A = \begin{bmatrix} 1 & -1 & 2 \\ 0 & 1 & 2 \\ 1 & -3 & -4 \end{bmatrix}$
11. $A = \begin{bmatrix} 1 & -4 & 8 \\ 1 & -3 & 2 \\ 2 & -7 & 10 \end{bmatrix}$
12. $A = \begin{bmatrix} -2 & 5 & 3 \\ 4 & -1 & 3 \\ 7 & -2 & 5 \end{bmatrix}$
13. $A = \begin{bmatrix} 1 & 2 & 3 & 4 \\ 0 & 1 & 3 & -5 \\ 0 & 0 & 1 & -2 \\ 0 & 0 & 0 & -1 \end{bmatrix}$
14. $A = \begin{bmatrix} -2 & -3 & 4 & 1 \\ 0 & 1 & 1 & 0 \\ 0 & 4 & -6 & 1 \\ -2 & -2 & 5 & 1 \end{bmatrix}$

15–28. Find A^{-1} of each matrix in Exercises 1–14. Use the Gauss-Jordan reduction method.

For Exercises 29 and 30, write a matrix equation equivalent to the system. Find the inverse of the coefficient matrix. Use the inverse of the coefficient matrix to solve each system.

29. $7x - 2y = -3,$
    $9x + 3y = 4$
30. $5x_1 + 3x_2 = -2,$
    $4x_1 - x_2 = 1$
31. $x_1 + x_3 = 1,$
    $2x_1 + x_2 = 3,$
    $x_1 - x_2 + x_3 = 4$
32. $x + 2y + 3z = -1,$
    $2x - 3y + 4z = 2,$
    $-3x + 5y - 6z = 4$

☆

33. Let

$$A = \begin{bmatrix} a & b & c \\ d & e & f \\ g & h & i \end{bmatrix} \text{ and } I = \begin{bmatrix} 1 & 0 & 0 \\ 0 & 1 & 0 \\ 0 & 0 & 1 \end{bmatrix}.$$

Show that $AI = IA = A$.

In each of the following state the conditions under which A^{-1} exists. Then find a formula for A^{-1}.

34. $A = [x]$
35. $A = \begin{bmatrix} x & 0 \\ 0 & y \end{bmatrix}$

36. $A = \begin{bmatrix} 0 & 0 & x \\ 0 & y & 0 \\ z & 0 & 0 \end{bmatrix}$

37. $A = \begin{bmatrix} x & 1 & 1 & 1 \\ 0 & y & 0 & 0 \\ 0 & 0 & z & 0 \\ 0 & 0 & 0 & w \end{bmatrix}$

★

38. Consider:

$$a_{11}x + a_{12}y = c_1,$$
$$a_{21}x + a_{22}y = c_2.$$

Use the cofactor method and the equivalent matrix equation $AX = C$ to prove Cramer's rule.

CHAPTER 6 REVIEW

Let $A = \begin{bmatrix} 1 & -1 & 0 \\ 2 & 3 & -2 \\ -2 & 0 & 1 \end{bmatrix}$, $B = \begin{bmatrix} -1 & 0 & 6 \\ 1 & -2 & 0 \\ 0 & 1 & -3 \end{bmatrix}$, and $C = \begin{bmatrix} -2 & 0 \\ 1 & 3 \end{bmatrix}$.

Find each of the following, if possible.

1. $A + B$ **2.** $-3A$ **3.** $-A$ **4.** AB
5. $B + C$ **6.** A^{-1} **7.** $A - B$ **8.** B^{-1}

Find A^{-1}, if it exists.

9. $A = \begin{bmatrix} -2 & 0 \\ 1 & 3 \end{bmatrix}$

10. $A = \begin{bmatrix} 0 & 0 & 3 \\ 0 & -2 & 0 \\ 4 & 0 & 0 \end{bmatrix}$

11. $A = \begin{bmatrix} 1 & 0 & 0 & 0 \\ 0 & 4 & -5 & 0 \\ 0 & 2 & 2 & 0 \\ 0 & 0 & 0 & 1 \end{bmatrix}$

12. Write a matrix equation equivalent to this system of equations.

$$3x - 2y + 4z = 13,$$
$$x + 5y - 3z = 7,$$
$$2x - 3y + 7z = -8$$

Evaluate.

13. $\begin{vmatrix} -4 & \sqrt{3} \\ \sqrt{3} & 7 \end{vmatrix}$

14. $\begin{vmatrix} 1 & -1 & 2 \\ -1 & 2 & 0 \\ -1 & 3 & 1 \end{vmatrix}$

15. $\begin{vmatrix} 0 & a & b \\ -a & 0 & c \\ -b & -c & 0 \end{vmatrix}$

16. $\begin{vmatrix} 4 & -7 & 6 & 7 \\ 0 & -3 & 9 & -8 \\ 0 & 0 & -2 & 6 \\ 0 & 0 & 0 & 5 \end{vmatrix}$

17. On the basis of Exercise 16, conjecture and prove a theorem regarding determinants.

18. Without expanding, show that,
$$\begin{vmatrix} 5a & 5b & 5c \\ 3a & 3b & 3c \\ d & e & f \end{vmatrix} = 0.$$

Factor.

19. $\begin{vmatrix} 1 & a & bc \\ 1 & b & ac \\ 1 & c & ab \end{vmatrix}$

20. $\begin{vmatrix} 1 & x^2 & x^3 \\ 1 & y^2 & y^3 \\ 1 & z^2 & z^3 \end{vmatrix}$

21. $\begin{vmatrix} 1 & a & a^2 & a^3 \\ 1 & b & b^2 & b^3 \\ 1 & c & c^2 & c^3 \\ 1 & d & d^2 & d^3 \end{vmatrix}$

7

Exponential and Logarithmic Functions

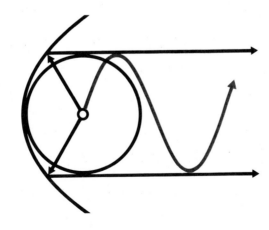

7.1 EXPONENTIAL AND LOGARITHMIC FUNCTIONS

Irrational Exponents

We have defined exponential notation for cases in which the exponent is a rational number. For example $x^{2.34}$, or $x^{234/100}$, means to take the 100th root of x and raise the result to the 234th power. We will now consider irrational exponents, such as π or $\sqrt{2}$.

Let us consider 2^π. We know that π has an unending decimal representation,

$$3.1415926535\ldots.$$

Now consider this sequence of numbers:

$$3,\ 3.1,\ 3.14,\ 3.141,\ 3.1415,\ 3.14159,\ldots.$$

Each of these numbers is an approximation to π, the more decimal places the better the approximation. Let us use these (rational) numbers to form a sequence as follows:

$$2^3,\ 2^{3.1},\ 2^{3.14},\ 2^{3.141},\ 2^{3.1415},\ 2^{3.14159},\ldots.$$

Each of the numbers in this sequence is already defined, the exponent being rational. The numbers in this sequence get closer and closer to some real number. We define that number to be 2^π.

We can define exponential notation for any irrational exponent in a similar way. Thus any exponential expression a^x now has meaning, whether the exponent is rational or irrational. The usual laws of exponents still hold, in case exponents are irrational. We will not attempt to prove that fact here, however.

Exponential Functions

Exponential functions are defined using exponential notation.

DEFINITION

The function $f(x) = a^x$, where a is some positive constant different from 1, is called *the exponential function, base a.*

Example 1 Graph $y = 2^x$. Use the graph to approximate 2^π.

x	0	1	2	3	−1	−2	−3
y	1	2	4	8	$\frac{1}{2}$	$\frac{1}{4}$	$\frac{1}{8}$

We find some solutions (see Fig. 1), plot them, and then draw the graph. (Note that as x increases, the function values increase. Check this on your calculator. As x decreases, the function values decrease toward 0. Check this on your calculator.)

To approximate 2^π, we locate π on the x-axis, at about 3.1. Then we find the corresponding function value. It is about 8.8.

Let us now look at some other exponential functions. We will make comparisons, using transformations.

Example 2 Graph: $y = 4^x$.

We could plot points and connect them, but let us be more clever. We note that $4^x = (2^2)^x = 2^{2x}$. Thus the function we wish to graph is

$$y = 2^{2x}.$$

Compare this with $y = 2^x$, graphed in Fig. 1. The graph of $y = 2^{2x}$ is a compression, in the x-direction, of the graph of $y = 2^x$. Knowing this allows us to graph $y = 2^{2x}$ at once (see Fig. 2). Each point on the graph of 2^x is moved half the distance to the y-axis.

Example 3 Graph: $y = (\tfrac{1}{2})^x$.

We could plot points and connect them, but again let us be more clever. We note that $(\tfrac{1}{2})^x = 1/2^x = 2^{-x}$. Thus the function we wish to graph is

$$y = 2^{-x}.$$

Compare this with the graph of $y = 2^x$ in Fig. 1. The graph of $y = 2^{-x}$ is a reflection, across the y-axis, of the graph of $y = 2^x$. Knowing this allows us to graph $y = 2^{-x}$ at once. See Fig. 3.

The preceding examples and exercises illustrate exponential functions of various bases. If $a = 1$, then $f(x) = a^x = 1^x = 1$ and the graph is a horizontal line. This is why we exclude 1 as a base for an exponential function.

We summarize.

Figure 1

Figure 2

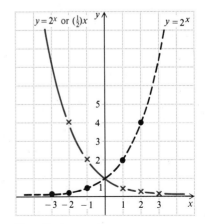

Figure 3

THEOREM 1

1. When $a > 1$, the function $f(x) = a^x$ is an increasing function. The greater the value of a, the faster the function increases.
2. When $0 < a < 1$, the function $f(x) = a^x$ is a decreasing function. The greater the value of a, the more slowly the function decreases.

Figure 4

Figure 5

It should be noted that for any value of a, the y-intercept is 1, as shown in Fig. 4.

As an application of exponential functions, consider the compound interest formula, $A = P(1 + i)^t$, considered in Section 2.2. Suppose principal P of \$1000 is invested at an interest rate of 18%, compounded annually. Then the amount A, in the account after time t, in years, is given by the exponential function $A = \$1000(1.18)^t$.

Logarithmic Functions

The inverse of an exponential function, for $a > 0$ and $a \neq 1$, is a function. It is called a *logarithmic* or *logarithm function*. Thus one way to describe a logarithm function is to interchange variables in $y = a^x$:

$$x = a^y.$$

The most useful and interesting logarithmic functions are those for which $a > 1$. The graph of such a function is shown in Fig. 5. It is a reflection of $y = a^x$ across the line $y = x$. Note that the domain of a logarithm function is the set of all positive real numbers.

For logarithm functions we use the notation $\log_a (x)$ or $\log_a x$.* That is, we use the symbol $\log_a x$ to denote the second coordinates of a function $x = a^y$. To say this another way, a logarithmic function can be described as $y = \log_a x$.

DEFINITION

The following are equivalent:
1. $x = a^y$; and
2. $y = \log_a x$ (read "y equals the logarithm, or simply log, base a, of x").

Thus $\log_a x$ represents the exponent in the equation $x = a^y$, so the logarithm, base a, of a number x is the power to which a is raised to get x.

It is important to be able to convert from an exponential equation to a logarithmic equation.

Examples Convert to logarithmic equations.

    4. $8 = 2^x \rightarrow x = \log_2 8$ It helps, in such conversions, to remember that the *logarithm is the exponent*.

---

*The parentheses in $\log_a (x)$ are like those in $f(x)$. In the case of logarithm functions we usually omit the parentheses.

5. $y^{-1} = 4 \to -1 = \log_y 4$
6. $a^b = c \to b = \log_a c$

It is also important to be able to convert from a logarithmic equation to an exponential equation.

Examples Convert to exponential equations.

7. $y = \log_3 5 \to 3^y = 5$ Again, it helps to remember that the *logarithm is the exponent*.
8. $-2 = \log_a 7 \to a^{-2} = 7$
9. $a = \log_b d \to b^a = d$

Certain equations containing logarithmic notation can be solved by first converting to exponential notation.

Examples

10. Solve: $\log_2 x = -3$.
    If $\log_2 x = -3$, then $2^{-3} = x$. So $x = \frac{1}{8}$.
11. Find: $\log_{27} 3$.
    Let $x = \log_{27} 3$. Then $27^x = 3$. Also $27^{1/3} = 3$ so $x = \frac{1}{3}$.
12. Solve: $\log_x 4 = \frac{1}{2}$.
    Now $\log_x 4 = \frac{1}{2}$ means $x^{1/2} = 4$. Then $(x^{1/2})^2 = 4^2$, or $x = 16$.

The exponential and logarithm functions are inverses of each other. Let us recall an important fact about functions and their inverses (p. 140). If the domains are suitable, then for any x,

$$f(f^{-1}(x)) = x$$

and

$$f^{-1}(f(x)) = x.$$

We apply this fact to exponential and logarithm functions. Suppose f is the exponential function, base a:

$$f(x) = a^x.$$

Then f^{-1} is the logarithm function, base a:

$$f^{-1}(x) = \log_a x.$$

Now let us find $f(f^{-1}(x))$:

$$f(f^{-1}(x)) = a^{f^{-1}(x)} = a^{\log_a x} = x.$$

Thus for any suitable base a, $a^{\log_a x} = x$ for any positive number x (negative numbers and 0 do not have logarithms).

Next, let us find $f^{-1}(f(x))$:
$$f^{-1}(f(x)) = \log_a f(x) = \log_a a^x = x.$$

Thus for any suitable base a, $\log_a a^x = x$ for any number x whatever.

These facts are important in simplification and should be learned well.

THEOREM 2

For any number a, suitable as a logarithm base,
1. $a^{\log_a x} = x$, for any positive number x; and
2. $\log_a a^x = x$, for any number x.

Examples Simplify.

13. $2^{\log_2 5} = 5$
14. $10^{\log_{10} t} = t$
15. $\log_e e^{-3} = -3$
16. $\log_{10} 10^{5.6} = 5.6$

EXERCISE SET 7.1

Graph.

1. a) $y = 4^x$ b) $y = \left(\frac{1}{4}\right)^x$ c) $y = \log_4 x$
2. a) $y = 3^x$ b) $y = \left(\frac{2}{3}\right)^x$ c) $y = \log_3 x$
3. $y = 5^x$
4. $y = 2^x$
5. $y = \log_2 x$
6. $y = \log_{10} x$

Write equivalent exponential equations.

7. $\log_2 32 = 5$
8. $\log_{10} 1000 = 3$
9. $\log_{10} 0.01 = -2$
10. $\log_{\sqrt{5}} 5 = 2$
11. $\log_6 6 = 1$
12. $\log_b M = N$

Write equivalent logarithmic equations.

13. $6^0 = 1$
14. $10^{-3} = 0.001$
15. $\left(\frac{6}{5}\right)^{-2} = \frac{25}{36}$
16. $5^4 = 625$
17. $5^{-2} = \frac{1}{25}$
18. $8^{1/3} = 2$
19. $e^{0.08} = 1.0833$
20. $10^{0.4771} = 3$

Solve.

21. $\log_{10} x = 4$ **22.** $\log_3 x = 2$ **23.** $\log_x \frac{1}{32} = 5$ **24.** $\log_8 x = \frac{1}{3}$

Find.

25. $\log_2 16$ **26.** $\log_3 3$ **27.** $\log_4 2$ **28.** $\log_2 64$
29. $\log_{10} 10^2$ **30.** $\log_3 3^4$ **31.** $\log_\pi \pi$ **32.** $\log_a a$
33. $\log_{10} 0.001$ **34.** $\log_{10} 1000$

Simplify.

35. $3^{\log_3 4x}$ **36.** $5^{\log_5(4x-5)}$ **37.** $\log_Q Q^{\sqrt{5}}$ **38.** $\log_e e^{|x-4|}$

☆ ───

Graph.

39. $y = \log_2(x + 3)$ **40.** $y = \log_3(x - 2)$ **41.** $y = 2^x - 1$
42. $y = 2^{x-3}$ **43.** $f(x) = 2^{|x|}$ **44.** $f(x) = \log_3 |x|$
45. $f(x) = 2^x + 2^{-x}$ **46.** $f(x) = 2^{-(x-1)}$

What is the domain of each function?

47. $f(x) = 3^x$ **48.** $f(x) = \log_{10} x$ **49.** $f(x) = \log_a x^2$
50. $f(x) = \log_4 x^3$ **51.** $f(x) = \log_{10}(3x - 4)$ **52.** $f(x) = \log_5 |x|$
53. $f(x) = \log_6(x^2 - 9)$

Solve using graphing.

54. $2^x > 1$ **55.** $3^x \leq 1$ **56.** $\log_2 x < 0$ **57.** $\log_2 x \geq 4$
58. $\log_2(x - 3) \leq 5$ **59.** $2^{x+3} > 1$

60. Approximate each of the following to six decimal places.

$$2^3, \quad 2^{3.1}, \quad 2^{3.14}, \quad 2^{3.141}, \quad 2^{3.1415}, \quad 2^{3.14159}$$

61. Which is larger, 5^π or π^5? **62.** Which is larger, $\sqrt{8}^3$ or $8^{\sqrt{3}}$?

63. Graph: $y = (0.745)^x$.

★ ───

Graph.

64. $y = 2^{-x^2}$ **65.** $y = 3^{-(x+1)^2}$ **66.** $y = |2^{x^2} - 8|$

7.2 PROPERTIES OF LOGARITHMIC FUNCTIONS

Let us now establish some basic properties of logarithmic functions.

THEOREM 3

For any positive numbers x and y,

$$\log_a (x \cdot y) = \log_a x + \log_a y,$$

where a is any positive number different from 1.

Theorem 3 says that the logarithm of a *product* is the *sum* of the logarithms of the factors. Note that the base a must remain constant. The logarithm of a sum is *not* the sum of the logarithms of the summands.

Proof of Theorem 3. Since a is positive and different from 1, it can serve as a logarithm base. Since x and y are assumed positive, they are in the domain of the function $y = \log_a x$. Now let $b = \log_a x$ and $c = \log_a y$.

Writing equivalent exponential equations, we have

$$x = a^b \quad \text{and} \quad y = a^c.$$

Next we multiply, to obtain

$$xy = a^b a^c = a^{b+c}.$$

Now writing an equivalent logarithmic equation, we obtain

$$\log_a (xy) = b + c, \quad \text{or}$$
$$\log_a (xy) = \log_a x + \log_a y,$$

which was to be shown. □

Example 1 Express as a sum of logarithms and simplify.

$$\log_2 (4 \cdot 16) = \log_2 4 + \log_2 16$$
$$= 2 + 4 = 6$$

THEOREM 4

For any positive number x and any number p,

$$\log_a x^p = p \cdot \log_a x,$$

where a is any logarithm base.

Theorem 4 says that the logarithm of a power of a number is the exponent times the logarithm of the number.

Proof of Theorem 4. Let $b = \log_a x$. Then, writing an equivalent exponential equation we have $x = a^b$. Next we raise both sides of the latter equation to the pth power. This gives us

$$x^p = (a^b)^p, \quad \text{or} \quad a^{bp}.$$

Now we can write an equivalent logarithmic equation,

$$\log_a x^p = \log_a a^{bp} = bp.$$

But $b = \log_a x$, so we have

$$\log_a x^p = p \cdot \log_a x,$$

which was to be shown. □

Examples Express as products.

2. $\log_b 9^{-5} = -5 \cdot \log_b 9$
3. $\log_a \sqrt[4]{5} = \log_a 5^{1/4} = \frac{1}{4} \log_a 5$

THEOREM 5

For any positive numbers x and y,

$$\log_a \frac{x}{y} = \log_a x - \log_a y,$$

where a is any logarithm base.

Theorem 5 says that the logarithm of a quotient is the difference of the logarithms (i.e., the logarithm of the dividend minus the logarithm of the divisor).

Proof of Theorem 5. $x/y = x \cdot y^{-1}$, so

$$\log_a \frac{x}{y} = \log_a (xy^{-1}).$$

By Theorem 2,

$$\log_a (xy^{-1}) = \log_a x + \log_a y^{-1},$$

and by Theorem 3,

$$\log_a y^{-1} = -1 \cdot \log_a y,$$

so we have

$$\log_a \frac{x}{y} = \log_a x - \log_a y,$$

which was to be shown. □

Example 4 Express as sums and differences of logarithms and without exponential notation or radicals.

$$\log_a \frac{\sqrt{17}}{5\pi} = \log_a \sqrt{17} - \log_a 5\pi$$

$$= \log_a 17^{1/2} - (\log_a 5 + \log_a \pi)$$

$$= \frac{1}{2} \log_a 17 - \log_a 5 - \log_a \pi$$

Example 5 Express in terms of logarithms of x, y, and z.

$$\log_a \sqrt[4]{\frac{xy}{z^3}} = \log_a \left(\frac{xy}{z^3}\right)^{1/4} = \frac{1}{4} \cdot \log_a \frac{xy}{z^3}$$

$$= \frac{1}{4} [\log_a xy - \log_a z^3]$$

$$= \frac{1}{4} [\log_a x + \log_a y - 3 \log_a z]$$

$$= \frac{1}{4} \log_a x + \frac{1}{4} \log_a y - \frac{3}{4} \log_a z$$

Example 6 Express as a single logarithm.

$$\frac{1}{2} \log_a x - 7 \log_a y + \log_a z = \log_a \sqrt{x} - \log_a y^7 + \log_a z$$

$$= \log_a \frac{\sqrt{x}}{y^7} + \log_a z$$

$$= \log_a \frac{z\sqrt{x}}{y^7}$$

Example 7 Given that $\log_a 2 = 0.301$ and $\log_a 3 = 0.477$, find:

a) $\log_a 6 = \log_a 2 \cdot 3 = \log_a 2 + \log_a 3$

$$= 0.301 + 0.477 = 0.778;$$

b) $\log_a \sqrt{3} = \log_a 3^{1/2} = \frac{1}{2} \cdot \log_a 3 = \frac{1}{2} \cdot 0.477$

$$= 0.2385;$$

c) $\log_a \frac{2}{3} = \log_a 2 - \log_a 3 = 0.301 - 0.477$

$$= -0.176;$$

d) $\log_a 5$; *no way to find, using these properties* ($\log_a 5 \neq \log_a 2 + \log_a 3$);

e) $\dfrac{\log_a 2}{\log_a 3} = \dfrac{0.301}{0.477} = 0.63.$ Note that we could not use any of the properties; we simply divided.

For any base a, $\log_a a = 1$. This is easily seen by writing an equivalent exponential equation, $a^1 = a$. Similarly, for any base a, $\log_a 1 = 0$. These facts are important and should be remembered well.

THEOREM 6

For any base a,

$$\log_a a = 1 \quad \text{and} \quad \log_a 1 = 0.$$

EXERCISE SET 7.2

Express in terms of logarithms of x, y, and z.

1. $\log_a x^2 y^3 z$
2. $\log_a 5xy^4 z^3$
3. $\log_b \dfrac{xy^2}{z^3}$
4. $\log_c \sqrt[3]{\dfrac{x^4}{y^3 z^2}}$

Express as a single logarithm and simplify if possible.

5. $\dfrac{2}{3} \log_a 64 - \dfrac{1}{2} \log_a 16$
6. $\dfrac{1}{2} \log_a x + 3 \log_a y - 2 \log_a x$
7. $\log_a 2x + 3(\log_a x - \log_a y)$
8. $\log_a x^2 - 2 \log_a \sqrt{x}$
9. $\log_a \dfrac{a}{\sqrt{x}} - \log_a \sqrt{ax}$
10. $\log_a (x^2 - 4) - \log_a (x - 2)$
11. $\log_a (x^3 + y^3) - \log_a (x + y)$
12. $\log_a (x - y) + \log_a (x^2 + xy + y^2)$

Express as a sum and/or difference of logarithms.

13. $\log_a \sqrt{1 - x^2}$
14. $\log_a \dfrac{x + t}{\sqrt{x^2 - t^2}}$

Given $\log_{10} 2 = 0.301$, $\log_{10} 3 = 0.477$, and $\log_{10} 10 = 1$, find:

15. $\log_{10} 4$
16. $\log_{10} 5$ $\left(\text{Hint: } 5 = \dfrac{10}{2}.\right)$
17. $\log_{10} 50$ $\left(\text{Hint: } 50 = \dfrac{100}{2}.\right)$
18. $\log_{10} 12$
19. $\log_{10} 60$
20. $\log_{10} \dfrac{1}{3}$
21. $\log_{10} \sqrt{\dfrac{2}{3}}$
22. $\log_{10} \sqrt[5]{12}$
23. $\log_{10} 90$
24. $\log_{10} \dfrac{9}{8}$
25. $\log_{10} \dfrac{9}{10}$
26. $\log_{10} \dfrac{1}{4}$

Which of the following are false?

27. $\dfrac{\log_a M}{\log_a N} = \log_a M - \log_a N$
28. $\dfrac{\log_a M}{\log_a N} = \log_a \dfrac{M}{N}$
29. $\dfrac{\log_a M}{c} = \log_a M^{1/c}$
30. $\log_N (M \cdot N)^x = x \log_N M + x$
31. $\log_a 2x = 2 \log_a x$
32. $\log_a 2x = \log_a 2 + \log_a x$
33. $\log_a (M + N) = \log_a M + \log_a N$
34. $\log_a x^3 = 3 \log_a x$

Solve.

35. $\log_\pi \pi^{2x+3} = 4$
36. $3^{\log_3 (8x-4)} = 5$
37. $4^{2 \log_4 x} = 7$
38. $8^{2 \log_8 x + \log_8 x} = 27$
39. $(x + 3) \cdot \log_a a^x = x$
40. $\log_a x^2 = 2 \log_a x$
41. $\log_a 5x = \log_a 5 + \log_a x$
42. $\log_b \dfrac{5}{x+2} = \log_b 5 - \log_b (x + 2)$

43. If $\log_a x = 2$, what is $\log_a \left(\dfrac{1}{x}\right)$?

44. If $\log_a x = 2$, what is $\log_{1/a} x$?

Prove the following for any base a and any positive number x.

45. $\log_a \left(\dfrac{1}{x}\right) = -\log_a x$
46. $\log_a \left(\dfrac{1}{x}\right) = \log_{1/a} x$

47. Show that
$$\log_a \left(\dfrac{x + \sqrt{x^2 - 5}}{5}\right) = -\log_a (x - \sqrt{x^2 - 5})$$

48. Graph and compare: $y = \log_2 |x|$ and $y = |\log_2 x|$.

7.3 COMMON LOGARITHMS

Base ten logarithms are known as *common logarithms*. Tables for these logarithms are readily available (Table 2 at the back of this book).

Logarithms in Computation

Before calculators and computers became so readily available, common logarithms were used extensively to do certain kinds of calculations. In fact, this is why logarithms were developed. Today, computations with logarithms are mainly of historical interest; the logarithm *functions* are of modern importance. The study of computations with logarithms can, however, help the student to fix his or her ideas with respect to properties of logarithm functions.

7.3 COMMON LOGARITHMS

Figure 6 contains a list of powers of 10, or *logarithms* base 10. The exponents are approximate, but accurate to four decimal places. To illustrate how logarithms can be used for computation we will use this table and do some easy calculations.

$$
\begin{aligned}
1 &= 10^{0.0000}, & \text{or} && \log_{10} 1 &= 0.0000 \\
2 &= 10^{0.3010}, & \text{or} && \log_{10} 2 &= 0.3010 \\
3 &= 10^{0.4771}, & \text{or} && \log_{10} 3 &= 0.4771 \\
4 &= 10^{0.6021}, & \text{or} && \log_{10} 4 &= 0.6021 \\
5 &= 10^{0.6990}, & \text{or} && \log_{10} 5 &= 0.6990 \\
6 &= 10^{0.7782}, & \text{or} && \log_{10} 6 &= 0.7782 \\
7 &= 10^{0.8451}, & \text{or} && \log_{10} 7 &= 0.8451 \\
8 &= 10^{0.9031}, & \text{or} && \log_{10} 8 &= 0.9031 \\
9 &= 10^{0.9542}, & \text{or} && \log_{10} 9 &= 0.9542 \\
10 &= 10^{1.0000}, & \text{or} && \log_{10} 10 &= 1.0000 \\
11 &= 10^{1.0414}, & \text{or} && \log_{10} 11 &= 1.0414 \\
12 &= 10^{1.0792}, & \text{or} && \log_{10} 12 &= 1.0792 \\
13 &= 10^{1.1139}, & \text{or} && \log_{10} 13 &= 1.1139 \\
14 &= 10^{1.1461}, & \text{or} && \log_{10} 14 &= 1.1461 \\
15 &= 10^{1.1761}, & \text{or} && \log_{10} 15 &= 1.1761 \\
16 &= 10^{1.2041}, & \text{or} && \log_{10} 16 &= 1.2041
\end{aligned}
$$

Figure 6

Example 1 Find 3×4 using the table of exponents.

$$
\begin{aligned}
3 \times 4 &= 10^{0.4771} \times 10^{0.6021} \\
&= 10^{1.0792} \qquad \text{Adding exponents}
\end{aligned}
$$

From the table we see that $10^{1.0792} = 12$, so

$$3 \times 4 = 12.$$

Note from Example 1 that we can find a product by adding the logarithms of the factors and then finding the number having the result as its logarithm. That is, we found the number $10^{1.0792}$. This number is often referred to as the *antilogarithm* of 1.0792. To state this another way, if

$$f(x) = \log_{10} x,$$

then

$$f^{-1}(x) = \text{antilog}_{10} x = 10^x.$$

In other words, an antilogarithm function is simply an exponential function.

Example 2 Find $\frac{14}{2}$, using base 10 logarithms.

$$\log_{10} \frac{14}{2} = \log_{10} 14 - \log_{10} 2$$

$$\log_{10} \frac{14}{2} = 1.1461 - 0.3010$$

$$\log_{10} \frac{14}{2} = 0.8451$$

$$\frac{14}{2} = \text{antilog}_{10} \, 0.8451 = 7$$

Example 3 Find $\sqrt[4]{16}$, using base 10 logarithms.

$$\log_{10} \sqrt[4]{16} = \log_{10} 16^{1/4} = \frac{1}{4} \cdot \log_{10} 16 = \frac{1}{4} \cdot 1.2041$$
$$\log_{10} \sqrt[4]{16} = 0.3010 \quad \text{Rounded to four decimal places}$$
$$\sqrt[4]{16} = \text{antilog}_{10} \, 0.3010 = 2$$

Example 4 Find 2^3, using base 10 logarithms.

$$\log_{10} 2^3 = 3 \cdot \log_{10} 2 = 3 \cdot 0.3010 = 0.9030$$
$$\log_{10} 2^3 = 0.9030$$
$$2^3 = \text{antilog}_{10} \, 0.9030 \approx 8$$

Note the rounding error.

We will often omit the base, 10, when working with common logarithms.

Table 2 at the back of the book contains logarithms of numbers from 1 to 10. Part of that table is shown below. To illustrate the use of the table, let us find log 5.24. We locate the row headed 5.2, then move across to the column headed 4. We find log 5.24 as the colored entry in the table.

x	0	1	2	3	4	5	6	7	8	9
5.0	0.6990	0.6998	0.7007	0.7016	0.7024	0.7033	0.7042	0.7050	0.7059	0.7067
5.1	0.7076	0.7084	0.7093	0.7101	0.7110	0.7118	0.7126	0.7135	0.7143	0.7152
5.2	0.7160	0.7168	0.7177	0.7185	0.7193	0.7202	0.7210	0.7218	0.7226	0.7235
5.3	0.7234	0.7251	0.7259	0.7267	0.7275	0.7284	0.7292	0.7300	0.7308	0.7316
5.4	0.7324	0.7332	0.7340	0.7348	0.7356	0.7364	0.7372	0.7380	0.7388	0.7396

We can find antilogarithms by reversing this process. For example, antilog $0.7193 = 10^{0.7193} = 5.24$. Similarly, antilog $0.7292 = 5.36$.

Using Table 2 and scientific notation* we can find logarithms of numbers that are not between 1 and 10. First recall the following:

$$\log_a a^k = k \text{ for any number } k. \quad \text{Theorem 2}$$

Thus

$$\log_{10} 10^k = k \text{ for any number } k.$$

Examples

5. $\log 52.4 = \log (5.24 \times 10^1)$
    $= \log 5.24 + \log 10^1$
    $= 0.7193 + 1$

6. $\log 0.524 = \log (5.24 \times 10^{-1})$
    $= \log 5.24 + \log 10^{-1}$
    $= 0.7193 + (-1)$

7. $\log 52{,}400 = \log (5.24 \times 10^4)$
    $= \log 5.24 + \log 10^4$
    $= 0.7193 + 4$

8. $\log 0.00524 = \log (5.24 \times 10^{-3})$
    $= \log 5.24 + \log 10^{-3}$
    $= 0.7193 + (-3)$

The preceding examples illustrate the importance of using the base 10 for computation. It allows great economy in the printing of tables. If we know the logarithm of a number from 1 to 10 we can multiply that number by any power of ten and easily determine the logarithm of the resulting number. For any base other than 10 this would not be the case. In each of these examples, the integer part of the logarithm is the exponent in the scientific notation. This integer is called the *characteristic* of the logarithm. The other part of the logarithm, a number between 0 and 1, is called the *mantissa* of the logarithm. Table 2 contains only mantissas.

Example 9 Find log 0.0538, indicating the characteristic and mantissa.

We first write scientific notation for the number:

$$5.38 \times 10^{-2}.$$

Then we find log 5.38. This is the mantissa:

$$\log 5.38 = 0.7308.$$

---

*It may be helpful to review scientific notation in Chapter 1.

The characteristic is the exponent -2. Now
$$\log 0.0538 = 0.7308 + (-2), \text{ or } -1.2692.$$

When negative characteristics occur, it is often best to name the logarithm in such a way that the characteristic and mantissa are preserved. In the preceding example, we have
$$\log 0.0538 = 0.7308 + (-2) = -1.2692,$$
but the latter notation displays neither the characteristic nor the mantissa. We can rename the characteristic, -2, as $8 - 10$, and then add the mantissa, to obtain
$$8.7308 - 10,$$
thus preserving both mantissa and characteristic.

The characteristic and mantissa are useful when working with logarithmic tables, but are not needed on a calculator. For example, on a calculator with a ten-digit readout,
$$\log 0.0538 = -1.269217724,$$
but this shows neither the characteristic nor the mantissa. Check this on your calculator. How can you find the characteristic and mantissa?

Example 10 Find $\log 0.00687$.

We write scientific notation (or at least visualize it):
$$0.00687 = 6.87 \times 10^{-3}.$$

The characteristic is -3, or $7 - 10$. The mantissa, from the table, is 0.8370. Thus $\log 0.00687 = 7.8370 - 10$.

Antilogarithms

To find antilogarithms, we reverse the procedure for finding logarithms.

Example 11 Find antilog 2.6085 (or find x such that $\log x = 2.6085$).
$$\text{antilog } 2.6085 = 10^{2.6085} = 10^{(2+0.6085)}$$
$$= 10^2 \cdot 10^{0.6085}$$

From the table we can find $10^{0.6085}$, or antilog 0.6085. It is 4.06. Hence we have
$$\text{antilog } 2.6085 = 10^2 \times 4.06, \text{ or } 406.$$

Note in this example, we in effect separate the number 2.6085 into an integer and a number between 0 and 1. We use the latter with the table, after which we have scientific notation for our answer.

Example 12 Find antilog 3.7118.

From the table we find antilog 0.7118 = 5.15. Thus

$$\text{antilog } 3.7118 = 5.15 \times 10^3 \qquad \text{Note that 3 is the characteristic.}$$
$$= 5150.$$

Example 13 Find antilog (7.7143 − 10).

The characteristic is −3 and the mantissa is 0.7143. From the table we find that antilog 0.7143 = 5.18. Thus

$$\text{antilog } (7.7143 - 10) = 5.18 \times 10^{-3}$$
$$= 0.00518.$$

Example 14 Find antilog −2.2857.

We are to find the antilog of a number, but the number is named so the mantissa is not apparent. To find the mantissa we add 0, naming it 10 − 10:

$$-2.2857 = -2.2857 + (10 - 10) = (-2.2857 + 10) - 10 = 7.7143 - 10.$$

Then we proceed as in Example 13. The answer is 0.00518.

Calculations with Logarithms

The kinds of calculations in which logarithms were helpful historically are multiplication, division, taking powers, and taking roots.

Example 15 Find $(0.0578 \times 32.7)/8460$.

We write a *plan* for the use of logarithms, then look up all the mantissas at one time. Let

$$N = \frac{0.0578 \times 32.7}{8460}.$$

Then

$$\log N = \log 0.0578 + \log 32.7 - \log 8460.$$

This gives us the plan.
We use a straight line to indicate addition and a wavy line to indicate subtraction.

Completion of plan:
$$\log 0.0578 = 8.7619 - 10$$
$$\log 32.7 = 1.5145$$
$$\log \text{numerator} = 10.2764 - 10$$
$$\log 8460 = 3.9274$$
$$\log \text{fraction} = 6.3490 - 10$$
$$\text{fraction} = 0.000223 \quad \text{Taking antilog}$$

Example 16 Use logarithms to find $\sqrt[4]{0.325} \times (4.23)^2$.

If
$$N = \sqrt[4]{0.325} \times (4.23)^2,$$
then
$$\log N = \frac{1}{4} \log 0.325 + 2 \log 4.23;$$

this yields the plan.

Note that powers and roots are involved. In such cases it is easier to find the following logs first:

$$\log 0.325 = 9.5119 - 10$$
$$\log 4.23 = 0.6263.$$

Consider $9.5119 - 10$. Since we will be dividing this number by 4, but -10 is not divisible by 4, we rename this number:

$$\log 0.325 = 39.5119 - 40.$$

Now the negative term is divisible by 4. We divide $\log 0.325$ by 4 and multiply $\log 4.23$ by 2. Then we have

$$\log \sqrt[4]{0.325} = 9.8780 - 10$$
$$\log (4.23)^2 = 1.2526$$
$$\log \text{product} = 11.1306 - 10$$
$$= 1.1306$$
$$\text{product} = 13.5.$$

EXERCISE SET 7.3

Use Table 2 at the back of the book to find each of the following. Where appropriate, write so that (positive) mantissas are preserved.

1. log 2.46
2. log 7.5
3. log 347
4. log 8720
5. log 52.5
6. log 20.8
7. log 624,000
8. log 13,400
9. log 0.0702
10. log 0.640
11. log 0.000216
12. log 0.173

13. antilog 2.3674
14. antilog 1.9222
15. antilog 8.2553 − 10
16. antilog 6.6294 − 10
17. antilog −5.9788
18. antilog −2.2628
19. $10^{1.4014}$
20. $10^{3.6590}$
21. $10^{8.9881-10}$
22. $10^{7.5391-10}$

Solve for x.

23. $\log x = 0.6522$
24. $\log x = 4.8156$
25. $\log x = 8.1239 - 10$
26. $\log x = -1.0218$

Use logarithms to compute. When exact values do not occur in the table, use the nearest values.

27. 3.14×60.4
28. 541×0.0152
29. $286 \div 1.05$
30. $12.8 \div 81.6$
31. $\sqrt{76.9}$
32. $\sqrt[3]{56.9}$
33. $(1.36)^{4.2}$
34. $(0.727)^{3.6}$
35. $\dfrac{70.7 \times (10.6)^2}{18.6 \times \sqrt{276}}$
36. $\sqrt[3]{\dfrac{3.24 \times (3.16)^2}{78.4 \times 24.6}}$

☆

Find each of the following. Round to six decimal places.

37. 🖩 log 56,789
38. 🖩 log 0.0111347
39. 🖩 log (log 3)

Find each antilogarithm. Since $\mathrm{antilog}_{10} x = 10^x$, you can find the antilogarithm, base 10, of a number x by raising 10 to the power x.

40. 🖩 $10^{0.4356}$
41. 🖩 $\mathrm{antilog}_{10}\ 7.8943$
42. 🖩 $\mathrm{antilog}_{10}(-7.5689)$
43. 🖩 $10^{-3.23445678}$

7.4 INTERPOLATION   SKIP

We just developed procedures for using Table 2 using three-digit precision. By using a procedure called *interpolation* we can find values between those listed in the table, obtaining four-digit precision.* Interpolation can be done in various ways, the simplest and most common being *linear* interpolation. We describe it now. What we say applies to a table for any continuous function.

Let us consider how a table of values for any function is made. We select members of the domain x_1, x_2, x_3, and so on. Then we compute or somehow determine the corresponding function values, $f(x_1), f(x_2), f(x_3)$, and so on. Then we tabulate the results. We might also graph the results. See Fig. 7.

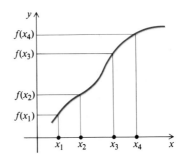

Figure 7

x	x_1	x_2	x_3	x_4	...
f(x)	$f(x_1)$	$f(x_2)$	$f(x_3)$	$f(x_4)$	...

---

*If a calculator or a five-place logarithmic table were used, there could be some variance in the last decimal place.

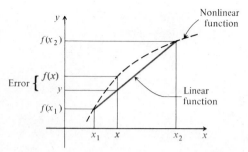

Figure 8

Suppose we want to find the function value $f(x)$ for a value of x not in the table. If x is halfway between x_1 and x_2, then we can take the number halfway between $f(x_1)$ and $f(x_2)$ as an approximation to $f(x)$. If x is one-fifth of the way between x_2 and x_3, we take the number that is one-fifth of the way between $f(x_2)$ and $f(x_3)$ as an approximation to $f(x)$. What we do is divide the length from x_2 to x_3 in a certain ratio, and then divide the length from $f(x_2)$ to $f(x_3)$ in the same ratio. This is *linear interpolation*.

We show this geometrically in Fig. 8. The length from x_1 to x_2 is divided in a certain ratio by x. The length from $f(x_1)$ to $f(x_2)$ is divided in the same ratio by y. The number y approximates $f(x)$ with the noted error. Note the slanted line in the figure. The approximation y comes from this line. This explains the use of the term *linear interpolation*. Let us apply linear interpolation to Table 2 of common logarithms.

Example 1 Find log 34,870.

a) Find the characteristic. Since $34{,}870 = 3.487 \times 10^4$, the characteristic is 4.

b) Find the mantissa. From Table 2 we have:

$$0.01 \left\{ 0.007 \left\{ \begin{array}{l} \log 3.480 = 0.5416 \\ \log 3.487 = 0.54?? \\ \log 3.490 = 0.5428 \end{array} \right. \right\}$$ The tabular difference is 0.0012.

The tabular difference (difference between consecutive values in the table) is 0.0012. Now 3.487 is $\tfrac{7}{10}$ of the way from 3.480 to 3.490. So we take 0.7 of 0.0012, which is 0.00084, and round it to 0.0008. We add this to 0.5416. The mantissa is 0.5424.

c) Add the characteristic and mantissa:

$$\log 34{,}870 = 4.5424.$$

With practice you will take 0.7 of 12, forgetting the zeros, but adding in the same way.

Example 2 Find log 0.009543.

a) Find the characteristic. Since $0.009543 = 9.543 \times 10^{-3}$, the characteristic is -3, or $7 - 10$.

b) Find the mantissa. From Table 2 we have:

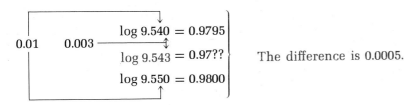
The difference is 0.0005.

Now 9.543 is $\frac{3}{10}$ of the way from 9.540 to 9.550, so we take 0.3 of 0.0005, which is 0.00015, and round it to 0.0002. We add this to 0.9795. The mantissa that results is 0.9797.

c) Add the characteristic and the mantissa:
$$\log 0.009543 = 7.9797 - 10.$$

Antilogarithms

We interpolate when finding antilogarithms, using the table in reverse.

Example 3 Find antilog 4.9164.

a) The characteristic is 4. The mantissa is 0.9164.

b) Find the antilog of the mantissa, 0.9164. From Table 2, we have

$$
\left.\begin{array}{l}
0.0006 \quad 0.0005 \left\{\begin{array}{l}
\text{antilog } 0.9159 = 8.240 \\
\text{antilog } 0.9164 = 8.24? \\
\text{antilog } 0.9165 = 8.250
\end{array}\right.
\end{array}\right\} \text{The difference is 0.010.}
$$

The difference between 0.9159 and 0.9165 is 0.0006. Thus 0.9164 is $\frac{0.0005}{0.0006}$, or $\frac{5}{6}$, of the way between 0.9159 and 0.9165. Then antilog 0.9164 is $\frac{5}{6}$ of the way between 8.240 and 8.250, or $\frac{5}{6}(0.010)$, which is $0.00833\ldots$, or 0.008 rounded. Thus the antilog of the mantissa is 8.248.

Thus antilog $4.9164 = 8.248 \times 10^4 = 82{,}480$.

Example 4 Find antilog $(7.4122 - 10)$.

a) The characteristic is -3. The mantissa is 0.4122.

b) Find the antilog of the mantissa, 0.4122. From Table 2 we have

The difference is 0.010.

The difference between 0.4116 and 0.4133 is 0.0017. Thus 0.4122 is $\frac{0.0006}{0.0017}$, or $\frac{6}{17}$ of the way between 0.4116 and 0.4133. Then antilog 0.4122 is $\frac{6}{17}$ of the way between 2.580 and 2.590, or $\frac{6}{17}(0.010)$, which is 0.0035, to four places. We round it to 0.004. Thus the antilog of the mantissa is 2.584.

So antilog $(7.4122 - 10) = 2.584 \times 10^{-3} = 0.002584$.

Calculations

Calculations with logarithms, using Table 2, can be done with four-digit precision when interpolation is used, in finding both logarithms and antilogarithms.

EXERCISE SET 7.4

Find each of the following logarithms using interpolation and Table 2.

1. log 41.63
2. log 472.1
3. log 2.944
4. log 21.76
5. log 650.2
6. log 37.37
7. log 0.1425
8. log 0.0904
9. log 0.004257
10. log 4518
11. log 0.1776
12. log 0.08356
13. log 600.6
14. log (log 3)
15. log (log 5)

Find each of the following antilogarithms using interpolation and Table 2.

16. antilog 1.6350
17. antilog 2.3512
18. antilog 0.6478
19. antilog 1.1624
20. antilog 0.0342
21. antilog 4.8453
22. antilog 9.8561 − 10
23. antilog 8.9659 − 10
24. antilog 7.4128 − 10
25. antilog 9.7278 − 10
26. antilog 8.2010 − 10
27. antilog 7.8630 − 10

☆

Use logarithms and interpolation to do the following calculations. Use four-digit precision. Answers may be checked using a calculator.

28. $\dfrac{35.24 \times (16.77)^3}{12.93 \times \sqrt{276.2}}$

29. $\sqrt[5]{\dfrac{16.79 \times (4.234)^3}{18.81 \times 175.3}}$

7.5 EXPONENTIAL AND LOGARITHMIC EQUATIONS

An equation with variables in exponents, such as $3^{2x-1} = 4$, is called an *exponential equation*. We can often solve such equations by taking the logarithm on both sides and then using Theorem 4, p. 258.*

Example 1 Solve: $3^x = 8$.

$$\log 3^x = \log 8 \quad \text{Taking log on both sides}$$

$$x \log 3 = \log 8 \quad \text{Using Theorem 4}$$

$$x = \frac{\log 8}{\log 3}$$

$$x \approx \frac{0.9031}{0.4771} \approx 1.8929 \quad \text{We look up the logs, or find them on a calculator, and divide.}$$

Example 2 Solve: $2^{3x-5} = 16$.

Method 1.

$$\log 2^{3x-5} = \log 16 \quad \text{Taking log on both sides}$$

$$(3x - 5) \log 2 = \log 16 \quad \text{Using Theorem 4}$$

$$3x - 5 = \frac{\log 16}{\log 2}$$

$$x = \frac{\frac{\log 16}{\log 2} + 5}{3} \approx \frac{\frac{1.2041}{0.3010} + 5}{3}$$

$$x \approx 3.0001$$

The answer is approximate because the logarithms are approximate.

Method 2. Note that $16 = 2^4$. Then we have

$$2^{3x-5} = 2^4.$$

Since the base is the same, 2, on both sides, the exponents must be the same. Thus

$$3x - 5 = 4.$$

We solve this equation to get $x = 3$. This answer is exact.

---

*The use of a calculator is recommended for the rest of this chapter.

Example 3 Solve $\dfrac{e^x + e^{-x}}{2} = t$, for x.

$$e^x + e^{-x} = 2t \quad \text{Multiplying by 2}$$

$$e^x + \frac{1}{e^x} = 2t \quad \text{Rewriting to eliminate the minus sign in an exponent}$$

$$e^{2x} + 1 = 2t \cdot e^x \quad \text{Multiplying on both sides by } e^x$$

$$(e^x)^2 - 2t \cdot e^x + 1 = 0$$

This equation is reducible to quadratic, with $u = e^x$. Using the quadratic formula, we obtain

$$e^x = \frac{2t \pm \sqrt{4t^2 - 4}}{2} = t \pm \sqrt{t^2 - 1}$$

$$\log e^x = \log(t \pm \sqrt{t^2 - 1}) \quad \text{Taking log on both sides}$$

$$x \log e = \log(t \pm \sqrt{t^2 - 1}) \quad \text{Using Theorem 4}$$

$$x = \frac{\log(t \pm \sqrt{t^2 - 1})}{\log e}$$

If e should be a number suitable as a logarithm base, we might also take base e logarithms on both sides. The denominator would then be $\log_e e$, or 1, and we would have

$$x = \log_e(t \pm \sqrt{t^2 - 1}).$$

Logarithmic Equations

Equations that contain logarithmic expressions are called *logarithmic equations*. To solve such equations we try to obtain a single logarithmic expression on one side of the equation and then take the antilogarithm on both sides.

Example 4 Solve: $\log x + \log(x - 3) = 1$.

$$\log x(x - 3) = 1 \quad \text{Using Theorem 3, p. 258, to obtain a single logarithm}$$

$$x(x - 3) = 10^1 \quad \text{Taking the antilog on both sides}$$

$$x^2 - 3x - 10 = 0$$

$$(x + 2)(x - 5) = 0 \quad \text{Factoring and principle of zero products}$$

$$x = -2 \quad \text{or} \quad x = 5$$

Possible solutions to logarithmic equations must be checked because domains of logarithmic functions are restricted to positive numbers.

Check:

$$\begin{array}{c|c}
\log x + \log (x - 3) = 1 & \log x + \log (x - 3) = 1 \\
\hline
\log (-2) + \log (-2 - 3) \quad | \quad 1 & \begin{array}{c} \log 5 + \log (5 - 3) \\ \log 5 + \log 2 \\ \log 10 \\ 1 \end{array} \quad \Bigg| \quad 1
\end{array}$$

The number -2 is not a solution because negative numbers do not have logarithms. The solution is 5.

Applications

Exponential and logarithmic functions and equations have many applications. We shall consider a few of them.

Example 5 (*Compound interest*). The amount A that principal P will be worth after t years at interest rate i, compounded annually, is given by the formula $A = P(1 + i)^t$. Suppose \$4000 principal is invested at 6% interest and yields \$5353. How many years was it invested?

Using the formula $A = P(1 + i)^t$, we have

$$5353 = 4000(1 + 0.06)^t, \quad \text{or} \quad 5353 = 4000(1.06)^t.$$

Solving for t, we have

$$\log 5353 = \log 4000(1.06)^t$$

$$\log 5353 = \log 4000 + t \log 1.06$$

$$\frac{\log 5353 - \log 4000}{\log 1.06} = t$$

$$\frac{3.7286 - 3.6021}{0.0253} = t$$

$$5 = t.$$

The money was invested for five years.

Example 6 (*Loudness of sound*). The sensation of loudness of sound is not proportional to the energy intensity, but rather is a logarithmic function. *Loudness*, in Bels (after Alexander Graham Bell), of a sound of intensity I is defined to be

$$L = \log \frac{I}{I_0},$$

where I_0 is the minimum intensity detectable by the human ear (such as the tick of a watch at 20 ft under quiet conditions). When a sound is 10 times as intense as another, its loudness is 1 Bel greater. If a sound is 100 times as intense as another, it is louder by 2 Bels, and so on. The Bel is a large unit, so a subunit, a *decibel*, is usually used. For L in decibels, the formula is

$$L = 10 \log \frac{I}{I_0}.$$

a) Find the loudness, in decibels, of the sound in a radio studio, for which the intensity I is 199 times I_0.

We substitute into the formula and calculate, using Table 2:

$$L = 10 \log \frac{199 \cdot I_0}{I_0} = 10 \log 199$$

$$= 10(2.2989)$$

$$= 23 \text{ decibels.}$$

b) Find the loudness of the sound of a heavy truck, for which the intensity is 10^9 times I_0.

$$L = 10 \log \frac{10^9 \cdot I_0}{I_0} = 10 \log 10^9$$

$$= 10 \cdot 9$$

$$= 90 \text{ decibels.}$$

Example 7 (*Earthquake magnitude*). The magnitude R (on the Richter scale) of an earthquake of intensity I is defined as follows:

$$R = \log \frac{I}{I_0},$$

where I_0 is a minimum intensity used for comparison.

The Mexico City earthquake of 1978 had an intensity $10^{7.85}$ times I_0. What is its magnitude on the Richter scale?

We substitute into the formula:

$$R = \log \frac{10^{7.85} \cdot I_0}{I_0} = \log 10^{7.85} = 7.85.$$

7.5 EXPONENTIAL AND LOGARITHMIC EQUATIONS

Example 8 (*Forgetting*). Here is a mathematical model from psychology. A group of people take a test and make an average score of S. After a time t they take an equivalent form of the same test. At that time the average score is $S(t)$. According to this model, $S(t)$ is given by the following function,

$$S(t) = A - B \log(t + 1),$$

where t is in months and the constants A and B are determined by experiment in various kinds of learning situations. The model is appropriate only over the interval $[0, 10^{A/B} - 1]$.

Students in a zoology class took a final exam, and took equivalent forms of the test at monthly intervals thereafter. The average scores were found to be given by the function

$$S(t) = 78 - 15 \log(t + 1).$$

What was the average score (a) when they took the test originally? (b) after four months?

We substitute into the equation defining the function:

a) $S(0) = 78 - 15 \log(0 + 1)$
$= 78 - 15 \log 1$
$= 78 - 0 = 78.$

b) $S(4) = 78 - 15 \log(4 + 1)$
$= 78 - 15 \log 5$
$= 78 - 15 \cdot 0.6990$
$= 78 - 10.49 = 67.51.$

EXERCISE SET 7.5

Solve.

1. $2^x = 32$
2. $3^{x-1} + 3 = 30$
3. $4^{2x} = 8^{3x-4}$
4. $3^{x^2+4x} = \dfrac{1}{27}$
5. $3^{5x} \cdot 9^{x^2} = 27$
6. $4^x = 7$
7. $2^x = 3^{x-1}$
8. $3^{x+2} = 5^{x-1}$
9. $\log x + \log(x - 9) = 1$
10. $\log x - \log(x + 3) = -1$
11. $\log(x + 9) - \log x = 1$
12. $\log(2x + 1) - \log(x - 2) = 1$
13. $\log_4(x + 3) + \log_4(x - 3) = 2$
14. $\log_8(x + 1) - \log_8 x = \log_8 4$
15. $\log x^2 = (\log x)^2$
16. $(\log_3 x)^2 - \log_3 x^2 = 3$
17. $\log_3(\log_4 x) = 0$
18. $\log(\log x) = 2$
19. Solve for x. (*Hint:* Use $\log_e$ in the last step.)

$$\dfrac{e^x - e^{-x}}{2} = t$$

20. Solve for x.

$$3^x + 3^{-x} = t$$

21. Solve for x.
$$\frac{5^x - 5^{-x}}{5^x + 5^{-x}} = t$$

22. Solve for x.
$$\frac{e^x + e^{-x}}{e^x - e^{-x}} = t$$

23. (*Doubling time*). How many years will it take an investment of $1000 to double itself when interest is compounded annually at 6%?

24. (*Tripling time*). How many years will it take an investment of $1000 to triple itself when interest is compounded annually at 5%?

25. Find the loudness of the sound of an automobile, having an intensity 3,100,000 times I_0.

26. Find the loudness of the sound of a dishwasher, having an intensity 2,500,000 times I_0.

27. Find the loudness of the threshold of sound pain, for which the intensity is 10^{14} times I_0.

28. Find the loudness of a jet aircraft, having an intensity 10^{12} times I_0.

29. The Los Angeles earthquake of 1971 had an intensity $10^{6.7}$ times I_0. What was its magnitude on the Richter scale?

30. The San Francisco earthquake of 1906 had an intensity $10^{8.25}$ times I_0. What was its magnitude on the Richter scale?

31. An earthquake has a magnitude of 5 on the Richter scale. What is its intensity?

32. An earthquake has a magnitude of 7 on the Richter scale. What is its intensity?

33. Students in an industrial mathematics course take a final exam and are then retested at monthly intervals. The forgetting function is given by $S(t) = 82 - 18 \log(t + 1)$.

a) What was the average score originally on the final exam?

b) What was the average score after five months had elapsed?

34. Students graduating from a cosmetology curriculum take a final exam and are then retested at monthly intervals. The forgetting function is given by $S(t) = 75 - 20 \log(t + 1)$.

a) What was the average score originally on the final exam?

b) What was the average score after six months had elapsed?

35. Refer to Exercise 33. How much time will elapse before the average score has decreased to 64?

36. Refer to Exercise 34. How much time will elapse before the average score has decreased to 61?

37. In chemistry, pH is defined as follows:
$$\text{pH} = -\log [\text{H}^+],$$
where $[\text{H}^+]$ is the hydrogen ion concentration in moles per liter. For example, the hydrogen ion concentration in milk is 4×10^{-7} moles per liter, so pH $= -\log(4 \times 10^{-7}) = -[\log 4 + (-7)] \approx 6.4$. For tomatoes, $[\text{H}^+]$ is about 6.3×10^{-5}. Find the pH.

38. For eggs, $[\text{H}^+]$ is about 1.6×10^{-8}. Find the pH.

☆ ─────────────────────────────

Solve.

39. $\log \sqrt{x} = \sqrt{\log x}$

40. $\log_5 \sqrt{x^2 + 1} = 1$

41. $(\log_a x)^{-1} = \log_a x^{-1}$

42. $|\log_5 x| = 2$

43. $\log_3 |x| = 2$

44. $\dfrac{(e^{3x+1})^2}{e^4} = e^{10x}$

45. $\dfrac{\sqrt{(e^{2x} \cdot e^{-5x})^{-4}}}{e^x \div e^{-x}} = e^7$

46. $\log x^{\log x} = 4$

47. Solve $y = ax^n$, for n. Use $\log_x$.

48. Solve $y = ke^{at}$, for t. Use $\log_e$.

49. Solve for t. Use $\log_e$.
$$P = P_0 e^{rt/100}$$

50. Solve for t. Use $\log_e$.
$$I = \frac{E}{R}(1 - e^{-(Rt/L)})$$

51. Solve for t.
$$T = T_0 + (T_1 - T_0)10^{-kt}$$

52. Solve for n. Use $\log_V$.
$$PV^n = c$$

53. Solve for Q.
$$\log_a Q = \frac{1}{3}\log_a y + b$$

54. Solve for y.
$$\log_a y = 2x + \log_a x$$

Solve for x.

55. $x^{\log x} = \dfrac{x^3}{100}$

56. $x^{\log x} = 100x$

★

Solve.

57. $|\log_5 x| + 3\log_5 |x| = 4$

58. $|\log_a x| = \log_a |x|$

59. $(0.5)^x < \dfrac{4}{5}$

60. $8x^{0.3} - 8x^{-0.3} = 63$

61. Solve the system of equations.
$$5^{x+y} = 100,$$
$$3^{2x-y} = 1000$$

62. Given that
$$\log_2[\log_3(\log_4 x)] = \log_3[\log_2(\log_4 y)]$$
$$= \log_4[\log_3(\log_2 z)] = 0,$$
find $x + y + z$.

63. If
$$2\log_3(x - 2y) = \log_3 x + \log_3 y,$$
find $\dfrac{x}{y}$.

64. Find the ordered pair (x, y) for which
$$4^{\log_{16} 27} = 2^x 3^y.$$

7.6 THE NUMBER e; NATURAL LOGARITHMS; CHANGE OF BASE

The Number e

One of the most important numbers is a certain irrational number, with a nonrepeating decimal representation. This number is usually named e, and it arises in a number of different ways, one of which is the following.

Consider the compound interest formula
$$A = p(1 + r)^t,$$

where A is the amount that the investment is worth, p is the principal, r is the rate of interest, and t is the time in years. Interest is usually compounded more often than once a year, say n times per year. In this event we alter the formula. The rate of interest for an *interest period* is r/n and the number of interest periods is $n \cdot t$, for t years. Thus we have

$$A = p\left(1 + \frac{r}{n}\right)^{nt}.$$

Suppose you invest one dollar for one year at 100%. Since $r = 100\%$, or 1, this makes the formula simple:

$$A = \left(1 + \frac{1}{n}\right)^n.$$

The amount A is now a function of n. The more often interest is compounded, the greater A becomes. Let us look at some values of this function in the table below.

Compounding	n	A
Annually	1	$(1 + \frac{1}{1})^1 = 2$
Semiannually	2	$(1 + \frac{1}{2})^2 = 2.25$
Quarterly	4	$(1 + \frac{1}{4})^4 = 2.44\ldots$

According to this table, for semiannual compounding the dollar grows to \$2.25 in one year. For quarterly compounding it grows to \$2.44. What would you expect to happen if interest were compounded daily, or once each minute?

Let us look at some more function values, listed in the table below.

Compounding	n	A
Daily	365	$2.71456\ldots$
Once per hour	8,760	$2.71812\ldots$
Once per minute	525,600	$2.71828\ldots$

This result may be surprising. The function values approach a limit as n increases. No matter how often interest is compounded your investment will not amount to more than \$2.72 for the year. The limit that these function values approach is the number called e:

$$e = 2.718281828459\ldots.$$

Natural Logarithms and Exponential Functions

The exponential function, base e, is important in many applications, and its inverse, the logarithm function base e, is also important in

7.6 THE NUMBER e; NATURAL LOGARITHMS; CHANGE OF BASE

mathematical theory and in applications as well. Logarithms to the base e are called *natural* logarithms. Most scientific calculators have an $\boxed{e^x}$ key. Table 4 at the back of the book gives function values for e^x and e^{-x}. Using a calculator or this table we can construct graphs of $y = e^x$ and $y = \log_e x$ as shown in Fig. 9.

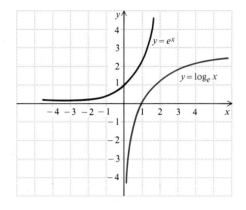

Figure 9

Natural Logarithms

The number $\log_e x$ is abbreviated $\ln x$; that is, $\ln x = \log_e x$. The following is a restatement of the basic theorems of logarithms in terms of natural logarithms.

$\ln xy = \ln x + \ln y$	Theorem 3
$\ln x^p = p \ln x$	Theorem 4
$\ln \dfrac{x}{y} = \ln x - \ln y$	Theorem 5
$\ln e^k = k$	Theorem 2

If you have an $\boxed{\ln}$ key on your calculator, you can find such logarithms directly. For example, to six decimal places,

$$\ln 5.24 = 1.656321, \quad \ln 52.4 = 3.958907, \quad \text{and}$$
$$\ln 0.001277 = -6.663242.$$

From $\ln 5.24$ and $\ln 52.4$ we note again that logarithms with bases other than ten do not have characteristics and mantissas. This is because our numeration system is based on ten.

If you do not have a calculator with a natural logarithm key, you can use Table 7 at the back of the book. Part of Table 7 is shown below. It shows some values of $\ln x$.

x	0.00	0.01	0.02	0.03	0.04	0.05	0.06	0.07	0.08	0.09
5.0	1.6094	1.6114	1.6134	1.6154	1.6174	1.6194	1.6214	1.6233	1.6253	1.6273
5.1	1.6292	1.6312	1.6332	1.6351	1.6371	1.6390	1.6409	1.6429	1.6448	1.6467
5.2	1.6487	1.6506	1.6525	1.6544	1.6563	1.6582	1.6601	1.6620	1.6639	1.6658
5.3	1.6677	1.6696	1.6715	1.6734	1.6752	1.6771	1.6790	1.6808	1.6827	1.6845
5.4	1.6864	1.6882	1.6901	1.6919	1.6938	1.6956	1.6974	1.6993	1.7011	1.7029

Example 1 Find $\ln 5.24$. Use Table 7.

To find $\ln 5.24$, locate the row headed 5.2, then move across to the column headed 0.04. Note the colored number in the table above. Thus

$$\ln 5.24 = 1.6563.$$

We find natural logarithms of numbers not in the table as follows. We first write scientific notation for the number.

Example 2 Find ln 5240. Use Table 7.

$$\begin{aligned}\ln 5240 &= \ln(5.24 \times 10^3) \\ &= \ln 5.24 + \ln 10^3 \quad \text{Theorem 3} \\ &= \ln 5.24 + 3 \ln 10 \quad \text{Theorem 4} \\ &= 1.6563 + 6.9078 \quad \text{Find ln 5.24 in the body of Table 7 and 3 ln 10 at the bottom.} \\ &= 8.5641\end{aligned}$$

Example 3 Find ln 0.000524. Use Table 7.

$$\begin{aligned}\ln 0.000524 &= \ln(5.24 \times 10^{-4}) \\ &= \ln 5.24 + \ln 10^{-4} \quad \text{Theorem 3} \\ &= \ln 5.24 - 4 \ln 10 \quad \text{Theorem 4} \\ &= 1.6563 - 9.2103 \quad \text{Find ln 5.24 in the body of Table 7 and 4 ln 10 at the bottom.} \\ &= -7.5540\end{aligned}$$

Exponential Equations

An equation in which a variable occurs in an exponent is called *exponential*. Logarithms can be used to manipulate or solve exponential equations.

Example 4 Solve: $e^t = 40$.

We could take the common logarithm on both sides as we did in Section 7.5, but taking the natural logarithm makes the simplification on the left easier.

$$\begin{aligned}e^t &= 40 \\ \ln e^t &= \ln 40 \quad \text{Taking the natural log on both sides} \\ t &= \ln 40 \quad \text{Theorem 2} \\ t &\approx 3.7 \quad \text{Use a calculator or Table 7.}\end{aligned}$$

Applications

There are many applications of exponential functions, base e. We consider a few.

Example 5 (*Population growth*). One mathematical model for describing population growth is the formula

$$P = P_0 e^{kt},$$

where P_0 is the number of people at time 0, P is the number of people at time t, and k is a positive constant depending on the situation. See the graph in Fig. 10.

The population of the United States in 1970 was 208 million. In 1980 it was 225 million. Use these data to find the value of k and then use the model to predict the population in 2000.

Time t begins with 1970; that is, $t = 0$ corresponds to 1970, $t = 10$ corresponds to 1980, and so on. Substituting the data into the formula, we get

$$225 = 208 e^{k \cdot 10}.$$

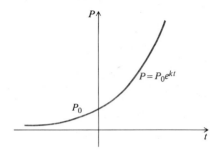

Figure 10

We solve for k:

$\ln 225 = \ln 208 e^{10k}$ Taking the natural logarithm on both sides

$\ln 225 = \ln 208 + \ln e^{10k}$

$\ln 225 = \ln 208 + 10k$

$k = \dfrac{\ln 225 - \ln 208}{10}$

$k = \dfrac{5.4161 - 5.3375}{10}$ Using a calculator or Table 7

$k = 0.008.$

To find the population in the year 2000, we will use $P_0 = 208$ (population in the year 1970). We then have

$$P = 208 e^{0.008 t}.$$

In 2000, t will be 30, so we have

$$P = 208 e^{0.008(30)} = 208 e^{0.24}.$$

Using a calculator or Table 4, we see that $e^{0.24} = 1.2712$. Multiplying by 208 gives us about 264 million. This is our prediction for the population of the United States in the year 2000.

Example 6 (*Radioactive decay*). In a radioactive substance such as radium, some of the atoms are always ceasing to become radioactive. Thus the amount of a radioactive substance decreases. This is called radioactive *decay*. A model for radioactive decay is

$$N = N_0 e^{-kt},$$

where N_0 is the amount of a radioactive substance at time 0, N is the amount at time t, and k is a positive constant depending on the situation. See the graph in Fig. 11.

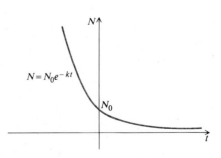

Figure 11

Strontium 90 has a *half-life* of 25 years. This means that half of a sample of the substance will cease to become radioactive in 25 yr. Find k in the formula and then use the formula to find how much of a 36-gram sample will remain after 100 years.

When $t = 25$ (half-life), N will be half of N_0, so we have

$$\tfrac{1}{2}N_0 = N_0 e^{-25k} \quad \text{or} \quad \tfrac{1}{2} = e^{-25k}.$$

We take the natural log on both sides:

$$\ln \frac{1}{2} = \ln e^{-25k} = -25k.$$

Thus

$$k = -\frac{\ln 0.5}{25} \approx 0.0277.$$

Now to find the amount remaining after 100 years, we use the following formula:

$$N = 36e^{-0.0277 \cdot 100}$$
$$= 36e^{-2.77}.$$

From a calculator or Table 4 we find that $e^{-2.8} = 0.0608$, and thus

$$N \approx 2.2 \text{ grams}.$$

Example 7 (*Atmospheric pressure*). Under standard conditions of temperature, the atmospheric pressure at height h is given by

$$P = P_0 e^{-kh},$$

where P is the pressure, P_0 is the pressure where $h = 0$, and k is a positive constant.

Standard sea level pressure is 1013 millibars. Suppose that the pressure at 18,000 ft is half that at sea level. Then find k and find the pressure at 1000 ft.

At 18,000 ft P is half of P_0, so we have

$$\frac{P_0}{2} = P_0 e^{-18,000k}.$$

Taking natural logarithms on both sides, we obtain

$$\ln \tfrac{1}{2} = -18,000k, \quad \text{or} \quad -\ln 2 = -18,000k.$$

Then

$$k = \frac{\ln 2}{18,000} = 3.85 \times 10^{-5}.$$

7.6 THE NUMBER e; NATURAL LOGARITHMS; CHANGE OF BASE

To find the pressure at 1000 ft we use the fact that $P_0 = 1013$ and we let $h = 1000$:

$$P = 1013e^{-3.85 \times 10^{-5} \times 1000}$$
$$= 1013e^{-0.0385}.$$

Rounding the exponent to -0.04 and using a calculator or Table 4, we calculate the pressure to be 973 millibars at 1000 ft.

Logarithm Tables and Change of Base

Tables of common and natural logarithms have been discussed. The following theorem shows how we can change from base 10, base e, or any base to any other base.

THEOREM 7

For any bases a and b and any positive number M,

$$\log_b M = \frac{\log_a M}{\log_a b}.$$

Proof. Let $x = \log_b M$. Then $b^x = M$, so that

$$\log_a M = \log_a b^x, \quad \text{or} \quad x \log_a b.$$

We now solve for x:

$$x = \log_b M = \frac{\log_a M}{\log_a b}.$$

This is the desired formula. □

Example 8 Find $\log_3 81$.

$$\log_3 81 = \frac{\log_{10} 81}{\log_{10} 3} = \frac{1.9085}{0.4771} \approx 4.0$$

Example 9 Find $\log_5 346$.

$$\log_5 346 = \frac{\log_{10} 346}{\log_{10} 5} = \frac{2.5391}{0.6990} \approx 3.6325$$

If, in Theorem 7, we let $a = 10$ and $b = e$, we obtain a formula for changing from common to natural logarithms:

$$\log_e M = \frac{\log_{10} M}{\log_{10} e}.$$

Now $\log_{10} e \approx 0.4343$, so we have:

THEOREM 8

$$\log_e M = \frac{\log_{10} M}{0.4343}$$

Example 10 Find $\log_e 257$.

$$\log_e 257 = \frac{\log_{10} 257}{0.4343} = \frac{2.4099}{0.4343} = 5.5489$$

EXERCISE SET 7.6

Graph.

1. $y = e^{2x}$
2. $y = e^{0.5x}$
3. $y = e^{-2x}$
4. $y = e^{-0.5x}$
5. $y = 1 - e^{-x}$, for $x \geq 0$
6. $y = 2(1 - e^{-x})$, for $x \geq 0$

Find each natural logarithm to four decimal places. Use a calculator or Table 7. Answers have been found using Table 7. If you use a calculator, there may be some variance in the last decimal place.

7. ln 1.88
8. ln 18.8
9. ln 0.0188
10. ln 0.188
11. ln 2.13
12. ln 213
13. ln 0.213
14. ln 0.00213
15. ln 4500
16. ln 81,000
17. ln 0.00056
18. ln 0.999
19. ln 0.08
20. ln 0.0471
21. ln 980,000
22. ln 765,000,000

Solve.

23. $e^t = 100$
24. $e^t = 1000$
25. $e^x = 60$
26. $e^k = 90$
27. $e^{-t} = 0.1$
28. $e^{-t} = 0.01$
29. $e^{-0.02k} = 0.06$
30. $e^{0.07t} = 2$

31. (*Population growth*). The population of Dallas was 680,000 in 1960. In 1969 it was 815,000. Find k in the growth formula and estimate the population in 1990.

32. (*Population growth*). The population of Kansas City was 475,000 in 1960. In 1970 it was 507,000. Find k in the growth formula and estimate the population in 2000.

33. (*Radioactive decay*). The half-life of polonium is 3 minutes. After 30 minutes, how much of a 410-gram sample will remain radioactive?

34. (*Radioactive decay*). The half-life of a lead isotope is 22 yr. After 66 yr, how much of a 1000-gram sample will remain radioactive?

35. (*Radioactive decay*). A certain radioactive substance decays from 66,560 grams to 6.5 grams in 16 days. What is its half-life?

36. Ten grams of uranium will decay to 2.5 grams in 496,000 years. What is its half-life?

7.6 THE NUMBER e; NATURAL LOGARITHMS; CHANGE OF BASE

37. (*Radiocarbon dating*). Carbon-14, an isotope of carbon, has a half-life of 5750 years. Organic objects contain carbon-14, as well as nonradioactive carbon, in known proportions. When a living organism dies, it takes in no more carbon. The carbon-14 decays, thus changing the proportions of the kinds of carbon in the organism. By determining the amount of carbon-14 it is possible to determine how long the organism has been dead, hence how old it is.

a) How old is an animal bone that has lost 30% of its carbon-14?

b) A mummy discovered in the pyramid Khufu in Egypt had lost 46% of its carbon-14. Determine its age.

38. (*Radiocarbon dating*).

a) How old is an animal bone that has lost 20% of its carbon-14?

b) The Statue of Zeus at Olympia in Greece is one of the seven wonders of the world. It is made of gold and ivory. The ivory was found to have lost 35% of its carbon-14. Determine the age of the statue.

39. (*Atmospheric pressure*). What is the pressure at the top of Mt. Shasta in California, 14,162 ft high?

40. (*Atmospheric pressure*). Blood will boil when atmospheric pressure goes below about 62 millibars. At what altitude, in an unpressurized vehicle, will a pilot's blood boil?

41. (*Consumer price index*). The *consumer price index* compares the costs of goods and services over various years. The base year is 1967. The same goods and services that cost $100 ($P_0$) in 1967 cost $184.50 in 1977. Assume the exponential model.

a) Find the value k and write the equation.

b) Estimate what the same goods and services will cost in 1987.

c) When did the same goods and services cost double those of 1967?

42. (*Cost of a double-dip ice cream cone*). In 1970 the cost of a double-dip ice cream cone was 52¢. In 1978 it was 66¢. Assume the exponential model.

a) Find the value k and write the equation.

b) Estimate the cost of a cone in 1986.

c) When will the cost of a cone be twice that of 1978?

Find.

43. $\log_4 20$ **44.** $\log_8 0.99$ **45.** $\log_5 0.78$ **46.** $\log_{12} 15{,}000$

47. $\log_e 12$ (Do not use Table 7.) **48.** $\log_e 0.77$ (Do not use Table 7.)

☆

Solve for t.

49. $P = P_0 e^{kt}$ **50.** $P = P_0 e^{-kt}$

Verify each of the following.

51. $\ln x = \dfrac{\log x}{\log e} \approx 2.3026 \log x$

52. $\log x = \dfrac{\ln x}{\ln 10} \approx 0.4343 \ln x$

53. ▦ Which is larger, e^π or π^e?

54. ▦ Which is larger, $e^{\sqrt{\pi}}$ or $\sqrt{e^\pi}$?

55. ▦ Given $f(x) = (1 + x)^{1/x}$, find $f(1)$, $f(0.5)$, $f(0.2)$, $f(0.1)$, $f(0.01)$, and $f(0.001)$ to six decimal places. This sequence of numbers approaches the number e.

56. ▦ Given $f(t) = t^{1/(t-1)}$, find $f(0.5)$, $f(0.9)$, $f(0.99)$, $f(0.999)$, and $f(0.9999)$ to six decimal places. This sequence of numbers approaches the number e.

57. Find a (simple) formula for radioactive decay involving H, the half-life.

58. The time required for a population to double is called the *doubling time*.
   a) Find an expression relating T, the doubling time, to k, the constant in the growth equation.
   b) Find a (simple) formula for population growth involving T.

Prove the following for any logarithm bases a and b.

59. $\log_a b = \dfrac{1}{\log_b a}$

60. $a^{(\log_b M) \div (\log_b a)} = M$

61. $a^{(\log_b M)(\log_b a)} = M^{(\log_b a)^2}$

62. $\log_a (\log_a x) = \log_a (\log_b x) - \log_a (\log_b a)$

CHAPTER 7 REVIEW

Graph.

1. $y = \log_2 (x - 1)$
2. $y = e^{0.4x}$
3. $y = e^{-x} - 3$

4. Write an exponential equation equivalent to $\log_8 \frac{1}{4} = -\frac{2}{3}$.

5. Write a logarithmic equation equivalent to $7^{2.3} = x$.

Solve.

6. $\log_x 64 = 3$
7. $\log_{16} 4 = x$

8. Write an equivalent expression containing a single logarithm.

$$\frac{1}{2} \log_b a + \frac{3}{2} \log_b c - 4 \log_b d$$

9. Solve for x: $\log_a a^{x^2-x-4} = 2$.

Given that $\log_a 2 = 0.301$, $\log_a 3 = 0.477$, and $\log_a 7 = 0.845$, find:

10. $\log_a 18$.
11. $\log_a \dfrac{7}{2}$.
12. $\log_a \dfrac{1}{4}$.
13. $\log_a \sqrt{3}$.

14. Express in terms of logarithms of M and N: $\log \sqrt[3]{M^2/N}$.

Using Table 2, find:

15. $\log 26.3$.
16. $\log 0.00806$.
17. $10^{1.8686}$.
18. $\log_5 290$ (round to the nearest tenth).
19. antilog $(8.4409 - 10)$.
20. Use logarithms to compute: $(0.0524)^2 \cdot \sqrt{0.0638}$.

Using Table 7, find:

21. $\ln 8.9$.   **22.** $\ln 560$.   **23.** $\ln 0.0462$.

24. Simplify: $\log_{12} 12^{x^2+1}$.

Solve.

25. $3^{1-x} = 9^{2x}$

26. $\log(x^2 - 1) - \log(x - 1) = 1$

27. How many years will it take an investment of $1000 to double if interest is compounded annually at 13%?

28. What is the loudness, in decibels, of a sound whose intensity is 1000 times I_0?

29. The half-life of a radioactive substance is 15 days. How much of a 25-gram sample will remain radioactive after 30 days?

Solve.

30. $|\log_4 x| = 3$

31. $\log x^2 = \log x$

32. $\log_2(x - 1) + \log_2(x + 1) = 3$

33. $\log 2 + 2 \log x = \log(5x + 3)$

34. $\log x = \ln x$

35. Graph: $y = |\log_3 x|$.

36. Graph: $y = |e^x - 4|$.

Find the domain.

37. $f(x) = \dfrac{17}{\sqrt{5 \ln x - 6}}$

38. $f(x) = \dfrac{8}{e^{4x} - 10}$

The Circular Functions

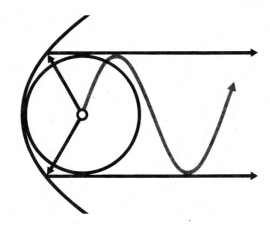

8.1 TRIANGLES*

In this chapter we consider an important class of functions called *trigonometric* functions, or *circular* functions. They are based, historically, on certain properties of triangles. In this section we consider some of those properties. We focus our attention on right triangles.

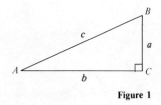

Figure 1

The angles and sides of a right triangle are often labeled as shown in Fig. 1, with $\angle C$ a right angle and c the length of the hypotenuse. Side a is opposite $\angle A$ and side b is opposite $\angle B$. In any triangle the sum of the angle measures is 180°. Thus in a right triangle the measures of the two acute angles add up to 90°.

Example 1 In the triangle in Fig. 2, find $m \angle B$ (the measure of $\angle B$).

We know that $m \angle A + m \angle B = 90°$ and that $m \angle A = 32°$, so

$$m \angle B = 90° - 32° = 58°.$$

Figure 2

Similar Triangles

Triangles are similar if their corresponding angles have the same measure. In the triangles in Fig. 3, if we know that $\angle A$ and $\angle A'$ are the same size then we know that *all* corresponding angles have the same measure, hence the triangles are similar.

In similar triangles corresponding sides are in the same ratio (are proportional). For the triangles in Fig. 3, this means the following:

$$\frac{a}{a'} = \frac{b}{b'} = \frac{c}{c'}.$$

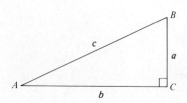

Figure 3

Similar triangles can be used to determine distances without measuring directly.

Example 2 Find the height of the flagpole using a rod and a measuring device as shown in Fig. 4.

Place the rod of length r so that the top of its shadow coincides with the shadow of the top of the flagpole. We know that we have similar triangles, because the acute angle θ at the ground is the same in both trian-

---

*The reader who wishes to approach trigonometry from the functional point of view first, and then consider triangles, can skip this section. The reader who wishes a brief introduction to triangles before studying the functions can read this section and then proceed.

There are several paths through the trigonometry. For example, it is possible to treat triangles thoroughly before delving deeply into analytic trigonometry. For a thorough discussion of this, see the preface.

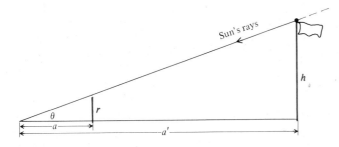

Figure 4

gles. We measure and find that $a = 3$ ft, that $a' = 30$ ft, and that $r = 2$ ft. Now $h/r = a'/a$ by similar triangles, so

$$h = r\frac{a'}{a} = 2\left(\frac{30}{3}\right) = 20 \text{ ft.}$$

Trigonometric Ratios

In a right triangle as shown in Fig. 5, the ratio a/c of the side *opposite* θ to the hypotenuse depends on the size of θ. In other words, this ratio is a *function* of θ.

This function is called the *sine* function. Another such function, called the *cosine* function, is the ratio b/c, the ratio of the side *adjacent* to θ to the hypotenuse. There are other such ratios, or trigonometric functions, defined as follows.

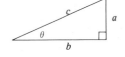

Figure 5

FUNCTION	ABBREVIATION FOR FUNCTION VALUE	DEFINING RATIO
sine	$\sin(\theta)$* or $\sin\theta$	$\dfrac{\text{side opposite }\theta}{\text{hypotenuse}}$
cosine	$\cos(\theta)$ or $\cos\theta$	$\dfrac{\text{side adjacent }\theta}{\text{hypotenuse}}$
tangent	$\tan(\theta)$ or $\tan\theta$	$\dfrac{\text{side opposite }\theta}{\text{side adjacent }\theta}$
cotangent	$\cot(\theta)$ or $\cot\theta$	$\dfrac{\text{side adjacent }\theta}{\text{side opposite }\theta}$
secant	$\sec(\theta)$ or $\sec\theta$	$\dfrac{\text{hypotenuse}}{\text{side adjacent }\theta}$
cosecant	$\csc(\theta)$ or $\csc\theta$	$\dfrac{\text{hypotenuse}}{\text{side opposite }\theta}$

*For function values we generally write $f(x)$. For these functions we often omit parentheses, as in $\sin\theta$.

Figure 6

Note that the function values do not depend on the size of a triangle, but only on the size of the angle θ.

Example 3 In the triangle in Fig. 6, find (a) the trigonometric function values for θ; (b) the trigonometric function values for ϕ.

$$\sin \theta = \frac{\text{side opposite } \theta}{\text{hypotenuse}} = \frac{3}{5}, \qquad \sin \phi = \frac{\text{side opposite } \phi}{\text{hypotenuse}} = \frac{4}{5},$$

$$\cos \theta = \frac{\text{side adjacent } \theta}{\text{hypotenuse}} = \frac{4}{5}, \qquad \cos \phi = \frac{\text{side adjacent } \phi}{\text{hypotenuse}} = \frac{3}{5},$$

$$\tan \theta = \frac{\text{side opposite } \theta}{\text{side adjacent } \theta} = \frac{3}{4}, \qquad \tan \phi = \frac{\text{side opposite } \phi}{\text{side adjacent } \phi} = \frac{4}{3},$$

$$\cot \theta = \frac{\text{side adjacent } \theta}{\text{side opposite } \theta} = \frac{4}{3}, \qquad \cot \phi = \frac{\text{side adjacent } \phi}{\text{side opposite } \phi} = \frac{3}{4},$$

$$\sec \theta = \frac{\text{hypotenuse}}{\text{side adjacent } \theta} = \frac{5}{4}, \qquad \sec \phi = \frac{\text{hypotenuse}}{\text{side adjacent } \phi} = \frac{5}{3},$$

$$\csc \theta = \frac{\text{hypotenuse}}{\text{side opposite } \theta} = \frac{5}{3}, \qquad \csc \phi = \frac{\text{hypotenuse}}{\text{side opposite } \phi} = \frac{5}{4}.$$

Values for the trigonometric functions have been worked out in great detail and are given in tables and by some calculators. When these values are available we can use them to help determine distances or angles without actually measuring them directly.

Example 4 An observer stands on level ground, 200 meters from the base of a TV tower, and looks up at an angle of 26.5° to see the top of the tower. How high is the tower above the observer's eye level?

We draw a diagram (Fig. 7) and see that a right triangle is formed. We use one of the trigonometric functions. This time the tangent function is most convenient. From the definition of the tangent function, we have

Figure 7

$$\frac{h}{200} = \tan 26.5°.$$

Then $h = 200 \tan 26.5°$. We find, from a table, that $\tan 26.5° = 0.499$, approximately. Thus $h = 200 \times 0.499 = 99.8$ m.

Calculators

Scientific calculators contain tables of trigonometric functions. It is recommended that students not use them until after the trigonometric tables have been studied (Section 9.5). In the meantime, calculators will be most useful for arithmetic calculations.

Example 5 A kite flies at a height of 60 ft when 130 ft of string is out. If we assume that the string is in a straight line, what is the angle that it makes with the ground?

Figure 8

We draw a diagram (Fig. 8) and then use the most convenient trigonometric function. A portion of a sine table is given below.

sin 26.5° = 0.4462	sin 28.5° = 0.4772
sin 27° = 0.4540	sin 29° = 0.4848
sin 27.5° = 0.4617	sin 29.5° = 0.4924
sin 28° = 0.4695	sin 30° = 0.5000

From the definition of the sine function, we have
$$\frac{60}{130} = \sin \theta = 0.4615.$$
From the table, we find that θ is about 27.5°.

---

1. Find $m \angle B$.

2. Find $m \angle A$.

3. Find b.

4. Find c.

5. Find the trigonometric function values for θ.

6. Find the trigonometric function values for ϕ.

7. 🖩 Find the trigonometric function values for θ. Round to four decimal places.

8. 🖩 Find the trigonometric function values for ϕ. Round to four decimal places.

296 THE CIRCULAR FUNCTIONS

For Exercises 9–12, use the following function values.

sin 36° = 0.5878	cos 36° = 0.8090	tan 36° = 0.7265	cot 36° = 1.376
sin 36.5° = 0.5948	cos 36.5° = 0.8039	tan 36.5° = 0.7400	cot 36.5° = 1.351
sin 37° = 0.6018	cos 37° = 0.7986	tan 37° = 0.7536	cot 37° = 1.327
sin 37.5° = 0.6088	cos 37.5° = 0.7934	tan 37.5° = 0.7673	cot 37.5° = 1.303
sin 38° = 0.6157	cos 38° = 0.7880	tan 38° = 0.7813	cot 38° = 1.280

9. a) Find ∠B.*
b) Find a.
c) Find b.

10. a) Find ∠B.
b) Find a.
c) Find b.

11. a) Find ∠A.
b) Find a.
c) Find c.

12. a) Find ∠A.
b) Find a.
c) Find c.

☆ ──────────────────────────

13. Find h.

14. Find a.

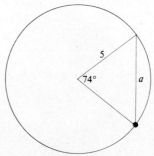

15. A 45° right triangle is half of a square. Use the Pythagorean theorem to find the length of a diagonal and then find:
a) sin 45°;
b) cos 45°;
c) tan 45°.

───────────────
*Instead of saying "find m ∠B," we often say "find ∠B."

16. A 30°-60° right triangle is half of an equilateral triangle. Using this fact, find:
   a) sin 30°;
   b) cos 30°;
   c) sin 60°;
   d) cos 60°.

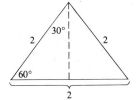

8.2 THE UNIT CIRCLE

To develop the important circular functions we will use a circle with a radius of length 1. Such a circle is called a *unit circle*. When its center is at the origin of a coordinate system, it has an equation

$$u^2 + v^2 = 1.$$

(We use u and v for the variables here because we wish to reserve x and y for other purposes later.)

Any circle centered at the origin is symmetric with respect to the u-axis, the v-axis, and the origin.

Example 1 Suppose the point $(1, 2)$ is on a circle centered at the origin. Find its reflection across the u-axis, the v-axis, and the origin.

The reflection across the u-axis is $(1, -2)$. (See Fig. 9.) The reflection across the v-axis is $(-1, 2)$. The reflection across the origin is $(-1, -2)$. By symmetry, all these points are on the circle.

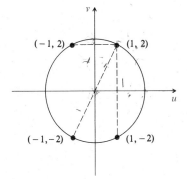

Figure 9

The intercepts (points of intersection with the axes) of the unit circle A, B, C, and D have coordinates as shown in Fig. 10. Let us find the coordinates of point E, which is the intersection of the circle with the line $u = v$. Note that this point divides the arc in the first quadrant in half.

Since for any point (u, v) on the circle $u^2 + v^2 = 1$, and for point E we have $u = v$, we know that the coordinates of E satisfy $u^2 + u^2 = 1$ (substituting u for v in the equation of the circle). Let us solve this equation:

$$u^2 + u^2 = 1, \quad \text{or} \quad 2u^2 = 1$$

$$u^2 = \frac{1}{2}$$

$$u = \sqrt{\frac{1}{2}}, \quad \text{or} \quad \frac{\sqrt{1}}{\sqrt{2}}, \quad \text{or} \quad \frac{1}{\sqrt{2}}.$$

Figure 10

Figure 11

Figure 12

Now
$$\frac{1}{\sqrt{2}} = \frac{1}{\sqrt{2}} \cdot \frac{\sqrt{2}}{\sqrt{2}} = \frac{\sqrt{2}}{2}.$$

Thus the coordinates of point E are $\left(\frac{\sqrt{2}}{2}, \frac{\sqrt{2}}{2}\right)$.

Distances on the Unit Circle

The circumference of a circle of radius r is $2\pi r$. Thus for the unit circle the circumference is 2π. If a point starts at A and travels counterclockwise around the circle, it will travel a distance of 2π (Fig. 11). If it travels halfway around the circle, it will travel a distance of π (Fig. 12).

If a point travels $\frac{1}{8}$ of the way around the circle, it will travel a distance of $(\frac{1}{8}) \cdot 2\pi$, or $\pi/4$. Note that the stopping point is $\frac{1}{4}$ of the way from A to C and $\frac{1}{2}$ of the way from A to B (Fig. 13).

If a point travels $\frac{1}{6}$ of the way around the circle ($\frac{2}{12}$ of the way, as shown in Fig. 14), it will travel a distance of $(\frac{1}{6}) \cdot 2\pi$, or $\pi/3$. Note that the stopping point is $\frac{1}{3}$ of the way from A to C and $\frac{2}{3}$ of the way from A to B.

A point may travel completely around the circle and then continue. For example, if it goes around once and then continues $\frac{1}{4}$ of the way around, it will have traveled a distance of $2\pi + (\pi/2)$, or $(\frac{5}{2})\pi$ (Fig. 15).

Any positive number thus determines a point on the unit circle. For the number 35, for example, we start at A and travel counterclockwise a distance of 35. The point at which we stop is the point "determined" by the number 35.

Negative numbers also determine points on the unit circle. For a negative number, we move clockwise around the circle. Points for $-\pi/4$ and $-3\pi/2$ are shown in Fig. 16. Any negative number thus determines a point on the unit circle. The same is true for any positive

Figure 13

Figure 14

Figure 15

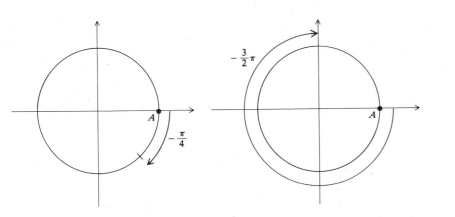

Figure 16

Figure 17

number. The number 0 determines the point A. Hence there is a point on the unit circle for every real number.

Points on the unit circle determined by two numbers that are additive inverses are situated as in Fig. 17. The points are reflections of each other across the u-axis. Thus their first coordinates are the same and their second coordinates are additive inverses of each other.

EXERCISE SET 8.2

1. The point $(-5, 2)$ is on a circle centered at the origin. Find the coordinates of its reflection across (a) the u-axis, (b) the v-axis, and (c) the origin. (d) Are these points on the circle? Why?

2. The point $(3, -5)$ is on a circle centered at the origin. Find the coordinates of its reflection across (a) the u-axis, (b) the v-axis, and (c) the origin. (d) Are these points on the circle? Why?

3. Find the coordinates of these points on a unit circle.

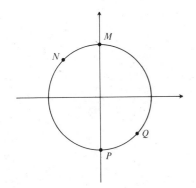

4. Find the coordinates of these points on a unit circle.

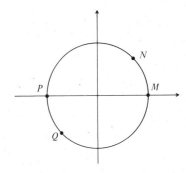

For each of Exercises 5–10, sketch a unit circle such as the one shown below.

5. Mark the points determined by (a) $\frac{\pi}{4}$, (b) $\frac{3}{2}\pi$, (c) $\frac{3}{4}\pi$, (d) π, (e) $\frac{11}{4}\pi$, and (f) $\frac{17}{4}\pi$.

6. Mark the points determined by (a) $\frac{\pi}{2}$, (b) $\frac{5}{4}\pi$, (c) 2π, (d) $\frac{9}{4}\pi$, (e) $\frac{13}{4}\pi$, and (f) $\frac{23}{4}\pi$.

7. Mark the points determined by (a) $\frac{\pi}{6}$, (b) $\frac{2}{3}\pi$, (c) $\frac{7}{6}\pi$, (d) $\frac{10}{6}\pi$, (e) $\frac{14}{6}\pi$, and (f) $\frac{23}{4}\pi$.

8. Mark the points determined by (a) $\frac{\pi}{3}$, (b) $\frac{5}{6}\pi$, (c) $\frac{11}{6}\pi$, (d) $\frac{13}{6}\pi$, (e) $\frac{23}{6}\pi$, and (f) $\frac{33}{6}\pi$.

9. Mark the points determined by (a) $-\frac{\pi}{2}$, (b) $-\frac{3}{4}\pi$, (c) $-\frac{5}{6}\pi$, (d) $-\frac{5}{2}\pi$, (e) $-\frac{17}{6}\pi$, and (f) $-\frac{9}{4}\pi$.

10. Mark the points determined by (a) $-\frac{3}{2}\pi$, (b) $-\frac{\pi}{3}$, (c) $-\frac{7}{6}\pi$, (d) $-\frac{13}{6}\pi$, (e) $-\frac{19}{6}\pi$, and (f) $-\frac{11}{4}\pi$.

11. Find two real numbers between -2π and 2π that determine each of these points.

12. Find two real numbers between -2π and 2π that determine each of these points.

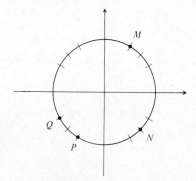

13. For the points in Exercise 11, find a real number between 2π and 4π that determines each point.

14. For the points in Exercise 12, find a real number between -2π and -4π that determines each point.

15. A point on the unit circle has u-coordinate $-\frac{1}{2}$. What is its v-coordinate?

16. A point on the unit circle has v-coordinate $-\sqrt{3}/2$. What is its u-coordinate?

17. The number $\pi/6$ determines a point on the unit circle with coordinates
$$\left(\frac{\sqrt{3}}{2}, \frac{1}{2}\right).$$
What are the coordinates of the point determined by $-\pi/6$?

18. The number $\pi/3$ determines a point on the unit circle with coordinates
$$\left(\frac{1}{2}, \frac{\sqrt{3}}{2}\right).$$
What are the coordinates of the point determined by $-\pi/3$?

19. ▨ A point on the unit circle has u-coordinate 0.25671. What is its v-coordinate?

20. ▨ A point on the unit circle has v-coordinate −0.88041. What is its u-coordinate?

8.3 THE SINE AND COSINE FUNCTIONS

The Sine Function

Every real number s determines a point T on the unit circle, as shown. To the real number s we assign the second coordinate of point T as a function value. The function thus defined is called the *sine function*. Function values are denoted by "sine s" or, more briefly, "sin s." *
By considering the unit circle, we see that

$$\sin 0 = 0,$$

$$\sin \frac{\pi}{2} = 1, \quad \text{(Fig. 18b)}$$

$$\sin \frac{\pi}{4} = \sin \frac{3}{4}\pi = \frac{\sqrt{2}}{2}. \quad \text{(Fig. 18c)}$$

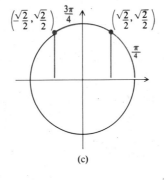

Figure 18

The sine function also has the following values, as we shall later verify:

$$\sin \frac{\pi}{6} = \frac{1}{2}, \quad \sin \frac{\pi}{3} = \frac{\sqrt{3}}{2}.$$

---

*The sine function is a generalization of what was done in Section 8.1. To help visualize this, we could draw a triangle as shown here. Its hypotenuse has length 1, so $\sin \theta = v/1$, or v. Note also that the arc length s determines the angle θ, so we think of sin s rather than sin θ.

The sine function

Figure 19

Figure 20

Figure 21

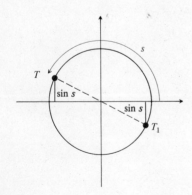

Figure 22

Function values increase to a maximum of 1 at $\pi/2$, then decrease to 0 at π, decrease further to -1 at $(\frac{3}{2})\pi$, then increase to 0 at 2π, and so on. The graph of the sine function is shown in Fig. 19.

From the graph of the sine function, certain properties are apparent.

1. **The sine function is periodic, with period 2π.**
2. **It is an odd function, by symmetry with respect to the origin. Thus we know that $\sin(-s) = -\sin s$ for all real numbers s.**
3. **It is continuous everywhere.**
4. **The domain is the set of all real numbers. The range is the set of all real numbers from -1 to 1, inclusive.**

The unit circle can be used to verify some of the above, as well as to determine other important properties of the sine function.

Let us verify that the period is 2π. Consider any real number s and its point T on the unit circle (Fig. 20). Now we increase s by 2π. The point for $s + 2\pi$ is again the point T. Hence for any real number s, $\sin s = \sin(s + 2\pi)$. There is no number smaller than 2π for which this is true. Hence the period is 2π. Actually, we know that $\sin s = \sin(s + 2k\pi)$, where k is any integer.

Consider any real number s and its additive inverse $-s$. These numbers determine points on the unit circle that are symmetric with respect to the u-axis (Fig. 21), hence their second coordinates are additive inverses of each other. Thus we know for any number s that $\sin(-s) = -\sin s$, so the sine function is odd.

Consider any real number s and its point T on the unit circle. Then $s + \pi$ and $s - \pi$ both determine a point T_1 symmetric with T with respect to the origin (Fig. 22). Thus the sines of $s + \pi$ and $s - \pi$ are the same, and are the additive inverse of $\sin s$. This gives us another important property. For any real number s, $\sin(s + \pi) = -\sin s$, and $\sin(s - \pi) = -\sin s$.

Let us add to this list a property concerning $\sin(\pi - s)$. Since $\pi - s$ is the additive inverse of $s - \pi$, $\sin(\pi - s) = \sin[-(s - \pi)]$. Since the sine function is odd, this is equal to $-\sin(s - \pi)$. By the second property in the preceding paragraph, this is equal to $-(-\sin s)$ or $\sin s$. Thus our last property is

$$\sin(\pi - s) = \sin s.$$

Equations that hold for all sensible replacements for the variables are known as *identities*. We shall use the sign $\equiv$ instead of $=$ to indicate that an equation is an identity. We now have the following identities concerning the sine function.

$$\sin s \equiv \sin(s + 2k\pi), \quad k \text{ any integer}$$
$$\sin(-s) \equiv -\sin s$$
$$\sin(s \pm \pi) \equiv -\sin s$$
$$\sin(\pi - s) \equiv \sin s$$

The *amplitude* of a periodic function is half the difference between its maximum and minimum function values. It is always positive. The maximum value of the sine function can be seen to be 1, either from the graph or the unit circle, while the minimum value is -1. Thus the amplitude of the sine function is 1.

The Cosine Function

Every real number s determines a point T on the unit circle. To the real number s we assign the first coordinate of point T as a function value. This defines the function known as the *cosine function*. Function values are denoted by "cosine s" or, more briefly, "$\cos s$."* By considering the unit circle, we see that

$$\cos 0 = 1, \quad \text{(Fig. 23b)}$$
$$\cos \frac{\pi}{2} = 0, \quad \text{(Fig. 23b)}$$
$$\cos \frac{\pi}{4} = \frac{\sqrt{2}}{2}, \quad \text{(Fig. 23c)}$$
$$\cos \frac{3}{4}\pi = -\frac{\sqrt{2}}{2}. \quad \text{(Fig. 23d)}$$

Figure 23

(a)

(b)

(c)   (d)

---

*The cosine function is also a generalization of what was done in Section 8.1. The triangle here has a hypotenuse of length 1, so $\cos \theta = u/1$, or u. Since the arc length s determines the angle θ, we think of $\cos s$ rather than $\cos \theta$.

The cosine function also has the following values, as we shall later verify:

$$\cos \frac{\pi}{6} = \frac{\sqrt{3}}{2},$$

$$\cos \frac{\pi}{3} = \frac{1}{2}.$$

It should be noted that for any point on the unit circle, the coordinates are (cos s, sin s).

The graph of the cosine function is shown in Fig. 24. From the graph, certain properties are apparent.

The cosine function

Figure 24

1. **The cosine function is periodic, with period 2π.**
2. **It is an even function, by symmetry with respect to the y-axis. Thus we know that $\cos(-s) = \cos s$ for all real numbers s.**
3. **It is continuous everywhere.**
4. **The domain is the set of all real numbers. The range is the set of all real numbers from -1 to 1, inclusive.**
5. **The amplitude is 1.**

The unit circle can be used to verify that the period is 2π and, in fact, that $\cos s = \cos(s + 2k\pi)$ for any integer k.

Consider any real number s and its additive inverse. They determine points symmetric with respect to the u-axis, hence they have the same first coordinates (see Fig. 25). Thus we know that for any number s, $\cos(-s) = \cos s$, and that the cosine function is even.

Let us consider $\cos s$ and $\cos(s \pm \pi)$. Recall that s and $s \pm \pi$ determine points symmetric with respect to the origin. Thus their coordinates are additive inverses of each other. This verifies that $\cos(s \pm \pi) = -\cos s$ for all real numbers s.

Let us consider $\cos(\pi - s)$. This is equal to $\cos[-(s - \pi)]$. Since the cosine function is even, this is equal to $\cos(s - \pi)$, and by the above this is in turn equal to $-\cos s$. Hence we have $\cos(\pi - s) =$

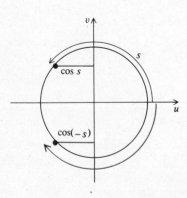

Figure 25

$-\cos s$. We now list the identities that we have developed for the cosine function.

$$\cos s \equiv \cos (s + 2k\pi), \quad k \text{ any integer}$$
$$\cos (-s) \equiv \cos s$$
$$\cos (s \pm \pi) \equiv -\cos s$$
$$\cos (\pi - s) \equiv -\cos s$$

Instead of the letter s, we can use any letter. We usually use x, as is the custom with most functions.

EXERCISE SET 8.3

1. Construct a graph of the sine function. The lower set of axes is shown here at half size, so on other paper, copy the axes at twice the size shown below. Then from the unit circle (shown full size here) transfer vertical distances with a compass.

2. Construct a graph of the cosine function. Follow the instructions for Exercise 1, except that horizontal distances on the unit circle, rather than vertical distances, will be transferred with a compass.

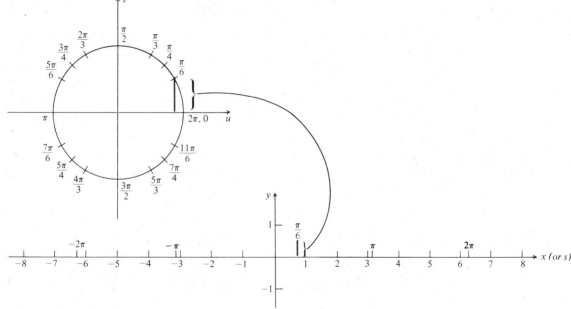

3. a) Sketch a graph of $y = \sin x$.
b) By reflecting the graph in (a), sketch a graph of $y = \sin (-x)$.
c) By reflecting the graph in (a), sketch a graph of $y = -\sin x$.
d) How do the graphs in (b) and (c) compare?

4. a) Sketch a graph of $y = \cos x$.
b) By reflecting the graph in (a), sketch a graph of $y = \cos (-x)$.
c) By reflecting the graph in (a), sketch a graph of $y = -\cos x$.
d) How do the graphs in (a) and (b) compare?

5. a) Sketch a graph of $y = \sin x$.
   b) By translating, sketch a graph of $y = \sin(x + \pi)$.
   c) By reflecting the graph of (a), sketch a graph of $y = -\sin x$.
   d) How do the graphs of (b) and (c) compare?

6. a) Sketch a graph of $y = \sin x$.
   b) By translating, sketch a graph of $y = \sin(x - \pi)$.
   c) By reflecting the graph of (a), sketch a graph of $y = -\sin x$.
   d) How do the graphs of (b) and (c) compare?

7. a) Sketch a graph of $y = \cos x$.
   b) By translating, sketch a graph of $y = \cos(x + \pi)$.
   c) By reflecting the graph of (a), sketch a graph of $y = -\cos x$.
   d) How do the graphs in (b) and (c) compare?

8. a) Sketch a graph of $y = \cos x$.
   b) By translating, sketch a graph of $y = \cos(x - \pi)$.
   c) By reflecting the graph of (a), sketch a graph of $y = -\cos x$.
   d) How do the graphs in (b) and (c) compare?

Complete.

9. $\cos(-x) \equiv$ _____

10. $\sin(-x) \equiv$ _____

11. $\sin(x + \pi) \equiv$ _____

12. $\sin(x - \pi) \equiv$ _____

13. $\cos(\pi - x) \equiv$ _____

14. $\sin(\pi - x) \equiv$ _____

15. $\cos(x + 2k\pi) \equiv$ _____

16. $\sin(x + 2k\pi) \equiv$ _____

17. $\cos(x - \pi) \equiv$ _____

18. $\cos(x + \pi) \equiv$ _____

19. Find: a) $\sin\frac{\pi}{4}$; b) $\sin\pi$; c) $\sin\frac{\pi}{6}$; d) $\sin\frac{5}{4}\pi$; e) $\sin\frac{3}{2}\pi$; f) $\sin\frac{5}{6}\pi$; g) $\sin\left(-\frac{\pi}{4}\right)$; h) $\sin(-\pi)$; j) $\sin\left(-\frac{5}{4}\pi\right)$; k) $\sin\left(-\frac{3}{2}\pi\right)$; m) $\sin\left(-\frac{\pi}{3}\right)$.

20. Find: a) $\sin\frac{\pi}{2}$; b) $\sin\frac{\pi}{3}$; c) $\sin\frac{3}{4}\pi$; d) $\sin\frac{7}{4}\pi$; e) $\sin 2\pi$; f) $\sin\left(-\frac{\pi}{2}\right)$; g) $\sin\left(-\frac{3}{4}\pi\right)$; h) $\sin\left(-\frac{7}{4}\pi\right)$; j) $\sin(-2\pi)$; k) $\sin\left(-\frac{2}{3}\pi\right)$; m) $\sin\left(-\frac{\pi}{6}\right)$.

21. Find: a) $\cos\frac{\pi}{4}$; b) $\cos\pi$; c) $\cos\frac{5}{4}\pi$; d) $\cos\frac{\pi}{6}$; e) $\cos\frac{3}{2}\pi$; f) $\cos\frac{5}{6}\pi$; g) $\cos\left(-\frac{\pi}{4}\right)$; h) $\cos(-\pi)$; j) $\cos\left(-\frac{5}{4}\pi\right)$; k) $\cos\left(-\frac{3}{2}\pi\right)$; m) $\cos\left(-\frac{\pi}{3}\right)$.

22. Find: a) $\cos\frac{\pi}{2}$; b) $\cos\frac{\pi}{3}$; c) $\cos\frac{3}{4}\pi$; d) $\cos\frac{7}{4}\pi$; e) $\cos 2\pi$; f) $\cos\left(-\frac{\pi}{2}\right)$; g) $\cos\left(-\frac{3}{4}\pi\right)$; h) $\cos\left(-\frac{\pi}{6}\right)$; j) $\cos\left(-\frac{7}{4}\pi\right)$; k) $\cos\left(-\frac{2}{3}\pi\right)$.

☆ ─────────────────────────────

23. For which numbers is
    a) $\sin x = 1$?
    b) $\sin x = -1$?

24. For which numbers is
    a) $\cos x = 1$?
    b) $\cos x = -1$?

25. Solve for x: $\sin x = 0$.

26. Solve for x: $\cos x = 0$.

Graph.

27. $y = 3 \sin x$

28. $y = \sin x + \cos x$

What are the domain, range, period, and amplitude of each function?

29. $y = \sin^2 x$

30. $y = |\cos x| + 1$

31. Consider $(\sin x)/x$ when x is between 0 and $\pi/2$. What can you say about this function? As x approaches 0, this function approaches a limit. What is it?

32. Find $f \circ g$ and $g \circ f$, where $f(x) = x^2 + 2x$ and $g(x) = \cos x$.

33. Solve, in the interval $[-\pi, \pi]$: $\sin x < \cos x$.

Which of the following have solutions for x? Why?

34. $8 \sin x = 10$

35. $3 \cos x = -2$

36. ▦ Calculate the following. Do not use the trigonometric function keys.

 a) $\sin \dfrac{\pi}{4}$ b) $\sin \dfrac{\pi}{3}$

37. ▦ Calculate the following. Do not use the trigonometric function keys.

 a) $\cos \dfrac{\pi}{6}$ b) $\cos \dfrac{\pi}{4}$

38. Show that the sine function does not have the *linearity* property, i.e., that it is not true that $\sin(x + y) = \sin x + \sin y$ for all numbers x and y.

8.4 THE OTHER CIRCULAR FUNCTIONS

There are four other circular functions, all defined in terms of the sine and cosine functions. Their definitions are as follows.

DEFINITIONS

The tangent function:	$\tan s \equiv \dfrac{\sin s}{\cos s}$	
The cotangent function:	$\cot s \equiv \dfrac{\cos s}{\sin s}$	
The secant function:	$\sec s \equiv \dfrac{1}{\cos s}$	
The cosecant function:	$\csc s \equiv \dfrac{1}{\sin s}$	

The Tangent Function

The tangent of a number is the quotient of its sine by its cosine.

Example 1 Find $\tan 0$.

$$\tan 0 = \frac{\sin 0}{\cos 0} = \frac{0}{1} = 0$$

Example 2 Find $\tan \pi/4$.

$$\tan \frac{\pi}{4} = \frac{\sin \dfrac{\pi}{4}}{\cos \dfrac{\pi}{4}} = \frac{\dfrac{\sqrt{2}}{2}}{\dfrac{\sqrt{2}}{2}} = 1$$

Example 3 Find $\tan \pi/3$.

$$\tan \frac{\pi}{3} = \frac{\sin \dfrac{\pi}{3}}{\cos \dfrac{\pi}{3}} = \frac{\dfrac{\sqrt{3}}{2}}{\dfrac{1}{2}} = \sqrt{3}$$

Not every real number has a tangent.

Example 4 Show that $\tan \pi/2$ does not exist.

$$\tan \frac{\pi}{2} = \frac{\sin \dfrac{\pi}{2}}{\cos \dfrac{\pi}{2}} = \frac{1}{0},$$

but division by 0 is undefined. Thus $\pi/2$ is not in the domain of the tangent function, and $\tan \pi/2$ does not exist.

The tangent function is undefined for any number whose cosine is 0. Thus it is undefined for $(\pi/2) + k\pi$, where k is any integer. In the first quadrant the function values are positive. In the second quadrant, however, the sine and cosine values have opposite signs, so the tangent values are negative. Tangent values are also negative in the fourth quadrant, and they are positive in quadrant three.

A graph of the tangent function is shown in Fig. 26. Note that the function value is 0 when $x = 0$, and the values increase as x increases toward $\pi/2$. As we approach $\pi/2$, the denominator becomes very small, so the tangent values become very large. In fact, they increase without bound. The dashed vertical line at $\pi/2$ is not a part of the graph. The graph approaches this vertical line closely, however. The vertical dashed lines are known as *asymptotes* to the curve.

1. The tangent function is periodic, with a period of π.
2. Its domain is the set of all real numbers except $\pi/2 + k\pi$, k any integer.
3. Its range is the set of all real numbers.
4. It is continuous except where it is not defined.

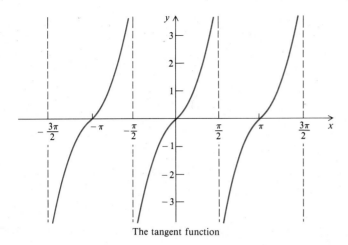

The tangent function

Figure 26

The Cotangent Function

The cotangent of a number is the quotient of its cosine by its sine. Thus the cotangent function is undefined for any number whose sine is 0. The graph of this function is shown in Fig. 27.

1. **The cotangent function is periodic, with a period of π.**
2. **Its domain is the set of all real numbers except $k\pi$, k any integer.**
3. **Its range is the set of all real numbers.**
4. **It is continuous except where it is not defined.**

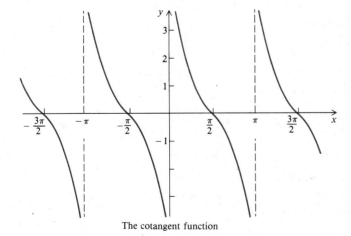

The cotangent function

Figure 27

Like the tangent function, the cotangent has positive function values in quadrants I and III and negative function values in quadrants II and IV. The tangent and cotangent functions are reciprocals, hence the following identities hold.

$$\tan x \equiv \frac{1}{\cot x}, \quad \cot x \equiv \frac{1}{\tan x}$$

The Secant Function

The secant and cosine functions are reciprocals.

Example 5 Find $\sec \pi/4$.

$$\sec \frac{\pi}{4} = \frac{1}{\cos \frac{\pi}{4}} = \frac{1}{\frac{\sqrt{2}}{2}} = \sqrt{2}$$

This function is undefined for those numbers x for which $\cos x = 0$. A graph of the secant function is shown in Fig. 28. The cosine graph is also shown for reference, since these functions are reciprocals.

The secant function

Figure 28

1. The secant function is periodic, with period 2π.
2. Its domain is the set of all real numbers except $\pi/2 + k\pi$, k any integer.
3. Its range consists of all real numbers 1 and greater, in addition to all real numbers -1 and less.
4. It is continuous wherever it is defined.

The Cosecant Function

Since the cosecant and sine functions are reciprocals, the cosecant will be undefined wherever the sine function is 0. The functions will be positive together and negative together. A graph of the cosecant function is shown in Fig. 29. The sine graph is shown for reference, since these functions are reciprocals.

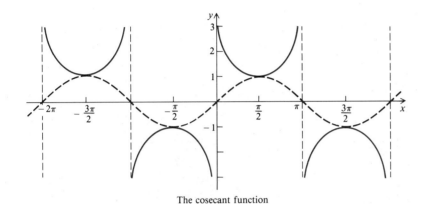

The cosecant function

Figure 29

1. The cosecant function is periodic, with period 2π.
2. Its domain is the set of all real numbers except $k\pi$, k any integer.
3. Its range is the same as for the secant function.
4. It is continuous wherever it is defined.

Signs of the Functions

Suppose a real number s determines a point on the unit circle in the first quadrant. Since both coordinates are positive there, the function values for all the circular functions will be positive. In the second quadrant the first coordinate is negative and the second coordinate is positive. Thus the sine function has positive values and the cosine has negative values. The secant, being the reciprocal of the cosine, also has negative values, and the cosecant, being the reciprocal of the sine, has positive values. The tangent and cotangent both have negative values in the second quadrant. Figure 30 summarizes, showing the quadrants in which the circular function values are positive. This diagram need not be memorized, because the signs can readily be determined by referring to a unit circle.

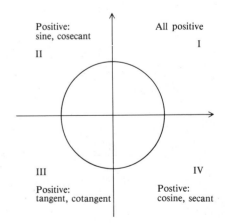

Figure 30

Example 6 Given that $\sin \pi/9 \approx 0.34202$ and $\cos \pi/9 \approx 0.93969$, calculate the other function values for $\pi/9$.

$$\tan \frac{\pi}{9} \approx \frac{\sin \frac{\pi}{9}}{\cos \frac{\pi}{9}} \approx \frac{0.34202}{0.93969} \approx 0.36397$$

$$\cot \frac{\pi}{9} \approx \frac{\cos \frac{\pi}{9}}{\sin \frac{\pi}{9}} \approx \frac{0.93969}{0.34202} \approx 2.74747$$

$$\sec \frac{\pi}{9} \approx \frac{1}{\cos \frac{\pi}{9}} \approx \frac{1}{0.93969} \approx 1.06418$$

$$\csc \frac{\pi}{9} \approx \frac{1}{\sin \frac{\pi}{9}} \approx \frac{1}{0.34202} \approx 2.92381$$

Example 7 Given that $\sin \frac{2}{3}\pi \approx 0.86603$ and $\cos \frac{2}{3}\pi = -0.50000$, calculate the other function values for $\frac{2}{3}\pi$.

$$\tan \frac{2}{3}\pi \approx \frac{\sin \frac{2}{3}\pi}{\cos \frac{2}{3}\pi} \approx \frac{0.86603}{-0.50000} \approx -1.73206$$

$$\cot \frac{2}{3}\pi \approx \frac{\cos \frac{2}{3}\pi}{\sin \frac{2}{3}\pi} \approx \frac{-0.50000}{0.86603} \approx -0.57735$$

$$\sec \frac{2}{3}\pi \approx \frac{1}{\cos \frac{2}{3}\pi} \approx \frac{1}{-0.50000} \approx -2.0000$$

$$\csc \frac{2}{3}\pi \approx \frac{1}{\sin \frac{2}{3}\pi} \approx \frac{1}{0.86603} \approx 1.15469$$

Some Identities

The tangent function appears to be odd. Let us investigate by finding $\tan(-x)$. By definition of the tangent function,

$$\tan(-x) \equiv \frac{\sin(-x)}{\cos(-x)}.$$

But we know that $\sin(-x) \equiv -\sin x$ and $\cos(-x) \equiv \cos x$. Hence

$$\tan(-x) \equiv \frac{\sin(-x)}{\cos(-x)}$$

$$\equiv \frac{-\sin x}{\cos x}$$

$$\equiv -\tan x.$$

Thus the tangent function is indeed odd. Similarly, it can be shown that the cotangent function is odd, the secant function is even, and the cosecant function is odd. Each such demonstration gives us an identity.

By use of the previously developed identities involving the sine and cosine functions we can develop a number of other identities.

Example 8 Show that $\tan(x + \pi) \equiv \tan x$.

$$\tan(x + \pi) \equiv \frac{\sin(x + \pi)}{\cos(x + \pi)} \quad \text{By definition of tangent function}$$

$$\equiv \frac{-\sin x}{-\cos x} \quad \text{Using } \sin(x + \pi) \equiv -\sin x \text{ and } \cos(x + \pi) \equiv -\cos x$$

$$\equiv \frac{\sin x}{\cos x} \equiv \tan x$$

Example 9 Show that $\sec(\pi - x) \equiv -\sec x$.

$$\sec(\pi - x) \equiv \frac{1}{\cos(\pi - x)} \quad \text{By definition of secant function}$$

$$\equiv \frac{1}{-\cos x} \quad \text{Using } \cos(\pi - x) \equiv -\cos x$$

$$\equiv -\frac{1}{\cos x}$$

$$\equiv -\sec x$$

Example 10 Show that $\csc(x - \pi) \equiv -\csc x$.

$$\csc(x - \pi) \equiv \frac{1}{\sin(x - \pi)} \quad \text{By definition of cosecant function}$$

$$\equiv \frac{1}{-\sin x} \quad \text{Using } \sin(x - \pi) \equiv -\sin x$$

$$\equiv -\csc x$$

EXERCISE SET 8.4

1. Complete this table of approximate function values. Do not use the trigonometric function keys. Round to five decimal places.

	$\frac{\pi}{16}$	$\frac{\pi}{8}$	$\frac{\pi}{6}$	$\frac{\pi}{4}$	$\frac{3}{8}\pi$	$\frac{7}{16}\pi$
sin	0.19509	0.38268	.3500	.707	0.92388	0.98079
cos	0.98079	0.92388	.8660	.707	0.38268	0.19509
tan	0.198					
cot	5.027					
sec	1.0196					
csc	5.126					

2. Complete this table of approximate function values. Do not use the trigonometric function keys. Round to five decimal places.

	$-\frac{\pi}{16}$	$-\frac{\pi}{8}$	$-\frac{\pi}{6}$	$-\frac{\pi}{4}$	$-\frac{\pi}{3}$
sin	−0.19509	−0.38268	−0.50000	−0.70711	−0.86603
cos	0.98079	0.92388	0.86603	0.70711	0.50000
tan					
cot					
sec					
csc					

In Exercises 3–6 use graph paper. Use the tables above, plus your other knowledge of the functions, to make graphs.

3. Graph the tangent function between -2π and 2π.
4. Graph the cotangent function between -2π and 2π.
5. Graph the secant function between -2π and 2π.
6. Graph the cosecant function between -2π and 2π.
7. Of the six circular functions, which are even?
8. Of the six circular functions, which are odd?
9. Which of the six circular functions have period 2π?
10. Which of the six circular functions have period π?
11. In which quadrants is the tangent function positive? negative?
12. In which quadrants is the cotangent function positive? negative?
13. In which quadrants is the secant function positive? negative?
14. In which quadrants is the cosecant function positive? negative?

Verify the following identities.

15. $\sec(-x) \equiv \sec x$
16. $\csc(-x) \equiv -\csc x$
17. $\cot(x + \pi) \equiv \cot x$
18. $\cot(x - \pi) \equiv \cot x$
19. $\sec(x + \pi) \equiv -\sec x$
20. $\tan(\pi - x) \equiv -\tan x$
21. Verify the identity $\sec(x - \pi) \equiv -\sec x$ graphically.
22. Verify the identity $\tan(x + \pi) \equiv \tan x$ graphically.

23. Describe how the graphs of the tangent and cotangent functions are related.

24. Describe how the graphs of the secant and cosecant functions are related.

25. For which pairs of circular functions do the same inputs give outputs of 0?

26. Describe how the asymptotes of the tangent, cotangent, secant, and cosecant functions are related to the inputs that produce outputs of 0.

27. Graph $f(x) = |\tan x|$.

28. Solve: $\cos x \leq \sec x$.

29. Solve: $\sin x > \csc x$.

8.5 SOME RELATIONS AMONG THE CIRCULAR FUNCTIONS

We have already seen that there are some relations among the six circular functions. In fact the sine and cosine functions are used to define the others. There are certain other important relations among these functions. These are given by identities.

The Pythagorean Identities

Recall that the equation of the unit circle in a *uv*-plane is

$$u^2 + v^2 = 1.$$

For any point on the unit circle, the coordinates u and v satisfy this equation. Suppose a real number s determines a point T on the unit circle, with coordinates (u, v). Then $u = \cos s$ and $v = \sin s$. Substituting these in the equation of the unit circle gives us the identity

$$\sin^2 s + \cos^2 s \equiv 1.$$

This is true because the coordinates of a point on the unit circle are $(\cos s, \sin s)$. When exponents are used with the circular functions, we write $\sin^2 s$ instead of $(\sin s)^2$. This identity relates the sine and cosine of any real number s. It is an important identity, known as one of the *Pythagorean* identities. We now develop another. We will divide the above identity by $\sin^2 s$:

$$\frac{\sin^2 s}{\sin^2 s} + \frac{\cos^2 s}{\sin^2 s} \equiv \frac{1}{\sin^2 s}.$$

This simplifies to

$$1 + \cot^2 s \equiv \csc^2 s.$$

This relation is valid for any number s for which $\sin^2 s \neq 0$, since we divided by $\sin^2 s$. But the numbers for which $\sin^2 s = 0$ (or $\sin s = 0$)

are exactly the ones for which the cotangent and cosecant functions are undefined. Hence our new equation holds for all numbers s for which cot s and csc s are defined, and is thus an identity.

The third Pythagorean identity, obtained by dividing the first by $\cos^2 s$, is

$$1 + \tan^2 s \equiv \sec^2 s.$$

The Pythagorean identities should be memorized. Certain variations of them, although most useful, need not be memorized, because they are so easily obtained from these three.

Example 1 Derive identities that give $\cos^2 x$ and $\cos x$ in terms of $\sin x$.

We begin with $\sin^2 x + \cos^2 x \equiv 1$, and solve for $\cos^2 x$. We have $\cos^2 x \equiv 1 - \sin^2 x$. This gives $\cos^2 x$ in terms of $\sin x$. If we now take the principal square root, we obtain

$$|\cos x| \equiv \sqrt{1 - \sin^2 x}.$$

We can also express this as

$$\cos x \equiv \pm \sqrt{1 - \sin^2 x},$$

with the understanding that the sign must be determined by the quadrant in which the point for x lies.

The Cofunction Identities

The sine and cosine are called *cofunctions* of each other. Similarly, the tangent and cotangent are cofunctions and the secant and cosecant are cofunctions. Another class of identities gives functions in terms of their cofunctions at numbers differing by $\pi/2$. We consider first the sine and cosine. Consider the graph in Fig. 31. The graph of $y = \sin x$ has been translated to the left a distance of $\pi/2$, to obtain the graph of $y = \sin(x + \pi/2)$. The latter is also a graph of the cosine function.

Figure 31

Thus we obtain the identity

$$\sin\left(x + \frac{\pi}{2}\right) \equiv \cos x.$$

By means similar to that above, we obtain the identity

$$\cos\left(x - \frac{\pi}{2}\right) \equiv \sin x.$$

Consider the graph in Fig. 32. The graph of $y = \sin x$ has been translated to the right a distance of $\pi/2$, to obtain the graph of $y = \sin(x - \pi/2)$. The latter is a reflection of the cosine function across the x-axis. In other words, it is a graph of $y = -\cos x$. We thus obtain the following identity:

$$\sin\left(x - \frac{\pi}{2}\right) \equiv -\cos x.$$

Figure 32

By a means similar to that above, we obtain the identity

$$\cos\left(x + \frac{\pi}{2}\right) \equiv -\sin x.$$

We have now established the following cofunction identities.*

$$\sin\left(x \pm \frac{\pi}{2}\right) \equiv \pm\cos x$$

$$\cos\left(x \pm \frac{\pi}{2}\right) \equiv \mp\sin x$$

These should be learned. The other cofunction identities can be obtained easily, using these and the definitions of the other circular functions.

---

*There are four identities in this list. Two of them are obtained by taking the top signs, the other two by taking the bottom signs.

Example 2 Derive an identity relating $\sin(\pi/2 - x)$ to $\cos x$.

Since the sine function is odd, we know that

$$\sin\left(\frac{\pi}{2} - x\right) \equiv \sin\left[-\left(x - \frac{\pi}{2}\right)\right] \equiv -\sin\left(x - \frac{\pi}{2}\right).$$

Now consider the identity already established,

$$\sin\left(x - \frac{\pi}{2}\right) \equiv -\cos x.$$

This is equivalent to

$$-\sin\left(x - \frac{\pi}{2}\right) \equiv \cos x,$$

and we now have

$$\sin\left(\frac{\pi}{2} - x\right) \equiv \cos x.$$

Example 3 Find an identity for $\tan(x + \pi/2)$.

By definition of the tangent function,

$$\tan\left(x + \frac{\pi}{2}\right) \equiv \frac{\sin\left(x + \frac{\pi}{2}\right)}{\cos\left(x + \frac{\pi}{2}\right)}.$$

Using the cofunction identities above, we obtain

$$\frac{\sin\left(x + \frac{\pi}{2}\right)}{\cos\left(x + \frac{\pi}{2}\right)} \equiv \frac{\cos x}{-\sin x} \equiv -\frac{\cos x}{\sin x} \equiv -\cot x.$$

Thus the identity we seek is

$$\tan\left(x + \frac{\pi}{2}\right) \equiv -\cot x.$$

Example 4 Find an identity for $\sec(\pi/2 - x)$.

By definition of the secant function,

$$\sec\left(\frac{\pi}{2} - x\right) \equiv \frac{1}{\cos\left(\frac{\pi}{2} - x\right)}.$$

Proceeding as in Example 2, we can establish that $\cos(\pi/2 - x) \equiv \cos(x - \pi/2)$. Then we obtain

$$\frac{1}{\cos\left(\dfrac{\pi}{2} - x\right)} \equiv \frac{1}{\sin x} \equiv \csc x.$$

Thus the identity we seek is

$$\sec\left(\frac{\pi}{2} - x\right) \equiv \csc x.$$

EXERCISE SET 8.5

1. From the identity $1 + \cot^2 x \equiv \csc^2 x$, obtain two other identities.

2. From the identity $1 + \tan^2 x \equiv \sec^2 x$, obtain two other identities.

3. a) Derive an identity for csc x in terms of cot x.
   b) Derive an identity for cot x in terms of csc x.

4. a) Derive an identity for tan x in terms of sec x.
   b) Derive an identity for sec x in terms of tan x.

5. Find an identity for $\tan\left(x - \dfrac{\pi}{2}\right)$.

6. Find an identity for $\cot\left(x + \dfrac{\pi}{2}\right)$.

7. Find an identity for $\sec\left(\dfrac{\pi}{2} - x\right)$.

8. Find an identity for $\csc\left(\dfrac{\pi}{2} - x\right)$.

9. For the six circular functions, list the cofunction identities involving $x \pm \dfrac{\pi}{2}$.

☆

10. In which quadrant is $\sin x = \sqrt{1 - \cos^2 x}$?

11. Given that $\sin x = 0.1425$ in quadrant II, find the other circular function values.

12. Given that $\tan x = 0.8012$ in quadrant III, find the other circular function values.

13. Given that $\cot x = 0.7534$ and $\sin x > 0$, find the other circular function values.

14. Given that $\sin \pi/8 = 0.38268$, use identities to find:

    a) $\cos \dfrac{\pi}{8}$;
    b) $\cos \dfrac{5\pi}{8}$;
    c) $\sin \dfrac{5\pi}{8}$;
    d) $\sin \dfrac{-3\pi}{8}$;
    e) $\cos \dfrac{-3\pi}{8}$.

15. Using sines, cosines, and a determinant, write an equation of the unit circle.

16. The coordinates of a point on the unit circle are (cos s, sin s) for some suitable s. Show that for a circle with center at the origin of radius r, the coordinates of any point on the circle are (r cos s, r sin s).

8.6 GRAPHS

We will consider graphs of some variations of the sine and cosine functions. In particular, we are interested in the following:

$$y = A + B \sin(Cx - D),$$

and

$$y = A + B \cos(Cx - D),$$

where A, B, C, and D are constants, some of which may be 0. These constants have the effect of translating, stretching, reflecting, or shrinking the basic graphs. Let us examine an equation to see what each of these constants does.* We first change the form a bit.

| Translates, upward if positive. | Stretches, vertically if greater than 1. | Stretches, horizontally if between 0 and 1. | Translates, to the right if positive. |

The amplitude of the sine or cosine function is multiplied by B, when that constant is present. The period will be affected by the constant C. If $C = 2$, for example, the period will be divided by 2. If $C = \frac{1}{2}$, the period will be multiplied by 2. In general, the new period will be $2\pi/C$.

Example 1 The graph of $y = 3 + \sin x$ is a translation of $y = \sin x$ upward three units ($A = 3$, $B = C = 1$, and $D = 0$). (See Fig. 33.)

Figure 33

---

*You may wish to review pp. 119–124.

Example 2 The graph of $y = 2 \sin x$ is a vertical stretching of the graph of $y = \sin x$. The amplitude of this function is 2 ($A = 0$, $B = 2$, $C = 1$, and $D = 0$). (See Fig. 34.)

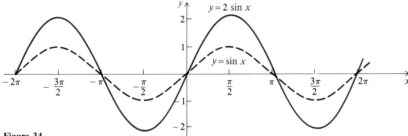

Figure 34

If the constant B in $y = B \sin x$ is negative, there will also be a reflection across the x-axis. If the absolute value of B is less than 1, then there will be a vertical shrinking; $|B|$ will be the amplitude.

Example 3 The graph of $y = -(\frac{1}{2}) \sin x$ is a vertical shrinking and a reflection of the graph of $y = \sin x$. The amplitude is $\frac{1}{2}$. (See Fig. 35.)

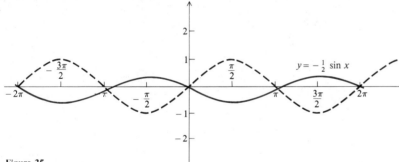

Figure 35

Example 4 The graph of $y = \sin 2x$ is a horizontal shrinking of the graph of $y = \sin x$. The period of this function is π (half the period of $y = \sin x$). (See Fig. 36.)

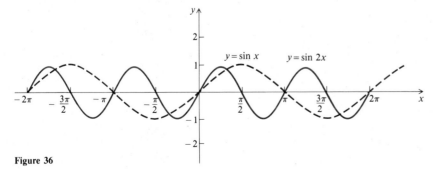

Figure 36

Example 5 The graph of $y = \sin(2x - \pi)$ is the same as that of $y = \sin 2(x - \pi/2)$. We merely factored the 2 out of the parentheses. The $\pi/2$ translates the graph of $y = \sin 2x$ a distance of $\pi/2$ to the right. The 2 in front of the parentheses shrinks by half, making the period π. (See Fig. 37.)

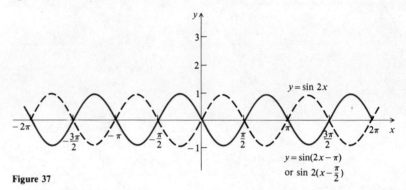

Figure 37

The quantity D/C translates to the right or left, and is called the *phase shift*. In Example 5, the phase shift is $\pi/2$. This means that the graph of $\sin 2x$ has been shifted forward a distance of $\pi/2$.

Example 6 The graph of $y = 3 \sin[2x + (\pi/2)]$ has an amplitude of 3. The period is π and the phase shift is $-\pi/4$. This means that the graph of $3 \sin 2x$ has been shifted back a distance of $\pi/4$. (See Fig. 38.)

Figure 38

Graphs of Sums

A function that is a sum of two functions can often be graphed by a method called *addition of ordinates*. We graph the two functions separately and then add the second coordinates (called *ordinates*) graphically. A compass may be helpful.

Example 7 Graph: $y = 2 \sin x + \sin 2x$.

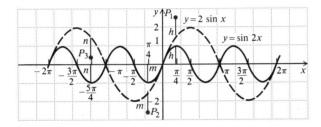

Figure 39

We graph $y = 2 \sin x$ and $y = \sin 2x$ using the same axes (see Fig. 39). Now we graphically add some ordinates to obtain points on the graph we seek. At $x = \pi/4$ we transfer the distance h, which is the value of $\sin 2x$, upward to add it to the value of $2 \sin x$. Point P_1 is on the graph we seek. At $x = -\pi/4$ we do a similar thing, but the distance m is negative. Point P_2 is on our graph. At $x = -(\frac{5}{4})\pi$ the distance n is negative, so we in effect subtract it from the value of $2 \sin x$. Point P_3 is on the graph we seek. We continue to plot points in this fashion and then connect them to get the desired graph, shown in Fig. 40.

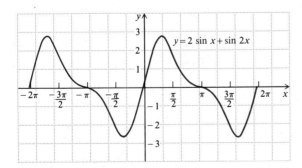

Figure 40

EXERCISE SET 8.6

In Exercises 1–14, use graph paper. Sketch graphs of the functions.

1. $y = 2 + \sin x$
2. $y = -3 + \cos x$
3. $y = 2 \cos x$
4. $y = 3 \cos x$
5. $y = -\frac{1}{2} \cos x$
6. $y = -2 \sin x$
7. $y = \cos(-2x)$
8. $y = \cos\left(-\frac{1}{3}x\right)$
9. $y = 2 \cos(2x - \pi)$
10. $y = -2 \sin\left(2x + \frac{\pi}{2}\right)$
11. $y = 2 \cos x + \cos 2x$
12. $y = 2 \sin x + \cos 2x$
13. $y = 3 \cos x - \sin 2x$
14. $y = 3 \sin x - \cos 2x$

In Exercises 15–18, determine the amplitude, period, and phase shift.

15. $y = 3 \cos\left(3x - \frac{\pi}{2}\right)$
16. $y = -5 \cos\left(4x + \frac{\pi}{3}\right)$
17. $y = -4 \sin\left(5x + \frac{\pi}{2}\right)$
18. $y = \frac{1}{2} \sin(2\pi x + \pi)$

☆

19. Graph $y = \sec^2 x$.
20. Graph $y = \tan x \csc x$.
21. Graph $y = 3 \cos 2(x + \pi/4)$ by first converting to a sine.

★

22. Graph $y = x + \sin x$. (*Hint:* Use addition of ordinates. You might also wish to use a calculator to find some function values to plot.)
23. Graph $y = 2^{-x} \sin x$, for $0 \leq x \leq 4\pi$. Functions like this have applications in the theory of damped oscillations. (*Hint:* First graph $y = 2^{-x}$ and also $y = -(2^{-x})$. Then graph $y = \sin x$ and consider multiplying the ordinates. You may wish to use a calculator to find some function values to plot.)

8.7 ALGEBRAIC AND TRIGONOMETRIC MANIPULATIONS

The circular functions and variations of them are called *trigonometric functions*. Trigonometric expressions such as $\sin 2x$ or $\tan(x - \pi)$ represent numbers, just as algebraic expressions such as $3x$ or $3x^2 - 2$ represent numbers. Thus we can manipulate expressions containing trigonometric expressions in much the same way as we manipulate purely algebraic expressions.

Example 1 Multiply and simplify: $\cos y\, (\tan y - \sec y)$.

$$\cos y\, (\tan y - \sec y) = \cos y \tan y - \cos y \sec y$$

The multiplication has now been accomplished, but we can simplify:

$$\cos y \tan y - \cos y \sec y = \cos y \frac{\sin y}{\cos y} - \cos y \frac{1}{\cos y}$$

$$= \sin y - 1.$$

In Example 1 it has been illustrated how we can use certain identities to accomplish simplification. There is no rule for doing this, but it is often helpful to put everything in terms of sines and cosines, as we did in Example 1.

Example 2 Factor and simplify: $\sin^2 x \cos^2 x + \cos^4 x$.

$$\sin^2 x \cos^2 x + \cos^4 x = \cos^2 x\, (\sin^2 x + \cos^2 x)$$

The factoring has been accomplished, but we can now simplify, using the Pythagorean identity $\sin^2 x + \cos^2 x \equiv 1$:

$$\cos^2 x\, (\sin^2 x + \cos^2 x) = \cos^2 x.$$

Example 3 Factor and simplify: $\tan x + \cos(x - \pi/2)$.

$$\tan x + \cos\left(x - \frac{\pi}{2}\right) = \sin x \frac{1}{\cos x} + \sin x.$$

Here we have used the identity $\cos(x - \pi/2) \equiv \sin x$, as well as the definition of the tangent function. Now we factor as follows:

$$\sin x \left(\frac{1}{\cos x} + 1\right) \quad \text{or} \quad \sin x\, (\sec x + 1).$$

Example 4 Simplify: $\dfrac{\sin x - \sin x \cos x}{\sin x + \sin x \tan x}$.

$$\frac{\sin x - \sin x \cos x}{\sin x + \sin x \tan x} = \frac{\sin x (1 - \cos x)}{\sin x (1 + \tan x)} \quad \text{Factoring}$$

$$= \frac{1 - \cos x}{1 + \tan x} \quad \text{Simplifying}$$

Example 5 Subtract and simplify: $\dfrac{2}{\sin x - \cos x} - \dfrac{3}{\sin x + \cos x}$.

$$\frac{2}{\sin x - \cos x} - \frac{3}{\sin x + \cos x} = \frac{2}{\sin x - \cos x} \cdot \frac{\sin x + \cos x}{\sin x + \cos x}$$

$$- \frac{3}{\sin x + \cos x} \cdot \frac{\sin x - \cos x}{\sin x - \cos x}$$

$$= \frac{2(\sin x + \cos x) - 3(\sin x - \cos x)}{\sin^2 x - \cos^2 x}$$

$$= \frac{-\sin x + 5\cos x}{\sin^2 x - \cos^2 x}$$

Example 6 Add and simplify: $\dfrac{2}{\tan^3 x - 2\tan^2 x} + \dfrac{2}{\tan x - 2}$.

$$\frac{2}{\tan^3 x - 2\tan^2 x} + \frac{2}{\tan x - 2} = \frac{2}{\tan^2 x (\tan x - 2)} + \frac{2}{\tan x - 2} \cdot \frac{\tan^2 x}{\tan^2 x}$$

$$= \frac{2 + 2\tan^2 x}{\tan^2 x (\tan x - 2)} = \frac{2(1 + \tan^2 x)}{\tan^2 x (\tan x - 2)}$$

Now we use the Pythagorean identity $1 + \tan^2 x \equiv \sec^2 x$ in the numerator, and obtain for that numerator,

$$2\sec^2 x, \quad \text{or} \quad 2\frac{1}{\cos^2 x}.$$

Thus we have

$$\frac{2}{\cos^2 x \dfrac{\sin^2 x}{\cos^2 x} (\tan x - 2)} \quad \text{or} \quad \frac{2}{\sin^2 x (\tan x - 2)}.$$

When radicals occur, the use of absolute value is sometimes necessary, but its consideration can be complex. In the examples and exercises that follow, we shall assume that all quantities are nonnegative.

Example 7 Multiply and simplify: $\sqrt{\sin^3 x \cos x} \cdot \sqrt{\cos x}$.

$$\sqrt{\sin^3 x \cos x} \cdot \sqrt{\cos x} = \sqrt{\sin^3 x \cos^2 x}$$
$$= \sin x \cos x \sqrt{\sin x}$$

Example 8 Rationalize the denominator: $\sqrt{2/\tan x}$.

$$\sqrt{\frac{2}{\tan x}} = \sqrt{\frac{2}{\tan x} \cdot \frac{\tan x}{\tan x}} = \sqrt{\frac{2\tan x}{\tan^2 x}}$$

$$= \frac{\sqrt{2\tan x}}{\tan x}$$

Example 9 Solve for tan x: $\tan^2 x + \tan x = 56$.

$$\tan^2 x + \tan x - 56 = 0 \qquad \text{Reducible to quadratic;}$$
$$\text{let } u = \tan x$$
$$(\tan x + 8)(\tan x - 7) = 0 \qquad \text{Factoring}$$
$$\tan x + 8 = 0 \quad \text{or} \quad \tan x - 7 = 0$$
$$\tan x = -8 \quad \text{or} \quad \tan x = 7$$

Example 10 Solve for sec x: $\sec^2 x - \frac{3}{4}\sec x = \frac{1}{2}$.

$$4\sec^2 x - 3\sec x - 2 = 0$$

We now use the quadratic formula, and obtain

$$\sec x = \frac{3 \pm \sqrt{3^2 - 4 \cdot 4 \cdot (-2)}}{2 \cdot 4}$$
$$= \frac{3 \pm \sqrt{41}}{8}.$$

Example 11 Solve for sin x: $8\sin^2 x + 2\cos(x + \pi/2) - 3 = 0$.

We first use the identity $\cos(x + \pi/2) \equiv -\sin x$, to get everything in terms of sin x:

$$8\sin^2 x - 2\sin x - 3 = 0.$$

Now we factor:

$$(4\sin x - 3)(2\sin x + 1) = 0$$
$$4\sin x - 3 = 0 \quad \text{or} \quad 2\sin x + 1 = 0$$
$$\sin x = \frac{3}{4} \quad \text{or} \quad \sin x = -\frac{1}{2}.$$

EXERCISE SET 8.7

In Exercises 1–8, multiply and simplify.

1. $(\sin x - \cos x)(\sin x + \cos x)$
2. $(\sin x + \cos x)(\sec x + \csc x)$
3. $\tan x(\cos x - \csc x)$
4. $\tan y \sin y(\cot y - \csc y)$
5. $(\sin y + \cos y)^2$
6. $(1 + \tan x)^2$
7. $(\sin x + \csc x)(\sin^2 x + \csc^2 x - 1)$
8. $(\cot x - \tan x)(\cot^2 x + 1 + \tan^2 x)$

In Exercises 9–18, factor and simplify.

9. $\sin^2 y - \cos^2 y$
10. $\tan^4 x - \sec^4 x$

11. $\tan x + \sin(\pi - x)$

12. $\cot x - \cos(\pi - x)$

13. $3\cot^2 y + 6\cot y + 3$

14. $8 + \tan^4 x - 2\tan^2 x$

15. $\sin^3 y + 27$

16. $\cos^3 x - \sec^3 x$

17. $\cos\left(x - \dfrac{\pi}{2}\right)\tan y + \sin x \cot y$

18. $\cos(\pi - x) + \cot x \sin\left(x - \dfrac{\pi}{2}\right)$

In Exercises 19–40, simplify.

19. $\dfrac{\sin^2 x \cos x}{\cos^2 x \sin x}$

20. $\dfrac{30 \sin^3 x \cos x}{6 \cos^2 x \sin x}$

21. $\dfrac{\cos^2 x - 2\cos x + 1}{\cos x - 1}$

22. $\dfrac{6 \tan x \sin x - 3 \sin x}{9 \sin^2 x + 3 \sin x}$

23. $\dfrac{\cos^2 x - 1}{\sin\left(\dfrac{\pi}{2} - x\right) - 1}$

24. $\dfrac{\cos x - \sin\left(\dfrac{\pi}{2} - x\right)\sin x}{\cos x - \cos(\pi - x)\tan x}$

25. $\dfrac{\cos^2 x + 2\sin\left(x - \dfrac{\pi}{2}\right) + 1}{\sin\left(\dfrac{\pi}{2} - x\right) - 1}$

26. $\dfrac{\cos^2 y \sin\left(y + \dfrac{\pi}{2}\right)}{\sin^2 y \sin\left(\dfrac{\pi}{2} - y\right)}$

27. $\dfrac{2\sin^2 x}{\cos^3 x} \cdot \left(\dfrac{\cos x}{2\sin x}\right)^2$

28. $\dfrac{5\cos x}{\sin^2 x} \cdot \dfrac{\sin^2 x - \sin x \cos x}{\sin^2 x - \cos^2 x}$

29. $\dfrac{\tan^2 y}{\sec y} \div \dfrac{3\tan^3 y}{\sec y}$

30. $\dfrac{\cot^3 y}{\csc y} \div \dfrac{4\cot^2 y}{\csc y}$

31. $\dfrac{1}{\sin^2 y - \cos^2 y} - \dfrac{2}{\cos y + \sin y}$

32. $\left(\dfrac{\cot x}{\csc x}\right)^2 + \dfrac{1}{\csc^2 x}$

33. $\dfrac{\sin^2 x - 9}{2\cos x + 1} \cdot \dfrac{10\cos x + 5}{3\sin x + 9}$

34. $\dfrac{9\cos^2 x - 25}{2\cos x - 2} \cdot \dfrac{\cos^2 x - 1}{6\cos x - 10}$

35. $\sqrt{\sin^2 x \cos x} \cdot \sqrt{\cos x}$

36. $\sqrt{\cos y \sin^2 y} - \sqrt{\cos^3 y}$

37. $\sqrt{\sin^2 x + 2\cos x \sin x + \cos^2 x}$

38. $\sqrt{\tan^2 x - 2\tan x \sin x + \sin^2 x}$

39. $(1 - \sqrt{\sin y})(\sqrt{\sin y} + 1)$

40. $\sqrt{\cos x}(\sqrt{3\cos x} - \sqrt{\sin x \cos x})$

In Exercises 41–44, rationalize the denominator.

41. $\sqrt{\dfrac{\sin x}{\cos x}}$

42. $\sqrt{\dfrac{\cos^2 x}{2\sin^2 x}}$

43. $\sqrt{\dfrac{1 + \sin x}{1 - \sin x}}$

44. $\sqrt{\dfrac{1 - \cos x}{1 + \cos x}}$

45. Solve for $\tan x$: $\tan^2 x + 4\tan x = 21$.

46. Solve for $\sec x$: $\sec^2 x - 7\sec x = -10$.

47. Solve for $\cot x$: $\cot^2 x + 9\cot x - 10 = 0$.

48. Solve for $\csc x$: $\csc^2 x + 3\csc x - 10 = 0$.

49. Solve for $\sin x$: $\sin^2 x - \cos\left(x + \dfrac{\pi}{2}\right) = 6$.

50. Solve for $\sin x$: $2\cos^2\left(x - \dfrac{\pi}{2}\right) - 3\cos\left(x + \dfrac{\pi}{2}\right) - 2 = 0$.

51. Solve for $\tan x$: $\tan^2 x - 6\tan x = 4$.

52. Solve for $\cot x$: $\cot^2 x + 8\cot x = 5$.

CHAPTER 8 REVIEW

The point (3, −2) is on a circle centered at the origin. Find the coordinates of its reflections across:

1. The y-axis.
2. The origin.
3. The x-axis.

On a unit circle, mark the points determined by:

4. $\dfrac{7\pi}{6}$.
5. $\dfrac{3\pi}{4}$.
6. $\dfrac{\pi}{6}$.
7. $\dfrac{9\pi}{4}$.

8. Sketch a graph of $y = \sin x$.
9. What is the domain of the sine function?
10. Sketch a graph of $y = \cos x$.
11. What is the period of the cosine function?
12. Sketch a graph of $y = \sin\left(x + \dfrac{\pi}{2}\right)$. Use the axes of Exercise 8.
13. Complete the following table.

	$\dfrac{\pi}{6}$	$\dfrac{\pi}{4}$	$\dfrac{\pi}{3}$	$\dfrac{\pi}{2}$	$\dfrac{3\pi}{4}$	$\dfrac{5\pi}{4}$
sin x						
cos x						

14. Sketch a graph of $y = \csc x$.
15. What is the period of the cosecant function?
16. What is the range of the cosecant function?
17. In which quadrants are the signs of the sine and the tangent the same?
18. Verify the following identity: $\cot(x - \pi) \equiv \cot x$.

Complete these Pythagorean identities.

19. $\sin^2 x + \cos^2 x \equiv$ _____
20. $1 + \cot^2 x \equiv$ _____

Complete these cofunction identities.

21. $\cos\left(x + \dfrac{\pi}{2}\right) \equiv$ _____
22. $\cos\left(\dfrac{\pi}{2} - x\right) \equiv$ _____
23. $\sin\left(x - \dfrac{\pi}{2}\right) \equiv$ _____
24. Express tan x in terms of sec x.

25. Sketch a graph of $y = 3 + \cos\left(x - \dfrac{\pi}{4}\right)$.
26. What is the phase shift of the function in Exercise 25?
27. What is the period of the function in Exercise 25?
28. Sketch a graph of $y = 3 \cos x + \sin x$ for values of x between 0 and 2π.

Simplify.

29. $\cos x (\tan x + \cot x)$
30. $\dfrac{\csc x (\sin^2 x + \cos^2 x \tan x)}{\sin x + \cos x}$

31. Rationalize the denominator: $\sqrt{\dfrac{\tan x}{\sec x}}$

32. Solve for $\tan x$: $3\tan^2 x - 2\tan x - 2 = 0$.

33. Does $5 \sin x = 7$ have a solution for x? Why or why not?

34. For what values of x in $(0, \pi/2]$ is the following true?
$$\sin x < x$$

Trigonometric Functions, Rotations, and Angles

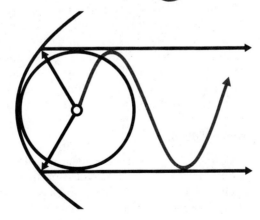

9.1 ROTATIONS AND ANGLES

We shall consider rotations abstractly. That is, instead of considering a specific rotating object, such as a wheel or part of a machine, we will consider a rotating ray, with its endpoint at the origin of an xy-plane. The ray starts in position along the positive half of the x-axis. A counterclockwise rotation will be called positive. Clockwise rotations will be called negative. Note in Fig. 1 that the rotating ray and the positive half of the x-axis form an angle. Thus we often speak of "rotations" and "angles" interchangeably. The rotating ray is often called the *terminal side* of the angle, and the positive half of the x-axis is called the *initial side*.

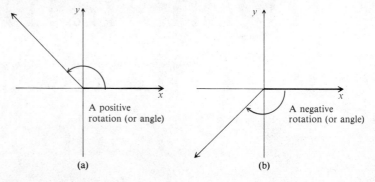

Figure 1

Measures of Rotations or Angles

The measure of an angle, or rotation, may be given in degrees. For example, a complete revolution has a measure of 360°, half a revolution has a measure of 180°, and so on.* We also speak of an *angle* of 90°, or 720°, or −240°, and so on. An angle between 0° and 90° has its terminal side in the first quadrant. An angle between 90° and 180° has its terminal side in the second quadrant. An angle between 0° and −90° has its terminal side in the fourth quadrant, and so on.

Radian Measure

Consider a circle with its center at the origin and radius of length 1 (a *unit* circle). The distance around this circle from the initial side of an angle to the terminal side can be used as a measure of the rotation. This kind of angle measure is very useful. The unit is called a *radian* (Fig. 2).

Since the circumference of a circle is $2\pi r$ and in this case $r = 1$, a rotation of 360° (one revolution) has a measure of 2π radians. Half a

---

*Subunits are *minutes* (60 minutes = 1°) and *seconds* (60 seconds = 1 minute).

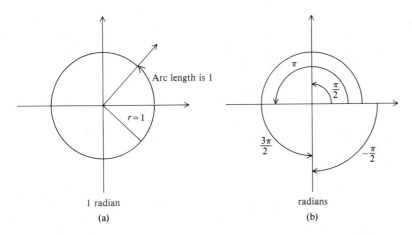

Figure 2

revolution is a rotation of 180° or π radians. A quarter revolution is a rotation of 90°, or $\pi/2$ radians, and so on (Fig. 2b). To convert between degrees and radians we can use the notion of "multiplying by one" (see Section 1.11). Note the following:

$$\frac{1 \text{ revolution}}{1 \text{ revolution}} = 1 = \frac{2\pi \text{ radians}}{360 \text{ degrees}} = \frac{\pi \text{ radians}}{180 \text{ degrees}};$$

also

$$\frac{180 \text{ degrees}}{\pi \text{ radians}} = 1.$$

When a rotation is given in radians, the word "radians" is optional and is most often omitted. Thus if no unit is given for a rotation, it is understood to be in radians.

Example 1 Convert 60° to radians.

$$60° = 60° \cdot \frac{\pi \text{ radians}}{180°} = \frac{60°}{180°} \pi \text{ radians}$$

$$= \frac{\pi}{3} \text{ radians, or } \frac{\pi}{3}$$

Using 3.14 for π, we find that $\pi/3$ is about 1.047.

Example 2 Convert $3\pi/4$ radians to degrees.

$$\frac{3\pi}{4} \text{ radians} = \frac{3\pi}{4} \text{ radians} \cdot \frac{180°}{\pi \text{ radians}}$$

$$= \frac{3\pi}{4\pi} \cdot 180° = 135°$$

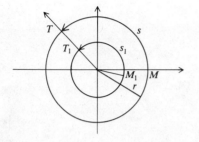

Figure 3

Arc Length and Central Angles

Radian measure can be determined using a circle other than a unit circle. In Fig. 3, a unit circle is shown along with another circle. The angle shown is a central angle of both circles, hence the arcs that it intercepts have their lengths in the same ratio as the radii of the circles. The radii of the circles are r and 1.

The corresponding arc lengths are MT and M_1T_1 or, more simply, s and s_1. We therefore have the proportion

$$\frac{s}{r} = \frac{s_1}{1}.$$

Now s_1 is the radian measure of the rotation in question. It is more common to use a Greek letter, such as θ, for the measure of an angle or rotation. We commonly use the letter s for arc length. If we adopt this convention, the above proportion becomes $\theta = s/r$. In any circle, arc length, central angle, and length of radius are related in this fashion. Or, in general, the following is true.

The radian measure θ* of a rotation is the ratio of the distance s traveled by a point at a radius r from the center of rotation to the length of the radius r:

$$\theta = \frac{s}{r}.$$

Example 3 Find the length of an arc of a circle of 5-cm radius associated with a central angle of $\pi/3$ radians.

$$\theta = \frac{s}{r}, \quad \text{or} \quad s = r\theta$$

Therefore, $s = 5 \cdot \dfrac{\pi}{3}$ cm or, using 3.14 for π, about 5.23 cm.

Example 4 Find the measure of a rotation in radians in which a point 2 m from the center of rotation travels 4 m.

$$\theta = \frac{s}{r} = \frac{4 \text{ m}}{2 \text{ m}} = 2 \quad \text{The unit is understood to be radians.}$$

A look at Examples 3 and 4 will show why the word "radian" is most often omitted. In Example 4, we have the division 4 m/2 m, which

---

*When using this formula, you must make sure that θ is in radians and that s and r are expressed in the same unit.

simplifies to the number 2, since m/m = 1. From this point of view it would seem preferable to omit the word "radians." In Example 3, had we used the word "radians" all the way through, our answer would have come out to be 5.23 cm-radians. It is a distance we seek, hence we know the unit should be centimeters. Thus we must omit the word "radians." A measure in radians is simply a number, so it is usually preferable to omit the word "radians."

EXERCISE SET 9.1

For angles of the following measures, state in which quadrant the terminal side lies.

1. $34°$
2. $320°$
3. $-120°$
4. $175°$
5. $\dfrac{\pi}{3}$
6. $-\dfrac{3}{4}\pi$
7. $\dfrac{11}{4}\pi$
8. $\dfrac{19}{4}\pi$

Convert to radian measure. Leave answers in terms of π.

9. $30°$
10. $15°$
11. $60°$
12. $200°$
13. $75°$
14. $300°$
15. $37.71°$
16. $12.73°$
17. $214.6°$
18. $73.87°$

Convert to radian measure. Do not leave answers in terms of π. Use 3.14 for π.

19. $120°$
20. $240°$
21. $320°$
22. $75°$
23. $200°$
24. $300°$
25. ▦ $117.8°$
26. ▦ $231.2°$
27. ▦ $1.354°$
28. ▦ $327.9°$

Convert these radian measures to degree measure.

29. 1 radian
30. 2 radians
31. 8π
32. -12π
33. $\dfrac{3}{4}\pi$
34. $\dfrac{5}{4}\pi$
35. ▦ 1.303
36. ▦ 2.347
37. ▦ 0.7532π
38. ▦ -1.205π

39. In a circle with 120-cm radius, an arc 132 cm long subtends a central angle of how many radians? how many degrees, to the nearest degree?

40. In a circle with 200-cm radius, an arc 65 cm long subtends a central angle of how many radians? how many degrees, to the nearest degree?

41. Through how many radians does the minute hand of a clock rotate in 50 min?

42. A wheel on a car has a 14-in. radius. Through what angle (in radians) does the wheel turn while the car travels 1 mi?

43. In a circle with 10-m radius, how long is an arc associated with a central angle of 1.6 radians?

44. In a circle with 5-m radius, how long is an arc associated with a central angle of 2.1 radians?

45. Certain positive angles are marked here in degrees. Find the corresponding radian measures.

46. Certain negative angles are marked here in degrees. Find the corresponding radian measures.

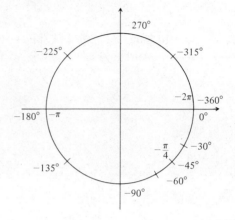

☆

47. The *grad* is a unit of angle measure similar to a degree. A right angle has a measure of 100 grads. Convert the following to grads.

a) 48°   b) 153°   c) $\frac{\pi}{8}$ radians   d) $\frac{5\pi}{7}$ radians

48. The *mil* is a unit of angle measure. A right angle has a measure of 1600 mils. Convert the following to degrees, minutes, and seconds.

a) 100 mils   b) 350 mils

49. On the earth, one degree of latitude is how many kilometers? how many miles? (Assume that the radius of the earth is 6400 km or 4000 miles, approximately.)

50. One minute of latitude on the earth is equivalent to one *nautical mile*. Find the circumference and radius of the earth in nautical miles.

51. An astronaut on the moon observes the earth, about 240,000 miles away. The diameter of the earth is about 8000 miles. Find the angle α.

52. The circumference of the earth was computed by Eratosthenes (276–195 B.C.). He knew the distance from Aswan to Alexandria to be about 500 miles. From each town he observed the sun at noon, finding the angular difference to be 7°12′. Do Eratosthenes' calculation.

★

53. What is the angle between the hands of a clock at 7:45?

54. At what time between noon and 1:00 P.M. are the hands of a clock first perpendicular?

9.2 ANGULAR SPEED

Speed is defined to be distance traveled per unit of time. Similarly, *angular speed* is defined to be amount of rotation per unit of time. For example, we might speak of the angular speed of a wheel as 150 revolutions per minute or the angular speed of the earth as 2π radians per day. The Greek letter ω (omega) is usually used for angular speed. Thus angular speed is defined as

$$\omega = \frac{\theta}{t}.$$

For many applications it is important to know a relationship between angular speed and linear speed. For example, we might wish to find the linear speed of a point on the earth, knowing its angular speed. Or we might wish to know the linear speed of an earth satellite, knowing its angular speed. To develop the relationship we seek, we recall the relation between angle and distance from the preceding section, $\theta = s/r$. This is equivalent to

$$s = r\theta.$$

We divide by the time, t, to obtain

$$\frac{s}{t} = r\frac{\theta}{t}.$$

Now s/t is linear speed v, and θ/t is angular speed ω. We thus have the relation we seek.

The linear speed v of a point a distance of r from the center of rotation is given by

$$v = r\omega,$$

where ω is the angular speed in radians per unit of time.

In deriving this formula we used the equation $s = r\theta$, in which the units for s and r must be the same and θ must be in radians. So, for our new formula $v = r\omega$, the units of distance for v and r must be the same, ω must be in radians per unit of time, and the units of time must be the same for v and ω.

Example 1 An earth satellite in a circular orbit 1200 km high makes one complete revolution every 90 min. What is its linear speed? Use 6400 km for the length of a radius of the earth.

We will use the formula $v = r\omega$; thus we shall need r and ω.

$r = 6400 \text{ km} + 1200 \text{ km}$ Radius of earth plus height of satellite

$ = 7600 \text{ km}$

$\omega = \dfrac{2\pi \text{ radians}}{90 \text{ min}} = \dfrac{\pi}{45 \text{ min}}$ We have, as usual, omitted the word radians.

Now, using $v = r\omega$, we have

$$v = 7600 \text{ km} \cdot \frac{\pi}{45 \text{ min}} = \frac{7600\pi}{45} \cdot \frac{\text{km}}{\text{min}}.$$

Using 3.14 for π, we obtain $v = 530$ km/min.

Example 2 A car is traveling 45 mph. Its tires have a 20-in. radius. Find the angle through which a wheel turns in 5 sec.

Recall that $\omega = \theta/t$ or $\theta = \omega t$. Thus we can find θ if we know ω and t. To find ω we use $v = r\omega$. For convenience we will first convert 45 mph to ft/sec:

$$v = 45 \frac{\text{mi}}{\text{hr}} \cdot \frac{1 \text{ hr}}{60 \text{ min}} \cdot \frac{1 \text{ min}}{60 \text{ sec}} \cdot \frac{5280 \text{ ft}}{1 \text{ mi}} = 66 \frac{\text{ft}}{\text{sec}}.$$

Now $r = 20$ in. We shall convert to ft, since v is in ft/sec:

$$r = 20 \text{ in.} \cdot \frac{1 \text{ ft}}{12 \text{ in.}} = \frac{20}{12} \text{ ft} = \frac{5}{3} \text{ ft}.$$

Using $v = r\omega$, we have

$$66 \frac{\text{ft}}{\text{sec}} = \frac{5}{3} \text{ ft} \cdot \omega, \quad \text{so } \omega = 39.6 \frac{\text{radians}}{\text{sec}}.$$

Then

$$\theta = \omega t = 39.6 \frac{\text{radians}}{\text{sec}} \cdot 5 \text{ sec} = 198 \text{ radians.*}$$

EXERCISE SET 9.2

1. A flywheel is rotating 7 radians/sec. It has a 15-cm diameter. What is the speed of a point on its rim, in cm/min?

2. A $33\frac{1}{3}$-rpm record has a radius of 15 cm. What is the linear velocity of a point on the rim, in cm/sec?

3. The earth has a 4000-mile radius and rotates one revolution every 24 hours. What is the linear speed of a point on the equator, in mph?

4. The earth is 93,000,000 miles from the sun and traverses its orbit, which is nearly circular, every 365.25 days. What is the linear velocity of the earth in its orbit, in mph?

5. The wheels of a bicycle have a 24-in. diameter. When the bike is being ridden so that the wheels make 12 rpm, how far will the bike travel in 1 min?

6. A horse on a merry-go-round is 7 m from the center and travels at 10 km/h. What is its angular speed?

7. A car is traveling 40 mph. Its wheels have a 15-in. radius. Find the angle through which a wheel rotates in 12 sec.

8. (*Determining the speed of a river*). A water wheel has a 10-ft radius. To get a good approximation to the speed of the river, you count the revolutions of the wheel and find that it makes 14 revolutions per minute. What is the speed of the river?

---

*If you have trouble working with the units, review Section 1.11.

9. Two pulleys, 50 cm and 30 cm in diameter, respectively, are connected by a belt. The larger pulley makes 12 revolutions per minute. Find the angular speed of the smaller pulley in radians/sec.

10. One gear wheel turns another, the teeth being on the rims. The wheels have 40-cm and 50-cm radii and the smaller wheel rotates at 20 rpm. Find the angular speed of the larger wheel in radians/sec.

11. An airplane engine is idling at 800 rpm. When the throttle is opened, it takes 4.3 seconds for the speed to come up to 2500 rpm. What was the angular acceleration (a) in rpm per second? (b) in radians per second per second?

12. The linear speed of an airplane, flying at low level over the sea, is 175 knots (nautical miles per hour). It accelerates to 325 knots in 12 seconds.
   a) What was its linear acceleration in knots per second?
   b) What was its angular acceleration in radians per second per second? (*Hint:* One nautical mile is equivalent to one minute of latitude.)

9.3 TRIGONOMETRIC FUNCTIONS OF ANGLES OR ROTATIONS

We now extend the domains of the functions defined on p. 293 to angles or rotations of any size. To do this, we consider a triangle with one vertex at the origin of a coordinate system and one side along the x-axis. The point R can be in any of the four quadrants or on one of the axes. The sides of the triangle have lengths x, y, and r, as shown in Fig. 4, and the angle θ is always measured from the positive half of the x-axis. The numbers x and y can be positive, negative, or zero, but the number r will always be considered positive.

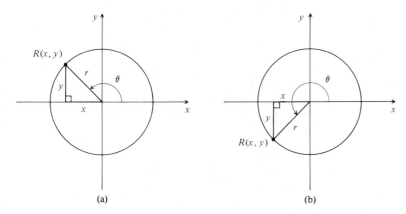

Figure 4

The values of the six trigonometric functions are as follows.

$$\sin \theta = \frac{y}{r} = \frac{\text{second coordinate}}{\text{radius}}$$

Figure 5

$$\cos \theta = \frac{x}{r} = \frac{\text{first coordinate}}{\text{radius}}$$

$$\tan \theta = \frac{y}{x} = \frac{\text{second coordinate}}{\text{first coordinate}}$$

$$\cot \theta = \frac{x}{y} = \frac{\text{first coordinate}}{\text{second coordinate}}$$

$$\sec \theta = \frac{r}{x} = \frac{\text{radius}}{\text{first coordinate}}$$

$$\csc \theta = \frac{r}{y} = \frac{\text{radius}}{\text{second coordinate}}$$

Example 1 Find the six trigonometric function values for the angle shown in Fig. 5.

$$\sin \theta = \frac{y}{r} = \frac{4}{5} = 0.8, \qquad \cos \theta = \frac{x}{r} = \frac{-3}{5} = -0.6,$$

$$\tan \theta = \frac{y}{x} = \frac{4}{-3} \approx -1.33, \qquad \cot \theta = \frac{x}{y} = \frac{-3}{4} = -0.75,$$

$$\sec \theta = \frac{r}{x} = \frac{5}{-3} \approx -1.67, \qquad \csc \theta = \frac{r}{y} = \frac{5}{4} = 1.25.$$

Figure 6

Function Values for Some Special Angles

Our knowledge of triangles enables us to determine function values for certain angles. Let us first recall the Pythagorean theorem about right triangles. It says that in any right triangle, as shown in Fig. 6, $a^2 + b^2 = c^2$, where c is the length of the hypotenuse.

A 45° right triangle is half of a square, so its legs are the same length. Let us consider such a triangle whose legs have length 1. Then its hypotenuse has length c given by

$$1^2 + 1^2 = c^2, \quad \text{or} \quad c^2 = 2, \quad \text{or} \quad c = \sqrt{2}.$$

Such a triangle is shown in Fig. 7. From this diagram we can easily determine the function values for 45° or $\pi/4$:

$$\sin \frac{\pi}{4} = \sin 45° = \frac{1}{\sqrt{2}} = \frac{\sqrt{2}}{2} \approx 0.707,$$

$$\cos \frac{\pi}{4} = \cos 45° = \frac{1}{\sqrt{2}} = \frac{\sqrt{2}}{2} \approx 0.707,$$

Figure 7

$$\tan\frac{\pi}{4} = \tan 45° = \frac{1}{1} = 1,$$

$$\cot\frac{\pi}{4} = \cot 45° = \frac{1}{1} = 1,$$

$$\sec\frac{\pi}{4} = \sec 45° = \frac{\sqrt{2}}{1} = \sqrt{2} \approx 1.414,$$

$$\csc\frac{\pi}{4} = \csc 45° = \frac{\sqrt{2}}{1} = \sqrt{2} \approx 1.414.$$

These function values should be memorized. The decimal values need not be, and it is sufficient to memorize the first three, since the others are their reciprocals.

$$\sin 45° = \frac{\sqrt{2}}{2}, \quad \cos 45° = \frac{\sqrt{2}}{2}, \quad \tan 45° = 1.$$

Next we consider an equilateral triangle, each of whose sides has length 2. If we take half of it, as shown in Fig. 8, we obtain a right triangle having a hypotenuse of length 2 and a leg of length 1. The other leg has length a, given by the Pythagorean theorem as follows:

$$a^2 + 1^2 = 2^2, \quad \text{or} \quad a^2 = 3, \quad \text{or} \quad a = \sqrt{3}.$$

The acute angles of this triangle have measures of 30° and 60° (see Fig. 9). We can now determine function values for these angles. The function values for 30° or $\pi/6$ are as follows:

$$\sin\frac{\pi}{6} = \sin 30° = \frac{1}{2},$$

$$\cos\frac{\pi}{6} = \cos 30° = \frac{\sqrt{3}}{2},$$

$$\tan\frac{\pi}{6} = \tan 30° = \frac{1}{\sqrt{3}} = \frac{\sqrt{3}}{3},$$

$$\cot\frac{\pi}{6} = \cot 30° = \sqrt{3},$$

$$\sec\frac{\pi}{6} = \sec 30° = \frac{2}{\sqrt{3}} = \frac{2\sqrt{3}}{3},$$

$$\csc\frac{\pi}{6} = \csc 30° = 2.$$

Figure 8

Figure 9

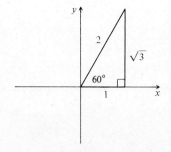

Figure 10

We obtain the function values for 60° or $\pi/3$ by repositioning the triangle as in Fig. 10:

$$\sin\frac{\pi}{3} = \sin 60° = \frac{\sqrt{3}}{2},$$

$$\cos\frac{\pi}{3} = \cos 60° = \frac{1}{2},$$

$$\tan\frac{\pi}{3} = \tan 60° = \sqrt{3},$$

$$\cot\frac{\pi}{3} = \cot 60° = \frac{1}{\sqrt{3}} = \frac{\sqrt{3}}{3},$$

$$\sec\frac{\pi}{3} = \sec 60° = 2,$$

$$\csc\frac{\pi}{3} = \csc 60° = \frac{2}{\sqrt{3}} = \frac{2\sqrt{3}}{3}.$$

The function values for 30° and 60° should be memorized, but again it is sufficient to learn only those for the sine, cosine, and tangent, since the others are their reciprocals.

$$\sin 30° = \frac{1}{2}, \qquad \cos 30° = \frac{\sqrt{3}}{2}, \qquad \tan 30° = \frac{\sqrt{3}}{3},$$

$$\sin 60° = \frac{\sqrt{3}}{2}, \qquad \cos 60° = \frac{1}{2}, \qquad \tan 60° = \sqrt{3}.$$

Figure 11

Signs of the Functions

Function values of the generalized trigonometric functions can be positive, negative, or zero, depending upon where the terminal side of the angle lies. In the first quadrant, all function values are positive. In the second quadrant, first coordinates are negative and second coordinates are positive (Fig. 11). Thus if θ is in quadrant II,

$\sin\theta = \dfrac{y}{r}$ is positive, because y and r are positive;

$\cos\theta = \dfrac{x}{r}$ is negative, because x is negative and r is positive;

$\tan\theta = \dfrac{y}{x}$ is negative, because y is positive and x is negative;

and so on.

Figure 12 summarizes, showing the quadrants in which the function values are positive. The diagram need not be memorized, because the signs can readily be determined as shown in Fig. 11.

Now let us suppose that the terminal side of an angle falls on one of the axes. In that case, one of the coordinates is zero. The definitions of the functions still apply, but in some cases functions will be undefined because a denominator will be 0.

Example 2 Find the trigonometric function values for 180°, or π radians. (See Fig. 13.)

We note that the first coordinate is negative, the second coordinate is 0, and x and r have the same absolute value (r always being positive). Thus we have

$$\sin 180° = \frac{0}{r} = 0,$$

$$\cos 180° = \frac{x}{r} = -1 \quad \text{Since } |x| = |r|, \text{ but } x \text{ and } r \text{ have opposite signs.}$$

$$\tan 180° = \frac{y}{x} = \frac{0}{x} = 0,$$

$$\cot 180° = \frac{x}{y} = \frac{x}{0}, \quad \text{Thus } \cot 180° \text{ is undefined.}$$

$$\sec 180° = \frac{r}{x} = -1, \quad \text{The reciprocal of } \cos 180°$$

$$\csc 180° = \frac{r}{0}. \quad \text{Thus } \csc 180° \text{ is undefined.}$$

Example 3 Find the trigonometric function values for 210°.

We draw a diagram showing the terminal side of a 210° angle (Fig. 14). We now have a right triangle with a 30° angle as shown. We can read off function values:

$$\sin 210° = \frac{-1}{2} = -\frac{1}{2},$$

$$\cos 210° = \frac{-\sqrt{3}}{2} = -\frac{\sqrt{3}}{2},$$

$$\tan 210° = \frac{-1}{-\sqrt{3}} = \frac{1}{\sqrt{3}}, \text{ or } \frac{\sqrt{3}}{3}.$$

Figure 12

Figure 13

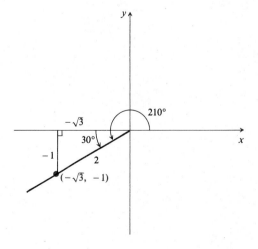

Figure 14

$$\cot 210° = \frac{-\sqrt{3}}{-1} = \sqrt{3},$$

$$\sec 210° = \frac{2}{-\sqrt{3}} = -\frac{2}{\sqrt{3}}, \text{ or } -\frac{2\sqrt{3}}{3},$$

$$\csc 210° = \frac{2}{-1} = -2.$$

We are now in a position to be able to determine the trigonometric function values for many other angles. If the terminal side of an angle falls on one of the axes, the function values are 0 or 1 or −1 or else are undefined. Thus we can determine function values for any multiple of 90° or $\pi/2$. We can also determine the function values for any angle whose terminal side makes a 30°, 45°, or 60° angle with the x-axis. Consider, for example, an angle of 150°, or $5\pi/6$ (Fig. 15). The terminal side makes a 30° angle with the x-axis, since 180° − 150° = 30°. As Fig. 15 shows, triangle ONR is congruent to triangle $ON'R'$, hence the ratios of the sides of the two triangles are the same except perhaps for sign. We could determine the function values directly from triangle ONR, but this is not necessary. If we remember that the sine is positive in quadrant II and that the cosine and tangent are negative, we can simply use the values for 30°, prefixing the appropriate sign. The triangle ONR is called a *reference* triangle and its acute angle at the origin is called a *reference* angle.

Figure 15

In general, to find the function values of an angle, we find them for the reference angle and prefix the appropriate sign.

Example 4 Find the sine, cosine, and tangent of 600°, or $10\pi/3$.

We find the multiple of 180° nearest 600°:

$$180° \times 2 = 360°, \quad 180° \times 3 = 540°, \quad \text{and} \quad 180° \times 4 = 720°.$$

The nearest multiple is 540°. The difference between 600° and 540° is 60°. This gives us the reference angle (Fig. 16). We recall that sin 60° = $\sqrt{3}/2$, cos 60° = $\frac{1}{2}$, and tan 60° = $\sqrt{3}$. We also note that the sine and cosine are negative in the third quadrant and that the tangent is positive. Hence we have

Figure 16

$$\sin \frac{10\pi}{3} = \sin 600° = -\frac{\sqrt{3}}{2},$$

$$\cos \frac{10\pi}{3} = \cos 600° = -\frac{1}{2},$$

$$\tan \frac{10\pi}{3} = \tan 600° = \sqrt{3}.$$

EXERCISE SET 9.3

Before beginning, make a list of the function values of 30°, 60°, and 45°. Then memorize your list. In Exercises 1-4, find the six trigonometric function values for the angle θ as shown.

1.
2.
3.
4.

In Exercises 5 and 6, use the fact that $\sqrt{2} \approx 1.414$ and $\sqrt{3} \approx 1.732$.

5. Find decimal values for the six trigonometric functions of 30°, or $\pi/6$.

6. Find decimal values for the six trigonometric functions of 60°, or $\pi/3$.

Find the following if they exist.

7. $\cos 180°$

8. $\sin 360°$

9. $\tan \dfrac{\pi}{2}$

10. $\cot \pi$

11. $\sec 720°$

12. $\csc 720°$

13. $\sin \left(-\dfrac{3\pi}{4}\right)$

14. $\cos \dfrac{3\pi}{4}$

15. $\sin \dfrac{5\pi}{6}$

16. $\cos \dfrac{5\pi}{6}$

17. $\tan 240°$

18. $\cot 240°$

19. $\sec \dfrac{7\pi}{4}$

20. $\csc \dfrac{7\pi}{4}$

21. $\tan (-315°)$

22. $\cot (-315°)$

23. $\sin (-210°)$

24. $\cos (-210°)$

25. $\tan \dfrac{7\pi}{6}$

26. $\cot \dfrac{7\pi}{6}$

27. $\sin \dfrac{11\pi}{4}$

28. $\cos \dfrac{11\pi}{4}$

29. $\tan \dfrac{11\pi}{6}$

30. $\cot \dfrac{11\pi}{6}$

31. Compile a list of function values for the sine function, from -2π to 2π. Then graph the function.

32. Is the sine function an even function? an odd function? periodic?

33. Compile a list of function values for the cosine function, from -2π to 2π. Then graph the function.

34. Is the cosine function an even function? an odd function? periodic?

35. This diagram shows a piston of a steam engine driving a drive wheel of a locomotive. The radius of the drive wheel (from the center of the wheel to the pin P) is R and the length of the rod is L.

Suppose that the drive wheel is rotating at a speed of ω radians per second (so that $\theta = \omega t$). Show that the distance of pin Q from O, the center of the wheel, is a function of time, given by

$$x = \sqrt{L^2 - R^2 \sin^2 \omega t} + R \cos \omega t.$$

9.4 RELATIONSHIPS AMONG FUNCTION VALUES

When we know one of the trigonometric function values for an angle, we can find the other five if we know the quadrant in which the terminal side lies. The idea is to sketch a triangle in the appropriate quadrant, use the Pythagorean theorem to find the lengths of its sides as needed, and then read off the ratios of its sides.

Example 1 Given that $\tan \theta = -\frac{2}{3}$ and that θ is in the second quadrant, find the other function values.

We first sketch a second quadrant triangle. Since $\tan \theta = -\frac{2}{3}$, we make the legs of lengths 2 and 3. The hypotenuse must then have length $\sqrt{13}$ (Fig. 17). Now we can read off the appropriate ratios:

$$\sin \theta = \frac{2}{\sqrt{13}}, \qquad \cos \theta = -\frac{3}{\sqrt{13}},$$

$$\tan \theta = -\frac{2}{3}, \qquad \cot \theta = -\frac{3}{2},$$

$$\sec \theta = -\frac{\sqrt{13}}{3}, \qquad \csc \theta = \frac{\sqrt{13}}{2}.$$

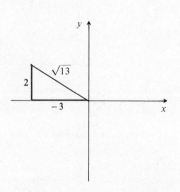

Figure 17

Example 2 Given that $\cot \theta = 2$ and θ is in the third quadrant, find $\sin \theta$, $\cos \theta$, and $\tan \theta$.

Again, we sketch a triangle and label the sides (Fig. 18), then read off the appropriate ratios:

$$\sin \theta = -\frac{1}{\sqrt{5}}, \qquad \cos \theta = -\frac{2}{\sqrt{5}}, \qquad \tan \theta = \frac{1}{2}.$$

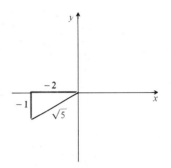

Figure 18

Cofunctions and Complements

In a right triangle the acute angles are complementary, since the sum of all three angle measures is 180° and the right angle accounts for 90° of this total. Thus if one acute angle of a right triangle is θ, the other is $90° - \theta$, or $\pi/2 - \theta$ (see Fig. 19). Note that the sine of $\angle A$ is also the cosine of $\angle B$, its complement (Fig. 19):

$$\sin \theta = \frac{a}{c}, \qquad \cos (90° - \theta) = \frac{a}{c}.$$

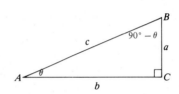

Figure 19

Similarly, the tangent of $\angle A$ is the cotangent of its complement and the secant of $\angle A$ is the cosecant of its complement.

These pairs of functions are called *cofunctions*. The name *cosine* originally meant the sine of the complement. The name *cotangent* meant the tangent of the complement and *cosecant* meant the secant of the complement. A complete list of the cofunction relations is as follows.*

$$\sin \theta \equiv \cos (90° - \theta), \qquad \cos \theta \equiv \sin (90° - \theta),$$
$$\tan \theta \equiv \cot (90° - \theta), \qquad \cot \theta \equiv \tan (90° - \theta),$$
$$\sec \theta \equiv \csc (90° - \theta), \qquad \csc \theta \equiv \sec (90° - \theta).$$

Example 3 Given that

$$\sin 18° = 0.3090, \qquad \cos 18° = 0.9511,$$
$$\tan 18° = 0.3249, \qquad \cot 18° = 3.078,$$
$$\sec 18° = 1.051, \qquad \csc 18° = 3.236,$$

find the six function values for 72°.

Since 72° and 18° are complements, we have $\sin 72° = \cos 18°$, etc., and the function values are

$$\sin 72° = 0.9511, \qquad \cos 72° = 0.3090,$$
$$\tan 72° = 3.078, \qquad \cot 72° = 0.3249,$$
$$\sec 72° = 3.236, \qquad \csc 72° = 1.051.$$

---

*Equations that hold for all sensible replacements for the variables are known as *identities*. We shall use the sign $\equiv$ instead of $=$ to indicate that an equation is an identity.

EXERCISE SET 9.4

In Exercises 1-6, a function value and a quadrant are given. Find the other five function values for each.

1. $\sin\theta = -\frac{1}{3}$, III

2. $\sin\theta = -\frac{1}{5}$, IV

3. $\cos\theta = \frac{3}{5}$, IV

4. $\cos\theta = -\frac{4}{5}$, II

5. $\cot\theta = -2$, IV

6. $\tan\theta = 5$, III

7. Given that

$$\sin 65° = 0.9063, \quad \cos 65° = 0.4226,$$
$$\tan 65° = 2.145, \quad \cot 65° = 0.4663,$$
$$\sec 65° = 2.366, \quad \csc 65° = 1.103,$$

find the six function values for 25°.

8. Given that

$$\sin 32° = 0.5299, \quad \cos 32° = 0.8480,$$
$$\tan 32° = 0.6249, \quad \cot 32° = 1.600,$$
$$\sec 32° = 1.179, \quad \csc 32° = 1.887,$$

find the six function values for 58°.

9. ▦ Given that $\sin 27° = 0.45399$,

a) find the other function values for 27°;

b) find the six function values for 63°.

10. ▦ Given that $\cot 54° = 0.72654$,

a) find the other function values for 54°;

b) find the six function values for 36°.

11. ▦ Given that $\cos 38° = 0.78801$, find the six function values for 128°.

12. ▦ Given that $\tan 73° = 3.27085$, find the six function values for 343°.

13. For any acute angle θ, we know that $\cos(90° - \theta) = \sin\theta$. Consider angles other than acute angles. Does this relation still hold, and if so to what extent?

14. Consider the equation $\sin(x + \pi/2) = \cos x$, where x is an angle in radians. Does this relation hold for acute angles? Does it hold for other than acute angles?

15. Consider a right triangle with sides a, b, and c, where c is the hypotenuse. We know that $a^2 + b^2 = c^2$. Divide by c^2 and interpret the results in terms of trigonometric functions.

9.5 TABLES OF TRIGONOMETRIC FUNCTIONS

By use of certain formulas, theoretically determined, tables of values for the trigonometric functions have been constructed. Table 3 at the back of the book is such a table. A portion of it is shown in Fig. 20. This table gives the function values for angles from 0° to 90° only. Values for other angles can always be determined from these, since any angle has an acute reference angle. Note that angle measures are given in degrees and minutes (60 minutes = 1°) as well as radians. Values in the table are given for intervals of 10′ (10 minutes) and are correct to four digits.

9.5 TABLES OF TRIGONOMETRIC FUNCTIONS

Degrees	Radians	Sin	Cos	Tan	Cot	Sec	Csc		
56°00′	.6283	.5878	.8090	.7265	1.376	1.236	1.701	.9425	54°00′
10	312	901	073	310	368	239	695	396	50
20	341	925	056	355	360	241	688	367	40
30	.6370	.5948	.8039	.7400	1.351	1.244	1.681	.9338	30
40	400	972	021	445	343	247	675	308	20
50	429	995	004	490	335	249	668	279	10
37°00′	.6458	.6018	.7986	.7536	1.327	1.252	1.662	.9250	53°00′
10	487	041	969	581	319	255	655	221	50
20	516	065	951	627	311	258	649	192	40
30	.6545	.6088	.7934	.7673	1.303	1.260	1.643	.9163	30
40	574	111	916	720	295	263	636	134	20
50	603	134	898	766	288	266	630	105	10
44°00′	.7679	.6947	.7193	.9657	1.036	1.390	1.440	.8029	46°00′
10	709	967	173	713	030	394	435	999	50
20	738	988	153	770	024	398	431	970	40
30	.7767	.7009	.7133	.9827	1.018	1.402	1.427	.7941	30
40	796	030	112	884	012	406	423	912	20
50	825	050	092	942	006	410	418	883	10
45°00′	.7854	.7071	.7071	1.000	1.000	1.414	1.414	.7854	45°00′
		Cos	Sin	Cot	Tan	Csc	Sec	Radians	Degrees

Figure 20

The headings on the left of Table 3 range from 0° to 45° only. Thus the table may seem to be only half of what it should be. Function values from 45° to 90° are the cofunction values of their complements, however. Hence these can be found without further entries. For angles from 45° to 90° the headings on the right are used, together with the headings at the bottom. For example, sin 37° is found to be 0.6018 using the top and left headings. The cosine of 53° (the complement of 37°) is found also to be 0.6018, using the bottom and right headings.

Example 1 Find cos 37°20′.

We find 37°20′ in the left column and then *cos* at the top (see Fig. 20). At the intersection of this row and column we find the entry we seek:

$$\cos 37°20′ = 0.7951.$$

Example 2 Find cot 0.9192.

Since no degree symbol is given, we know that the angle is in radians. We do not find 0.9192 on the left, under radians. This is because the angle is greater than 45° or $\pi/4$ radians. Therefore we will use the right and bottom headings. At the right, under radians, we find 0.9192. Next we find *cot* at the bottom. At the intersection of this row and column we find the entry we seek:

$$\cot 0.9192 = 0.7627.$$

Function Values for Small Angles

Figure 21 shows a unit circle and a small angle. Since we are assuming s to be small, the length of arc s is very nearly the same as y. Also x is very nearly 1. Since y = sin s, we see that $s \approx \sin s$. Since tan s = y/x and $x \approx 1$, we see that $\tan s \approx s$. These relations also hold if s is negative with small absolute value, and are useful in certain applications.

If $|s|$ is a small number, $s \approx \sin s \approx \tan s$ (a small angle—in radians—is approximately equal to its sine and its tangent).

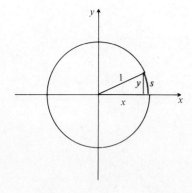

Figure 21

Minutes Versus Tenths and Hundredths

It is usual to express the measure of an angle in degrees and minutes, but it is often useful to have the measure in degrees, tenths, hundredths, etc. Thus we will need to know how to convert between these.

Example 3 Convert 16.35° to degrees and minutes.

$$16.35° = 16° + 0.35 \times 1°$$

Now

$$0.35 \times 1° = 0.35 \times 60'$$
$$= 21'.$$

So

$$16.35° = 16°21'.$$

Example 4 Convert 34°37′ to degrees and decimal parts of degrees.

$$34°37' = 34° + \frac{37°}{60},$$

$$\frac{37}{60} = 0.617$$

so

$$34°37' = 34.617°$$

Interpolation

The process of linear interpolation, described in Section 7.4, can be applied to tables of any function. In particular, we can use it with tables of the trigonometric functions.

Example 5 Find $\tan 27°43'$.

$$
10' \begin{cases} 3' \begin{cases} \tan 27°40' = 0.5243 \\ \tan 27°43' = 0.52?? \\ \tan 27°50' = 0.5280 \end{cases} \end{cases} \quad \text{The tabular difference is 0.0037.}
$$

The tabular difference is 0.0037, and 43' is $\frac{3}{10}$ of the way from 40' to 50', so we take $\frac{3}{10}$ of 0.0037. This is 0.0011. Thus

$$\tan 27°43' = 0.5243 + 0.0011, \quad \text{or} \quad 0.5254.$$

Interpolation for decreasing functions is about the same as for increasing functions, but there is a slight difference as the next example shows.

Example 6 Find $\cot 29°44'$.

$$
10' \begin{cases} 4' \begin{cases} \cot 29°40' = 1.756 \\ \cot 29°44' = 1.7?? \\ \cot 29°50' = 1.744 \end{cases} \end{cases} \quad \text{The tabular difference is 0.012.}
$$

We take 0.4 of the tabular difference, 0.012, and obtain 0.0048, or 0.005. Because the cotangent function is *decreasing* in the interval $(0°, 90°]$, we *subtract* 0.005 from 1.756. Thus $\cot 29°44' = 1.751$.

Once the process of interpolation is understood, it will not be necessary to write as much as we did in the above examples. After a bit of practice, you will find that interpolation is rather easy, and you can accomplish some of the steps without writing.

Let us look at an example of using the tables in reverse, that is, given a function value, to find the measure of an angle.

Example 7 Given $\tan B = 0.3727$, find B (between $0°$ and $90°$).

$$
10' \begin{cases} \tan 20°20' = 0.3706 \\ \tan B = 0.3727 \\ \tan 20°30' = 0.3739 \end{cases} \; 0.0021 \quad \text{The tabular difference is 0.0033.}
$$

We find that B is $\frac{21}{33}$ or $\frac{7}{11}$ of the way from $20°20'$ to $20°30'$. The tabular difference is $10'$, so we take $\frac{7}{11}$ of $10'$ and obtain $6'$. Thus $B = 20°26'$.

Function Values for Angles of Any Size

Table 3 at the back of the book gives function values for angles from $0°$ to $90°$. For any other angle we can still use the table to find the function values. To do this, we first find the reference angle, which is the angle that the terminal side makes with the x-axis. We then look up the function values for the reference angle in the table and use the appropriate sign, depending on the quadrant in which the terminal side lies.

Example 8 Find $\sin 285°40'$.

We first find the reference angle. To do this we find the multiple of $180°$ that is nearest $285°40'$. The multiples of $180°$ are $180°$, $360°$, and so on. Now $285°40'$ is nearest $360°$. We subtract to find the reference angle.

$$\begin{array}{r} 359°60' \\ -285°40' \\ \hline 74°20' \end{array}$$

Now we find that $\sin 74°20' = 0.9628$. The terminal side is in quadrant IV, where the sine is negative. Hence

$$\sin 285°40' = -0.9628.$$

It is helpful to make a diagram, like Fig. 22, showing the reference angle.

Figure 22

The Use of Calculators

Scientific calculators contain tables of trigonometric functions, usually the sine, cosine, and tangent only. One can enter an angle and obtain any of these function values and, by taking their reciprocals, find the other three trigonometric function values.

Important: Some calculators require angles to be entered in degrees. On others either degrees or radians can be used. Be sure to use the unit appropriate to the calculator you are using.

You can use the tables in reverse—for example, to find an angle when its sine is given, enter the sine value first. Then press the key marked $\boxed{\sin^{-1}}$ (on some calculators). On other calculators the key is marked $\boxed{\text{arcsin}}$. On still others, two keys must be pressed in sequence, $\boxed{\text{arc}}$ and then $\boxed{\sin}$. Be sure to read the instructions for the calculator you are using.

EXERCISE SET 9.5

In Exercises 1 and 2, use Table 3 at the back of the book.

1. How big must an angle be in order that it first differs from its tangent in the fourth decimal place?

2. How big must an angle be in order that it first differs from its tangent in the third decimal place?

Convert to degrees and minutes.

3. 46.38°
4. 85.21°
5. 67.84°
6. 38.48°

Convert to degrees and decimal parts of degrees.

7. 45°45′
8. 36°17′
9. 76°53′
10. 12°23′

Use Table 3 to find the following.

11. sin 13°20′
12. cos 71°50′
13. tan 0.3956
14. cot 1.0443
15. $\sin \frac{5\pi}{7}$
16. $\cos \frac{4\pi}{5}$
17. sin 292°40′
18. cos 514°30′
19. tan (−237°50′)
20. sec 653°50′
21. sin 28°13′
22. 1 36°42′
23. cos 53°53′
24.
25. tan 349°28′
26. cot 253°34′
27. sin 17.43°
28. sin 38.72°
29. cos 75.15°
30. cos 81.91°

For each of the following, find θ in degrees and minutes between 0° and 90°.

31. sin θ = 0.2363
32. sin θ = 0.3854
33. cos θ = 0.3719
34. cos θ = 0.6361
35. cos θ = 0.3538
36. cos θ = 0.9678

For each of the following, find θ in radians between 0 and $\pi/2$.

37. tan θ = 0.4699
38. cot θ = 1.621
39. sec θ = 1.457
40. csc θ = 1.173

41. Angles are measured in degrees, minutes (60 minutes = 1 degree), and seconds (60 seconds = 1 minute). Minutes are denoted ′ and seconds are denoted ″. Convert radian measure of $\pi/7$ to degrees, minutes, and seconds.

42. Convert 61°38′22″ to degrees and decimal parts of degrees.

43. The formula sin x = x − (x³/3!) + (x⁵/5!) gives an approximation for sine values when x is in radians. Calculate sin 0.5 and compare your answer with Table 3.

44. Graph the tangent function, using values from the table, between −90° and 90°. What is the domain of the tangent function?

45. Use the fact that $\sin\theta \approx 0$ when θ is small to calculate the diameter of the sun. The sun is 93 million miles from the earth and the angle it subtends at the earth's surface is about $18''$.

46. When light passes from one substance to another, rays are bent, depending upon the speed of light in those substances. For example, light traveling in air will be bent as it enters water. Angle i (see the figure) is called the angle of *incidence* and angle r is called the angle of *refraction*. Snell's law states that $\sin i/\sin r$ is constant. The constant is called the *index of refraction*.

The index of refraction of a certain crystal is 1.52. Light strikes it, making an angle of incidence of 27°. What will be the angle of refraction?

★

47. ▦ The function $(\sin x)/x$ (with x in radians) approaches a limit as x approaches 0. What is it?

48. ▦ Does the function $(x + \tan x)/(\sin x)$ approach a limit as x approaches 0? If so, what is it?

49. a) Find $\log \sin 57°$ (the logarithm of the sine of 57°).
b) What is the domain of the log sin function?

50. For what real numbers x is $\sin x > x$ (x in radians)?

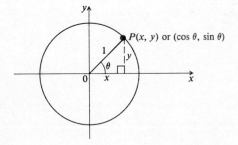

Figure 23

9.6 FURTHER RELATIONSHIPS AMONG THE FUNCTIONS (OPTIONAL)*

Consider the unit circle, whose equation is $x^2 + y^2 = 1$. For any point (x, y) on the circle, the second coordinate is $\sin\theta$ and the first coordinate is $\cos\theta$, since the length of the hypotenuse is 1 (see Fig. 23). Thus we have the following.

For any point P on the unit circle, the coordinates are $(\cos\theta, \sin\theta)$, where θ is any angle having $\overrightarrow{OP}$ for its terminal side.

If we substitute $\cos\theta$ and $\sin\theta$ for x and y in the equation of the unit circle, we obtain the following important identity.

$$\sin^2\theta + \cos^2\theta \equiv 1$$

---

*This section should be omitted, except in the "minimal course." (See the preface.)

For any point P on the unit circle, $\tan\theta = y/x$ and $\cot\theta = x/y$, provided denominators do not become 0. Since $x = \cos\theta$ and $y = \sin\theta$, we have

$$\tan\theta = \frac{y}{x} = \frac{\sin\theta}{\cos\theta} \quad \text{and} \quad \cot\theta = \frac{x}{y} = \frac{\cos\theta}{\sin\theta}.$$

Since the secant and cosine values are reciprocals and the cosecant and sine values are reciprocals, we have:

$$\sec\theta = \frac{1}{\cos\theta} \quad \text{and} \quad \csc\theta = \frac{1}{\sin\theta}.$$

We now list four useful identities:

$$\tan\theta \equiv \frac{\sin\theta}{\cos\theta}, \quad \cot\theta \equiv \frac{\cos\theta}{\sin\theta},$$

$$\sec\theta \equiv \frac{1}{\cos\theta}, \quad \csc\theta \equiv \frac{1}{\sin\theta}.$$

Example 1 Given that $\cos\theta = -0.61$ and that θ is in the second quadrant, use identities to find the other function values.

We begin with $\sin^2\theta + \cos^2\theta \equiv 1$, solving for $\sin\theta$.

$\sin\theta \equiv \pm\sqrt{1 - \cos^2\theta}$ The sign is to be chosen according to the quadrant, in this case, $+$.

$\sin\theta = \sqrt{1 - (-0.61)^2} = 0.7924$

We proceed to calculate, using the appropriate identities:

$$\tan\theta = \frac{\sin\theta}{\cos\theta} = \frac{0.7924}{-0.61} = -1.2990,$$

$$\cot\theta = \frac{\cos\theta}{\sin\theta} = \frac{-0.61}{0.7924} = -0.7698,$$

$$\sec\theta = \frac{1}{\cos\theta} = \frac{1}{-0.61} = -1.6393,$$

$$\csc\theta = \frac{1}{\sin\theta} = \frac{1}{0.7924} = 1.2620.$$

Some More Identities

The identity $\sin^2\theta + \cos^2\theta \equiv 1$ is one of the so-called *Pythagorean identities*, because it can be derived from the Pythagorean property of right triangles. There are two other Pythagorean identities, obtained from this one by dividing. In one case we divide by $\sin^2 \theta$ and in the other we divide by $\cos^2 \theta$.

$$\frac{\sin^2 \theta}{\sin^2 \theta} + \frac{\cos^2 \theta}{\sin^2 \theta} \equiv \frac{1}{\sin^2 \theta} \qquad \text{Dividing by } \sin^2 \theta$$

Simplifying, we obtain $1 + \cot^2 \theta \equiv \csc^2 \theta$. We list the Pythagorean identities thus derived:

$$1 + \tan^2 \theta \equiv \sec^2 \theta, \qquad 1 + \cot^2 \theta \equiv \csc^2 \theta.$$

Some of the trigonometric functions are even and some of them are odd. This fact gives us some other useful identities. We determine evenness or oddness for the sine and cosine functions by considering an angle θ and its inverse $-\theta$. Figure 24 shows only special cases, but it is true in general that $\sin(-\theta) = -\sin \theta$ and $\cos(-\theta) = \cos \theta$. Thus the cosine function is even and the sine function is odd.

$$\sin(-\theta) \equiv -\sin \theta, \qquad \cos(-\theta) \equiv \cos \theta$$

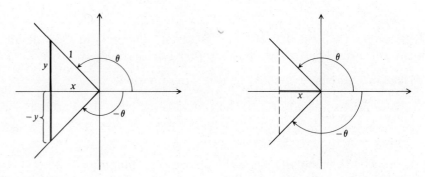

Figure 24

Example 2 Determine whether the tangent function is even or odd.

We know that $\tan \theta \equiv \sin \theta / \cos \theta$. We substitute from the identities just derived.

$$\tan(-\theta) \equiv \frac{\sin(-\theta)}{\cos(-\theta)} \equiv \frac{-\sin \theta}{\cos \theta}$$

$$\equiv -\frac{\sin \theta}{\cos \theta} \equiv -\tan \theta$$

The tangent function is therefore odd.

A similar derivation shows that the cotangent function is also odd. We thus have the following:

$$\tan(-\theta) \equiv -\tan \theta, \qquad \cot(-\theta) \equiv -\cot \theta.$$

Graphs of the Functions

Graphs of the trigonometric functions can be constructed by plotting and connecting points, using values obtained from the tables and keeping in mind the signs of the functions in the various quadrants. Figures 25–28 are graphs of four of the functions. The angles are in radians, and we have chosen to use x instead of θ, as is customary when graphing functions.

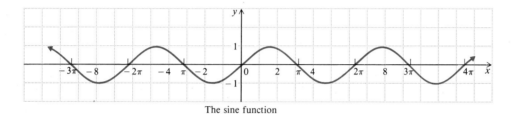

The sine function

Figure 25

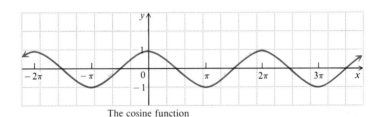

The cosine function

Figure 26

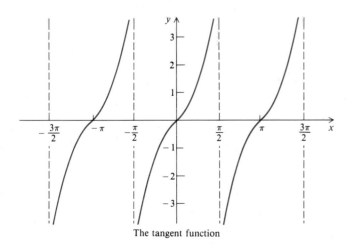

The tangent function

Figure 27

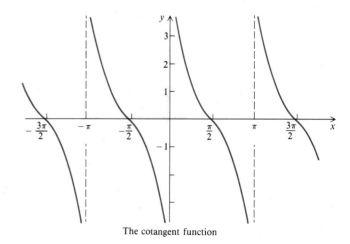

The cotangent function

Figure 28

The Cofunction Identities

Another class of identities gives functions in terms of their cofunctions at numbers differing by $\pi/2$. We consider first the sine and cosine. Consider the graph in Fig. 29. The graph of $y = \sin x$ has been translated to the left a distance of $\pi/2$, to obtain the graph of $y = \sin(x + \pi/2)$.

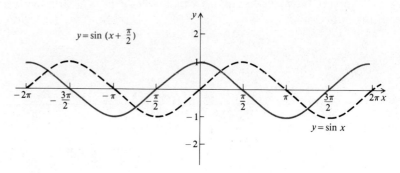

Figure 29

The latter is also a graph of the cosine function. Thus we obtain the identity

$$\sin\left(x + \frac{\pi}{2}\right) \equiv \cos x.$$

By means similar to that above, we obtain the identity

$$\cos\left(x - \frac{\pi}{2}\right) \equiv \sin x.$$

Consider the graph in Fig. 30. The graph of $y = \sin x$ has been translated to the right a distance of $\pi/2$, to obtain the graph of $y = \sin(x - \pi/2)$. The latter is a reflection of the cosine function across the

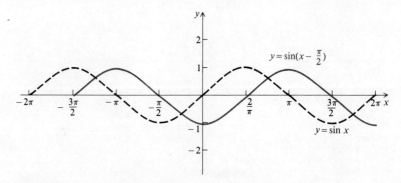

Figure 30

x-axis. In other words, it is a graph of $y = -\cos x$. We thus obtain the following identity:

$$\sin\left(x - \frac{\pi}{2}\right) \equiv -\cos x.$$

By a means similar to that above, we obtain the identity

$$\cos\left(x + \frac{\pi}{2}\right) \equiv -\sin x.$$

We have now established the following cofunction identities:*

$$\sin\left(x \pm \frac{\pi}{2}\right) \equiv \pm\cos x,$$

$$\cos\left(x \pm \frac{\pi}{2}\right) \equiv \mp\sin x.$$

These should be learned. The other cofunction identities can be obtained easily from them.

Example 3 Derive an identity relating $\sin(\pi/2 - x)$ to $\cos x$.

Since the sine function is odd, we know that

$$\sin\left(\frac{\pi}{2} - x\right) \equiv \sin-\left(x - \frac{\pi}{2}\right) \equiv -\sin\left(x - \frac{\pi}{2}\right).$$

Now consider the identity already established,

$$\sin\left(x - \frac{\pi}{2}\right) \equiv -\cos x.$$

This is equivalent to

$$-\sin\left(x - \frac{\pi}{2}\right) \equiv \cos x,$$

and we now have

$$\sin\left(\frac{\pi}{2} - x\right) \equiv \cos x.$$

EXERCISE SET 9.6

1. Given that $\sin x = 0.1425$ in quadrant II, find the other circular function values, using identities.

2. Given that $\tan x = 0.8012$ in quadrant III, find the other circular function values, using identities.

---

*There are four identities in this list. Two of them are obtained by taking the top signs, the other two by taking the bottom signs.

3. ▦ Given that cot x = 0.7534 and sin x > 0, find the other circular function values, using identities.

4. ▦ Given that sin $(\pi/8)$ = 0.38268, use identities to find:

a) $\cos \frac{\pi}{8}$;   b) $\cos \frac{5\pi}{8}$;

c) $\sin \frac{5\pi}{8}$;   d) $\sin \frac{-3\pi}{8}$;

e) $\cos \frac{-3\pi}{8}$.

5. Determine whether the cotangent function is even or odd.

6. Determine whether the secant and cosecant functions are even or odd.

7. Make a diagram to show that $\sin(-\theta) = -\sin\theta$ when θ is between 180° and 270°.

8. Make a diagram to show that $\cos(-\theta) = \cos\theta$ when θ is between −360° and −540°.

9. Using function values from a table, construct a graph of the sine function between -2π and 2π.

10. Using function values from a table, construct a graph of the cosine function between -2π and 2π.

11. For the sine function, determine the domain, range, and period.

12. For the cosine function, determine the domain, range, and period.

13. For the tangent function, determine the domain, range, and period.

14. For the cotangent function, determine the domain, range, and period.

15. Find an identity for $\cos(\pi/2 - x)$.

16. Find an identity for $\tan(\pi/2 - x)$.

17. Find an identity for $\tan(x - \pi/2)$.

18. Find an identity for $\tan(x + \pi/2)$.

19. Find an identity for $\sec(x - \pi/2)$.

20. Find an identity for $\csc(\pi/2 - x)$.

☆

21. Graph the secant function.

22. Graph the cosecant function.

23. Graph $y = \tan(x/2)$.

24. Graph $y = \cot 2x$.

25. The coordinates of any point on the unit circle are (cos s, sin s), for some suitable s. Show that for a circle with center at the origin of radius r, the coordinates of any point on the circle are (r cos s, r sin s).

CHAPTER 9 REVIEW

For angles of the following measures, state in which quadrant the terminal side lies, convert to radian measure in terms of π, and convert to radian measure not in terms of π. Use 3.14 for π.

1. 87°   **2.** 145°   **3.** 30°   **4.** −30°

Convert to degree measure.

5. $\frac{3\pi}{2}$ **6.** 4π

7. Find the length of an arc of a circle, given a central angle of $\pi/4$ and a radius of 7 cm.

8. An arc 18 m long on a circle of radius 8 m subtends an angle of how many radians? how many degrees, to the nearest degree?

9. A phonograph record revolves at 45 rpm. What is the linear velocity in cm/min of a point 4 cm from the center?

10. An automobile wheel has a diameter of 26 in. If the car travels 30 mph, what is the angular velocity in radians/hr of a point on the edge of the wheel?

11. Find the six trigonometric function values for the angle θ as shown.

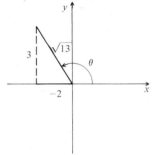

12. Complete the following table.

θ	0°	30°	45°	60°	90°	270°
$\sin \theta$						
$\cos \theta$						
$\tan \theta$						
$\cot \theta$						
$\sec \theta$						
$\csc \theta$						

Find the following exactly. Do not use the table.

13. sin 495° **14.** tan −315° **15.** cot 210°

16. Graph the sine function and state its domain, range, and period.

Given that $\tan \theta = 2/\sqrt{5}$ and that the terminal side is in quadrant III, find the following.

17. $\sin \theta$ **18.** $\cos \theta$ **19.** $\cot \theta$ **20.** $\sec \theta$ **21.** $\csc \theta$

Given $\sin \theta = 0.6820$, $\cos \theta = 0.7314$, $\tan \theta = 0.9325$, $\cot \theta = 1.0724$, $\sec \theta = 1.3673$, and $\csc \theta = 1.4663$, find the following.

22. $\sin (90° - \theta)$ **23.** $\cos (90° - \theta)$ **24.** $\tan (90° - \theta)$
25. $\cot (90° - \theta)$ **26.** $\sec (90° - \theta)$ **27.** $\csc (90° - \theta)$

28. Convert to degrees and minutes: 22.20°.

29. Convert to degrees and decimal parts of degrees: 47°33′.

Use Table 3 at the back of the book to find the following.

30. cos 8°20′ **31.** tan 27°14′ **32.** sec 44°30′

33. $\cot \theta = 3.450$ and $0° < \theta < 90°$. Find θ.

34. $\sin \theta = 0.6293$ and $0° < \theta < 90°$. Find θ.

Trigonometric Identities, Inverse Functions, and Equations

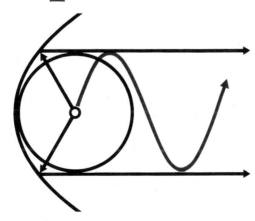

10.1 SUM AND DIFFERENCE FORMULAS

We now develop some important identities involving sums or differences of two numbers (or angles), first an identity for the cosine of the difference of two numbers. We shall use the Greek letters α (alpha) and β (beta) for these numbers. Let us consider a real number α in the interval $[\pi/2, \pi]$ and a real number β in the interval $[0, \pi/2]$. These determine points A and B on the unit circle (Fig. 1). The arc length s is $\alpha - \beta$, and we know that $0 \leq s \leq \pi$. Recall that the coordinates of A are $(\cos \alpha, \sin \alpha)$, and the coordinates of B are $(\cos \beta, \sin \beta)$. Using the distance formula, we can write an expression for the square of the distance AB:

$$AB^2 = (\cos \alpha - \cos \beta)^2 + (\sin \alpha - \sin \beta)^2.$$

Figure 1

This can be simplified as follows:

$$\begin{aligned} AB^2 &= \cos^2 \alpha - 2 \cos \alpha \cos \beta + \cos^2 \beta \\ &\quad + \sin^2 \alpha - 2 \sin \alpha \sin \beta + \sin^2 \beta \\ &= (\sin^2 \alpha + \cos^2 \alpha) + (\sin^2 \beta + \cos^2 \beta) \\ &\quad - 2(\cos \alpha \cos \beta + \sin \alpha \sin \beta) \\ &= 2 - 2(\cos \alpha \cos \beta + \sin \alpha \sin \beta). \end{aligned}$$

Now let us imagine rotating the circle above so that point B is at $(1, 0)$, as in Fig. 2. The coordinates of point A are now $(\cos s, \sin s)$. The distance AB has not changed. Again we use the distance formula to write an expression for the square of AB:

$$AB^2 = (\cos s - 1)^2 + (\sin s - 0)^2.$$

This simplifies as follows:

$$\begin{aligned} AB^2 &= \cos^2 s - 2 \cos s + 1 + \sin^2 s \\ &= (\sin^2 s + \cos^2 s) + 1 - 2 \cos s = 2 - 2 \cos s. \end{aligned}$$

Figure 2

Equating our two expressions for AB^2 and simplifying, we obtain

$$\cos s = \cos \alpha \cos \beta + \sin \alpha \sin \beta. \tag{1}$$

But $s = \alpha - \beta$, so we have the equation

$$\cos (\alpha - \beta) = \cos \alpha \cos \beta + \sin \alpha \sin \beta. \tag{2}$$

This formula holds for any numbers α and β for which $\alpha - \beta$ is the length of the shortest arc from A to B; in other words, when $0 \leq \alpha - \beta \leq \pi$.

Example 1 Find $\cos(3\pi/4 - \pi/3)$.

$$\cos\left(\frac{3\pi}{4} - \frac{\pi}{3}\right) = \cos\frac{3\pi}{4}\cos\frac{\pi}{3} + \sin\frac{3\pi}{4}\sin\frac{\pi}{3}$$

$$= -\frac{\sqrt{2}}{2}\cdot\frac{1}{2} + \frac{\sqrt{2}}{2}\cdot\frac{\sqrt{3}}{2}$$

$$= \frac{\sqrt{2}}{4}(\sqrt{3} - 1) \quad \text{or} \quad \frac{\sqrt{6} - \sqrt{2}}{4}$$

Example 2 Find $\cos 15°$.

$$\cos 15° = \cos(45° - 30°)$$

$$= \cos 45° \cos 30° + \sin 45° \sin 30°$$

$$= \frac{\sqrt{2}}{2}\cdot\frac{\sqrt{3}}{2} + \frac{\sqrt{2}}{2}\cdot\frac{1}{2}$$

$$= \frac{\sqrt{2}}{4}(\sqrt{3} + 1) \quad \text{or} \quad \frac{\sqrt{6} + \sqrt{2}}{4}$$

Formula (1) above holds when s is the length of the shortest arc from A to B. Given any real numbers α and β, the length of the shortest arc from A to B is not always $\alpha - \beta$. However, $\cos s$ is always equal to $\cos(\alpha - \beta)$. We will not prove this in detail, but Fig. 3 helps to show how a proof would be done. The length of the shortest arc from A to B always turns out to be $\alpha - \beta$ or $\beta - \alpha$, plus some multiple of 2π.

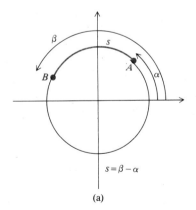
(a) $s = \beta - \alpha$

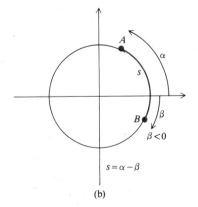
(b) $s = \alpha - \beta$, $\beta < 0$

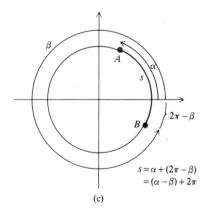
(c) $s = \alpha + (2\pi - \beta) = (\alpha - \beta) + 2\pi$

Figure 3

Since $s = |\alpha - \beta| + 2k\pi$, where k is an integer, we have

$$\cos s = \cos(|\alpha - \beta| + 2k\pi)$$
$$= \cos|\alpha - \beta|,$$

since the cosine function has a period of 2π. Now let us recall the identity $\cos(-s) \equiv \cos s$. From this identity we see that we can drop the absolute value signs and then we know that $\cos s = \cos(\alpha - \beta)$ for any real numbers α and β. Thus Formula (2) holds for all real numbers α and β. That formula is thus the identity we sought:

$$\cos(\alpha - \beta) \equiv \cos\alpha\cos\beta + \sin\alpha\sin\beta.$$

Example 3 Simplify:

$$\sin\left(-\frac{5\pi}{2}\right)\sin\frac{\pi}{2} + \cos\frac{\pi}{2}\cos\left(-\frac{5\pi}{2}\right).$$

This is equal to

$$\cos\frac{\pi}{2}\cos\left(-\frac{5\pi}{2}\right) + \sin\frac{\pi}{2}\sin\left(-\frac{5\pi}{2}\right).$$

By the identity above we can simplify to

$$\cos\left[\frac{\pi}{2} - \left(-\frac{5\pi}{2}\right)\right], \quad \text{or} \quad \cos\frac{6\pi}{2}, \quad \text{or} \quad \cos 3\pi,$$

which is -1.

Other Formulas

The other sum and difference formulas we seek follow easily from the one we have just derived. Let us consider $\cos(\alpha + \beta)$. This is equal to $\cos[\alpha - (-\beta)]$, and by the identity above, we have

$$\cos(\alpha + \beta) \equiv \cos\alpha\cos(-\beta) + \sin\alpha\sin(-\beta).$$

But $\cos(-\beta) \equiv \cos\beta$ and $\sin(-\beta) \equiv -\sin\beta$, so the identity we seek is

$$\cos(\alpha + \beta) \equiv \cos\alpha\cos\beta - \sin\alpha\sin\beta.$$

To develop an identity for the sine of a sum, we recall the cofunction identity $\sin\theta \equiv \cos(\pi/2 - \theta)$. In this identity we shall substitute $\alpha + \beta$ for θ, obtaining $\sin(\alpha + \beta) \equiv \cos[\pi/2 - (\alpha + \beta)]$. We can now use the identity for the cosine of a difference. We get

$$\sin(\alpha + \beta) \equiv \cos\left[\frac{\pi}{2} - (\alpha + \beta)\right]$$

$$\equiv \cos\left[\left(\frac{\pi}{2} - \alpha\right) - \beta\right]$$

$$\equiv \cos\left(\frac{\pi}{2} - \alpha\right)\cos\beta + \sin\left(\frac{\pi}{2} - \alpha\right)\sin\beta$$

$$\equiv \sin\alpha\cos\beta + \cos\alpha\sin\beta.$$

Thus the identity we seek is

$$\sin(\alpha + \beta) \equiv \sin\alpha\cos\beta + \cos\alpha\sin\beta.$$

To find a formula for the sine of a difference, we can use the identity just derived, substituting $-\beta$ for β. We obtain

$$\sin(\alpha - \beta) \equiv \sin\alpha\cos\beta - \cos\alpha\sin\beta.$$

A formula for the tangent of a sum can be derived as follows, using identities already established:

$$\tan(\alpha + \beta) \equiv \frac{\sin(\alpha + \beta)}{\cos(\alpha + \beta)}$$

$$\equiv \frac{\sin\alpha\cos\beta + \cos\alpha\sin\beta}{\cos\alpha\cos\beta - \sin\alpha\sin\beta} \cdot \frac{\frac{1}{\cos\alpha\cos\beta}}{\frac{1}{\cos\alpha\cos\beta}}$$

$$\equiv \frac{\frac{\sin\alpha\cos\beta}{\cos\alpha\cos\beta} + \frac{\cos\alpha\sin\beta}{\cos\alpha\cos\beta}}{\frac{\cos\alpha\cos\beta}{\cos\alpha\cos\beta} - \frac{\sin\alpha\sin\beta}{\cos\alpha\cos\beta}}$$

$$\equiv \frac{\frac{\sin\alpha}{\cos\alpha} + \frac{\sin\beta}{\cos\beta}}{1 - \frac{\sin\alpha\sin\beta}{\cos\alpha\cos\beta}} \equiv \frac{\tan\alpha + \tan\beta}{1 - \tan\alpha\tan\beta}.$$

Similarly, a formula for the tangent of a difference can be established. Following is a list of the sum and difference formulas. These should be memorized.

$$\cos(\alpha \mp \beta) \equiv \cos\alpha\cos\beta \pm \sin\alpha\sin\beta*$$

$$\sin(\alpha \mp \beta) \equiv \sin\alpha\cos\beta \mp \cos\alpha\sin\beta$$

$$\tan(\alpha \mp \beta) \equiv \frac{\tan\alpha \mp \tan\beta}{1 \pm \tan\alpha\tan\beta}$$

The identities involving sines and tangents can be used in the same way as those involving cosines in the earlier examples.

Example 4 Find $\tan 15°$.

$$\tan 15° = \tan(45° - 30°) = \frac{\tan 45° - \tan 30°}{1 + \tan 45° \tan 30°}$$

$$= \frac{1 - \sqrt{3}/3}{1 + \sqrt{3}/3} = \frac{3 - \sqrt{3}}{3 + \sqrt{3}}$$

Example 5 Simplify $\sin(\pi/3)\cos\pi + \sin\pi\cos(\pi/3)$.

This is equal to

$$\sin\frac{\pi}{3}\cos\pi + \cos\frac{\pi}{3}\sin\pi.$$

Thus by the fourth identity in the list above, we can simplify to

$$\sin\left(\frac{\pi}{3} + \pi\right), \quad \text{or} \quad \sin\frac{4\pi}{3}, \quad \text{or} \quad -\frac{\sqrt{3}}{2}.$$

Angles Between Lines

Recall that a nonvertical line has an equation $y = mx + b$, where m is the slope. Note that such a line makes an angle with the positive half of the x-axis whose tangent is the slope of the line. As in Fig. 4, $\tan\theta = m$.

One of the identities just developed gives us an easy way to find an angle formed by two lines. We shall consider the case in which neither line is vertical. Thus we shall consider two lines with equations as follows.

$$l_1: y = m_1x + b_1, \quad \text{and} \quad l_2: y = m_2x + b_2.$$

Figure 4

---

*There are six identities here, half of them obtained by taking the colored signs.

The slopes m_1 and m_2 are the tangents of the angles θ_1 and θ_2 that the lines form with the positive half of the x-axis (Fig. 5). Thus we have $m_1 = \tan \theta_1$ and $m_2 = \tan \theta_2$. We wish to find the measure of the smallest angle through which l_1 can be rotated in the positive direction to get to l_2. In this case that angle is $\theta_2 - \theta_1$, or ϕ. We proceed as follows:

$$\tan \phi = \tan(\theta_2 - \theta_1) = \frac{\tan \theta_2 - \tan \theta_1}{1 + \tan \theta_2 \tan \theta_1}$$

$$= \frac{m_2 - m_1}{1 + m_2 m_1}.$$

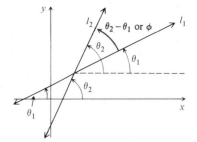

Figure 5

Suppose we had taken the lines in the reverse order. Then the smallest *positive* angle from l_1 to l_2 would be ϕ, as shown in Fig. 6. Note that $\tan \theta_1 = m_1$. Since $\theta_2 = \alpha + 180°$, $\tan \theta_2 = \tan \alpha$ and $\tan \theta_2 = m_2$. Thus the same formula for $\tan \phi$ derived above holds. In the first case ϕ is an acute angle, so $\tan \phi$ will be positive. In the second case ϕ is obtuse, so $\tan \phi$ will be negative. Thus we have a general formula for finding the angle from one line to another.

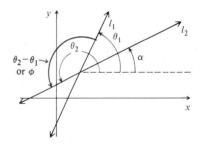

Figure 6

THEOREM 1

The smallest positive angle ϕ from a line l_1 to a line l_2 (where neither line is vertical) can be obtained from

$$\tan \phi = \frac{m_2 - m_1}{1 + m_2 m_1},$$

where m_1 and m_2 are the slopes of l_1 and l_2, respectively. If $\tan \phi$ is positive, ϕ is acute. If $\tan \phi$ is negative, then ϕ is obtuse.

Example 6 Two lines have slopes as follows: l_1 has slope $\sqrt{3}/3$, l_2 has slope $\sqrt{3}$. How large is the smallest positive angle from l_1 to l_2?

We have $m_1 = \sqrt{3}/3$ and $m_2 = \sqrt{3}$. Then

$$\tan \phi = \frac{\sqrt{3} - \frac{\sqrt{3}}{3}}{1 + \sqrt{3} \cdot \frac{\sqrt{3}}{3}} = \frac{\frac{3\sqrt{3} - \sqrt{3}}{3}}{\frac{3+3}{3}} = \frac{2\sqrt{3}}{6} = \frac{\sqrt{3}}{3}.$$

Since $\tan \phi$ is positive, ϕ is acute. Hence

$$\phi = \frac{\pi}{6} \quad \text{or} \quad 30°.$$

Example 7 Find the smallest positive angle from l_1 to l_2:

$$l_1: 2y = x + 3, \quad l_2: 3y + 4x + 6 = 0.$$

We first solve for y to obtain the slopes:

$$l_1: y = \frac{1}{2}x + \frac{3}{2}, \quad \text{so} \quad m_1 = \frac{1}{2};$$

$$l_2: y = -\frac{4}{3}x - 2, \quad \text{so} \quad m_2 = -\frac{4}{3};$$

$$\tan \phi = \frac{-\frac{4}{3} - \frac{1}{2}}{1 + \left(-\frac{4}{3}\right)\left(\frac{1}{2}\right)} = \frac{\frac{-8-3}{6}}{\frac{6-4}{6}} = -\frac{11}{2} = -5.5.$$

Since $\tan \phi$ is negative, we know that ϕ is obtuse. From the table we find an angle whose tangent is 5.5 to be about $79°40'$. This is a reference angle. Subtracting from $180°$, we obtain

$$\phi = 100°20', \quad \text{approximately}.$$

EXERCISE SET 10.1

Use the sum and difference identities to evaluate each of the following.

1. $\sin 75°$
 (*Hint*: $75° = 45° + 30°$.)
2. $\cos 75°$
3. $\sin 15°$
 (*Hint*: $15° = 45° - 30°$.)
4. $\cos 15°$
5. $\sin 105°$
 (*Hint*: Use the results of Exercises 1 and 2.)
6. $\cos 105°$
7. $\tan 75°$
8. $\cot 15°$

Assume that $\sin u = \frac{3}{5}$ and $\sin v = \frac{4}{5}$ and that u and v are between 0 and $\pi/2$. Then evaluate each of the following.

9. $\sin(u + v)$
10. $\sin(u - v)$
11. $\cos(u + v)$
12. $\cos(u - v)$
13. $\tan(u + v)$
14. $\tan(u - v)$

Assume that $\sin \theta = 0.6249$ and $\cos \phi = 0.1102$ and that θ and ϕ are both first-quadrant angles. Evaluate.

15. ▦ $\sin(\theta + \phi)$
16. ▦ $\cos(\theta + \phi)$

Simplify or evaluate.

17. $\sin 37° \cos 22° + \cos 37° \sin 22°$
18. $\cos 37° \cos 22° - \sin 22° \sin 37°$
19. $\dfrac{\tan 20° + \tan 32°}{1 - \tan 20° \tan 32°}$
20. $\dfrac{\tan 35° - \tan 12°}{1 + \tan 35° \tan 12°}$

21. $\sin(\alpha + \beta) + \sin(\alpha - \beta)$
22. $\sin(\alpha + \beta) - \sin(\alpha - \beta)$
23. $\cos(\alpha + \beta) + \cos(\alpha - \beta)$
24. $\cos(\alpha + \beta) - \cos(\alpha - \beta)$
25. $\cos(u + v)\cos v + \sin(u + v)\sin v$
26. $\sin(u - v)\cos v + \cos(u - v)\sin v$

Find the angle from l_1 to l_2.

27. $l_1: 3y = \sqrt{3}x + 2,$
    $l_2: 3y + \sqrt{3}x = -3$
28. $l_1: 3y = \sqrt{3}x + 3,$
    $l_2: y = \sqrt{3}x + 2$
29. $l_1: 2x = 3 - 2y,$
    $l_2: x + y = 5$
30. $l_1: 2x = 3 + 2y,$
    $l_2: x - y = 5$
31. $l_1: 2x - 5y + 1 = 0,$
    $l_2: 3x + y - 7 = 0$
32. $l_1: 2x + y - 4 = 0,$
    $l_2: y - 2x + 5 = 0$
33. $l_1: y = 3,$
    $l_2: x + y = 5$
34. $l_1: y = 5,$
    $l_2: x - y = 2$

---

35. Find an identity for $\sin 2\theta$.
    (Hint: $2\theta = \theta + \theta$.)
36. Find an identity for $\cos 2\theta$.
    (Hint: $2\theta = \theta + \theta$.)
37. Derive an identity for $\cot(\alpha + \beta)$ in terms of $\cot \alpha$ and $\cot \beta$.
38. Derive an identity for $\cot(\alpha - \beta)$ in terms of $\cot \alpha$ and $\cot \beta$.

The cofunction identities can be derived from the sum and difference formulas. Derive identities for the following.

39. $\sin\left(\dfrac{\pi}{2} - x\right)$
40. $\cos\left(\dfrac{\pi}{2} - x\right)$

Find the slope of line l_1, where m_2 is the slope of line l_2 and ϕ is the smallest positive angle from l_1 to l_2.

41. $m_2 = \dfrac{4}{3}$, $\phi = 45°$.
42. $m_2 = \dfrac{2}{3}$, $\phi = \text{Arctan}\dfrac{5}{2}$.
43. Line l_1 contains the points $(-2, 4)$ and $(5, -1)$. Find the slope of line l_2 such that the angle from l_1 to l_2 is $45°$.
44. Line l_1 contains $(3, -1)$ and $(-4, 2)$. Find the slope of line l_2 such that the angle from l_1 to l_2 is $-45°$.
45. Line l_1 contains the points $(-2.123, 3.899)$ and $(-4.892, -0.9012)$ and line l_2 contains $(0, -3.814)$ and $(5.925, 4.013)$. Find the smallest positive angle from l_1 to l_2.
46. Line l_1 contains the points $(-3, 7)$ and $(-3, -2)$. Line l_2 contains $(0, -4)$ and $(2, 6)$. Find the smallest positive angle from l_1 to l_2.

★

47. Find an identity for $\cos(\alpha + \beta)$ involving only cosines.
48. Show that if $\alpha + \beta = \pi/2$, then
$$\sin(\alpha - \beta) = \cos 2\beta = -\cos 2\alpha.$$
49. Find an identity for $\sin(\alpha + \beta + \gamma)$.
50. Show that if $\alpha + \beta + \gamma = \pi$, then
$$\cos\alpha \cos\beta + \cos\gamma - \sin\alpha \sin\beta = 0.$$

10.2 SOME IMPORTANT IDENTITIES

Two important classes of trigonometric identities are known as the half-angle identities and the double-angle identities.

Double-Angle Identities

To develop these identities we shall use the sum formulas from the preceding section. We first develop a formula for sin 2θ. Recall that

$$\sin(\alpha + \beta) \equiv \sin\alpha \cos\beta + \cos\alpha \sin\beta.$$

We shall consider a number θ and substitute it for both α and β in this identity. We obtain

$$\sin(\theta + \theta) \equiv \sin 2\theta \equiv \sin\theta \cos\theta + \cos\theta \sin\theta$$
$$\equiv 2\sin\theta \cos\theta.$$

The identity we seek is

$$\sin 2\theta \equiv 2\sin\theta \cos\theta.$$

Example 1 If $\sin\theta = \frac{3}{8}$ and θ is in the first quadrant, what is $\sin 2\theta$? From Fig. 7, we see that $\cos\theta = \sqrt{55}/8$. Thus

$$\sin 2\theta = 2\sin\theta \cos\theta = 2 \cdot \frac{3}{8} \cdot \frac{\sqrt{55}}{8} = \frac{3\sqrt{55}}{32}.$$

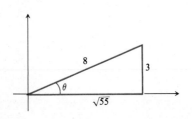

Figure 7

Double-angle identities for the cosine and tangent functions can be derived in much the same way as the above identity:

$$\cos(\alpha + \beta) \equiv \cos\alpha \cos\beta - \sin\alpha \sin\beta$$
$$\cos 2\theta \equiv \cos(\theta + \theta) \equiv \cos\theta \cos\theta - \sin\theta \sin\theta$$
$$\equiv \cos^2\theta - \sin^2\theta$$
$$\cos 2\theta \equiv \cos^2\theta - \sin^2\theta.$$

Now we derive an identity for tan 2θ:

$$\tan(\alpha + \beta) \equiv \frac{\tan\alpha + \tan\beta}{1 - \tan\alpha \tan\beta}$$

$$\tan 2\theta \equiv \tan(\theta + \theta) \equiv \frac{\tan\theta + \tan\theta}{1 - \tan\theta \tan\theta} \equiv \frac{2\tan\theta}{1 - \tan^2\theta}$$

$$\tan 2\theta \equiv \frac{2\tan\theta}{1 - \tan^2\theta}.$$

Example 2 Given that $\tan\theta = -\frac{3}{4}$ and θ is in the second quadrant, find sin 2θ, cos 2θ, tan 2θ, and the quadrant in which 2θ lies.

By drawing a diagram as shown in Fig. 8, we find that $\sin\theta = \frac{3}{5}$ and $\cos\theta = -\frac{4}{5}$. Now,

$$\sin 2\theta = 2\sin\theta \cos\theta$$
$$= 2 \cdot \frac{3}{5} \cdot \left(-\frac{4}{5}\right) = -\frac{24}{25};$$

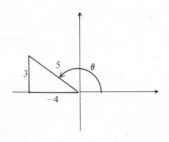

Figure 8

$$\cos 2\theta = \cos^2 \theta - \sin^2 \theta = \left(-\frac{4}{5}\right)^2 - \left(\frac{3}{5}\right)^2$$

$$= \frac{16}{25} - \frac{9}{25} = \frac{7}{25};$$

$$\tan 2\theta = \frac{2 \tan \theta}{1 - \tan^2 \theta} = \frac{2 \cdot \left(-\frac{3}{4}\right)}{1 - \left(-\frac{3}{4}\right)^2}$$

$$= -\frac{\frac{3}{2}}{1 - \frac{9}{16}} = -\frac{24}{7}.$$

Since $\sin 2\theta$ is negative and $\cos 2\theta$ is positive, we know that 2θ is in quadrant IV. Note that $\tan 2\theta$ could have been found more easily in this case by dividing the values of $\sin 2\theta$ and $\cos 2\theta$.

Two other useful identities for $\cos 2\theta$ can easily be derived, as follows:

$$\cos 2\theta \equiv \cos^2 \theta - \sin^2 \theta$$
$$\equiv (1 - \sin^2 \theta) - \sin^2 \theta \quad \text{Using } \sin^2 \theta + \cos^2 \theta \equiv 1$$
$$\equiv 1 - 2 \sin^2 \theta.$$

Similarly,

$$\cos 2\theta \equiv \cos^2 \theta - \sin^2 \theta$$
$$\equiv \cos^2 \theta - (1 - \cos^2 \theta)$$
$$\equiv 2 \cos^2 \theta - 1.$$

Solving these two identities for $\sin^2 \theta$ and $\cos^2 \theta$, respectively, we obtain two more identities that are often useful. Following is a list of double-angle identities. It should be memorized.

$$\sin 2\theta \equiv 2 \sin \theta \cos \theta,$$
$$\cos 2\theta \equiv \cos^2 \theta - \sin^2 \theta$$
$$\equiv 1 - 2 \sin^2 \theta$$
$$\equiv 2 \cos^2 \theta - 1,$$
$$\tan 2\theta \equiv \frac{2 \tan \theta}{1 - \tan^2 \theta},$$
$$\sin^2 \theta \equiv \frac{1 - \cos 2\theta}{2},$$
$$\cos^2 \theta \equiv \frac{1 + \cos 2\theta}{2}$$

By division and the last two identities, it is always easy to deduce the following identity, which is also often useful:

$$\tan^2 \theta \equiv \frac{1 - \cos 2\theta}{1 + \cos 2\theta}.$$

From the basic identities (in the lists to be memorized), others can be obtained.

Example 3 Find a formula for $\sin 3\theta$ in terms of function values of θ.

$$\sin 3\theta \equiv \sin (2\theta + \theta)$$
$$\equiv \sin 2\theta \cos \theta + \cos 2\theta \sin \theta$$
$$\equiv (2 \sin \theta \cos \theta) \cos \theta + (2 \cos^2 \theta - 1) \sin \theta$$
$$\equiv 2 \sin \theta \cos^2 \theta + 2 \sin \theta \cos^2 \theta - \sin \theta$$
$$\sin 3\theta \equiv 4 \sin \theta \cos^2 \theta - \sin \theta$$

Example 4 Find a formula for $\cos^3 x$ in terms of function values of x or 2x, raised only to the first power.

$$\cos^3 x \equiv \cos^2 x \cos x \equiv \frac{1 + \cos 2x}{2} \cos x$$

Half-Angle Identities

To develop these identities, we use three of the previously developed ones. As shown below, we take square roots and replace θ by $\phi/2$:

$$\sin^2 \theta \equiv \frac{1 - \cos 2\theta}{2} \rightarrow \left|\sin \frac{\phi}{2}\right| \equiv \sqrt{\frac{1 - \cos \phi}{2}},$$

$$\cos^2 \theta \equiv \frac{1 + \cos 2\theta}{2} \rightarrow \left|\cos \frac{\phi}{2}\right| \equiv \sqrt{\frac{1 + \cos \phi}{2}},$$

$$\tan^2 \theta \equiv \frac{1 - \cos 2\theta}{1 + \cos 2\theta} \rightarrow \left|\tan \frac{\phi}{2}\right| \equiv \sqrt{\frac{1 - \cos \phi}{1 + \cos \phi}}.$$

The half-angle formulas are those on the right above. We can eliminate the absolute value signs by introducing $\pm$ signs, with the understanding that we use $+$ or $-$ depending on the quadrant in which the angle $\phi/2$ lies. We thus obtain these formulas in the following form:

$$\sin \frac{\phi}{2} \equiv \pm \sqrt{\frac{1 - \cos \phi}{2}},$$

$$\cos \frac{\phi}{2} \equiv \pm \sqrt{\frac{1 + \cos \phi}{2}}, \qquad \tan \frac{\phi}{2} \equiv \pm \sqrt{\frac{1 - \cos \phi}{1 + \cos \phi}}.$$

10.2 SOME IMPORTANT IDENTITIES

These formulas should be memorized.

There are two other formulas for $\tan(\phi/2)$ that are often useful. They can be obtained as follows:

$$\left|\tan\frac{\phi}{2}\right| \equiv \sqrt{\frac{1-\cos\phi}{1+\cos\phi}} \equiv \sqrt{\frac{1-\cos\phi}{1+\cos\phi}\cdot\frac{1+\cos\phi}{1+\cos\phi}}$$

$$\equiv \sqrt{\frac{1-\cos^2\phi}{(1+\cos\phi)^2}}$$

$$\equiv \sqrt{\frac{\sin^2\phi}{(1+\cos\phi)^2}} \equiv \frac{|\sin\phi|}{|1+\cos\phi|}.$$

Now $1 + \cos\phi$ cannot be negative because $\cos\phi$ is never less than -1. Hence the absolute value signs are not necessary in the denominator. As the graph in Fig. 9 shows, $\tan(\phi/2)$ and $\sin\phi$ have the same sign for all ϕ for which $\tan(\phi/2)$ is defined. Thus we can dispense with the other absolute value signs, and obtain the formula we seek. A second formula can be obtained similarly:

$$\tan\frac{\phi}{2} \equiv \frac{\sin\phi}{1+\cos\phi}, \quad \tan\frac{\phi}{2} \equiv \frac{1-\cos\phi}{\sin\phi}.$$

These formulas have the advantage that they give the sign of $\tan(\phi/2)$ directly.

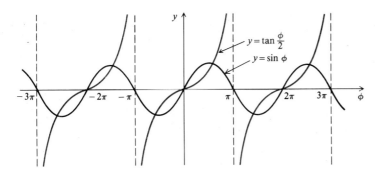

Figure 9

Example 5 Find $\sin 15°$.

$$\sin 15° = \sin\frac{30°}{2} = \pm\sqrt{\frac{1-\cos 30°}{2}}$$

$$= \pm\sqrt{\frac{1-(\sqrt{3}/2)}{2}} = \pm\sqrt{\frac{2-\sqrt{3}}{4}} = \frac{\sqrt{2-\sqrt{3}}}{2}$$

(The expression is positive, because $15°$ is in the first quadrant.)

Example 6 Find $\tan \frac{\pi}{8}$.

$$\tan \frac{\pi}{8} = \tan \frac{\frac{\pi}{4}}{2} = \frac{\sin \frac{\pi}{4}}{1 + \cos \frac{\pi}{4}} = \frac{\frac{\sqrt{2}}{2}}{1 + \frac{\sqrt{2}}{2}}$$

$$= \frac{\sqrt{2}}{2 + \sqrt{2}}$$

$$= \sqrt{2} - 1 \quad \text{Rationalizing the denominator}$$

Many simplifications of trigonometric expressions are possible through the use of the identities we have developed.

Example 7 Simplify: $(1 - \cos 2x)/(4 \sin x \cos x)$.

We search the list of identities, attempting to find some substitution that will simplify the expression. In this case, one might notice that the denominator is $2(2 \sin x \cos x)$ and thus find a simplification, since $2 \sin x \cos x \equiv \sin 2x$. We now have

$$\frac{1 - \cos 2x}{2 \sin 2x}.$$

Now

$$\frac{1 - \cos \phi}{\sin \phi} \equiv \tan \frac{\phi}{2}.$$

Using this we obtain

$$\frac{1}{2} \cdot \frac{1 - \cos 2x}{\sin 2x} \equiv \frac{1}{2} \tan x.$$

Sum-Product Identities

On occasion, it is convenient to convert a product of trigonometric expressions to a sum, or the reverse. The following identities are useful in this connection. Proofs are left as exercises.

$$\sin u \cdot \cos v \equiv \frac{1}{2}[\sin (u + v) + \sin (u - v)],$$

$$\cos u \cdot \sin v \equiv \frac{1}{2}[\sin (u + v) - \sin (u - v)],$$

$$\cos u \cdot \cos v \equiv \frac{1}{2}[\cos (u - v) + \cos (u + v)],$$

$$\sin u \cdot \sin v \equiv \frac{1}{2}[\cos(u-v) - \cos(u+v)],$$

$$\sin x + \sin y \equiv 2 \sin \frac{x+y}{2} \cos \frac{x-y}{2},$$

$$\sin x - \sin y \equiv 2 \cos \frac{x+y}{2} \sin \frac{x-y}{2},$$

$$\cos y + \cos x \equiv 2 \cos \frac{x+y}{2} \cos \frac{x-y}{2},$$

$$\cos y - \cos x \equiv 2 \sin \frac{x+y}{2} \sin \frac{x-y}{2}.$$

EXERCISE SET 10.2

In Exercises 1–6, find $\sin 2\theta$, $\cos 2\theta$, $\tan 2\theta$, and the quadrant in which 2θ lies.

1. $\sin \theta = \frac{4}{5}$ (θ in quadrant I)
2. $\sin \theta = \frac{5}{13}$ (θ in quadrant I)
3. $\cos \theta = -\frac{4}{5}$ (θ in quadrant III)
4. $\cos \theta = -\frac{3}{5}$ (θ in quadrant III)
5. $\tan \theta = \frac{4}{3}$ (θ in quadrant III)
6. $\tan \theta = \frac{3}{4}$ (θ in quadrant III)
7. Find a formula for $\sin 4\theta$ in terms of function values of θ.
8. Find a formula for $\cos 4\theta$ in terms of function values of θ.
9. Find a formula for $\sin^4 \theta$ in terms of function values of θ or 2θ or 4θ, raised only to the first power.
10. Find a formula for $\cos^4 \theta$ in terms of function values of θ or 2θ or 4θ, raised only to the first power.
11. Find $\sin 75°$ without using tables. (*Hint:* $75 = 150/2$.)
12. Find $\cos 75°$ without using tables.
13. Find $\tan 75°$ without using tables.
14. Find $\tan 67.5°$ without using tables. (*Hint:* $67.5 = 135/2$.)
15. Find $\sin \frac{5\pi}{8}$ without using tables.
16. Find $\cos \frac{5\pi}{8}$ without using tables.

Given that $\sin \theta = 0.3416$ and that θ is in the first quadrant, find:

17. ▦ $\sin 2\theta$.
18. ▦ $\cos 2\theta$.
19. ▦ $\sin 4\theta$.
20. ▦ $\cos 4\theta$.
21. ▦ $\sin \frac{\theta}{2}$.
22. ▦ $\cos \frac{\theta}{2}$.

Simplify.

23. $\dfrac{\sin 2x}{2 \sin x}$
24. $\dfrac{\sin 2x}{2 \cos x}$
25. $1 - 2 \sin^2 \dfrac{x}{2}$
26. $2 \cos^2 \dfrac{x}{2} - 1$

27. $2 \sin \frac{x}{2} \cos \frac{x}{2}$

28. $2 \sin 2x \cos 2x$

29. $\cos^2 \frac{x}{2} - \sin^2 \frac{x}{2}$

30. $\cos^4 x - \sin^4 x$

31. $(\sin x + \cos x)^2 - \sin 2x$

32. $(\sin x - \cos x)^2 + \sin 2x$

33. $2 \sin^2 \frac{x}{2} + \cos x$

34. $2 \cos^2 \frac{x}{2} - \cos x$

35. $(-4 \cos x \sin x + 2 \cos 2x)^2 + (2 \cos 2x + 4 \sin x \cos x)^2$

36. $(-4 \cos 2x + 8 \cos x \sin x)^2 + (8 \sin x \cos x + 4 \cos 2x)^2$

37. $2 \sin x \cos^3 x + 2 \sin^3 x \cos x$

38. $2 \sin x \cos^3 x - 2 \sin^3 x \cos x$

☆

39. Prove the first four of the sum–product identities. (*Hint:* They follow from the sum and difference formulas of Section 10.1.)

40. Prove the last four of the sum–product identities. (*Hint:* Use the results of Exercise 39.)

41. Show that $|\sin 2x| \leq |2 \sin x|$ for all real numbers x.

Graph.

42. $f(x) = \cos^2 x - \sin^2 x$

43. $y = |\sin x \cos x|$

44. $y = 2 \sin x \cos x$

10.3 PROVING IDENTITIES

One should remember certain trigonometric identities. Then as the occasion arises, other identities can be proved using those memorized. Following is a minimal list of identities that should be learned. Note that most formulas involving cotangents, secants, and cosecants are not in this list. This is because these functions are reciprocals of the sine, cosine, and tangent, and thus formulas for the former can easily be derived from those for the latter.

A LIST OF BASIC IDENTITIES

$$\sin(-x) \equiv -\sin x,$$
$$\cos(-x) \equiv \cos x,$$
$$\tan(-x) \equiv -\tan x$$

Pythagorean identities

$$\sin^2 x + \cos^2 x \equiv 1,$$
$$1 + \tan^2 x \equiv \sec^2 x,$$
$$1 + \cot^2 x \equiv \csc^2 x$$

Cofunction identities

$$\sin\left(x \pm \frac{\pi}{2}\right) \equiv \pm\cos x,$$

$$\cos\left(x \pm \frac{\pi}{2}\right) \equiv \mp\sin x$$

SUM AND DIFFERENCE IDENTITIES

$$\sin(\alpha \pm \beta) \equiv \sin\alpha\cos\beta \pm \cos\alpha\sin\beta,$$
$$\cos(\alpha \pm \beta) \equiv \cos\alpha\cos\beta \mp \sin\alpha\sin\beta,$$
$$\tan(\alpha \pm \beta) \equiv \frac{\tan\alpha \pm \tan\beta}{1 \mp \tan\alpha\tan\beta}$$

DOUBLE-ANGLE IDENTITIES

$$\sin 2x \equiv 2\sin x \cos x,$$
$$\cos 2x \equiv \cos^2 x - \sin^2 x \equiv 1 - 2\sin^2 x \equiv \cos^2 x - 1,$$
$$\tan 2x \equiv \frac{2\tan x}{1 - \tan^2 x},$$
$$\sin^2 x \equiv \frac{1 - \cos 2x}{2},$$
$$\cos^2 x \equiv \frac{1 + \cos 2x}{2}$$

HALF-ANGLE IDENTITIES

$$\sin\frac{x}{2} \equiv \pm\sqrt{\frac{1 - \cos x}{2}},$$
$$\cos\frac{x}{2} \equiv \pm\sqrt{\frac{1 + \cos x}{2}},$$
$$\tan\frac{x}{2} \equiv \pm\sqrt{\frac{1 - \cos x}{1 + \cos x}} \equiv \frac{\sin x}{1 + \cos x} \equiv \frac{1 - \cos x}{\sin x}$$

In proving other identities, using these, it is wise to consult the list to gain an idea of how to proceed. It is usually helpful to express all function values in terms of sines and cosines. *It is most important* in proving an identity to work with one side at a time. The idea is to simplify each side separately until the same expression is obtained in the two cases.

Example 1 Prove the following identity.

$$\tan^2 x - \sin^2 x \equiv \sin^2 x \tan^2 x$$

$$\begin{array}{c|c} \dfrac{\sin^2 x}{\cos^2 x} - \sin^2 x & \sin^2 x \dfrac{\sin^2 x}{\cos^2 x} \\[2mm] \dfrac{\sin^2 x - \sin^2 x \cos^2 x}{\cos^2 x} & \dfrac{\sin^4 x}{\cos^2 x} \\[2mm] \dfrac{\sin^2 x (1 - \cos^2 x)}{\cos^2 x} & \\[2mm] \dfrac{\sin^2 x \sin^2 x}{\cos^2 x} & \\[2mm] \dfrac{\sin^4 x}{\cos^2 x} & \end{array}$$

In this proof, a vertical line has been drawn as a reminder to work separately with the two sides. Everything has been expressed in terms of sines and cosines and then the two sides are simplified.

We work with each side separately because, as we saw in Chapter 2, "doing the same thing to both sides of an equation" does not always yield equivalent equations.

Example 2 Prove the following identity.

$$\frac{\sin 2\theta}{\sin \theta} - \frac{\cos 2\theta}{\cos \theta} \equiv \sec \theta$$

$$\begin{array}{c|c} \dfrac{2 \sin \theta \cos \theta}{\sin \theta} - \dfrac{\cos^2 \theta - \sin^2 \theta}{\cos \theta} & \dfrac{1}{\cos \theta} \\[2mm] \dfrac{2 \cos^2 \theta - \cos^2 \theta + \sin^2 \theta}{\cos \theta} & \\[2mm] \dfrac{\cos^2 \theta + \sin^2 \theta}{\cos \theta} & \\[2mm] \dfrac{1}{\cos \theta} & \end{array}$$

HINTS FOR PROVING IDENTITIES

1. Work with one side at a time.
2. Work with the more complex side first.

3. Do the algebraic manipulations, such as adding and subtracting.
4. Converting all expressions to sines and cosines is often helpful.
5. Try something! Identities are never proved by staring.

An Identity for $c \sin ax + d \cos ax$

An expression $c \sin ax + d \cos ax$, where a, c, and d are constants, is equivalent to an expression $A \sin (ax + b)$, where A and b are also constants. To prove this we shall look for the numbers A and b. First, let us consider the numbers $c/\sqrt{c^2 + d^2}$ and $d/\sqrt{c^2 + d^2}$. These numbers are the coordinates of a point on the unit circle because

$$\left(\frac{c}{\sqrt{c^2 + d^2}}\right)^2 + \left(\frac{d}{\sqrt{c^2 + d^2}}\right)^2 = 1.$$

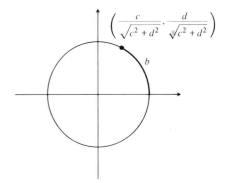

Figure 10

(See Fig. 10.) Thus there is a number b for which these numbers are the cosine and sine, respectively. In other words,

$$\frac{c}{\sqrt{c^2 + d^2}} = \cos b,$$

and

$$\frac{d}{\sqrt{c^2 + d^2}} = \sin b,$$

for some number b. Now we will consider the expression $c \sin ax + d \cos ax$, and multiply it by 1, as follows:

$$\frac{\sqrt{c^2 + d^2}}{\sqrt{c^2 + d^2}} [c \sin ax + d \cos ax]$$

$$\equiv \sqrt{c^2 + d^2} \left[(\sin ax) \cdot \frac{c}{\sqrt{c^2 + d^2}} + (\cos ax) \frac{d}{\sqrt{c^2 + d^2}} \right]$$

$$\equiv \sqrt{c^2 + d^2} \, [\sin ax \cos b + \cos ax \sin b].$$

Now, by the identity for the sine of a sum, we obtain

$$\sqrt{c^2 + d^2} \sin (ax + b).$$

Thus we have an identity as follows.

THEOREM 2

For any numbers a, c, and d,

$$c \sin ax + d \cos ax \equiv A \sin (ax + b),$$

where $A = \sqrt{c^2 + d^2}$ and b is a number whose cosine is c/A and whose sine is d/A.

Example 3 Find an expression equivalent to $\sqrt{3}\sin x + \cos x$ that involves only the sine function.

In this expression, referring to Theorem 2, we see that $c = \sqrt{3}$, $d = 1$, and $a = 1$. From this we find A and b:

$$A = \sqrt{c^2 + d^2} = \sqrt{3 + 1} = 2,$$

$$\cos b = \frac{\sqrt{3}}{2}, \quad \text{and} \quad \sin b = \frac{1}{2}.$$

We can use 30° or $\pi/6$ for b. Thus

$$\sqrt{3}\sin x + \cos x \equiv 2\sin\left(x + \frac{\pi}{6}\right).$$

Example 4 Find an expression equivalent to $\sqrt{3}\sin(\pi/4)t + \cos(\pi/4)t$ involving only the sine function.

In this case $c = \sqrt{3}$, $d = 1$, and $a = \frac{\pi}{4}$:

$$A = \sqrt{3 + 1} = 2,$$

$$\cos b = \frac{\sqrt{3}}{2}, \quad \text{and} \quad \sin b = \frac{1}{2}.$$

Thus we have

$$\sqrt{3}\sin\frac{\pi}{4}t + \cos\frac{\pi}{4}t \equiv 2\sin\left(\frac{\pi}{4}t + \frac{\pi}{6}\right).$$

Example 5 Find an expression equivalent to $3\sin 2x - 4\cos 2x$ involving only the sine function.

In this case $c = 3$, $d = -4$, and $a = 2$:

$$A = \sqrt{3^2 + (-4)^2} = 5,$$

$$\cos b = \frac{3}{5}, \quad \text{and} \quad \sin b = -\frac{4}{5}.$$

Thus we have $3\sin 2x - 4\cos 2x \equiv 5\sin(2x + b)$, where b is a number whose cosine is $\frac{3}{5}$ and whose sine is $-\frac{4}{5}$.

Example 6 Graph: $y = \sin x - \cos x$.

We first transform the right-hand side to an expression involving only the sine function. Since $c = 1$ and $d = -1$, we have

$$A = \sqrt{1^2 + (-1)^2} = \sqrt{2},$$

$$\cos b = \frac{1}{\sqrt{2}}, \quad \text{and} \quad \sin b = -\frac{1}{\sqrt{2}}.$$

So $b = -\pi/4$ and we have $y = \sqrt{2} \sin(x - \pi/4)$. This can be graphed (Fig. 11) by translating the sine graph $\pi/4$ units to the right and stretching it vertically.

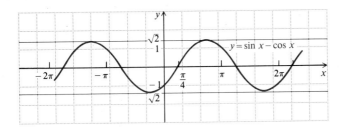

Figure 11

Skill in making graphs like this is important in such fields as engineering, electricity, and physics.

EXERCISE SET 10.3

Prove the following identities.

1. $\csc x - \cos x \cot x \equiv \sin x$

2. $\sec x - \sin x \tan x \equiv \cos x$

3. $\dfrac{1 + \cos \theta}{\sin \theta} + \dfrac{\sin \theta}{\cos \theta} \equiv \dfrac{\cos \theta + 1}{\sin \theta \cos \theta}$

4. $\dfrac{1}{\sin \theta \cos \theta} - \dfrac{\cos \theta}{\sin \theta} \equiv \dfrac{\sin \theta \cos \theta}{1 - \sin^2 \theta}$

5. $\dfrac{1 - \sin x}{\cos x} \equiv \dfrac{\cos x}{1 + \sin x}$

6. $\dfrac{1 - \cos x}{\sin x} \equiv \dfrac{\sin x}{1 + \cos x}$

7. $\dfrac{1 + \tan \theta}{1 + \cot \theta} \equiv \dfrac{\sec \theta}{\csc \theta}$

8. $\dfrac{\cot \theta - 1}{1 - \tan \theta} \equiv \dfrac{\csc \theta}{\sec \theta}$

9. $\dfrac{\sin x + \cos x}{\sec x + \csc x} \equiv \dfrac{\sin x}{\sec x}$

10. $\dfrac{\sin x - \cos x}{\sec x - \csc x} \equiv \dfrac{\cos x}{\csc x}$

11. $\dfrac{1 + \tan \theta}{1 - \tan \theta} + \dfrac{1 + \cot \theta}{1 - \cot \theta} \equiv 0$

12. $\dfrac{\cos^2 \theta + \cot \theta}{\cos^2 \theta - \cot \theta} \equiv \dfrac{\cos^2 \theta \tan \theta + 1}{\cos^2 \theta \tan \theta - 1}$

13. $\dfrac{1 + \cos 2\theta}{\sin 2\theta} \equiv \cot \theta$

14. $\dfrac{2 \tan \theta}{1 + \tan^2 \theta} \equiv \sin 2\theta$

15. $\sec 2\theta \equiv \dfrac{\sec^2 \theta}{2 - \sec^2 \theta}$

16. $\cot 2\theta \equiv \dfrac{\cot^2 \theta - 1}{2 \cot \theta}$

17. $\dfrac{\sin(\alpha + \beta)}{\cos \alpha \cos \beta} \equiv \tan \alpha + \tan \beta$

18. $\dfrac{\cos(\alpha - \beta)}{\cos \alpha \sin \beta} \equiv \tan \alpha + \cot \beta$

19. $1 - \cos 5\theta \cos 3\theta - \sin 5\theta \sin 3\theta \equiv 2 \sin^2 \theta$

20. $2 \sin \theta \cos^3 \theta + 2 \sin^3 \theta \cos \theta \equiv \sin 2\theta$

21. $\dfrac{\tan \theta + \sin \theta}{2 \tan \theta} \equiv \cos^2 \dfrac{\theta}{2}$

22. $\dfrac{\tan \theta - \sin \theta}{2 \tan \theta} \equiv \sin^2 \dfrac{\theta}{2}$

23. $\cos^4 x - \sin^4 x \equiv \cos 2x$

24. $\dfrac{\cos^4 x - \sin^4 x}{1 - \tan^4 x} \equiv \cos^4 x$

25. $\dfrac{\tan 3\theta - \tan \theta}{1 + \tan 3\theta \tan \theta} \equiv \dfrac{2 \tan \theta}{1 - \tan^2 \theta}$

26. $\left(\dfrac{1 + \tan \theta}{1 - \tan \theta}\right)^2 \equiv \dfrac{1 + \sin 2\theta}{1 - \sin 2\theta}$

27. $\dfrac{\cos^3 x - \sin^3 x}{\cos x - \sin x} \equiv \dfrac{2 + \sin 2x}{2}$

28. $\dfrac{\sin^3 t + \cos^3 t}{\sin t + \cos t} \equiv \dfrac{2 - \sin 2t}{2}$

29. $\sin(\alpha + \beta)\sin(\alpha - \beta) \equiv \sin^2 \alpha - \sin^2 \beta$

30. $\cos(\alpha + \beta)\cos(\alpha - \beta) \equiv \cos^2 \alpha - \sin^2 \beta$

31. $\cos(\alpha + \beta) + \cos(\alpha - \beta) \equiv 2 \cos \alpha \cos \beta$

32. $\sin(\alpha + \beta) + \sin(\alpha - \beta) \equiv 2 \sin \alpha \cos \beta$

Find an equivalent expression involving the sine function only.

33. $\sin 2x + \sqrt{3} \cos 2x$

34. $\sqrt{3} \sin 3x - \cos 3x$

35. $4 \sin x + 3 \cos x$

36. $4 \sin 2x - 3 \cos 2x$

37. $6.75 \sin 0.374x + 4.08 \cos 0.374x$

38. $97.81 \sin 0.8081x - 4.89 \cos 0.8081x$

39. Graph $y = \sin 2x - \cos 2x$

40. Graph $y = \sin x + \sqrt{3} \cos x$

41. Show that

$$\log(\cos x - \sin x) + \log(\cos x + \sin x) \equiv \log \cos 2x.$$

42. The following equation occurs in the study of mechanics:

$$\sin \theta = \dfrac{I_1 \cos \phi}{\sqrt{(I_1 \cos \phi)^2 + (I_2 \sin \phi)^2}}.$$

It can happen that $I_1 = I_2$. Assuming that this happens, simplify the equation.

43. In the theory of alternating current, the following equation is used:

$$R = \dfrac{1}{\omega C (\tan \theta + \tan \phi)}.$$

Show that this equation is equivalent to

$$R = \dfrac{\cos \theta \cos \phi}{\omega C \sin(\theta + \phi)}.$$

44. Show that

$$\begin{vmatrix} \cos x & \sin x \\ -\sin x & \cos x \end{vmatrix} \equiv \begin{vmatrix} \cos x & -\sin x \\ \sin x & \cos x \end{vmatrix} \equiv 1.$$

45. Show that

$$\begin{bmatrix} \cos x & \sin x \\ -\sin x & \cos x \end{bmatrix} \begin{bmatrix} \cos y & \sin y \\ -\sin y & \cos y \end{bmatrix}$$

$$\equiv \begin{bmatrix} \cos(x+y) & \sin(x+y) \\ -\sin(x+y) & \cos(x+y) \end{bmatrix}.$$

★

46. In electrical theory, the following equations are used:

$$E_1 = \sqrt{2} E_t \cos\left(\theta + \dfrac{\pi}{P}\right),$$

$$E_2 = \sqrt{2} E_t \cos\left(\theta - \dfrac{\pi}{P}\right).$$

Assuming that these equations hold, show that

$$\dfrac{E_1 + E_2}{2} = \sqrt{2} E_t \cos \theta \cos \dfrac{\pi}{P}$$

and

$$\dfrac{E_1 - E_2}{2} = -\sqrt{2} E_t \sin \theta \sin \dfrac{\pi}{P}.$$

47. In the theory of electrical machinery, the following equation occurs:

$$L_1 = I^2 \cos^2(\omega t + \alpha) L_2 \sin 2\theta.$$

Show that this equation is equivalent to

$$L_1 = \dfrac{1}{2} I^2 L_2 \sin 2\theta + \dfrac{1}{2} I^2 L_2 \sin 2\theta \cos(2\omega t + 2\alpha).$$

10.4 INVERSES OF THE TRIGONOMETRIC FUNCTIONS

Recall that to obtain the inverse of any relation we interchange the first and second members of each ordered pair in the relation. If a relation is defined by an equation, say in x and y, interchanging x and y produces an equation of the inverse relation. The graphs of a relation and its inverse are reflections of each other across the line y = x. The x- and y-axes are interchanged in such a reflection. Let us consider the inverse of the sine function, y = sin x. The inverse may be denoted in several ways:

$$x = \sin y, \qquad y = \sin^{-1} x, \qquad y = \arcsin x.$$

Thus $\sin^{-1} x$ is a number whose sine is x.

The notation arcsin x arises because it is the length of an arc on the unit circle for which the sine is x. The notation $\sin^{-1} x$ is not exponential notation. It does *not* mean 1/sin x! Either of the latter two kinds of notation above can be read "the inverse sine of x" or "the arc sine of x" or "the number (or angle) whose sine is x." Notation is chosen similarly for the inverses of the other trigonometric functions: $\cos^{-1} x$ or arccos x, $\tan^{-1} x$ or arctan x, and so on.

Graphs of $y = \sin^{-1} x$ and $y = \tan^{-1} x$ are shown in Figs. 12 and 13. Note that these relations are not functions.

Figure 12

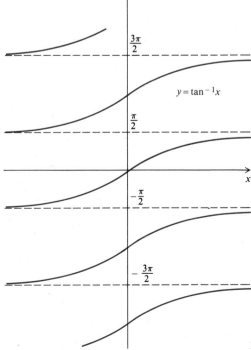

Figure 13

Finding Inverse Values

We illustrate finding inverse values using both a graph and the unit circle. In practice the unit circle is easier to use.

Example 1 Find all values of arcsin $\frac{1}{2}$.

On the graph of $y = \arcsin x$, we draw a vertical line at $x = \frac{1}{2}$ (Fig. 14). It intersects the graph at points whose y-value is arcsin $\frac{1}{2}$. Some of the numbers whose sine is $\frac{1}{2}$ are seen to be $\pi/6$, $5\pi/6$, $-7\pi/6$, and so on. From the graph we can see that $\pi/6$ plus any multiple of 2π is such a number. Also $5\pi/6$ plus any multiple of 2π is such a number. The complete set of values is given by $\pi/6 + 2k\pi$, k an integer, and $5\pi/6 + 2k\pi$, k an integer.

Figure 14

Example 2 Find all values of arcsin $\frac{1}{2}$.

On the unit circle there are two points at which the sine is $\frac{1}{2}$ (Fig. 15). The arc length for the point in the first quadrant is $\pi/6$ plus any multiple of 2π. The arc length for the point in the second quadrant is $5\pi/6$ plus any multiple of 2π. Hence we obtain all values of arcsin $\frac{1}{2}$ as follows:

$$\frac{\pi}{6} + 2k\pi \quad \text{or} \quad \frac{5\pi}{6} + 2k\pi, \quad k \text{ an integer.}$$

Example 3 Find all values of $\cos^{-1}(-0.9397)$ in degrees.

From Table 3 at the back of the book we find that the angle whose cosine is 0.9397 is 20°. This is the reference angle. We sketch this on a unit circle to find the two points where the cosine is -0.9397 (Fig. 16). The angles are 160° and 200°, plus any multiple of 360°. Thus the values of $\cos^{-1}(-0.9397)$ are

$$160° + k \cdot 360° \quad \text{or} \quad 200° + k \cdot 360°,$$

where k is any integer.

Figure 15

Figure 16

Example 4 Find all values of arctan 1 (see Fig. 17).

We find the two points on the unit circle at which the tangent is 1 (Fig. 17). These points are opposite ends of a diameter. Hence the arc lengths differ by π. Thus we have for all values of arctan 1,

$$\frac{\pi}{4} + k\pi, \quad k \text{ an integer.}$$

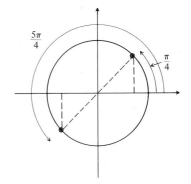

Figure 17

Principal Values

The inverses of the trigonometric functions are not themselves functions. However, if we restrict the ranges of these relations, we can obtain functions. Figure 18 shows how this restriction is made. These relations with their ranges so restricted are functions, and the values in these restricted ranges are called *principal values*. To denote principal values we shall capitalize, as follows:

$$\text{Arcsin } x, \quad \text{Sin}^{-1} x, \quad \text{Arccos } x, \quad \text{Cos}^{-1} x,$$

and so on. Thus, whereas arcsin $(\tfrac{1}{2})$ represents an infinite set of numbers, Arcsin $(\tfrac{1}{2})$ represents the single number $\pi/6$.

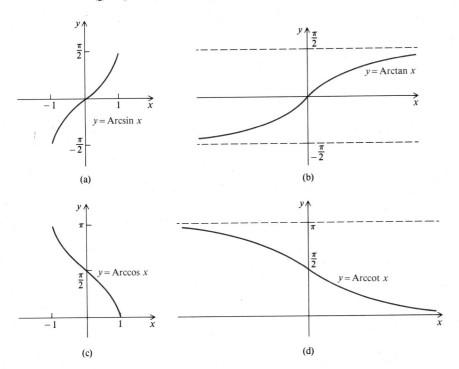

Figure 18

Note that for the function y = Arcsin x the range is the interval $[-\pi/2, \pi/2]$. For the function y = Arctan x the range is $(-\pi/2, \pi/2)$. For the function y = Arccos x the range is $[0, \pi]$, and for y = Arccot x the range is $(0, \pi)$. Figure 19 shows where principal values are found on the unit circle.

Example 5 Find Arcsin $\sqrt{2}/2$ and $\text{Cos}^{-1}(-\frac{1}{2})$.

In the restricted range as shown in Fig. 19, the only number whose sine is $\sqrt{2}/2$ is $\pi/4$. Hence Arcsin $\sqrt{2}/2 = \pi/4$.

The only number whose cosine is $-\frac{1}{2}$ in the restricted range is $2\pi/3$. Hence $\text{Cos}^{-1}(-\frac{1}{2}) = 2\pi/3$.

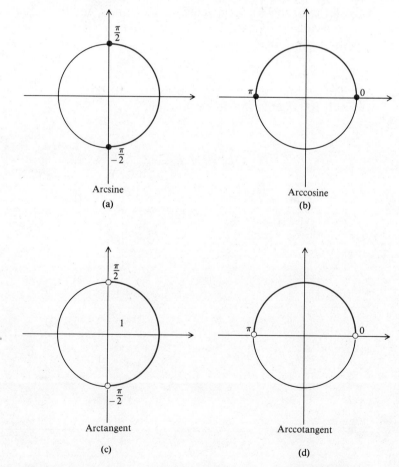

Figure 19

10.4 INVERSES OF THE TRIGONOMETRIC FUNCTIONS

The Use of Calculators

Scientific calculators can be used to find inverses of trigonometric function values. The values obtained from a calculator are *principal* values, and the typical calculator provides values for only the Arcsin, Arccos, and Arctan functions.

Some calculators give inverse function values in either radians or degrees. Some give values only in degrees. The key strokes involved in finding inverse function values vary with the calculator. Be sure to read the instructions for the calculator you are using.

EXERCISE SET 10.4

Find all values of the following without using tables.

1. $\arcsin \frac{\sqrt{2}}{2}$
2. $\arcsin \frac{\sqrt{3}}{2}$
3. $\cos^{-1} \frac{\sqrt{2}}{2}$
4. $\cos^{-1} \frac{\sqrt{3}}{2}$
5. $\sin^{-1}\left(-\frac{\sqrt{2}}{2}\right)$
6. $\sin^{-1}\left(-\frac{\sqrt{3}}{2}\right)$
7. $\arccos\left(-\frac{\sqrt{2}}{2}\right)$
8. $\arccos\left(-\frac{\sqrt{3}}{2}\right)$
9. $\arctan \sqrt{3}$
10. $\arctan \frac{\sqrt{3}}{3}$
11. $\cot^{-1} 1$
12. $\cot^{-1} \sqrt{3}$
13. $\arctan\left(-\frac{\sqrt{3}}{3}\right)$
14. $\arctan(-\sqrt{3})$
15. $\text{arccot}(-1)$
16. $\text{arccot}(-\sqrt{3})$
17. $\text{arcsec } 1$
18. $\text{arcsec } 2$
19. $\csc^{-1} 1$
20. $\csc^{-1} 2$

Use tables to find, in degrees, all values of the following.

21. $\arcsin 0.3907$
22. $\arcsin 0.9613$
23. $\sin^{-1} 0.6293$
24. $\sin^{-1} 0.8746$
25. $\arccos 0.7990$
26. $\arccos 0.9265$
27. $\cos^{-1} 0.9310$
28. $\cos^{-1} 0.2735$
29. $\tan^{-1} 0.3673$
30. $\tan^{-1} 1.091$
31. $\cot^{-1} 1.265$
32. $\cot^{-1} 0.4770$
33. $\sec^{-1} 1.167$
34. $\sec^{-1} 1.440$
35. $\text{arccsc } 6.277$
36. $\text{arccsc } 1.111$

Find the following without using tables.

37. $\text{Arcsin } \frac{\sqrt{2}}{2}$
38. $\text{Arcsin } \frac{1}{2}$
39. $\text{Cos}^{-1} \frac{1}{2}$
40. $\text{Cos}^{-1} \frac{\sqrt{2}}{2}$
41. $\text{Sin}^{-1}\left(-\frac{\sqrt{3}}{2}\right)$
42. $\text{Sin}^{-1}\left(-\frac{1}{2}\right)$
43. $\text{Arccos}\left(-\frac{\sqrt{2}}{2}\right)$
44. $\text{Arccos}\left(-\frac{\sqrt{3}}{2}\right)$
45. $\text{Tan}^{-1}\left(-\frac{\sqrt{3}}{3}\right)$
46. $\text{Tan}^{-1}(-\sqrt{3})$
47. $\text{Arccot}\left(-\frac{\sqrt{3}}{3}\right)$
48. $\text{Arccot}(-\sqrt{3})$

Find the following, in degrees, using tables.

49. $\text{Arcsin } 0.2334$
50. $\text{Arcsin } 0.4514$
51. $\text{Sin}^{-1}(-0.6361)$
52. $\text{Sin}^{-1}(-0.8192)$

53. Arccos (−0.8897)
54. Arccos (−0.2924)
55. Tan^{-1} (−0.4074)
56. Tan^{-1} (−0.2401)
57. Cot^{-1} (−5.396)
58. Cot^{-1} (−1.319)

☆

Do the following calculations.

59. ▦ Cos^{-1} (sin 48°35′)
60. ▦ sin (Tan^{-1} 1.405)
61. ▦ cos (Tan^{-1} (−3.333))
62. ▦ Tan^{-1} (sin (−37°20′))
63. ▦ cos (Cos^{-1} 0.7324)
64. ▦ Tan^{-1} (tan 1.0023)
65. ▦ $\sqrt{\ln \mathrm{Sin}^{-1} \sqrt[3]{0.96}}$

66. ▦ Graph $y = \mathrm{Sin}^{-1} x + \mathrm{Tan}^{-1} x$. What is the domain of this function?

67. The principal values of the inverses of the trigonometric functions are themselves functions. Determine which of them are even or odd and then write as many identities as you can involving the inverse functions.

68. Graph $y = \mathrm{Sin}^{-1}\left(x - \frac{\pi}{2}\right)$.

69. Graph $y = 2\,\mathrm{Cos}^{-1}\left(x + \frac{\pi}{2}\right)$.

70. ▦ The following gives an approximation to Tan^{-1} x when $|x| \leq 1$:

$$x - \frac{x^3}{3} + \frac{x^5}{5} - \frac{x^7}{7} + \frac{x^9}{9} - \cdots$$

Let $x = 1$ and thus approximate Tan^{-1} 1. What number are you approximating?

71. ▦ Determine whether the following function approaches a limit as x approaches 0 and, if possible, determine what that limit is, if it exists.

$$\frac{\mathrm{Sin}^{-1} 2x}{\mathrm{Sin}^{-1} x}$$

10.5 COMBINATIONS OF TRIGONOMETRIC FUNCTIONS AND THEIR INVERSES

In practice, various combinations of trigonometric functions and their inverses arise. We shall discuss evaluating and simplifying such expressions. Let us consider, for example,

sin Arcsin x   (the sine of a number whose sine is x).

Recall, from Chapter 3, that if a function f has an inverse that is also a function, then $f(f^{-1}(x)) = x$ and also $f^{-1}(f(x)) = x$ for all x in the domains of the functions. Since the sine and arcsine relations are inverses of each other, we might suspect that

sin Arcsin x = x

for all x in the domain of the Arcsin relation. However, we cannot apply the result of Chapter 3 here because the inverse of the sine function is not a function. Let us study sin Arcsin x using a unit circle. Consider a number a in the domain of the Arcsine function (a number from −1 to 1). Then Arcsin a is the length of an arc s as shown in Fig. 20. Its sine is a. In other words, s = Arcsin a. From the figure, we see

10.5 COMBINATIONS OF FUNCTIONS AND THEIR INVERSES

that sin s = a. Taking the sine on both sides of the last equation, we get

$$\sin \operatorname{Arcsin} a = a.$$

Thus it is true that sin Arcsin x ≡ x.

Similar identities hold for the other trigonometric functions:

sin Arcsin $x \equiv x$,   cot Arccot $x \equiv x$,
cos Arccos $x \equiv x$,   sec Arcsec $x \equiv x$,
tan Arctan $x \equiv x$,   csc Arccsc $x \equiv x$.

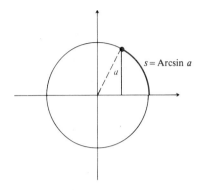

Figure 20

Now let us consider Arcsin sin x. We might also suspect that this is equal to x for any x, but this is not true unless x is in the range of the Arcsine function.

Example 1 Find Arcsin sin $3\pi/4$.

We first find sin $3\pi/4$. It is $\sqrt{2}/2$.
Next we find Arcsin $\sqrt{2}/2$. It is $\pi/4$.
So Arcsin sin $3\pi/4 = \pi/4$.

For any x in the range of the Arcsine function, we do have Arcsin sin x = x. Similar conditions hold for the other functions.

THEOREM 3

The following are true if x is in the range of the inverse function. Otherwise they are not.

Arcsin sin $x = x$,   Arccot cot $x = x$,
Arccos cos $x = x$,   Arcsec sec $x = x$,
Arctan tan $x = x$,   Arccsc csc $x = x$.

Now we shall consider some other kinds of combinations.

Example 2 Find sin Arctan (-1).

We first find Arctan (-1). It is $-\pi/4$. Now we find the sine of this, sin $(-\pi/4) = -\sqrt{2}/2$, so sin Arctan $(-1) = -\sqrt{2}/2$.

Example 3 Find Cos^{-1} sin $\pi/2$.

We first find sin $\pi/2$. It is 1. Now we find $\operatorname{Cos}^{-1} 1$. It is 0, so Cos^{-1} sin $\pi/2 = 0$.

Now let us consider

$$\cos \operatorname{Arcsin} \frac{3}{5}.$$

Figure 21

Without using tables, we cannot find Arcsin $\frac{3}{5}$. However, we can still evaluate the entire expression without using tables. In a case like this we sketch a triangle as in Fig. 21. The angle θ in this triangle is an angle whose sine is $\frac{3}{5}$ (it is Arcsin $\frac{3}{5}$). We wish to find the cosine of this angle. Since the triangle is a right triangle, we can find the length of the base, b. It is 4. Thus we know that $\cos \theta = b/5$ or $\frac{4}{5}$. Therefore

$$\cos \text{Arcsin} \frac{3}{5} = \frac{4}{5}.$$

Example 4 Find sin Arccot $x/2$.

We draw a right triangle (Fig. 22) whose sides have lengths x and 2, so that $\cot \theta = x/2$. We find the length of the hypotenuse and then read off the sine ratio.

$$\sin \text{Arccot} \frac{x}{2} = \frac{2}{\sqrt{x^2 + 2^2}}$$

Figure 22

In some cases the use of certain identities is needed to evaluate expressions.

Example 5 Evaluate $\sin (\text{Sin}^{-1} \frac{1}{2} + \text{Cos}^{-1} \frac{4}{5})$.

To simplify the use of an identity we shall make substitutions. Let

$$u = \text{Sin}^{-1} \frac{1}{2}, \quad \text{and} \quad v = \text{Cos}^{-1} \frac{4}{5}.$$

Then we have $\sin (u + v)$. By a sum formula, this is equivalent to

$$\sin u \cos v + \cos u \sin v.$$

Now we make the reverse substitutions and obtain

$$\sin \text{Sin}^{-1} \frac{1}{2} \cdot \cos \text{Cos}^{-1} \frac{4}{5} + \cos \text{Sin}^{-1} \frac{1}{2} \cdot \sin \text{Cos}^{-1} \frac{4}{5}.$$

This immediately simplifies to

$$\frac{1}{2} \cdot \frac{4}{5} + \cos \text{Sin}^{-1} \frac{1}{2} \cdot \sin \text{Cos}^{-1} \frac{4}{5}.$$

Now $\cos \text{Sin}^{-1} \frac{1}{2}$ readily simplifies to $\sqrt{3}/2$.

To find $\sin \text{Cos}^{-1} \frac{4}{5}$, we shall need a triangle (Fig. 23). We find that $\sin \text{Cos}^{-1} \frac{4}{5} = \frac{3}{5}$. Our expression is now simplified to

$$\frac{1}{2} \cdot \frac{4}{5} + \frac{\sqrt{3}}{2} \cdot \frac{3}{5},$$

which simplifies to

$$\frac{4 + 3\sqrt{3}}{10}.$$

Figure 23

EXERCISE SET 10.5

Evaluate or simplify.

1. $\sin \text{Arcsin } 0.3$
2. $\cos \text{Arccos } 0.2$
3. $\tan \text{Tan}^{-1}(-4.2)$
4. $\cot \text{Cot}^{-1}(-1.5)$
5. $\text{Arcsin } \sin \frac{2\pi}{3}$
6. $\text{Arccos } \cos \frac{3\pi}{2}$
7. $\text{Sin}^{-1} \sin\left(-\frac{3\pi}{4}\right)$
8. $\text{Cos}^{-1} \cos\left(-\frac{\pi}{4}\right)$
9. $\text{Sin}^{-1} \sin \frac{\pi}{5}$
10. $\text{Cos}^{-1} \cos \frac{\pi}{7}$
11. $\text{Tan}^{-1} \tan \frac{2\pi}{3}$
12. $\text{Cot}^{-1} \cot \frac{2\pi}{3}$
13. $\sin \text{Arctan } \sqrt{3}$
14. $\sin \text{Arctan } \frac{\sqrt{3}}{3}$
15. $\cos \text{Arcsin } \frac{\sqrt{3}}{2}$
16. $\cos \text{Arcsin } \frac{\sqrt{2}}{2}$
17. $\tan \text{Cos}^{-1} \frac{\sqrt{2}}{2}$
18. $\tan \text{Cos}^{-1} \frac{\sqrt{3}}{2}$
19. $\text{Cos}^{-1} \sin \frac{\pi}{3}$
20. $\text{Cos}^{-1} \sin \pi$
21. $\text{Arcsin } \cos \frac{\pi}{6}$
22. $\text{Arcsin } \cos \frac{\pi}{4}$
23. $\text{Sin}^{-1} \tan \frac{\pi}{4}$
24. $\text{Sin}^{-1} \tan\left(-\frac{\pi}{4}\right)$
25. $\sin \text{Arctan } \frac{x}{2}$
26. $\sin \text{Arctan } \frac{a}{3}$
27. $\cos \text{Tan}^{-1} \frac{\sqrt{2}}{3}$
28. $\cos \text{Tan}^{-1} \frac{\sqrt{3}}{4}$
29. $\tan \text{Arcsin } 0.1$
30. $\tan \text{Arcsin } 0.2$
31. $\cot \text{Cos}^{-1}(-0.2)$
32. $\cot \text{Cos}^{-1}(-0.3)$
33. $\sin \text{Arccot } y$
34. $\sin \text{Arccot } x$
35. $\tan\left(\frac{1}{2} \text{Arcsin } \frac{4}{5}\right)$
36. $\tan\left(\frac{1}{2} \text{Arcsin } \frac{1}{2}\right)$
37. $\cos\left(\frac{1}{2} \text{Arcsin } \frac{1}{2}\right)$
38. $\cos\left(\frac{1}{2} \text{Arcsin } \frac{\sqrt{3}}{2}\right)$
39. $\sin\left(2 \text{Cos}^{-1} \frac{3}{5}\right)$
40. $\sin\left(2 \text{Cos}^{-1} \frac{1}{2}\right)$
41. $\cos\left(2 \text{Sin}^{-1} \frac{5}{13}\right)$
42. $\cos\left(2 \text{Cos}^{-1} \frac{4}{5}\right)$
43. $\sin\left(\text{Sin}^{-1} \frac{1}{2} + \text{Cos}^{-1} \frac{3}{5}\right)$
44. $\sin\left(\text{Sin}^{-1} \frac{1}{2} - \text{Cos}^{-1} \frac{4}{5}\right)$
45. $\cos\left(\text{Sin}^{-1} \frac{\sqrt{2}}{2} + \text{Cos}^{-1} \frac{3}{5}\right)$
46. $\cos\left(\text{Sin}^{-1} \frac{4}{5} - \text{Cos}^{-1} \frac{1}{2}\right)$
47. ▦ $\sin(\text{Sin}^{-1} 0.6032 + \text{Cos}^{-1} 0.4621)$.
48. ▦ $\cos(\text{Sin}^{-1} 0.7325 - \text{Cos}^{-1} 0.4838)$.

★

49. ▦ Graph: $f(x) = \sin(\text{Tan}^{-1} x)$.

50. Prove the following identity:
$$\tan(\text{Tan}^{-1} x + \text{Tan}^{-1} 1) \equiv \frac{1+x}{1-x}.$$

51. Prove that for $x > 0$, $\text{Tan}^{-1} x + \text{Tan}^{-1} \frac{1}{x} = \frac{\pi}{2}$.

52. An observer's eye is at point A, observing a mural of height h, with the bottom of the mural y feet above the eye. The eye is x feet from the wall. Write an expression for θ in terms of x, y, and h.

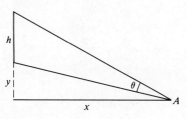

53. ▦ Evaluate the expression in Exercise 52 when $x = 20$ ft, $y = 7$ ft, and $h = 25$ ft.

10.6 TRIGONOMETRIC EQUATIONS

When an equation contains a trigonometric expression with a variable such as $\sin x$, it is called a *trigonometric equation*. To solve such an equation, we find all replacements for the variable that make the equation true.

Example 1 Solve: $2 \sin x = 1$.

We first solve for $\sin x$:

$$\sin x = \frac{1}{2}.$$

Now we note that the solutions are those numbers having a sine of $\frac{1}{2}$. We look for them. The unit circle is helpful. There are just two points on it for which the sine is $\frac{1}{2}$, as shown in Fig. 24. They are the points for $\pi/6$ and $5\pi/6$. These numbers, plus any multiple of 2π, are the solutions

$$\frac{\pi}{6} + 2k\pi \quad \text{or} \quad \frac{5\pi}{6} + 2k\pi,$$

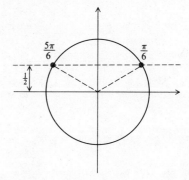

Figure 24

where k is any integer. In degrees, the solutions are

$$30° + k \cdot 360° \quad \text{or} \quad 150° + k \cdot 360°,$$

where k is any integer.

Example 2 Solve: $4 \cos^2 x = 1$.

We first solve for $\cos x$:

$$\cos^2 x = \frac{1}{4}$$

$$|\cos x| = \frac{1}{2} \quad \text{Taking principal square root}$$

$$\cos x = \pm\frac{1}{2}.$$

Now we use the unit circle to find those numbers having a cosine of $\frac{1}{2}$ or $-\frac{1}{2}$. (See Fig. 25.) The solutions are $\pi/3$, $2\pi/3$, $4\pi/3$, $5\pi/3$, plus any multiple of 2π.

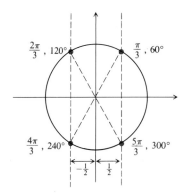

Figure 25

In solving trigonometric equations, it is usually sufficient to find just the solutions from 0 to 2π. We then remember that any multiple of 2π may be added to obtain the rest of the solutions.

The following example illustrates that when we look for solutions in the interval $[0, 2\pi)$ we must be cautious.

Example 3 Solve $2 \sin 2x = 1$ in the interval $[0, 2\pi)$.

We first solve for $\sin 2x$:

$$\sin 2x = \tfrac{1}{2}.$$

Using the unit circle we find points for which $\sin 2x = \frac{1}{2}$ (Fig. 26). However, since x is to run from 0 to 2π, we must let $2x$ run from 0 to 4π. These values of $2x$ are $\pi/6$, $5\pi/6$, $13\pi/6$, and $17\pi/6$. Thus the desired values of x in $[0, 2\pi)$ are half of these. Therefore,

$$x = \frac{\pi}{12}, \frac{5\pi}{12}, \frac{13\pi}{12}, \frac{17\pi}{12}.$$

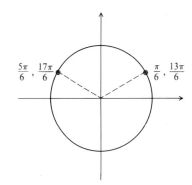

Figure 26

In solving some trigonometric equations, it is necessary to use tables. In such cases, we usually find answers in degrees.

Example 4 Solve $2 + \sin x = 2.5299$ in $[0, 360°)$.

We first solve for $\sin x$:

$$\sin x = 0.5299.$$

From the table we find the reference angle, x = 32°. Since sin x is positive, the solutions are to be found in the first and second quadrants. The solutions are

$$32° \quad \text{and} \quad 148°$$

Example 5 Solve $\sin x - 1 = -1.5299$ in $[0, 360°)$.

We first solve for sin x:

$$\sin x = -0.5299.$$

From the table we find the reference angle, 32°. Since sin x is negative, the solutions are to be found in the third and fourth quadrants. The solutions are

$$212° \quad \text{and} \quad 328°.$$

In solving trigonometric equations, we can expect to apply some algebra before concerning ourselves with the trigonometric part. In the next examples, we begin by factoring.

Example 6 Solve: $8 \cos^2 \theta - 2 \cos \theta = 1$.

Since we will be using the principle of zero products, we first obtain a 0 on one side of the equation.

$$8 \cos^2 \theta - 2 \cos \theta - 1 = 0$$

$(4 \cos \theta + 1)(2 \cos \theta - 1) = 0$ Factoring

$4 \cos \theta + 1 = 0$ or $2 \cos \theta - 1 = 0$ Principle of zero products

$\cos \theta = -\dfrac{1}{4} = -0.25$ or $\cos \theta = \dfrac{1}{2}$

$\theta = 104°30', 255°30'$ or $\theta = 60°, 300°$ $\left[\text{or} \quad \dfrac{\pi}{3}, \dfrac{5\pi}{3}\right]$

The solutions in $[0, 360°)$ are $104°30', 255°30', 60°$, and $300°$.

Example 7 Solve: $2 \sin^2 \phi + \sin \phi = 0$.

$$\sin \phi (2 \sin \phi + 1) = 0 \quad \text{Factoring}$$

$\sin \phi = 0$ or $2 \sin \phi + 1 = 0$ Principle of zero products

$\sin \phi = 0$ or $\sin \phi = -\dfrac{1}{2}$

$\phi = 0, \pi$ or $\phi = \dfrac{7\pi}{6}, \dfrac{11\pi}{6}$

The solutions in $[0, 2\pi)$ are $0, \pi, 7\pi/6$, and $11\pi/6$.

In case a trigonometric equation is quadratic but difficult or impossible to factor, we use the quadratic formula.

Example 8 Solve $10 \sin^2 x - 12 \sin x - 7 = 0$ in $[0, 360°)$.

$$\sin x = \frac{12 \pm \sqrt{144 + 280}}{20} \quad \text{Using the quadratic formula}$$

$$= \frac{12 \pm \sqrt{424}}{20} = \frac{12 \pm 2\sqrt{106}}{20} = \frac{6 \pm \sqrt{106}}{10}$$

$$= \frac{6 \pm 10.296}{10}$$

$$\sin x = 1.6296 \quad \text{or} \quad \sin x = -0.4296$$

Since sines are never greater than 1, the first of the equations on the preceding line has no solution. We look up the other number and find the reference angle to be $25°30'$ to the nearest $10'$. Thus the solutions are $205°30'$ and $334°30'$.

EXERCISE SET 10.6

Solve, finding all solutions.

1. $\sin x = \dfrac{\sqrt{3}}{2}$
2. $\cos x = \dfrac{\sqrt{3}}{2}$
3. $\cos x = \dfrac{1}{\sqrt{2}}$
4. $\tan x = \sqrt{3}$
5. $\sin x = 0.3448$
6. $\cos x = 0.6406$

Solve, finding all solutions in $[0, 2\pi)$ or $[0°, 360°)$.

7. $\cos x = -0.5495$
8. $\sin x = -0.4279$
9. $2 \sin x + \sqrt{3} = 0$
10. $\sqrt{3} \tan x + 1 = 0$
11. $2 \tan x + 3 = 0$
12. $4 \sin x - 1 = 0$
13. $4 \sin^2 x - 1 = 0$
14. $2 \cos^2 x = 1$
15. $\cot^2 x - 3 = 0$
16. $\csc^2 x - 4 = 0$
17. $2 \sin^2 x + \sin x = 1$
18. $2 \cos^2 x + 3 \cos x = -1$
19. $\cos^2 x + 2 \cos x = 3$
20. $2 \sin^2 x - \sin x = 3$
21. $4 \sin^3 x - \sin x = 0$
22. $2 \cos^2 x - \sqrt{3} \cos x = 0$
23. $2 \sin^2 \theta + 7 \sin \theta = 4$
24. $2 \sin^2 \theta - 5 \sin \theta + 2 = 0$
25. $6 \cos^2 \phi + 5 \cos \phi + 1 = 0$
26. $2 \sin^2 \phi + \sin \phi - 1 = 0$
27. $2 \sin t \cos t + 2 \sin t - \cos t - 1 = 0$
28. $2 \sin t \tan t + \tan t - 2 \sin t - 1 = 0$
29. $\cos 2x \sin x + \sin x = 0$
30. $\sin 2x \cos x - \cos x = 0$

Solve, restricting solutions to $[0, 2\pi)$ or $[0, 360°)$ where sensible to do so.

31. $|\sin x| = \dfrac{\sqrt{3}}{2}$

32. $|\cos x| = \dfrac{1}{2}$

33. $\sqrt{\tan x} = \sqrt[4]{3}$

34. $12 \sin x - 7\sqrt{\sin x} + 1 = 0$

35. $16 \cos^4 x - 16 \cos^2 x + 3 = 0$

36. $\ln \cos x = 0$

37. $e^{\sin x} = 1$

38. $\sin \ln x = -1$

39. $e^{\ln \sin x} = 0$

40. Solve this system of equations for x and y.

$$x \cos \theta - y \sin \theta = 0,$$
$$x \sin \theta + y \cos \theta = 1$$

41. ▦ Solve graphically.

$$y = 2 \sin \dfrac{x}{2},$$
$$y = \mathrm{Tan}^{-1} x$$

42. Solve graphically: $\sin x = \tan \dfrac{x}{2}$.

10.7 IDENTITIES IN SOLVING EQUATIONS

When a trigonometric equation involves more than one function, we may use identities to put it in terms of a single function. This can usually be done in several ways. In the following three examples, we illustrate by solving the same equation three different ways.

Example 1 Solve the equation $\sin x + \cos x = 1$.

To express $\cos x$ in terms of $\sin x$, we use the identity

$$\sin^2 x + \cos^2 x \equiv 1.$$

From this we obtain $\cos x \equiv \pm\sqrt{1 - \sin^2 x}$. Now we substitute in the original equation:

$$\sin x \pm \sqrt{1 - \sin^2 x} = 1.$$

This is a radical equation. We get the radical alone on one side and then square both sides:

$$1 - \sin^2 x = 1 - 2 \sin x + \sin^2 x.$$

This simplifies to

$$-2 \sin^2 x + 2 \sin x = 0.$$

Now we factor:

$$-2 \sin x (\sin x - 1) = 0.$$

Using the principle of zero products, we have

$$-2 \sin x = 0 \quad \text{or} \quad \sin x - 1 = 0$$
$$\sin x = 0 \quad \text{or} \quad \sin x = 1.$$

The values of x in $[0, 2\pi)$ satisfying these are

$$x = 0, \quad x = \pi, \quad \text{or} \quad x = \frac{\pi}{2}.$$

Now we check these in the original equation. We find that π does not check but the other values do. Thus the solutions are 0 and $\pi/2$.

It is important to check in examples such as these. Values are often obtained that are not solutions of the original equation.

Example 2 Solve: $\sin x + \cos x = 1$.

This time we will square both sides, obtaining

$$\sin^2 x + 2 \sin x \cos x + \cos^2 x = 1.$$

Since $\sin^2 x + \cos^2 x \equiv 1$, we can simplify to

$$2 \sin x \cos x = 0.$$

Now we use the identity $2 \sin x \cos x \equiv \sin 2x$, and obtain

$$\sin 2x = 0.$$

The values in $[0, 2\pi)$ satisfying this equation are 0, $\pi/2$, π, and $3\pi/2$. Only 0 and $\pi/2$ check.

Example 3 Solve: $\sin x + \cos x = 1$.

We recall that $c \sin ax + d \cos ax \equiv A \sin (ax + b)$, where $A = \sqrt{c^2 + d^2}$ and b is a number whose cosine is c/A and whose sine is d/A. In this case $c = d = a = 1$, so $A = \sqrt{2}$ and $b = \pi/4$. Thus our equation is equivalent to

$$\sqrt{2} \sin \left(x + \frac{\pi}{4} \right) = 1.$$

Then

$$\sin \left(x + \frac{\pi}{4} \right) = \frac{1}{\sqrt{2}} = \frac{\sqrt{2}}{2}$$

$$x + \frac{\pi}{4} = \frac{\pi}{4} \quad \text{or} \quad x + \frac{\pi}{4} = \frac{3\pi}{4}$$

$$x = 0 \quad \text{or} \quad x = \frac{\pi}{2}.$$

Both values check.

Example 4 Solve: $2\cos^2 x \tan x - \tan x = 0$.

First, we factor:
$$\tan x\,(2\cos^2 x - 1) = 0.$$

Now we use the identity $2\cos^2 x - 1 \equiv \cos 2x$:

$\tan x \cos 2x = 0$

$\tan x = 0$ or $\cos 2x = 0$ Principle of zero products

$x = 0, \pi$ or $2x = \dfrac{\pi}{2}, \dfrac{3\pi}{2}, \dfrac{5\pi}{2}, \dfrac{7\pi}{2}$

$x = 0, \pi$ or $x = \dfrac{\pi}{4}, \dfrac{3\pi}{4}, \dfrac{5\pi}{4}, \dfrac{7\pi}{4}.$

All values check. The solutions in $[0, 2\pi)$ are $0, \pi, \pi/4, 3\pi/4, 5\pi/4$, and $7\pi/4$.

Example 5 Solve: $\cos 2x + \sin x = 1$.

We first use the identity $\cos 2x \equiv 1 - 2\sin^2 x$, to get

$1 - 2\sin^2 x + \sin x = 1$
$-2\sin^2 x + \sin x = 0$
$\sin x\,(1 - 2\sin x) = 0$ Factoring

$\sin x = 0$ or $1 - 2\sin x = 0$ Principle of zero products

$\sin x = 0$ or $\sin x = \dfrac{1}{2}$

$x = 0, \pi$ or $x = \dfrac{\pi}{6}, \dfrac{5\pi}{6}.$

All values check. The solutions in $[0, 2\pi)$ are $0, \pi, \pi/6$, and $5\pi/6$.

Example 6 Solve: $\sin 5\theta \cos 2\theta - \cos 5\theta \sin 2\theta = \sqrt{2}/2$.

We use the identity
$$\sin(\alpha - \beta) \equiv \sin \alpha \cos \beta - \cos \alpha \sin \beta$$
to get $\sin 3\theta = \sqrt{2}/2$. Then

$3\theta = \dfrac{\pi}{4}, \dfrac{3\pi}{4}, \dfrac{9\pi}{4}, \dfrac{11\pi}{4}, \dfrac{17\pi}{4}, \dfrac{19\pi}{4}$

and

$\theta = \dfrac{\pi}{12}, \dfrac{\pi}{4}, \dfrac{3\pi}{4}, \dfrac{11\pi}{12}, \dfrac{17\pi}{12}, \dfrac{19\pi}{12}.$

Let us check $\pi/12$:

$$\sin\frac{5\pi}{12}\cos\frac{\pi}{6} - \cos\frac{5\pi}{12}\sin\frac{\pi}{6}.$$

We again use the difference identity to obtain $\sin(5\pi/12 - \pi/6)$. This is equal to $\sin 3\pi/12$ or $\sin \pi/4$, and this is equal to $\sqrt{2}/2$. Hence $\pi/12$ checks. All of the above answers check.

Example 7 Solve: $\tan^2 x + \sec x - 1 = 0$.

We use the identity $1 + \tan^2 x \equiv \sec^2 x$. Substituting, we get

$$\sec^2 x - 1 + \sec x - 1 = 0,$$

or

$$\sec^2 x + \sec x - 2 = 0$$

$$(\sec x + 2)(\sec x - 1) = 0 \quad \text{Factoring}$$

$\sec x = -2 \quad$ or $\quad \sec x = 1 \quad$ Principle of zero products

$$x = \frac{2}{3}\pi, \frac{4}{3}\pi \quad \text{or} \quad x = 0.$$

All of these values check. The solutions in $[0, 2\pi)$ are 0, $2\pi/3$, and $4\pi/3$.

Example 8 Solve: $\sin x = \sin(2x - \pi)$.

We can use either the identity $\sin(y - \pi) \equiv -\sin y$ or the identity $\sin(\alpha - \beta) \equiv \sin \alpha \cos \beta - \cos \alpha \sin \beta$ with the right-hand side. We get

$$\sin x = -\sin 2x.$$

Next we use the identity $\sin 2x \equiv 2 \sin x \cos x$, and obtain

$$\sin x = -2 \sin x \cos x, \quad \text{or} \quad 2 \sin x \cos x + \sin x = 0$$

$$\sin x (2 \cos x + 1) = 0 \quad \text{Factoring}$$

$\sin x = 0 \quad$ or $\quad \cos x = -\dfrac{1}{2} \quad$ Principle of zero products

$$x = 0, \pi \quad \text{or} \quad x = \frac{2\pi}{3}, \frac{4\pi}{3}.$$

All of these values check, so the solutions in $[0, 2\pi)$ are 0, π, $2\pi/3$, and $4\pi/3$.

EXERCISE SET 10.7

Find all solutions of the following equations in $[0, 2\pi)$.

1. $\tan x \sin x - \tan x = 0$

2. $2 \sin x \cos x + \sin x = 0$

3. $2 \sec x \tan x + 2 \sec x + \tan x + 1 = 0$
4. $2 \csc x \cos x - 4 \cos x - \csc x + 2 = 0$
5. $\sin 2x - \cos x = 0$
6. $\cos 2x - \sin x = 1$
7. $\sin 2x \sin x - \cos x = 0$
8. $\sin 2x \cos x - \sin x = 0$
9. $\sin 2x + 2 \sin x \cos x = 0$
10. $\cos 2x \sin x + \sin x = 0$
11. $\cos 2x \cos x + \sin 2x \sin x = 1$
12. $\sin 2x \sin x - \cos 2x \cos x = -\cos x$
13. $\sin 4x - 2 \sin 2x = 0$
14. $\sin 4x + 2 \sin 2x = 0$
15. $\sin 2x + 2 \sin x - \cos x - 1 = 0$
16. $\sin 2x + \sin x + 2 \cos x + 1 = 0$
17. $\sec^2 x = 4 \tan^2 x$
18. $\sec^2 x - 2 \tan^2 x = 0$
19. $\sec^2 x + 3 \tan x - 11 = 0$
20. $\tan^2 x + 4 = 2 \sec^2 x + \tan x$
21. $\cot x = \tan(2x - 3\pi)$
22. $\tan x = \cot(2x + \pi)$
23. $\cos(\pi - x) + \sin\left(x - \dfrac{\pi}{2}\right) = 1$
24. $\sin(\pi - x) + \cos\left(\dfrac{\pi}{2} - x\right) = 1$
25. $\dfrac{\cos^2 x - 1}{\sin\left(\dfrac{\pi}{2} - x\right) - 1} = \dfrac{\sqrt{2}}{2} + 1$
26. $\dfrac{\sin^2 x - 1}{\cos\left(\dfrac{\pi}{2} - x\right) + 1} = \dfrac{\sqrt{2}}{2} - 1$
27. $2 \cos x + 2 \sin x = \sqrt{6}$
28. $2 \cos x + 2 \sin x = \sqrt{2}$
29. $\sqrt{3} \cos x - \sin x = 1$
30. $\sqrt{2} \cos x - \sqrt{2} \sin x = 2$

☆

Solve.

31. $\text{Arc} \cos x = \text{Arccos}\, \dfrac{3}{5} - \text{Arcsin}\, \dfrac{4}{5}$
32. $\text{Sin}^{-1} x = \text{Tan}^{-1} \dfrac{1}{3} + \text{Tan}^{-1} \dfrac{1}{2}$

33. Solve graphically: $\sin x - \cos x = \cot x$.

★

34. Solve.

$$\sin x + \cos y = 1,$$
$$2 \sin x = \sin 2y$$

CHAPTER 10 REVIEW

Use the sum and difference formulas to write equivalent expressions. You need not simplify.

1. $\cos\left(x + \dfrac{3\pi}{2}\right)$
2. $\tan(45° - 30°)$
3. Simplify: $\cos 27° \cos 16° + \sin 27° \sin 16°$.
4. Given that $\tan \alpha = \sqrt{3}$, $\sin \beta = \sqrt{2}/2$, and α and β are between 0 and $\pi/2$, evaluate $\tan(\alpha - \beta)$ exactly.
5. Find $\sin \dfrac{\pi}{8}$.
6. Simplify: $\dfrac{\sin 2\theta}{\sin^2 \theta}$.

7. Prove the identity: $\tan 2\theta \equiv \dfrac{2 \tan \theta}{1 - \tan^2 \theta}$

8. Find, in radians, all values of $\sin^{-1} \dfrac{1}{2}$.

9. Use a table to find, in degrees, all values of $\cot^{-1} 0.1584$.

10. Find $\text{Sin}^{-1}\left(-\dfrac{\sqrt{2}}{2}\right)$.

11. Evaluate or simplify $\tan\left(\text{Arctan}\,\dfrac{7}{8}\right)$.

12. Find the angle, to the nearest 10 minutes, from l_1 to l_2, given the equations

$$l_1: y = 2x - 4 \quad \text{and} \quad l_2: x - y = 2.$$

Determine whether the lines are parallel, perpendicular, or neither.

13. Graph: $y + 1 = 2 \cos^2 x$.

14. Graph: $f(x) = 2 \sin^{-1}\left(x + \dfrac{\pi}{2}\right)$.

15. Solve $\sin^2 x - 7 \sin x = 0$, solutions in $[0, 2\pi)$.

16. Find an equivalent expression involving the sine function only: $6 \sin 3x + 2 \cos 3x$.

Triangles, Vectors, and Applications

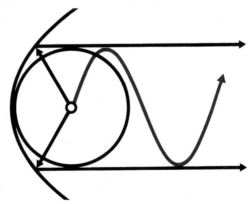

11.1 SOLVING RIGHT TRIANGLES

In Section 8.1 the trigonometric functions were defined and the solving of right triangles was introduced. Tables of the trigonometric functions were considered in Chapter 9. We continue consideration of solving right triangles, a topic important in many applications of trigonometry. The word *trigonometry* actually means "triangle measurement."

The precision to which we can find an angle, using a ratio of sides such as the sine ratio, depends upon how precisely we know the lengths of the sides. The table at the left shows the relationship. The table can also be read in reverse. For example, if we know an angle to the nearest ten minutes, then the number of significant digits in a value obtained from the table is 3, hence the number of significant digits obtained for the length of a side of a triangle is also 3.

NUMBER OF DIGITS IN RATIO	PRECISION OF ANGLE MEASURE
4	To nearest minute
3	To nearest ten minutes
2	To nearest degree

When we have four-digit precision, but a number itself has fewer than four digits, we often indicate the precision by writing 0's after the decimal point, as in 18.00 or 7.000.

In some of the exercises in this section, the precision is not realistic. For example, one would scarcely consider the distance from Los Angeles to Chicago to the nearest thousandth of a mile. The exercises may use unwarranted precision, from the standpoint of reality, in order to provide valid practice in calculating.

Example 1 Find the length b in the triangle of Fig. 1. Use four-digit precision.

The known side is the hypotenuse. The side we seek is adjacent to the known angle, so we shall use the cosine: $\cos A = b/70$. Thus

$$b = 70 \times \cos 19° = 70 \times 0.9455$$
$$b = 66.19.$$

Figure 1

When we solve a triangle, we find the *measures* of its sides and angles not already known. Note that we sometimes shorten this to saying that we "find the angles" or "find the sides."

Applications of Solving Triangles

In many applied problems, unknown parts of right triangles are to be found.

Example 2 *Finding cloud height.* A device for measuring cloud height at night consists of a vertical beam of light, which makes a spot on the clouds. The spot is viewed from a point 135 ft away. The angle of elevation is 67°40′. (The angle between the horizontal and a line of sight is called an *angle of elevation* or an *angle of depression*, the latter if the line of sight is below the horizontal.) Find the height of the clouds.

From Fig. 2 we see that

$$\frac{h}{135} = \tan 67°40'$$

$$h = 135 \times \tan 67°40' = 135 \times 2.434 = 329 \text{ ft.}$$

In the preceding example, it should be noted that distances are precise to three digits and the angle is precise to the nearest ten minutes.

Calculators

A calculator would be most convenient for use with Example 2. The same is true for most of the examples and exercises of this chapter, and the use of a calculator is recommended, although not essential, except for certain exercises marked 🖩.

If function values are obtained from a calculator rather than from Table 3 at the back of the book, it should be kept in mind that the number of decimal places will be different, hence answers may vary because of differences in rounding. The answers at the back of the book are computed using Table 3.

It is vital to note whether the calculator requires parts of degrees to be entered in tenths and hundredths, rather than minutes and seconds.

Figure 2

🖩 **Example 3** From an observation tower two markers are viewed on the ground. The markers and the base of the tower are on a line and the observer's eye is 65.3 ft above the ground. The angles of depression to the markers are 53°10′ and 27°50′. How far is it from one marker to the other?

From Fig. 3 we see that the distance we seek is d, which is $d_1 - d_2$. From the right triangles in the figure, we have

$$\frac{d_1}{65.3} = \cot \theta_1 \quad \text{and} \quad \frac{d_2}{65.3} = \cot \theta_2.$$

Then

$$d_1 = 65.3 \cot 27°50' \quad \text{and} \quad d_2 = 65.3 \cot 53°10',$$

and

$$\begin{aligned} d = d_1 - d_2 &= 65.3 \cot 27°50' - 65.3 \cot 53°10' \\ &= 65.3 \left(\cot 27°50' - \cot 53°10' \right) \\ &= 65.3 \left(\cot 27.83° - \cot 53.17° \right) \\ &= 65.3(1.894 - 0.7490) \\ &= 74.8 \text{ ft.} \end{aligned}$$

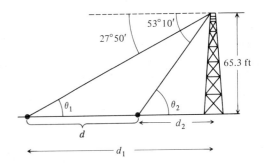

Figure 3

408 TRIGONOMETRY, VECTORS, AND APPLICATIONS

Figure 4

Example 4 In aerial navigation, directions are given in degrees, clockwise from north. Thus east is 90°, south is 180°, and so on. An airplane leaves an airport and travels for 100 mi in a direction 300°. How far north of the airport is the plane then? How far west?

The direction of flight is as shown in Fig. 4. Thus in the triangle shown in Fig. 5, d_1 is the northerly distance and d_2 is the westerly distance. Then

$$\frac{d_1}{100} = \sin 30° \quad \text{and} \quad \frac{d_2}{100} = \cos 30°,$$

$$d_1 = 100 \sin 30° = 100 \times 0.5$$
$$= 50 \text{ mi} \quad \text{(to the nearest mile),}$$

$$d_2 = 100 \cos 30° = 100 \times 0.866$$
$$= 87 \text{ mi.}$$

Figure 5

Figure 6

Example 5 In some applications, directions, or *bearings*, are given by reference to north or south using an acute angle. For example, N 40°W means 40° west of north and S 30°E means 30° east of south. A forest ranger at point A sights a fire directly south. A second ranger at a point B, 7 miles east, sights the fire at a bearing of S 27°20′W. How far from A is the fire?

From Fig. 6 we see that the desired distance d is part of a right triangle, as shown. We have

$$\frac{d}{7} = \tan 62°40'$$

$$d = 7 \tan 62°40' = 7 \times 1.935$$

$$= 13.5 \text{ mi.}$$

▪ EXERCISE SET 11.1

In Exercises 1–18, standard lettering for a right triangle will be used: A, B, and C are the angles, C being the right angle. The sides opposite A, B, and C are a, b, and c, respectively. Solve the triangles, using three-digit precision.

1. $A = 36°10'$, $a = 27.2$
2. $A = 87°40'$, $a = 9.73$
3. $B = 12°40'$, $b = 98.1$
4. $B = 69°50'$, $b = 127$
5. $A = 17°20'$, $b = 13.6$
6. $A = 78°40'$, $b = 1340$
7. $B = 23°10'$, $a = 0.0345$
8. $B = 69°20'$, $a = 0.00488$
9. $A = 47°30'$, $c = 48.3$
10. $A = 88°50'$, $c = 3950$

11. $B = 82°20'$, $c = 0.982$
12. $B = 56°30'$, $c = 0.0447$
13. $a = 12.0$, $b = 18.0$
14. $a = 10.0$, $b = 20.0$
15. $a = 16.0$, $c = 20.0$
16. $a = 15.0$, $c = 45.0$
17. $b = 1.80$, $c = 4.00$
18. $b = 100$, $c = 450$

19. A guy wire to a pole makes an angle of $73°10'$ with the level ground, and is 14.5 ft from the pole at the ground. How far above the ground is the wire attached to the pole?

20. A kite string makes an angle of $31°40'$ with the (level) ground and 455 ft of string is out. How high is the kite?

21. A road rises 3 m per 100 horizontal m. What angle does it make with the horizontal?

22. A kite is 120 ft high when 670 ft of string is out. What angle does the kite make with the ground?

23. What is the angle of elevation of the sun when a 6-ft man casts a 10.3-ft shadow?

24. What is the angle of elevation of the sun when a 35-ft mast casts a 20-ft shadow?

25. From a balloon 2500 ft high, a command post is seen with an angle of depression of $7°40'$. How far is it from a point on the ground below the balloon to the command post?

26. From a lighthouse 55 ft above sea level, the angle of depression to a small boat is $11°20'$. How far from the foot of the lighthouse is the boat?

27. An airplane travels at 120 km/h for 2 hr in a direction of $243°$ from Chicago. At the end of this time, how far south of Chicago is the plane?

28. An airplane travels at 150 km/h for 2 hr in a direction of $138°$ from Omaha. At the end of this time, how far east of Omaha is the plane?

29. Ship A is due west of a lighthouse. Ship B is 12 km south of ship A. From ship B the bearing to the lighthouse is N $63°20'$E. How far is ship A from the lighthouse?

30. Lookout station A is 15 km west of station B. The bearing from A to a fire directly south of B is S $37°50'$E. How far is the fire from B?

31. From a balloon two km high, the angles of depression to two towns in line with the balloon are $81°20'$ and $13°40'$. How far apart are the towns?

32. From a balloon 1000 m high, the angles of depression to two artillery posts, in line with the balloon, are $11°50'$ and $84°10'$. How far apart are the artillery posts?

33. A weather balloon is directly west of two observing stations 10 km apart. The angles of elevation of the balloon from the two stations are $17°50'$ and $78°10'$. How high is the balloon?

34. From two points south of a hill on level ground and 1000 ft apart, the angles of elevation of the hill are $12°20'$ and $82°40'$. How high is the hill?

35. Show that the area of a right triangle is $\frac{1}{4}c^2 \sin 2A$.

36. Show that the area of a right triangle is $(bc/2) \sin A$.

37. Find a formula for the distance to the horizon as a function of the height of the observer above the earth. Calculate the distance to the horizon from an airplane at an altitude of 1000 ft.

38. The base of this tetrahedron is an equilateral triangle. The other faces are isosceles triangles. Find b, s, and θ (a vertex angle).

11.2 THE LAW OF SINES

The trigonometric functions can be used to solve triangles that are not right triangles (oblique triangles). In order to solve oblique triangles we need to derive some properties, one of which is called the *law of sines*.

Figure 7

Figure 8

We shall consider any oblique triangle. It may or may not have an obtuse angle. We will consider both cases, but the derivations are essentially the same. The triangles are lettered in the standard way, with angles A, B, and C and the sides opposite them a, b, and c, respectively. We have drawn an altitude from vertex C. It has length h. In either triangle we now have, from triangle ADC,

$$\frac{h}{b} = \sin A \quad \text{or} \quad h = b \sin A.$$

From triangle DBC in Fig. 7 we have $h/a = \sin B$, or $h = a \sin B$. In Fig. 8 we have $h/a = \sin(180° - B) = \sin B$. So in either kind of triangle we now have

$$h = b \sin A \quad \text{and} \quad h = a \sin B.$$

Thus it follows that

$$b \sin A = a \sin B.$$

We divide by $\sin A \sin B$ to obtain

$$\frac{a}{\sin A} = \frac{b}{\sin B}.$$

There is no danger of dividing by 0 here because we are dealing with triangles whose angles are never 0° or 180°.

If we were to consider an altitude from vertex A in the triangles in Figs. 7 and 8 the same argument would give us

$$\frac{b}{\sin B} = \frac{c}{\sin C}.$$

We combine these results to obtain the law of sines, which holds for right triangles as well as oblique triangles.

THEOREM 1

The law of sines. In any triangle ABC,

$$\frac{a}{\sin A} = \frac{b}{\sin B} = \frac{c}{\sin C}$$

(the sides are proportional to the sines of the opposite angles).

Solving Triangles (AAS)

When two angles and a side of any triangle are known, the law of sines can be used to solve the triangle.

11.2 THE LAW OF SINES

Example 1 In triangle ABC, $a = 4.56$, $A = 43°$, and $C = 57°$. Solve the triangle.

We first draw a sketch (see Fig. 9). We find B, as follows:

$$B = 180° - (43° + 57°) = 80°.$$

We can now find the other two sides, using the law of sines:

$$\frac{c}{\sin C} = \frac{a}{\sin A}$$

$$c = \frac{a \sin C}{\sin A} = \frac{4.56 \sin 57°}{\sin 43°} = \frac{4.56 \times 0.8387}{0.6820} = 5.61;$$

$$\frac{b}{\sin B} = \frac{a}{\sin A}$$

$$b = \frac{a \sin B}{\sin A} = \frac{4.56 \sin 80°}{0.6820} = \frac{4.56 \times 0.9848}{0.6820} = 6.58.$$

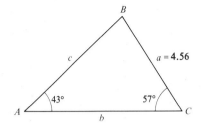

Figure 9

We have now found the unknown parts of the triangle, $B = 80°$, $c = 5.61$, and $b = 6.58$.

The Ambiguous Case (SSA)

When two sides of a triangle and an angle opposite one of them are known, the law of sines can be used to solve the triangle. However, there may be more than one solution. Thus this is known as the ambiguous case. Suppose a, b, and A are given. Then the various possibilities are shown in the four cases in Fig. 10. The following examples correspond to the four possibilities described in the figure.

Case 1

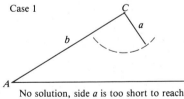

No solution, side a is too short to reach the base.

Case 2

One solution, side a just reaches the base and is perpendicular to it.

Case 3

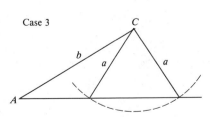

Two solutions, an arc of radius a meets the base at two points.

Case 4

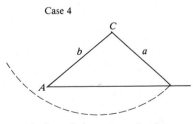

$a \geq b$. One solution, an arc of radius a meets the base at just one point, other than A.

Figure 10

Example 2 In triangle ABC, $a = 15$, $b = 25$, and $A = 47°$. Solve the triangle.

We look for B:
$$\frac{a}{\sin A} = \frac{b}{\sin B}.$$
Then
$$\sin B = \frac{b \sin A}{a} = \frac{25 \sin 47°}{15} = \frac{25 \times 0.7314}{15} = 1.219.$$

Since there is no angle having a sine greater than 1, there is no solution.

Example 3 In triangle ABC, $a = 12$, $b = 5$, and $B = 24°38'$ (see Fig. 11). Solve the triangle.

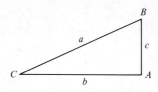

Figure 11

We look for A:
$$\frac{a}{\sin A} = \frac{b}{\sin B}$$
$$\sin A = \frac{a \sin B}{b} = \frac{12 \sin 24°38'}{5}$$
$$= \frac{12 \times 0.4168}{5} = 1.000$$
$$A = 90°.$$

Then $C = 90° - 24°38' = 65°22'$. Since $c/a = \cos B$,
$$c = a \cos B = 12 \times 0.9090 = 10.9.$$

Example 4 In triangle ABC, $a = 20$, $b = 15$, and $B = 30°$ Solve the triangle.

We look for A:
$$\frac{a}{\sin A} = \frac{b}{\sin B}$$
$$\sin A = \frac{a \sin B}{b} = \frac{20 \sin 30°}{15}$$
$$= \frac{20 \times 0.5}{15} = 0.6667.$$

There are two angles less than 180° having a sine of 0.6667. They are 42° and 138°, to the nearest degree. This gives us two possible solutions.

Possible solution 1. We know that $A = 42°$. Then $C = 180° - (30° + 42°) = 108°$ (Fig. 12).

Figure 12

We now find c:

$$\frac{c}{\sin C} = \frac{b}{\sin B}$$

$$c = \frac{b \sin C}{\sin B} = \frac{15 \sin 108°}{\sin 30°} = \frac{15 \times 0.9511}{0.5} = 28.5.$$

These parts make a triangle, as shown; hence we have a solution.

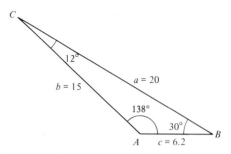

Figure 13

Possible solution 2. We know that $A = 138°$. Then $C = 12°$.

We now find c:

$$c = \frac{b \sin C}{\sin B} = \frac{15 \sin 12°}{\sin 30°} = \frac{15 \times 0.2079}{0.5} = 6.2.$$

These parts make a triangle; hence we have a second solution (Fig. 13).

Example 5 In triangle ABC, $a = 25$, $b = 10$, and $A = 42°$. Solve the triangle.

We look for B:

$$\frac{b}{\sin B} = \frac{a}{\sin A}$$

$$\sin B = \frac{b \sin A}{a} = \frac{10 \sin 42°}{25} = \frac{10 \times 0.6691}{25} = 0.2676.$$

Then $B = 15°30'$ or $B = 164°30'$. Since $a > b$, we know there is only one solution. If we had not noticed this, we could tell it now. An angle of $164°30'$ cannot be an angle of this triangle because it already has an angle of $42°$, and these two would total more than $180°$:

$$C = 180° - (42° + 15°30') = 122°30',$$

$$\frac{c}{\sin C} = \frac{a}{\sin A},$$

so

$$c = \frac{a \sin C}{\sin A} = \frac{25 \sin 122°30'}{\sin 42°} \quad \text{Solving for } c$$

$$c = \frac{25 \times 0.8434}{0.6691} = 31.5.$$

EXERCISE SET 11.2

In Exercises 1–12, solve the triangle ABC.

1. $A = 133°$, $B = 30°$, $b = 18$
2. $B = 120°$, $C = 30°$, $a = 16$
3. $B = 38°$, $C = 21°$, $b = 24$
4. $A = 131°$, $C = 23°$, $b = 10$

5. $A = 68°30'$, $C = 42°40'$, $c = 23.5$
6. $B = 118°20'$, $C = 45°40'$, $b = 42.1$
7. $A = 36°$, $a = 24$, $b = 34$
8. $C = 43°$, $c = 28$, $b = 27$
9. $A = 116°20'$, $a = 17.2$, $c = 13.5$
10. $A = 47°50'$, $a = 28.3$, $b = 18.2$
11. $C = 61°10'$, $c = 30.3$, $b = 24.2$
12. $B = 58°40'$, $a = 25.1$, $b = 32.6$

13. Points A and B are on opposite sides of a lunar crater. Point C is 50 meters from A. The measure of $\angle BAC$ is determined to be $112°$ and the measure of $\angle ACB$ is determined to be $42°$. What is the width of the crater?

14. A guy wire to a pole makes a $71°$ angle with level ground. At a point 25 ft farther from the pole than the guy wire, the angle of elevation of the top of the pole is $37°$. How long is the guy wire?

15. A pole leans away from the sun at an angle of $7°$ to the vertical. When the angle of elevation of the sun is $51°$, the pole casts a shadow 47 ft long on level ground. How long is the pole?

16. A vertical pole stands by a road that is inclined $10°$ to the horizontal. When the angle of elevation of the sun is $23°$, the pole casts a shadow 38 ft long directly downhill along the road. How long is the pole?

17. A reconnaissance airplane leaves its airport on the east coast of the United States and flies in a direction of $085°$. Because of bad weather it returns to another airport 230 km to the north of its home base. For the return it flies in a direction of $283°$. What was the total distance it flew?

18. Lookout station B is 10.2 km east of station A. The bearing of a fire from A is S $10°40'$W. The bearing of the fire from B is S $31°20'$W. How far is the fire from A? from B?

19. A boat leaves a lighthouse A and sails 5.1 km. At this time it is sighted from lighthouse B, 7.2 km west of A. The bearing of the boat from B is N $65°10'$E. How far is the boat from B?

20. An airplane leaves airport A and flies 200 km. At this time its bearing from airport B, 250 km to the west, is $120°$. How far is the airplane from B?

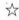

21. Prove that the area of a parallelogram is the product of two sides and the sine of the included angle.

22. Prove that in any triangle ABC, $\sin A + \sin B > \sin(A + B)$.

23. Prove that the area of a quadrilateral is half the product of the lengths of its diagonals and the sine of an angle between the diagonals.

24. Following is the *law of tangents*:

$$\tan \tfrac{1}{2}(A - B) = \frac{a - b}{a + b} \tan \tfrac{1}{2}(A + B)$$

in any triangle ABC. Prove this law.

25. When two objects, such as ships, airplanes, or runners, move in straight-line paths, if the distance between them is decreasing and if the bearing from one of them to the other is constant, they will collide. ("Constant bearing means collision," as mariners put it.) Prove that this statement is true.

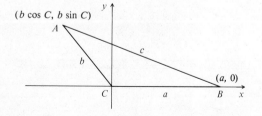

Figure 14

11.3 THE LAW OF COSINES

A second property of triangles important in solving oblique triangles is called the *law of cosines*. To derive this property we consider any triangle ABC placed on a coordinate system (Fig. 14). We place the origin at one of the vertices, say C, and the positive half of the x-axis along one of the sides, say CB. Then the coordinates of B are $(a, 0)$, and the coordinates of A are $(b \cos C, b \sin C)$.

We will next use the distance formula to determine c^2:

$$c^2 = (b \cos C - a)^2 + (b \sin C - 0)^2.$$

Now we multiply and simplify:

$$c^2 = b^2 \cos^2 C - 2ab \cos C + a^2 + b^2 \sin^2 C$$
$$c^2 = a^2 + b^2 (\sin^2 C + \cos^2 C) - 2ab \cos C$$
$$c^2 = a^2 + b^2 - 2ab \cos C.$$

Had we placed the origin at one of the other vertices, we would have obtained

$$a^2 = b^2 + c^2 - 2bc \cos A \quad \text{or} \quad b^2 = a^2 + c^2 - 2ac \cos B.$$

This result can be summarized as follows.

THEOREM 2

The law of cosines. **In any triangle ABC,**

$$a^2 = b^2 + c^2 - 2bc \cos A,$$
$$b^2 = a^2 + c^2 - 2ac \cos B, \quad \text{and}$$
$$c^2 = a^2 + b^2 - 2ab \cos C.$$

(In any triangle, the square of a side is the sum of the squares of the other two sides, minus twice the product of those sides and the cosine of the included angle.)

Only one of the above formulas need be memorized. The other two can be obtained by a change of letters.

Solving Triangles (SAS)

When two sides of a triangle and the included angle are known, we can use the law of cosines to find the third side. The law of cosines or the law of sines can then be used to finish solving the triangle.

Example 1 In triangle ABC, $a = 24$, $c = 32$, and $B = 115°$ (see Fig. 15). Solve the triangle.

We first find the third side. From the law of cosines,

$$b^2 = a^2 + c^2 - 2ac \cos B$$
$$= 24^2 + 32^2 - 2 \cdot 24 \cdot 32(-0.4226)$$
$$= 2249.$$

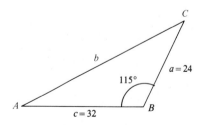

Figure 15

Then $b = \sqrt{2249} = 47.4$. Next, we use the law of sines to find a second angle:

$$\frac{a}{\sin A} = \frac{b}{\sin B}, \quad \text{and} \quad \sin A = \frac{a \sin B}{b};$$

$$\sin A = \frac{24 \sin 115°}{47.4} = \frac{24 \times 0.9063}{47.4} = 0.4589.$$

Thus
$$A = 27°20'.$$

The third angle is now easy to find:

$$C = 180° - (115° + 27°20') = 37°40'.$$

Solving Triangles (SSS)

When all three sides of a triangle are known, the law of cosines can be used to solve the triangle.

Example 2 In triangle ABC, $a = 18$, $b = 25$, and $c = 12$ (see Fig. 16). Solve the triangle.

Figure 16

Let us find angle B. We select the formula from the law of cosines that contains $\cos B$; in other words, $b^2 = a^2 + c^2 - 2ac \cos B$. We solve this for $\cos B$ and substitute:

$$\cos B = \frac{a^2 + c^2 - b^2}{2ac} = \frac{18^2 + 12^2 - 25^2}{2 \cdot 18 \cdot 12} = -0.3634.$$

Then $B = 111°20'$.

Similarly, we shall find angle A:*

$$a^2 = b^2 + c^2 - 2bc \cos A$$

$$\cos A = \frac{b^2 + c^2 - a^2}{2bc} = \frac{25^2 + 12^2 - 18^2}{2 \cdot 25 \cdot 12} = 0.7417.$$

Thus $A = 42°10'$. Then $C = 180° - (111°20' + 42°10') = 26°30'$.

EXERCISE SET 11.3

In Exercises 1–12, solve the triangles.

1. $A = 30°$, $b = 12$, $c = 24$
2. $C = 60°$, $a = 15$, $b = 12$
3. $A = 133°$, $b = 12$, $c = 15$
4. $B = 116°$, $a = 31$, $c = 25$
5. $B = 72°40'$, $c = 16$, $a = 78$
6. $A = 24°30'$, $b = 68$, $c = 14$

*The law of sines could be used at this point.

7. $a = 12$, $b = 14$, $c = 20$

8. $a = 22$, $b = 22$, $c = 35$

9. $a = 3.3$, $b = 2.7$, $c = 2.8$

10. $a = 16$, $b = 20$, $c = 32$

11. $a = 2.2$, $b = 4.1$, $c = 2.4$

12. $a = 3.6$, $b = 6.2$, $c = 4.1$

13. A ship leaves a harbor and sails 15 nautical mi east. It then sails 18 nautical mi in a direction of S 27°E. How far is it, then, from the harbor and in what direction?

14. An airplane leaves an airport and flies west 147 km. It then flies 200 km in a direction of 220°. How far is it, then, from the airport and in what direction?

15. Two ships leave harbor at the same time. The first sails N 15°W at 25 knots (a knot is one nautical mile per hour). The second sails N 32°E at 20 knots. After 2 hr, how far apart are the ships?

16. Two airplanes leave an airport at the same time. The first flies 150 km/h in a direction of 320°. The second flies 200 km/h in a direction of 200°. After 3 hr, how far apart are the planes?

17. A hill is inclined 5° to the horizontal. A 45-ft pole stands at the top of the hill. How long a rope will it take to reach from the top of the pole to a point 35 ft downhill from the base of the pole?

18. A hill is inclined 15° to the horizontal. A 40-ft pole stands at the top of the hill. How long a rope will it take to reach from the top of the pole to a point 68 ft downhill from the base of the pole?

19. A piece of wire 5.5 m long is bent into a triangular shape. One side is 1.5 m long and another is 2 m long. Find the angles of the triangle.

20. A triangular lot has sides 120 ft long, 150 ft long, and 100 ft long. Find the angles of the lot.

21. A slow-pitch softball diamond is a square 60 ft on a side. The pitcher's mound is 46 ft from home. How far is it from the pitcher's mound to first base?

22. A baseball diamond is a square 90 ft on a side. The pitcher's mound is 60.5 ft from home. How far does the pitcher have to run to cover first base?

23. The longer base of an isosceles trapezoid measures 14 ft. The nonparallel sides measure 10 ft, and the base angles measure 80°.
  a) Find the length of a diagonal.   b) Find the area.

24. An isosceles triangle has a vertex angle of 38° and this angle is included by two sides, each measuring 20 ft. Find the area of the triangle.

25. A field in the shape of a parallelogram has sides that measure 50 yd and 70 yd. One angle of the field measures 78°. Find the area of the field.

26. An aircraft takes off to fly a 180-mile trip. After flying 75 miles it is 10 miles off course. How much should the heading be corrected to then fly straight to the destination, assuming no wind correction?

27. From the top of a hill 20 ft above the surface of a lake, a tree across the lake is sighted. The angle of elevation is 11°. From the same position, the top of the tree is seen by reflection in the lake. The angle of depression is 14°. What is the height of the tree?

28. A bridge is being built across a canyon. The length of the bridge is 5045 ft. From the deepest point in the canyon, the angles of elevation of the ends of the bridge are 78° and 72°. How deep is the canyon?

29. Find a formula for the area of an isosceles triangle, in terms of the congruent sides and their included angle. Under what conditions will the area of a triangle with fixed congruent sides be a maximum?

30. Show that in any triangle ABC,

$$a^2 + b^2 + c^2 = 2(bc \cos A + ac \cos B + ab \cos C).$$

31. Show that in any triangle ABC,

$$\frac{\cos A}{a} + \frac{\cos B}{b} + \frac{\cos C}{c} = \frac{a^2 + b^2 + c^2}{2abc}.$$

32. Show that in any triangle,

$$\sin^2 A = \sin^2 B + \sin^2 C - 2 \sin B \sin C \cos A.$$

33. Hero, of Alexandria, derived the following formula for the area of any triangle. Prove this formula.

$$\text{Area} = \sqrt{s(s-a)(s-b)(s-c)},$$

where s is half the perimeter.

11.4 VECTORS

In many applications there arise certain quantities in which a direction is specified. Any such quantity having a *direction* and a *magnitude* is called a *vector quantity*, or *vector*. Here are some examples of vector quantities.

Displacement. An object moves a certain distance in a certain direction.
  A train travels 100 mi to the northeast.
  A person takes 5 steps to the west.
  A batter hits a ball 100 m along the left-field foul line.

Velocity. An object travels at a certain speed in a certain direction.
  The wind is blowing 15 mph from the northwest.
  An airplane is traveling 450 km/h in a direction of 243°.

Force. A push or pull is exerted on an object in a certain direction.
  A 15-kg force is exerted downward on the handle of a jack.
  A 25-lb upward force is required to lift a box.
  A wagon is pulled up a 30° incline, requiring an effort of 200 kg.

We shall represent vectors abstractly by directed line segments, or arrows. The length is chosen, according to some scale, to represent the magnitude of the vector, and the direction of the arrow represents the direction of the vector. For example, if we let 1 cm represent 5 km/h, then a 15-km/h wind from the northwest would be represented by an arrow 3 cm long, as in Fig. 17.

Figure 17

Vector Addition

To "add" vectors, we find a single vector that would have the same effect as the vectors combined. We shall illustrate vector addition for displacements, but it is done the same way for any vector quantities. Suppose a man takes 4 steps east and then 3 steps north. He will then be

5 steps from the starting point in the direction shown. The *sum* of the two vectors is the vector 5 steps in magnitude and in the direction shown in Fig. 18. The sum is also called the *resultant* of the two vectors.

In general, if we have two vectors **a** and **b**,* we can add them as in the above example. That is, we place the tail of one arrow at the head of the other and then find the vector that forms the third side of a triangle. Or, we can place the tails of the arrows together, complete a parallelogram, and find the diagonal of the parallelogram (see Fig. 19).

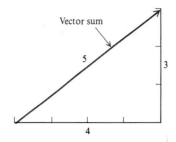

Figure 18

Example 1 Forces of 15 kg and 25 kg act on an object at right angles to each other. Find their *sum*, or *resultant*, giving the angle that it makes with the larger force.

We make a drawing (Fig. 20), this time a rectangle, using s for the length of vector **OB**. Since OAB is a right triangle, we have

$$\tan \theta = \frac{15}{25} = 0.6.$$

Thus θ, the angle the resultant makes with the larger force, is 31°. Now $s/15 = \csc \theta$, or $s = 15 \csc \theta = 15 \times 1.942 = 29.1$. Thus the resultant **OB** has a magnitude of 29.1 kg and makes an angle of 31° with the larger force.

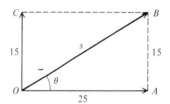

Figure 19

Example 2 An airplane heads in a direction of 100° at 180-km/h airspeed while a wind is blowing 40 km/h from 220°. Find the speed of the airplane over the ground and the direction of its track over the ground.

We first make a drawing (Fig. 21). The wind is represented by **OC** and the velocity vector of the airplane by **OA**. The resultant velocity is **v**, the sum of the two vectors. We denote the length of **v** by $|\mathbf{v}|$. By the law of cosines in $\triangle OAB$, we have

$$|\mathbf{v}|^2 = 40^2 + 180^2 - 2 \cdot 40 \cdot 180 \cos 120° = 41{,}200.$$

Thus $|\mathbf{v}|$ is 203 km/h. By the law of sines in the same triangle,

$$\frac{40}{\sin \theta} = \frac{203}{\sin 120°},$$

or

$$\sin \theta = \frac{40 \sin 120°}{203} = 0.1706.$$

Thus $\theta = 10°$ to the nearest degree. Therefore the ground speed of the airplane is 203 km/h and its track is in a direction of 90°.

Figure 20

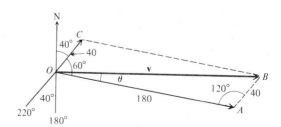

Figure 21

---

*It is usual to denote vectors using boldface type or, sometimes in handwriting, to write a bar over the letters.

EXERCISE SET 11.4

In Exercises 1–8, magnitudes of vectors **a** and **b** are given and the angle between the vectors is θ. Find the resultant, giving the direction by specifying to the nearest degree the angle it makes with vector **a**.

1. $|\mathbf{a}| = 45$, $|\mathbf{b}| = 35$, $\theta = 90°$
2. $|\mathbf{a}| = 54$, $|\mathbf{b}| = 43$, $\theta = 90°$
3. $|\mathbf{a}| = 10$, $|\mathbf{b}| = 12$, $\theta = 67°$
4. $|\mathbf{a}| = 25$, $|\mathbf{b}| = 30$, $\theta = 75°$
5. $|\mathbf{a}| = 20$, $|\mathbf{b}| = 20$, $\theta = 117°$
6. $|\mathbf{a}| = 30$, $|\mathbf{b}| = 30$, $\theta = 123°$
7. $|\mathbf{a}| = 23$, $|\mathbf{b}| = 47$, $\theta = 27°$
8. $|\mathbf{a}| = 32$, $|\mathbf{b}| = 74$, $\theta = 72°$

9. Two forces of 5 kg and 12 kg act on an object at right angles. Find the magnitude of the resultant and the angle it makes with the smaller force.

10. Two forces of 30 kg and 40 kg act on an object at right angles. Find the magnitude of the resultant and the angle it makes with the smaller force.

11. Forces of 420 kg and 300 kg act on an object. The angle between the forces is 50°. Find the resultant, giving the angle it makes with the larger force.

12. Forces of 410 kg and 600 kg act on an object. The angle between the forces is 47°. Find the resultant, giving the angle it makes with the smaller force.

13. A balloon is rising 12 ft/sec while the wind is blowing 18 ft/sec. Find the speed of the balloon and the angle it makes with the horizontal.

14. A balloon is rising 10 ft/sec while a wind is blowing 5 ft/sec. Find the speed of the balloon and the angle it makes with the horizontal.

15. A boat heads 35°, propelled by a force of 750 lb. A wind from 320° exerts a force of 150 lb on the boat. How large is the resultant force and in what direction is the boat moving?

16. A boat heads 220°, propelled by a 650-lb force. A wind from 080° exerts a force of 100 lb on the boat. How large is the resultant force, and in what direction is the boat moving?

17. A ship sails N 80°E for 120 nautical mi, then S 20°W for 200 nautical mi. How far is it then from the starting point, and in what direction?

18. An airplane flies 032° for 210 km, then 280° for 170 km. How far is it, then, from the starting point, and in what direction?

19. A motorboat has a speed of 15 km/h. It crosses a river whose current has a speed of 3 km/h. In order to cross the river at right angles, the boat should be pointed in what direction?

20. An airplane has an airspeed of 150 km/h. It is to make a flight in a direction of 080° while there is a 25-km/h wind from 350°. What should be the airplane's heading?

11.5 VECTORS AND COORDINATES

Components of Vectors

Given a vector, it is often convenient to reverse the addition procedure, that is, to find two vectors whose sum is the given vector. Usually the two vectors we seek will be perpendicular. The two vectors we find are called *components* of the given vector.

Example 1 A certain vector **a** has a magnitude of 130 and is inclined 40° with the horizontal. Resolve the vector into horizontal and vertical components.

Figure 22

We first make a drawing (Fig. 22), showing horizontal and vertical vectors whose sum is the given vector **a**. From $\triangle OAB$, we see

$$|\mathbf{h}| = 130 \cos 40° = 99.6, \quad \text{and} \quad |\mathbf{v}| = 130 \sin 40° = 83.6.$$

These are the components we seek.

Example 2 An airplane is flying at 200 km/h in a direction of 305°. Find the westerly and northerly components of its velocity.

We first make a drawing (Fig. 23), showing westerly and northerly vectors whose sum is the given velocity. From $\triangle OAB$, we see that

$$|\mathbf{n}| = 200 \cos 55° = 115 \text{ km/h},$$
$$|\mathbf{w}| = 200 \sin 55° = 164 \text{ km/h}.$$

Figure 23

Given a vector, we can resolve it into components. Given components of a vector, we can find the vector. We simply find the vector sum of the components.

Example 3 A vector **v** has a westerly component of 12 and a southerly component of 16. Find the vector.

From Fig. 24 we see that $\tan \theta = \frac{12}{16} = 0.75$. Thus $\theta = 36°50'$. Then $|\mathbf{v}| = 12 \csc 36°50' = 20$. Thus the vector has a magnitude of 20 and a direction of S 36°50'W.

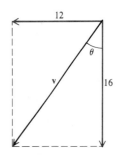

Figure 24

Adding Vectors Using Components

When we have two vectors resolved into components, we can add the two vectors by adding the components.

Example 4 A vector **v** has a westerly component of 3 and a northerly component of 5. A second vector **w** has an easterly component of 8 and a northerly component of 4. (See Fig. 25.) Find the components of the sum. Find the sum.

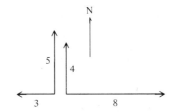

Figure 25

Adding the east–west components, we obtain 5 east. Adding the north–south components, we obtain 9 north. The components of **v** + **w** are 5 east and 9 north. We now add these components:

$$\tan \theta = \frac{9}{5} = 1.8,$$
$$\theta = 61°,$$
$$|\mathbf{v} + \mathbf{w}| = 9 \csc 61° = 10.3.$$

The sum **v** + **w** has a magnitude of 10.3 and a direction of N 29°E (Fig. 26).

Figure 26

Analytic Representation of Vectors

If we place a coordinate system so that the origin is at the tail of an arrow representing a vector, we say that the vector is in *standard position* (Fig. 27). Then if we know the coordinates of the other end of the vector, we know the vector. The coordinates will be the x-component and the y-component of the vector. Thus we can consider an ordered pair (a, b) to be a vector. When vectors are given in this form, it is easy to add them. We simply add the respective components.

Figure 27

Example 5 Find $\mathbf{u} + \mathbf{v}$, where $\mathbf{u} = (3, -7)$ and $\mathbf{v} = (4, 2)$.

The sum is found by adding the x-components and then adding the y-components:

$$\mathbf{u} + \mathbf{v} = (3, -7) + (4, 2) = (7, -5).$$

Vectors can also be specified by giving the length and direction. When this is done, we say that we have the vector in *polar form*. An example of polar form is $(15, 260°)$. The angle is measured from the positive half of the x-axis counterclockwise. In Section 11.6, we consider a coordinate system (called a *polar coordinate system*) in which polar notation is very natural.

Figure 28

Example 6 Find polar form for the vector $\mathbf{v}$, where $\mathbf{v} = (7, -2)$.

From Fig. 28 we see the reference angle ϕ:

$$\tan \phi = \frac{-2}{7} = -0.2857,$$

$$\phi = -16°,$$

$$|\mathbf{v}| = \sqrt{(-2)^2 + 7^2} = 7.28.$$

Thus polar form for $\mathbf{v}$ is

$$(7.28, 344°) \quad \text{or} \quad (7.28, -16°).$$

Given polar form for a vector we can also find the ordered-pair notation (rectangular form).

Example 7 Find rectangular form for the vector $\mathbf{w}$, where $\mathbf{w} = (14, 155°)$.

From Fig. 29 we have

$$x = 14 \cos 155° = -12.7,$$
$$y = 14 \sin 155° = 5.92.$$

Figure 29

Thus rectangular form for $\mathbf{w}$ is $(-12.7, 5.92)$.

Properties of Vectors and Vector Algebra

With vectors represented in rectangular form (a, b) it is easy to determine many of their properties. We consider the set of all ordered pairs of real numbers (a, b). This is the set of all vectors on a plane. Consider the sum of any two vectors:

$$\mathbf{u} = (a, b), \quad \mathbf{v} = (c, d),$$
$$\mathbf{u} + \mathbf{v} = (a + c, b + d), \quad \mathbf{v} + \mathbf{u} = (c + a, d + b).$$

We have just shown that addition of vectors is commutative. Other properties are also not difficult to show. The system of vectors is not a field, but it does have some of the familiar field properties.

PROPERTIES OF VECTORS

Commutativity. Addition of vectors is commutative.

Associativity. Addition of vectors is associative.

Identity. There is an additive identity, the vector $(0, 0)$, or $\mathbf{0}$.

Inverses. Every vector (a, b) has an additive inverse $(-a, -b)$.

Subtraction. We define $\mathbf{u} - \mathbf{v}$ to be the vector which when added to $\mathbf{v}$ gives $\mathbf{u}$. It follows that if $\mathbf{u} = (a, b)$ and $\mathbf{v} = (c, d)$, then $\mathbf{u} - \mathbf{v} = (a - c, b - d)$. Thus subtraction is always possible. Also, $\mathbf{u} - \mathbf{v} = \mathbf{u} + (-\mathbf{v})$.

Length, or absolute value. If $\mathbf{u} = (a, b)$, then $|\mathbf{u}| = \sqrt{a^2 + b^2}$.

A vector can be multiplied by a real number. A real number in this context is called a *scalar*. For example $2\mathbf{v}$ is a vector in the same direction as $\mathbf{v}$ but twice as long. Analytically, we define scalar multiplication as follows.

DEFINITION

Scalar multiplication. If r is any real number and $\mathbf{v} = (a, b)$, then $r\mathbf{v} = (ar, br)$.

Example 8 Do the following calculations, where $\mathbf{u} = (4, 3)$ and $\mathbf{v} = (-5, 8)$.

a) $\mathbf{u} + \mathbf{v} = (4, 3) + (-5, 8) = (-1, 11)$

b) $\mathbf{u} - \mathbf{v} = \mathbf{u} + (-\mathbf{v}) = (4, 3) + (5, -8) = (9, -5)$

c) $3\mathbf{u} - 4\mathbf{v} = 3(4, 3) - 4(-5, 8) = (12, 9) - (-20, 32) = (32, -23)$

d) $|3\mathbf{u} - 4\mathbf{v}| = \sqrt{32^2 + (-23)^2} = \sqrt{1553} \approx 39.41$

EXERCISE SET 11.5

In Exercises 1–14, vectors are given in rectangular form. In Exercises 1–6, add, giving the result in rectangular form.

1. $(3, 7) + (2, 9)$
2. $(5, -7) + (-3, 2)$
3. $(17, 7.6) + (-12.2, 6.1)$
4. $(-15.2, 37.1) + (7.9, -17.8)$
5. $(-650, -750) + (-12, 324)$
6. $(-354, -973) + (-75, 256)$

In Exercises 7–14, find polar form.

7. $(3, 4)$
8. $(4, 3)$
9. $(10, -15)$
10. $(17, -10)$
11. $(-3, -4)$
12. $(-4, -3)$
13. $(-10, 15)$
14. $(-17, 10)$

In Exercises 15–22, find rectangular form.

15. $(4, 30°)$
16. $(8, 60°)$
17. $(10, 235°)$
18. $(15, 210°)$
19. $(20, 330°)$
20. $(20, 200°)$
21. $(100, -45°)$
22. $(150, -60°)$

23. A vector **u** with magnitude 150 is inclined upward 52° from the horizontal. Find the horizontal and vertical components of **u**.

24. A vector **u** with magnitude 170 is inclined downward 63° from the horizontal. Find the horizontal and vertical components of **u**.

25. An airplane is flying 220° at 250 km/h. Find the southerly and westerly components of its velocity **v**.

26. A wind is blowing from 310° at 25 mph. Find the southerly and easterly components of the wind velocity **v**.

27. A force **f** has a westerly component of 25 kg and a southerly component of 35 kg. Find the magnitude and direction of **f**.

28. A force **f** has an upward component of 65 kg and a component to the left of 90 kg. Find the magnitude and direction of **f**.

29. Vector **u** has a westerly component of 15 and a northerly component of 22. Vector **v** has a northerly component of 6 and an easterly component of 8. Find (a) the components of **u** + **v**; (b) the magnitude and direction of **u** + **v**.

30. Vector **u** has a component of 18 upward and a component of 12 to the left. Vector **v** has a component of 5 downward and a component of 35 to the right. Find (a) the components of **u** + **v**; (b) the magnitude and direction of **u** + **v**.

In Exercises 31–40, do the calculations for the following vectors: $\mathbf{u} = (3, 4)$, $\mathbf{v} = (5, 12)$, $\mathbf{w} = (-6, 8)$.

31. $3\mathbf{u} + 2\mathbf{v}$
32. $3\mathbf{v} - 2\mathbf{w}$
33. $(\mathbf{u} + \mathbf{v}) - \mathbf{w}$
34. $\mathbf{u} - (\mathbf{v} + \mathbf{w})$
35. $|\mathbf{u}| + |\mathbf{v}|$
36. $|\mathbf{u}| - |\mathbf{v}|$
37. $|\mathbf{u} + \mathbf{v}|$
38. $|\mathbf{u} - \mathbf{v}|$
39. $2|\mathbf{u} + \mathbf{v}|$
40. $2|\mathbf{u}| + 2|\mathbf{v}|$

☆

41. If **PQ** is any vector, what is **PQ** + **QP**?

42. The *inner product* of vectors **u** • **v** is a scalar, defined as

$$\mathbf{u} \cdot \mathbf{v} = |\mathbf{u}| \, |\mathbf{v}| \cos \theta,$$

where θ is the angle between the vectors. Show that vectors **u** and **v** are perpendicular if and only if $\mathbf{u} \cdot \mathbf{v} = 0$.

★

43. Show that scalar multiplication is distributive over vector addition.

44. Let $\mathbf{u} = (3, 4)$. Find a vector that has the same direction as **u** but has length 1.

45. Prove that for any vectors **u** and **v**, $|\mathbf{u} + \mathbf{v}| \leq |\mathbf{u}| + |\mathbf{v}|$.

11.6 POLAR COORDINATES

In graphing we locate a point with an ordered pair of numbers (a, b). We may consider any such ordered pair to be a vector. From our work with vectors we know that we could also locate a point with a vector in polar form, if given a length and a direction. When we use vectors in rectangular form to locate points, we describe the coordinate system as *rectangular*.* When we use vectors in polar form to locate points, we describe the coordinate system as *polar*. As Fig. 30 shows, any point has rectangular coordinates (x, y) and has polar coordinates (r, θ). On a polar graph the origin is called the *pole* and the positive half of the x-axis is called the *polar axis*.

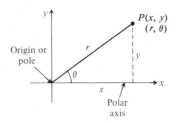

Figure 30

To plot points on a polar graph, we usually locate θ first, then move a distance r from the pole in that direction. If r is negative, we move in the opposite direction. Polar-coordinate graph paper, shown in Fig. 31 facilitates plotting. Point B illustrates that θ may be in radians. Point E illustrates that the polar coordinates of a point are not unique.

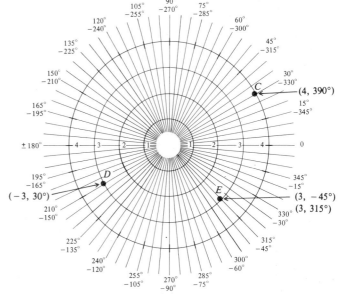

Figure 31

Polar and Rectangular Equations

Some curves have simpler equations in polar coordinates than in rectangular coordinates. For others the reverse is true. The following

---
*Also called *Cartesian*, after the French mathematician René Descartes.

relationships between the two kinds of coordinates allow us to convert an equation from rectangular to polar coordinates:

$$x = r \cos \theta, \quad y = r \sin \theta.$$

Example 1 Convert to a polar equation: $x^2 + y^2 = 25$.

$$\begin{aligned} (r \cos \theta)^2 + (r \sin \theta)^2 &= 25 \quad \text{Substituting for x and y} \\ r^2 \cos^2 \theta + r^2 \sin^2 \theta &= 25 \\ r^2 (\cos^2 \theta + \sin^2 \theta) &= 25 \\ r^2 &= 25 \\ r &= 5 \end{aligned}$$

Example 1 illustrates that the polar equation of a circle centered at the origin is much simpler than the rectangular equation.

Example 2 Convert to a polar equation: $2x - y = 5$.

$$\begin{aligned} 2(r \cos \theta) - (r \sin \theta) &= 5 \\ r(2 \cos \theta - \sin \theta) &= 5 \end{aligned}$$

The relationship that we need to convert from polar equations to rectangular equations can easily be determined from the triangle in Fig. 30. They are as follows:

$$r = \pm\sqrt{x^2 + y^2},$$

$$\sin \theta = \frac{\pm y}{\sqrt{x^2 + y^2}},$$

$$\cos \theta = \frac{\pm x}{\sqrt{x^2 + y^2}},$$

$$\tan \theta = \frac{y}{x}.$$

To convert to rectangular notation, we substitute as needed from either of the two lists, choosing the + or − signs as appropriate in each case.

Example 3 Convert to a rectangular (Cartesian) equation: $r = 4$.

$$\begin{aligned} +\sqrt{x^2 + y^2} &= 4 \quad \text{Substituting for r} \\ x^2 + y^2 &= 16 \quad \text{Squaring} \end{aligned}$$

In squaring like this, we must be careful not to introduce solutions of the equation that are not already present. In this example we did not, because the graph is a circle of radius 4 centered at the origin, in both cases.

Example 4 Convert to a rectangular equation: $r \cos \theta = 6$.

From our first list we know that $x = r \cos \theta$, so we have

$$x = 6.$$

Example 5 Convert to a rectangular equation: $r = 2 \cos \theta + 3 \sin \theta$.

Let us first multiply on both sides by r:

$$r^2 = 2r \cos \theta + 3r \sin \theta$$
$$x^2 + y^2 = 2x + 3y \quad \text{Substituting}$$

EXERCISE SET 11.6

Use polar coordinate paper. Graph the following points.

1. $(4, 30°)$
2. $(5, 45°)$
3. $(0, 37°)$
4. $(0, 48°)$
5. $(-6, 150°)$
6. $(-5, 135°)$
7. $(-8, 210°)$
8. $(-5, 270°)$
9. $(3, -30°)$
10. $(6, -45°)$
11. $(7, -315°)$
12. $(4, -270°)$
13. $(-3, -30°)$
14. $(-6, -45°)$
15. $(-3.2, 27°)$
16. $(-6.8, 34°)$
17. $\left(6, \frac{\pi}{4}\right)$
18. $\left(5, \frac{\pi}{6}\right)$
19. $\left(4, \frac{3\pi}{2}\right)$
20. $\left(3, \frac{3\pi}{4}\right)$
21. $\left(-6, \frac{\pi}{4}\right)$
22. $\left(-5, \frac{\pi}{6}\right)$
23. $\left(-4, -\frac{3\pi}{2}\right)$
24. $\left(-3, -\frac{3\pi}{4}\right)$

Find polar coordinates of the following points.

25. $(4, 4)$
26. $(5, 5)$
27. $(0, 5)$
28. $(0, -3)$
29. $(4, 0)$
30. $(-5, 0)$
31. $(3, 3\sqrt{3})$
32. $(-3, -3\sqrt{3})$
33. $(\sqrt{3}, 1)$
34. $(-\sqrt{3}, 1)$
35. $(3\sqrt{3}, 3)$
36. $(4\sqrt{3}, -4)$

Find Cartesian coordinates of the following points.

37. $(4, 45°)$
38. $(5, 60°)$
39. $(0, 23°)$
40. $(0, -34°)$
41. $(-3, 45°)$
42. $(-5, 30°)$
43. $(6, -60°)$
44. $(3, -120°)$
45. $\left(10, \frac{\pi}{6}\right)$
46. $\left(12, \frac{3\pi}{4}\right)$
47. $\left(-5, \frac{5\pi}{6}\right)$
48. $\left(-6, \frac{3\pi}{4}\right)$

Convert to polar equations.

49. $3x + 4y = 5$
50. $5x + 3y = 4$
51. $x = 5$
52. $y = 4$
53. $x^2 + y^2 = 36$
54. $x^2 + y^2 = 16$
55. $x^2 - 4y^2 = 4$
56. $x^2 - 5y^2 = 5$

Convert to rectangular equations.

57. $r = 5$
58. $r = 8$
59. $\theta = \frac{\pi}{4}$
60. $\theta = \frac{3\pi}{4}$

61. $r \sin \theta = 2$ **62.** $r \cos \theta = 5$ **63.** $r = 4 \cos \theta$ **64.** $r = -3 \sin \theta$

65. $r - r \sin \theta = 2$ **66.** $r + r \cos \theta = 3$ **67.** $r - 2 \cos \theta = 3 \sin \theta$ **68.** $r + 5 \sin \theta = 7 \cos \theta$

★

69. Convert to a rectangular equation: $r = \sec^2 \dfrac{\theta}{2}$.

70. The center of a regular hexagon is at the origin and one vertex is the point $(4, 0°)$. Find the coordinates of the other vertices.

71. Show that $(2, 90°)$, $(2, 210°)$, and $(2, 330°)$ are vertices of an equilateral triangle. Then find the length of a side.

72. Show that the law of cosines is a distance formula for polar coordinates.

11.7 FORCES IN EQUILIBRIUM

When several forces act through the same point on an object, their vector sum must be **0** in order for a balance to occur. When a balance occurs, then the object is either stationary or moving in a straight line without acceleration. The fact that the vector sum must be **0** for a balance, and vice versa, allows us to solve many applied problems involving forces. In these problems we usually resolve vectors into components. For a balance, the sum of the components must be **0** separately.*

Figure 32

Example 1 A 200-lb sign is hanging from the end of a horizontal hinged boom, supported by a cable from the end of the boom and inclined 35° with the horizontal (see Fig. 32). Find the force (tension) in the cable and the force (compression) in the boom.

At the end of the boom we have three forces acting at a point: the weight of the sign acting down, the cable pulling up and to the right, and the boom pushing to the left. We draw a force diagram showing these (see Fig. 33). We have called the tension in the cable **T**, the compression in the boom **F**, and the weight **W**. We resolve the tension in the cable into horizontal and vertical components. We have a balance, so $\mathbf{F} + \mathbf{T} + \mathbf{W} = \mathbf{0}$. In fact, the sum of the horizontal components must be **0**, and the sum of the vertical components must also be **0**. Thus we know that $|\mathbf{T_v}| = 200$ lb and that $|\mathbf{F}| = |\mathbf{T_h}|$. Then we have

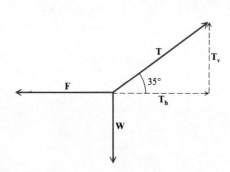

Figure 33

$$|\mathbf{T_v}| = 200 = |\mathbf{T}| \sin 35°$$

$$|\mathbf{T}| = \frac{|\mathbf{T_v}|}{\sin 35°} = \frac{200}{0.5736} = 349 \text{ lb.}$$

---

*For a balance, the rotational tendency of the forces must also be zero. This will always be the case if the forces are concurrent (act through the same point). In this section we consider only applications in which forces are concurrent.

Then $|\mathbf{F}| = |\mathbf{T_h}| = 349 \cos 35° = 286$ lb.

The answer is that the tension in the cable is 349 lb and the compression in the boom is 286 lb.

Figure 34

Example 2 A block weighing 100 kg rests on a 25° incline. Find the components of its weight perpendicular and parallel to the incline.

The weight **W** is 100 kg acting downward as shown in Fig. 34. We draw $\mathbf{F_1}$ and $\mathbf{F_2}$, the components. The angle at B has the same measure as the angle at A because their sides are respectively perpendicular. Thus we have

$$|\mathbf{F_1}| = 100 \cos 25° = 90.6 \text{ kg}, \quad \text{and} \quad |\mathbf{F_2}| = 100 \sin 25° = 42.3 \text{ kg}.$$

In a situation like that of Example 2, the force necessary to hold the block on the incline is a force the size of $\mathbf{F_2}$ but opposite in direction. The force with which the block pushes on the incline is $\mathbf{F_1}$.

Example 3 A 500-kg block is suspended by two ropes as shown in Fig. 35. Find the tension in each rope.

At point A there are three forces acting: the block is pulling down, and the two ropes are pulling upward and outward. We draw a force diagram showing these (Fig. 36).

We have called the forces exerted by the ropes **P** and **Q**. Their horizontal components have magnitudes $|\mathbf{P}| \cos 60°$ and $|\mathbf{Q}| \cos 30°$. Since there is a balance, these must be the same:

$$|\mathbf{P}| \cos 60° = |\mathbf{Q}| \cos 30°.$$

Figure 35

The vertical components in the ropes have magnitudes $|\mathbf{P}| \sin 60°$ and $|\mathbf{Q}| \sin 30°$. Since there is a balance, these must total 500 lb, the weight of the block:

$$|\mathbf{P}| \sin 60° + |\mathbf{Q}| \sin 30° = 500.$$

We now have two equations with unknowns $|\mathbf{P}|$ and $|\mathbf{Q}|$:

$$|\mathbf{P}| \cos 60° - |\mathbf{Q}| \cos 30° = 0, \tag{1}$$
$$|\mathbf{P}| \sin 60° + |\mathbf{Q}| \sin 30° = 500. \tag{2}$$

To solve, we begin by solving Eq. (1) for P:

$$|\mathbf{P}| = |\mathbf{Q}| \frac{\cos 30°}{\cos 60°}. \tag{3}$$

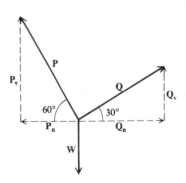

Figure 36

Next, we substitute in Eq. (2):

$$|\mathbf{Q}| \frac{\cos 30° \sin 60°}{\cos 60°} + |\mathbf{Q}| \sin 30° = 500$$

$$|\mathbf{Q}| \frac{\cos 30° \sin 60°}{\cos 60°} + |\mathbf{Q}| \frac{\sin 30° \cos 60°}{\cos 60°} = 500$$

$$|\mathbf{Q}| (\cos 30° \sin 60° + \sin 30° \cos 60°) = 500 \cos 60°$$

$$|\mathbf{Q}| (\sin 90°) = 500 \cos 60° \quad \text{Using an identity}$$

$$|\mathbf{Q}| = 500 \cdot \frac{1}{2} = 250 \text{ kg.}$$

Then substituting in Eq. (3), we obtain

$$|\mathbf{P}| = 250 \frac{\cos 30°}{\cos 60°} = 250 \frac{\sqrt{3}/2}{1/2} = 250\sqrt{3} = 433 \text{ kg.}$$

EXERCISE SET 11.7

1. A 150-lb sign is hanging from the end of a hinged boom, supported by a cable inclined 42° with the horizontal. Find the tension in the cable and the compression in the boom.

2. A 300-lb sign is hanging from the end of a hinged boom, supported by a cable inclined 51° from the horizontal. Find the tension in the cable and the compression in the boom (see the figure for Exercise 1).

3. A weight of 200 kg is supported by a frame made of two rods and hinged at A, B, and C (see the figure). Find the forces exerted by the two rods.

4. A weight of 300 kg is supported by the hinged frame of Exercise 3, but where the angle shown is 50°. Find the forces exerted by the two rods.

5. The force due to air movement on an airplane wing in flight is 2800 lb, acting at an angle of 28° with the vertical. What is the lift (vertical component) and what is the drag (horizontal component)?

6. The force due to air movement on an airplane wing in flight is 3500 lb, acting at an angle of 32° with the vertical. What is the lift (vertical component) and what is the drag (horizontal component)?

7. A 100-kg block of ice rests on a 37° incline. What force parallel to the incline is necessary to keep it from sliding down?

8. What force is necessary to pull a 6500-kg truck up a 7° incline?

9. The moon's gravity is $\frac{1}{6}$ that of the earth. To simulate the moon's gravity an incline is used, as shown in the figure. The astronaut is suspended in a sling and is free to move horizontally. The component of his weight perpendicular to the incline should be $\frac{1}{6}$ of his weight. What must the angle θ be?

10. To simulate gravity on a planet having $\frac{3}{8}$ of the earth's gravity, what should the angle θ be? (See the figure and explanation for Exercise 9.)

11. A 400-kg block is suspended by two ropes, as shown. Find the tension in each rope.

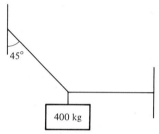

12. A 1000-lb block is suspended by two ropes, as shown. Find the tension in each rope.

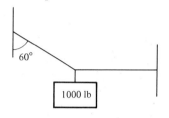

13. A 2000-kg block is suspended by two ropes, as shown. Find the tension in each rope.

14. A 1500-lb block is suspended by two ropes as shown. Find the tension in each rope.

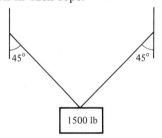

15. A 2500-kg block is suspended by two ropes, as shown. Find the tension in each rope.

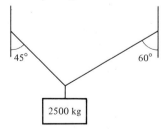

16. A 2000-kg block is suspended by two ropes as shown. Find the tension in each rope.

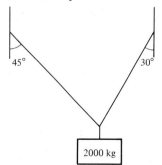

CHAPTER 11 REVIEW

In Exercises 1 and 2 solve the triangles. Angle C is a right angle. Use three-digit precision.

1. $a = 7.3$, $c = 8.6$

2. $a = 30.5$, $B = 51°10'$.

3. One leg of a right triangle bears east. The hypotenuse is 734 cm long and bears N 57°20'E. Find the perimeter.

4. Solve triangle ABC, where $B = 118°20'$, $C = 27°30'$, and $b = 0.974$.

5. In an isosceles triangle, the base angles each measure 52°20' and the base is 513 cm long. Find the lengths of the other two sides.

6. In triangle ABC, $a = 3.7$, $c = 4.9$, and $B = 135°$. Find b.

7. An observer's eye is 6 ft above the floor. A mural is being viewed. The bottom of the mural is at floor level. The observer looks downward 13° to see the bottom and upward 17° to see the top. How tall is the mural?

8. A parallelogram has sides of lengths 3.21 cm and 7.85 cm. One of its angles measures 147°. Find the area of the parallelogram.

9. A car is moving north at 45 mph. A ball is thrown from the car at 37 mph in a direction of N 45°E. Find the speed and direction of the ball over the ground.

10. A block weighing 150 lb rests on an inclined plane. The plane makes an angle of 45° with the ground. Find the components of the weight parallel and perpendicular to the plane.

11. Find rectangular form for the vector $(20, -200°)$.

12. Find polar form for the vector $(-2, 3)$.

13. Convert to a polar equation: $x^2 + y^2 + 2x - 3y = 0$.

14. Vector **u** has a component of 12 lb upward and a component of 18 lb to the right. Vector **v** has a component of 35 lb downward and a component of 35 lb to the left. Find the components of $\mathbf{u} + \mathbf{v}$ and the magnitude and direction of $\mathbf{u} + \mathbf{v}$.

15. Do the following calculations for $\mathbf{u} = (4, 3)$ and $\mathbf{v} = (-3, 4)$.
   a) $4\mathbf{u} - 3\mathbf{v}$
   b) $|\mathbf{u} + \mathbf{v}|$

16. A crane is supporting a 500-kg steel beam. The boom of the crane is inclined 30° from the vertical. Find the compression in the boom of the crane.

17. A parallelogram has sides of lengths 3.42 and 6.97. Its area is 18.4. Find the sizes of its angles.

Imaginary and Complex Numbers

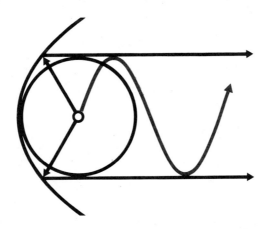

IMAGINARY AND COMPLEX NUMBERS

12.1 INTRODUCTION TO IMAGINARY AND COMPLEX NUMBERS

Negative numbers do not have square roots in the system of real numbers. Certain equations have no solutions. A new kind of number, called *imaginary*, was invented so that negative numbers would have square roots and certain equations would have solutions. These numbers were devised, starting with an imaginary unit, named i, with the agreement that $i^2 = -1$ or $i = \sqrt{-1}$. All other imaginary numbers can be expressed as a product of i and a real number.

Example 1 Express $\sqrt{-5}$ and $-\sqrt{-7}$ in terms of i.

$$\sqrt{-5} = \sqrt{-1 \cdot 5} = \sqrt{-1}\sqrt{5} = i\sqrt{5};$$
$$-\sqrt{-7} = -\sqrt{-1 \cdot 7} = -\sqrt{-1}\sqrt{7} = -i\sqrt{7}$$

It is also to be understood that the imaginary unit obeys the familiar laws of real numbers, such as the commutative and associative laws.

Example 2 Simplify: $\sqrt{-3}\sqrt{-7}$.

Important: We first express the two imaginary numbers in terms of i:

$$\sqrt{-3}\sqrt{-7} = i\sqrt{3} \cdot i\sqrt{7}.$$

Now, rearranging and combining, we have

$$i^2\sqrt{3}\sqrt{7} = -1 \cdot \sqrt{21} = -\sqrt{21}.$$

It is important, as in Example 2, to express imaginary numbers in terms of i before simplifying. Had we not done this in Example 2, we would have obtained $\sqrt{(-3)(-7)}$, or $\sqrt{21}$, instead of $-\sqrt{21}$.

Example 3 Simplify: $\dfrac{\sqrt{-20}}{\sqrt{-5}}$.

$$\frac{\sqrt{-20}}{\sqrt{-5}} = \frac{i\sqrt{20}}{i\sqrt{5}} = \frac{\sqrt{20}}{\sqrt{5}} = \sqrt{4} = 2$$

Example 4 Simplify: $\sqrt{-9} + \sqrt{-25}$.

$$\sqrt{-9} + \sqrt{-25} = i\sqrt{9} + i\sqrt{25} = 3i + 5i = (3 + 5)i = 8i$$

Powers of i

Let us look at the powers of i:

$i^2 = -1,$ $\qquad i^5 = i^4 \cdot i = 1 \cdot i = i,$
$i^3 = i^2 \cdot i = -1 \cdot i = -i,$ $\qquad i^6 = i^5 \cdot i = i \cdot i = -1,$
$i^4 = i^2 \cdot i^2 = -1(-1) = 1,$ $\qquad i^7 = i^6 \cdot i = -1 \cdot i = -i.$

The first four powers of i are all different, but thereafter there is a repeating pattern in cycles of four. Note that $i^4 = 1$ and that all powers of i^4, such as i^8, i^{16}, and so on, are 1. To find a higher power we express it in terms of the nearest power of 4 less than the given one.

Example 5 Simplify i^{17} and i^{23}.

$$i^{17} = i^{16} \cdot i$$

Since i^{16} is a power of i^4 and $i^4 = 1$, we know that $i^{16} = 1$ so

$$i^{17} = 1 \cdot i = i.$$

Also,

$$i^{23} = i^{20} \cdot i^3.$$

Since i^{20} is a power of i^4, $i^{20} = 1$, so $i^{23} = 1 \cdot i^3 = 1 \cdot (-i) = -i$.

Complex Numbers

The equation $x^2 + 1 = 0$ has no solution in real numbers, but it has the imaginary solutions i and $-i$. There are equations that do not have real or imaginary solutions. For example, $x^2 - 2x + 2 = 0$ does not. If we allow sums of real and imaginary numbers, this equation does have a solution, $1 + i$. Let us verify this by substitution:

$$x^2 - 2x + 2 = 0$$
$$(1 + i)^2 - 2(1 + i) + 2 = 0$$
$$1 + 2i + i^2 - 2 - 2i + 2 = 0$$
$$1 + 2i - 1 - 2 - 2i + 2 = 0.$$

In order that more equations will have solutions, we invent a new number system called the *system of complex numbers*.* A complex number is a sum of a real number and an imaginary number.

DEFINITION

The set of complex numbers consists of all numbers $a + bi$, where a and b are real numbers.

For the complex number $a + bi$ we say that the *real part* is a and the *imaginary part* is bi.

An Extension of the Real Numbers

The complex-number system is an extension of the real-number system. Any number $a + bi$, where a and b are real numbers, is a com-

---

*One may wonder why we do not invent a system of imaginary numbers. Since we do not even have closure ($i^2 = -1$, a real number) this would not make sense.

plex number. The number b can be 0, in which case we have $a + 0i$, which simplifies to the real number a. Thus the complex numbers include all of the real numbers. They also include all of the imaginary numbers, because any imaginary number bi is equal to $0 + bi$.

The familiar properties of real numbers (the *field* properties) hold also for complex numbers.* We list them.

Commutativity. **Addition and multiplication are commutative.**

Associativity. **Addition and multiplication are associative.**

Distributivity. **Multiplication is distributive over addition and also over subtraction.**

Identities. **The additive identity is $0 + 0i$, or 0. The multiplicative identity is $1 + 0i$, or 1.**

Additive inverses. **Every complex number $a + bi$ has the additive inverse $-a - bi$.**

Multiplicative inverses. **Every nonzero complex number has a multiplicative inverse, or reciprocal.**

Example 6 Simplify: $(8 + 6i) + (3 + 2i)$.

$$(8 + 6i) + (3 + 2i) = 8 + 3 + 6i + 2i$$
$$= 11 + 8i$$

Example 7 Simplify: $(1 + 2i)(1 + 3i)$.

$$(1 + 2i)(1 + 3i) = 1 + 6i^2 + 2i + 3i$$
$$= 1 - 6 + 2i + 3i \quad i^2 = -1$$
$$= -5 + 5i$$

Example 8 Simplify: $(3 + 2i) - (5 - 2i)$.

$$(3 + 2i) - (5 - 2i) = (3 + 2i) - 5 + 2i$$
$$= 3 - 5 + 2i + 2i$$
$$= -2 + 4i$$

Equality for Complex Numbers

An equation $a + bi = c + di$ will be true if and only if the real parts are the same and the imaginary parts are the same. In other words, we have the following:

$$a + bi = c + di \quad \text{if and only if} \quad a = c \text{ and } b = d.$$

---

* In a more rigorous treatment, addition and multiplication are defined for complex numbers. It is then proved that the familiar properties hold.

Example 9 Suppose that $3x + yi = 5x + 1 + 2i$. Find x and y.

We equate the real parts: $3x = 5x + 1$. Solving this equation, we obtain $x = -\frac{1}{2}$. We equate the imaginary parts: $yi = 2i$. Thus $y = 2$.

EXERCISE SET 12.1

Express in terms of i.

1. $\sqrt{-15}$
2. $\sqrt{-17}$
3. $\sqrt{-16}$
4. $\sqrt{-25}$
5. $-\sqrt{-12}$
6. $-\sqrt{-20}$

Simplify. Leave answers in terms of i in Exercises 7-10.

7. $\sqrt{-16} + \sqrt{-25}$
8. $\sqrt{-36} - \sqrt{-4}$
9. $\sqrt{-7} - \sqrt{-10}$
10. $\sqrt{-5} + \sqrt{-7}$
11. $\sqrt{-5}\sqrt{-11}$
12. $\sqrt{-7}\sqrt{-8}$
13. $-\sqrt{-4}\sqrt{-5}$
14. $-\sqrt{-9}\sqrt{-7}$
15. $\dfrac{\sqrt{-5}}{\sqrt{-2}}$
16. $\dfrac{\sqrt{-7}}{\sqrt{-5}}$
17. $\dfrac{\sqrt{-9}}{\sqrt{-4}}$
18. $\dfrac{\sqrt{-25}}{\sqrt{-16}}$
19. $\dfrac{-\sqrt{-36}}{\sqrt{-9}}$
20. $\dfrac{\sqrt{-25}}{-\sqrt{-16}}$

Simplify.

21. $(2 + 3i) + (4 + 2i)$
22. $(5 - 2i) + (6 + 3i)$
23. $(4 + 3i) + (4 - 3i)$
24. $(2 + 3i) + (-2 - 3i)$
25. $(8 + 11i) - (6 + 7i)$
26. $(9 - 5i) - (4 + 2i)$
27. $2i - (4 + 3i)$
28. $3i - (5 + 2i)$
29. $(1 + 2i)(1 + 3i)$
30. $(1 + 4i)(1 - 3i)$
31. $(1 + 2i)(1 - 3i)$
32. $(2 + 3i)(2 - 3i)$
33. $3i(4 + 2i)$
34. $5i(3 - 4i)$
35. $(2 + 3i)^2$
36. $(3 - 2i)^2$
37. i^{13}
38. i^{72}
39. Determine whether $1 + 2i$ is a solution of
$$x^2 - 2x + 5 = 0.$$
40. Determine whether $1 - 2i$ is a solution of
$$x^2 - 2x + 5 = 0.$$

Solve for x and y.

41. $4x + 7i = -6 + yi$
42. $-4 + (x + y)i = 2x - 5y + 5i$

☆

43. A function f is defined as follows: $f(z) = z^2 - 4z + i$. Find $f(3 + i)$.
44. A function g is defined as follows: $g(z) = 2z^2 + z - 2i$. Find $g(2 - i)$.
45. Show that the general rule for radicals, in real numbers, $\sqrt{a \cdot b} = \sqrt{a} \cdot \sqrt{b}$, does not hold for complex numbers.
46. Show that the general rule for radicals, in real numbers, $\sqrt{\dfrac{a}{b}} = \dfrac{\sqrt{a}}{\sqrt{b}}$, does not hold for complex numbers.
47. Multiply. $\begin{bmatrix} i & 0 \\ 0 & 3i \end{bmatrix} \begin{bmatrix} i & 2i \\ 1 - i & 4 - i \end{bmatrix}$

★

48. Solve $z^2 = -2i$. (*Hint*: Let $z = a + bi$.)

49. Solve $\dfrac{z^2}{4} = i$. (*Hint*: Let $z = a + bi$.)

50. A function R is defined as follows:

$R(z)$ is the real part of the complex number z.

In other words, if $z = a + bi$, the function value $R(z)$ is a. Determine whether this function has the linearity property, i.e., $R(z_1) + R(z_2) = R(z_1 + z_2)$, and prove your answer.

12.2 CONJUGATES AND DIVISION

We shall define the conjugate of a complex number as follows.

DEFINITION

The *conjugate* of a complex number $a + bi$ is $a - bi$, and the conjugate of $a - bi$ is $a + bi$.

We illustrate:

The conjugate of $3 + 4i$ is $3 - 4i$.
The conjugate of $5 - 7i$ is $5 + 7i$.
The conjugate of $5i$ is $-5i$.
The conjugate of 6 is 6.

Division

Fractional notation is useful for division of complex numbers. We also use the notion of conjugates.

Example 1 Divide $4 + 5i$ by $1 + 4i$.

We shall write fractional notation and then multiply by 1.

$$\frac{4 + 5i}{1 + 4i} = \frac{4 + 5i}{1 + 4i} \cdot \frac{1 - 4i}{1 - 4i} \quad \text{Note that } 1 - 4i \text{ is the conjugate of the divisor}$$

$$= \frac{(4 + 5i)(1 - 4i)}{1^2 - 4^2 i^2}$$

$$= \frac{24 - 11i}{1 + 16} = \frac{24 - 11i}{17} \quad i^2 = -1$$

$$= \frac{24}{17} - \frac{11}{17}i$$

The procedure used in Example 1 allows us always to find a quotient of two numbers and express it in the form $a + bi$. This is true because the product of a number and its conjugate is always a real number (we will prove this later), giving us a real-number denominator.

Reciprocals

We can find the reciprocal, or multiplicative inverse, of a complex number by division. The reciprocal of a complex number z is $1/z$.

Example 2 Find the reciprocal of $2 - 3i$ and express it in the form $a + bi$.

a) The reciprocal of $2 - 3i$ is $\dfrac{1}{2 - 3i}$.

b) We can express it in the form $a + bi$ as follows:

$$\frac{1}{2 - 3i} = \frac{1}{2 - 3i} \cdot \frac{2 + 3i}{2 + 3i} = \frac{2 + 3i}{2^2 - 3^2 i^2}$$

$$= \frac{2 + 3i}{4 + 9} = \frac{2}{13} + \frac{3}{13}i.$$

Properties of Conjugates

We may use a single letter for a complex number. For example, we could shorten $a + bi$ to z. To denote the conjugate of a number, we use a bar. The conjugate of z is $\bar{z}$. Or, the conjugate of $a + bi$ is $\overline{a + bi}$. Of course, by the definition of conjugates, $\overline{a + bi} = a - bi$ and $\overline{a - bi} = a + bi$. We have already noted that the product of a number and its conjugate is always a real number. Let us state this formally and prove it.

THEOREM 1

For any complex number z, $z \cdot \bar{z}$ is a real number.

Proof. Let $z = a + bi$. Then

$$z \cdot \bar{z} = (a + bi)(a - bi) = a^2 - b^2 i^2 = a^2 + b^2.$$

Since a and b are real numbers, so is $a^2 + b^2$. Thus $z \cdot \bar{z}$ is real. □

The sum of a number and its conjugate is also always real. We state this as the second property.

THEOREM 2

For any complex number z, $z + \bar{z}$ is a real number.

Proof. Let $z = a + bi$. Then
$$z + \bar{z} = (a + bi) + (a - bi) = 2a.$$
Since a is a real number, $2a$ is real. Thus $z + \bar{z}$ is real. □

Now we consider the conjugate of a sum and compare it with the sum of the conjugates.

Example 3 Compare $\overline{(2 + 4i) + (5 + i)}$ and $\overline{(2 + 4i)} + \overline{(5 + i)}$.

a) $\overline{(2 + 4i) + (5 + i)} = \overline{7 + 5i}$ Adding the complex numbers
$= 7 - 5i$ Taking the conjugate

b) $\overline{(2 + 4i)} + \overline{(5 + i)} = (2 - 4i) + (5 - i)$ Taking conjugates
$= 7 - 5i$ Adding

Taking the conjugate of a sum gives the same result as adding the conjugates. Let us state and prove this.

THEOREM 3

For any complex numbers z and w, $\overline{z + w} = \bar{z} + \bar{w}$.

Proof. Let $z = a + bi$ and $w = c + di$. Then
$$\overline{z + w} = \overline{(a + bi) + (c + di)} = \overline{(a + c) + (b + d)i}.$$
by adding. We now take the conjugate and obtain $(a + c) - (b + d)i$. Now
$$\bar{z} + \bar{w} = \overline{(a + bi)} + \overline{(c + di)} = (a - bi) + (c - di),$$
taking the conjugates. We will now add to obtain $(a + c) - (b + d)i$, the same result as before. Thus $\overline{z + w} = \bar{z} + \bar{w}$. □

Let us next consider the conjugate of a product.

Example 4 Compare $\overline{(3 + 2i)(4 - 5i)}$ and $\overline{(3 + 2i)} \cdot \overline{(4 - 5i)}$.

$\overline{(3 + 2i)(4 - 5i)} = \overline{22 - 7i}$ Multiplying
$= 22 + 7i$ Taking the conjugate

$\overline{(3 + 2i)} \cdot \overline{(4 - 5i)} = (3 - 2i)(4 + 5i)$ Taking conjugates
$= 22 + 7i$ Multiplying

The conjugate of a product is the product of the conjugates. This is our next result.

THEOREM 4

For any complex numbers z and w, $\overline{z \cdot w} = \overline{z} \cdot \overline{w}$.

Proof. Let $z = a + bi$ and $w = c + di$. Then
$$\overline{z \cdot w} = \overline{(a + bi)(c + di)}$$
$$= \overline{(ac - bd) + (bc + ad)i}. \quad \text{Multiplying}$$

Taking the conjugate, we obtain $(ac - bd) - (bc + ad)i$.

Now $\overline{z} \cdot \overline{w} = \overline{(a + bi)} \cdot \overline{(c + di)} = (a - bi)(c - di)$, taking conjugates. By multiplication we obtain $(ac - bd) - (bc + ad)i$, the same result as before. Thus $\overline{z \cdot w} = \overline{z} \cdot \overline{w}$. □

Let us now consider conjugates of powers, using the preceding result.

Example 5 Show that for any complex number z, $\overline{z^2} = \overline{z}^2$.

$$\overline{z^2} = \overline{z \cdot z} \quad \text{By definition of exponents}$$
$$= \overline{z} \cdot \overline{z} \quad \text{By Theorem 4}$$
$$= \overline{z}^2 \quad \text{By definition of exponents}$$

We now state our next result.

THEOREM 5

For any complex number z, $\overline{z^n} = \overline{z}^n$. **In other words, the conjugate of a power is the power of the conjugate. We understand that the exponent is a natural number.**

The conjugate of a real number $a + 0i$ is $a - 0i$, and both are equal to a. Thus a real number is its own conjugate. We state this as our next result.

THEOREM 6

If z is a real number, then $\overline{z} = z$.

Conjugates of Polynomials

Given a polynomial in z, where z is a variable for a complex number, we can find its conjugate in terms of $\bar{z}$.

Example 6 Find a polynomial in $\bar{z}$ that is the conjugate of $3z^2 + 2z - 1$.

We write an expression for the conjugate and then use the properties of conjugates.

$$\begin{aligned}\overline{3z^2 + 2z - 1} &= \overline{3z^2} + \overline{2z} - \overline{1} & \text{By Theorem 3} \\ &= \overline{3z^2} + \overline{2}\cdot\overline{z} - \overline{1} & \text{By Theorem 4} \\ &= \overline{3z^2} + 2\overline{z} - 1 & \text{The conjugate of a real number is the number itself} \\ &= 3\bar{z}^2 + 2\bar{z} - 1 & \text{By Theorem 5}\end{aligned}$$

EXERCISE SET 12.2

Find the reciprocal and express it in the form $a + bi$.

1. $4 + 3i$
2. $4 - 3i$
3. $5 - 2i$
4. $2 + 5i$
5. i
6. $-i$
7. $-4i$
8. $5i$

Simplify.

9. $\dfrac{4 + 3i}{1 - i}$
10. $\dfrac{2 - 3i}{5 - 4i}$
11. $\dfrac{\sqrt{2} + i}{\sqrt{2} - i}$
12. $\dfrac{\sqrt{3} + i}{\sqrt{3} - i}$
13. $\dfrac{3 + 2i}{i}$
14. $\dfrac{2 + 3i}{i}$
15. $\dfrac{i}{2 + i}$
16. $\dfrac{3}{5 - 11i}$
17. $\dfrac{1 - i}{(1 + i)^2}$
18. $\dfrac{1 + i}{(1 - i)^2}$
19. $\dfrac{3 - 4i}{(2 + i)(3 - 2i)}$
20. $\dfrac{(4 - i)(5 + i)}{(6 - 5i)(7 - 2i)}$
21. $\dfrac{1 + i}{1 - i} \cdot \dfrac{2 - i}{1 - i}$
22. $\dfrac{1 - i}{1 + i} \cdot \dfrac{2 + i}{1 + i}$
23. $\dfrac{3 + 2i}{1 - i} + \dfrac{6 + 2i}{1 - i}$
24. $\dfrac{4 - 2i}{1 + i} + \dfrac{2 - 5i}{1 + i}$

Find a polynomial in $\bar{z}$ that is the conjugate.

25. $3z^5 - 4z^2 + 3z - 5$
26. $7z^4 + 5z^3 - 12z$
27. $4z^7 - 3z^5 + 4z$
28. $5z^{10} - 7z^8 + 13z^2 - 4$

☆
29. Solve: $z + 6\bar{z} = 7$.
30. Solve: $5z - 4\bar{z} = 7 + 8i$.
31. Let $z = a + bi$. Find $\frac{1}{2}(z + \bar{z})$.
32. Let $z = a + bi$. Find $\frac{1}{2}i(\bar{z} - z)$.
33. Find $f(3 + i)$, where $f(z) = \dfrac{z}{z - 1}$.
34. Show that for any complex number z, $z + \bar{z}$ is real.

★

35. Let $z = a + bi$. Find a general expression for $1/z$.

36. Let $z = a + bi$ and $w = c + di$. Find a general expression for w/z.

37. State and prove a theorem about the conjugate of a polynomial.

38. Let R be the function that assigns the real part of z as $R(z)$ (see Exercise 49 on p. 438). Prove that for any complex numbers z_1 and z_2, not both zero,

$$R\left(\frac{z_1}{z_1 + z_2}\right) + R\left(\frac{z_2}{z_1 + z_2}\right) = 1.$$

39. Find the inverse of this matrix.

$$\begin{bmatrix} i & -1 \\ 2 & -1 \end{bmatrix}$$

40. Prove that

$$\overline{\begin{vmatrix} z_1 & z_2 \\ z_3 & z_4 \end{vmatrix}} = \begin{vmatrix} \overline{z_1} & \overline{z_2} \\ \overline{z_3} & \overline{z_4} \end{vmatrix},$$

for any complex numbers z_1, z_2, z_3, and z_4.

12.3 EQUATIONS AND COMPLEX NUMBERS

The Quadratic Formula

In Chapter 2 we derived the quadratic formula for equations with real-number coefficients. A glance at that derivation shows that all the steps can also be done with complex-number coefficients with one possible exception. We need to know whether the principle of zero products holds for complex numbers. This is the case, as will be shown later. Thus the quadratic formula holds for equations with complex coefficients.

THEOREM 7

Any equation $ax^2 + bx + c = 0$, where $a \neq 0$ and a, b, and c are complex numbers, has solutions

$$\frac{-b \pm \sqrt{b^2 - 4ac}}{2a}.$$

Example 1 Solve: $x^2 + (1 + i)x - 2i = 0$.

We note that $a = 1$, $b = 1 + i$, and $c = -2i$. Thus

$$x = \frac{-(1 + i) \pm \sqrt{(1 + i)^2 - 4 \cdot 1 \cdot (-2i)}}{2 \cdot 1} \quad \text{Substituting in the formula}$$

$$= \frac{-1 - i \pm \sqrt{10i}}{2}. \quad \text{Simplifying}$$

In Example 1 we have not attempted to evaluate $\sqrt{10i}$. This will be done later. Let us say, however, that $10i$ has two square roots that are additive inverses of each other. This is true of every nonzero complex number. There is no problem of negative numbers not having square roots, because in the system of complex numbers there are no positive or negative numbers. Whenever we speak of positive or negative numbers, we will be referring to the system of real numbers.

Quadratic Equations with Real Coefficients

Since any real number a is $a + 0i$, we can consider real numbers to be special kinds of complex numbers. If a quadratic equation has real coefficients, the quadratic formula always gives solutions in the system of complex numbers. Recall that in using the formula, we take the square root of $b^2 - 4ac$. If this *discriminant* is negative, the solutions will have nonzero imaginary parts.

Example 2 Solve: $x^2 + 2x + 5 = 0$.

We note that $a = 1$, $b = 2$, and $c = 5$. Thus

$$x = \frac{-2 \pm \sqrt{2^2 - 4 \cdot 1 \cdot 5}}{2 \cdot 1}$$

$$= \frac{-2 \pm \sqrt{-16}}{2} = \frac{-2 \pm 4i}{2}.$$

The solutions are $-1 + 2i$ and $-1 - 2i$.

It is easy to see from Example 2 that when there are nonzero imaginary parts, an equation has two solutions that are conjugates of each other.

Writing Equations with Specified Solutions

The principle of zero products for real numbers states that a product is 0 if and only if at least one of the factors is 0. This principle also holds for complex numbers, as will be shown in Section 12.4. Since the principle holds for complex numbers we can write equations having specified solutions.

Example 3 Find an equation having the numbers 1, i, and $-i$ as solutions.

12.3 EQUATIONS AND COMPLEX NUMBERS

The factors we use will be $x - 1$, $x - i$, and $x + i$. Next we set the product of these factors equal to 0:

$$(x - 1)(x - i)(x + i) = 0.$$

Now we multiply and simplify:

$$(x - 1)(x^2 - i^2) = 0$$
$$(x - 1)(x^2 + 1) = 0$$
$$x^3 - x^2 + x - 1 = 0.$$

Example 4 Find an equation having -1, i, and $1 + i$ as solutions.

$$(x + 1)(x - i)[x - (1 + i)] = 0$$
$$(x^2 + x - ix - i)(x - 1 - i) = 0$$
$$x^3 - 2ix^2 - ix - 2x - 1 + i = 0$$

Solving Equations

First-degree equations in complex numbers are solved very much like first-degree equations in real numbers.

Example 5 Solve: $3ix + 4 - 5i = (1 + i)x + 2i$.

$$3ix - (1 + i)x = 2i - (4 - 5i) \quad \text{Adding } -(1 + i)x \text{ and } -(4 - 5i)$$

$$(-1 + 2i)x = -4 + 7i \quad \text{Simplifying}$$

$$x = \frac{-4 + 7i}{-1 + 2i} \quad \text{Dividing}$$

$$x = \frac{-4 + 7i}{-1 + 2i} \cdot \frac{-1 - 2i}{-1 - 2i}$$

$$x = \frac{18 + i}{5} = \frac{18}{5} + \frac{1}{5}i$$

In Example 1 we used the quadratic formula, but did not evaluate $\sqrt{10i}$. We wish to find complex numbers z for which $z^2 = 10i$. We do this in the following example.

Example 6 Solve: $z^2 = 10i$.

Let $z = x + yi$. Then we have $(x + yi)^2 = 10i$, or $x^2 + 2xyi + y^2i^2 = 10i$, or $x^2 - y^2 + 2xyi = 0 + 10i$. Since the real parts must be the same and the imaginary parts the same, $x^2 - y^2 = 0$ and $2xy = 10$.

Here we have two equations in real numbers. We can find a solution by solving the second equation for y and substituting in the first.

Since
$$y = \frac{5}{x},$$
then
$$x^2 - \left(\frac{5}{x}\right)^2 = 0$$

$$x^2 - \frac{25}{x^2} = 0 \quad \text{or} \quad x^4 - 25 = 0$$

$(x^2 + 5)(x^2 - 5) = 0 \quad$ Factoring

$x^2 + 5 = 0 \quad \text{or} \quad x^2 - 5 = 0 \quad$ Principle of zero products

The real solutions are $\sqrt{5}$ and $-\sqrt{5}$. Now, going back to $2xy = 10$, we see that if $x = \sqrt{5}$, then $y = \sqrt{5}$, and if $x = -\sqrt{5}$, then $y = -\sqrt{5}$. Thus the solutions of our equation (and the square roots of $10i$) are $\sqrt{5} + \sqrt{5}i$ and $-\sqrt{5} - \sqrt{5}i$.

EXERCISE SET 12.3

Find an equation having the specified solutions.

1. $2i, -2i$
2. $3i, -3i$
3. $1 + i, 1 - i$
4. $2 + i, 2 - i$
5. $2 + 3i, 2 - 3i$
6. $4 + 3i, 4 - 3i$
7. $3, i$
8. $5, i$
9. $1, 3i, -3i$
10. $1, 2i, -2i$
11. $2, 1 + i, i$
12. $3, 1 - i, i$

Solve the equations.

13. $(3 + i)x + i = 5i$
14. $(2 + i)x - i = 5 + i$
15. $2ix + 5 - 4i = (2 + 3i)x - 2i$
16. $(1 + 2i)x + 3 - 2i = 4 - 5i + 3ix$
17. $x^2 - 2x + 5 = 0$
18. $x^2 - 4x + 13 = 0$
19. $x^2 + 3x + 4 = 0$
20. $3x^2 + x + 2 = 0$

Solve using the quadratic formula. You need not evaluate square roots.

21. $x^2 + (1 - i)x + i = 0$
22. $2x^2 + ix + 1 = 0$
23. $3x^2 + (1 + 2i)x + 1 - i = 0$
24. $3x^2 + (1 - 2i)x + 1 + i = 0$

Solve.

25. $z^2 = 4i$
26. $z^2 = -4i$
27. $z^2 = 3 + 4i$
28. $z^2 = 5 - 12i$

☆ ──────────────────────────────

Solve. (Hint: Factor first.)

29. $x^3 - 8 = 0$
30. $x^3 + 8 = 0$

Solve.

31. $2x + 3y = 7 - 7i$,
$3x - 2y = 4 + 9i$

32. $3x + 4y = 11 + 9i$,
$2x - y = -5i$

★

Solve for x and y.

33. $\log x + 2^y i = 2 + 16i$

34. $\sin^2 x + i \cos^2 y = 1 + \dfrac{i}{2}$

35. $e^{2x} + ie^y = 1 + i$

36. Solve. $\begin{vmatrix} z - 2 & 7 \\ 3i & 9 \end{vmatrix} = 0$

12.4 GRAPHICAL REPRESENTATION AND POLAR NOTATION

The real numbers can be graphed on a line. Complex numbers are graphed on a plane. We graph a complex number $a + bi$ in the same way that we graph an ordered pair of real numbers (a, b). In place of an x-axis we have a *real axis,* and in place of a y-axis we have an *imaginary axis.*

Examples Graph each of the following.

1. $3 + 2i$
2. $-4 + 5i$
3. $-5 - 4i$

See Fig. 1.

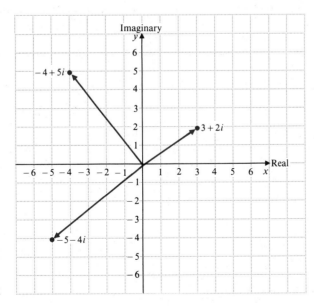

Figure 1

Horizontal distances correspond to the real part of a number. Vertical distances correspond to the imaginary part. Graphs of the numbers are shown above as vectors. The horizontal component of the vector for $a + bi$ is a and the vertical component is b. Complex numbers are sometimes used in the study of vectors.

Adding complex numbers is like adding vectors using components. For example, to add $3 + 2i$ and $5 + 4i$ we add the real parts and the imaginary parts to obtain $8 + 6i$. Graphically, then, the sum of two complex numbers looks like a vector sum. It is the diagonal of a parallelogram.

Example 4 Show graphically $2 + 2i$ and $3 - i$. Show also their sum.

See Fig. 2.

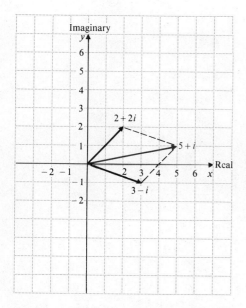

Figure 2

Polar Notation for Complex Numbers

In Chapter 11 we studied polar notation for vectors, in which we use the length of a vector and the angle that it makes with the x-axis. In a similar way we develop polar notation for complex numbers. From Fig. 3 we can see that the length of the vector for $a + bi$ is $\sqrt{a^2 + b^2}$. Note that this quantity is a nonnegative real number. It is called the *absolute value* of $a + bi$, and is denoted $|a + bi|$.

Figure 3

DEFINITION

The *absolute value* of a complex number $a + bi$ is denoted $|a + bi|$ and is defined to be $\sqrt{a^2 + b^2}$.

Example 5 Find $|3 + 4i|$.

$$|3 + 4i| = \sqrt{3^2 + 4^2} = \sqrt{9 + 16} = 5$$

Now let us consider any complex number $a + bi$. Suppose that its absolute value is r. Let us also suppose that the angle that the vector makes with the real axis is θ. As Fig. 4 shows, we have

$$a = r \cos \theta \quad \text{and} \quad b = r \sin \theta.$$

Thus

$$a + bi = r \cos \theta + ir \sin \theta$$
$$= r(\cos \theta + i \sin \theta).$$

Figure 4

This is polar notation for $a + bi$. The angle θ is called the *argument*. The notation is often abbreviated to $r \operatorname{cis} \theta$.

Polar notation for complex numbers is also called *trigonometric notation*.

Change of Notation

To change from polar notation to *binomial*, or *rectangular*, notation $a + bi$, we use components, noting that $a = r \cos \theta$ and $b = r \sin \theta$.

Example 6 Write binomial notation for $2(\cos 120° + i \sin 120°)$.

$$a = 2 \cos 120° = -1,$$
$$b = 2 \sin 120° = \sqrt{3}$$

Thus $2(\cos 120° + i \sin 120°) = -1 + i\sqrt{3}$.

Example 7 Write binomial notation for $\sqrt{8} \text{ cis } \frac{7\pi}{4}$.

$$a = \sqrt{8} \cos \frac{7\pi}{4} = \sqrt{8} \cdot \frac{1}{\sqrt{2}} = 2$$

$$b = \sqrt{8} \sin \frac{7\pi}{4} = \sqrt{8} \cdot \frac{-1}{\sqrt{2}} = -2$$

Thus $\sqrt{8} \text{ cis } \frac{7\pi}{4} = 2 - 2i$.

To change from binomial notation to polar notation, we remember that $r = \sqrt{a^2 + b^2}$ and that θ is an angle for which $\sin \theta = b/r$ and $\cos \theta = a/r$.

Example 8 Find polar notation for $1 + i$.

We note that $a = 1$ and $b = 1$. Then

$$r = \sqrt{1^2 + 1^2} = \sqrt{2},$$

$$\sin \theta = \frac{1}{\sqrt{2}} \quad \text{and} \quad \cos \theta = \frac{1}{\sqrt{2}}.$$

Thus $\theta = \frac{\pi}{4}$, or 45°, and we have

$$1 + i = \sqrt{2} \text{ cis } \frac{\pi}{4} \quad \text{or} \quad 1 + i = \sqrt{2} \text{ cis } 45°.$$

Example 9 Find polar notation for $\sqrt{3} - i$.

$$r = \sqrt{(\sqrt{3})^2 + (-1)^2} = 2,$$

$$\sin \theta = -\frac{1}{2} \quad \text{and} \quad \cos \theta = \frac{\sqrt{3}}{2}.$$

Thus $\theta = \dfrac{11\pi}{6}$, or $330°$, and we have

$$\sqrt{3} - i = 2 \text{ cis } \dfrac{11\pi}{6} = 2 \text{ cis } 330°.$$

When changing to polar notation, note that there are many angles satisfying the given conditions. We ordinarily choose the smallest positive angle.

Multiplication and Polar Notation

Multiplication of complex numbers is somewhat easier to do with polar notation than with rectangular notation. We simply multiply the absolute values and add the arguments. Let us state this more formally and prove it.

THEOREM 8

For any complex numbers $r_1 \text{ cis } \theta_1$ and $r_2 \text{ cis } \theta_2$,

$$(r_1 \text{ cis } \theta_1)(r_2 \text{ cis } \theta_2) = r_1 \cdot r_2 \text{ cis } (\theta_1 + \theta_2).$$

Proof. Let us first multiply $a_1 + b_1 i$ by $a_2 + b_2 i$:

$$(a_1 + b_1 i)(a_2 + b_2 i) = (a_1 a_2 - b_1 b_2) + (a_2 b_1 + a_1 b_2)i.$$

Recall that

$$a_1 = r_1 \cos \theta_1, \quad b_1 = r_1 \sin \theta_1,$$

and

$$a_2 = r_2 \cos \theta_2, \quad b_2 = r_2 \sin \theta_2.$$

We substitute these in the product above, to obtain

$$r_1(\cos \theta_1 + i \sin \theta_1) \cdot r_2(\cos \theta_2 + i \sin \theta_2)$$
$$= (r_1 r_2 \cos \theta_1 \cos \theta_2 - r_1 r_2 \sin \theta_1 \sin \theta_2)$$
$$+ (r_1 r_2 \sin \theta_1 \cos \theta_2 + r_1 r_2 \cos \theta_1 \sin \theta_2)i.$$

This simplifies to

$$r_1 r_2 (\cos \theta_1 \cos \theta_2 - \sin \theta_1 \sin \theta_2) + r_1 r_2 (\sin \theta_1 \cos \theta_2 + \cos \theta_1 \sin \theta_2)i.$$

Now, using identities for sums of angles, we can simplify this to

$$r_1 r_2 \cos (\theta_1 + \theta_2) + r_1 r_2 \sin (\theta_1 + \theta_2)i,$$

or

$$r_1 r_2 \text{ cis } (\theta_1 + \theta_2),$$

which was to be shown. □

12.4 GRAPHICAL REPRESENTATION AND POLAR NOTATION

To divide complex numbers we do the reverse of the above. We state this, but will omit the proof.

THEOREM 9

For any complex numbers $r_1 \text{ cis } \theta_1$ and $r_2 \text{ cis } \theta_2$ $(r_2 \neq 0)$,

$$\frac{r_1 \text{ cis } \theta_1}{r_2 \text{ cis } \theta_2} = \frac{r_1}{r_2} \text{ cis } (\theta_1 - \theta_2).$$

Example 10 Find the product of $3 \text{ cis } 40°$ and $7 \text{ cis } 20°$.

$$3 \text{ cis } 40° \cdot 7 \text{ cis } 20° = 3 \cdot 7 \text{ cis } (40° + 20°) = 21 \text{ cis } 60°$$

Example 11 Find the product of $2 \text{ cis } \pi$ and $3 \text{ cis } \left(-\frac{\pi}{2}\right)$.

$$2 \text{ cis } \pi \cdot 3 \text{ cis } \left(-\frac{\pi}{2}\right) = 2 \cdot 3 \text{ cis } \left(\pi - \frac{\pi}{2}\right) = 6 \text{ cis } \frac{\pi}{2}.$$

Example 12 Divide $2 \text{ cis } \pi$ by $4 \text{ cis } \frac{\pi}{2}$.

$$\frac{2 \text{ cis } \pi}{4 \text{ cis } \frac{\pi}{2}} = \frac{2}{4} \text{ cis } \left(\pi - \frac{\pi}{2}\right) = \frac{1}{2} \text{ cis } \frac{\pi}{2}$$

The Principle of Zero Products

In the system of complex numbers, when we multiply by 0 the result is 0. This is half of the principle of zero products. We have yet to show that the converse is true; that if a product is zero, at least one of the factors must be zero. With polar notation this is easy to do. The number 0 has polar notation $0 \text{ cis } \theta$, where θ can be any angle. Now consider a product $r_1 \text{ cis } \theta_1 \cdot r_2 \text{ cis } \theta_2$, where neither factor is 0. This means that $r_1 \neq 0$ and $r_2 \neq 0$. Since these are nonzero real numbers, their product is not zero, and the product of the complex numbers $r_1 r_2 \text{ cis } (\theta_1 + \theta_2)$ is not zero. Thus the principle of zero products holds in complex numbers.

EXERCISE SET 12.4

Graph each pair of complex numbers and their sum.

1. $3 + 2i, 2 - 5i$
2. $4 + 3i, 3 - 4i$
3. $-5 + 3i, -2 - 3i$
4. $-4 + 2i, -3 - 4i$

5. $2 - 3i, -5 + 4i$
6. $3 - 2i, -5 + 5i$
7. $-2 - 5i, 5 + 3i$
8. $-3 - 4i, 6 + 3i$

Find rectangular notation.

9. $3(\cos 30° + i \sin 30°)$
10. $6(\cos 150° + i \sin 150°)$
11. $10 \text{ cis } 270°$
12. $5 \text{ cis } (-60°)$
13. $\sqrt{8}\left(\cos \frac{\pi}{4} + i \sin \frac{\pi}{4}\right)$
14. $5\left(\cos \frac{\pi}{3} + i \sin \frac{\pi}{3}\right)$
15. $\sqrt{8} \text{ cis } \frac{5\pi}{4}$
16. $\sqrt{8} \text{ cis } \left(-\frac{\pi}{4}\right)$

Find polar notation.

17. $1 - i$
18. $\sqrt{3} + i$
19. $10\sqrt{3} - 10i$
20. $-10\sqrt{3} + 10i$
21. -5
22. $-5i$

Convert to polar notation and then multiply or divide.

23. $(1 - i)(2 + 2i)$
24. $(1 + i\sqrt{3})(1 + i)$
25. $(2\sqrt{3} + 2i)(2i)$
26. $(3\sqrt{3} - 3i)(2i)$
27. $\dfrac{1 + i}{1 - i}$
28. $\dfrac{1 - i}{\sqrt{3} - i}$
29. $\dfrac{2\sqrt{3} - 2i}{1 + \sqrt{3}i}$
30. $\dfrac{3 - 3\sqrt{3}i}{\sqrt{3} - i}$

31. Show that for any complex number z, $|z| = |-z|$. (Hint: Let $z = a + bi$.)
32. Show that for any complex number z, $|z| = |\bar{z}|$. (Hint: Let $z = a + bi$.)
33. Show that for any complex number z, $|z\bar{z}| = |z|^2$.
34. Show that for any complex number z, $|z^2| = |z|^2$.
35. Show that for any complex numbers z and w, $|z \cdot w| = |z| \cdot |w|$. (Hint: Let $z = r_1 \text{ cis } \theta_1$ and $w = r_2 \text{ cis } \theta_2$.)
36. Show that for any complex number z and any nonzero complex number w, $|z/w| = |z|/|w|$. (Use the hint for Exercise 35.)
37. On a complex plane, graph $|z| = 1$.
38. On a complex plane, graph $z + \bar{z} = 3$.

39. Find polar notation for $(\cos \theta + i \sin \theta)^{-1}$.
40. Let d be the function given by $d(z) = |z|$. Show that $d(z - w)$ is the distance between the vectors for z and w on the complex plane.
41. (See Exercise 40.) Show that for any real number a and any complex number z, $a \cdot d(z) = d(a \cdot z)$.
42. (See Exercise 40.) Find and prove a relationship between $d(u + v)$ and $d(u) + d(v)$.
43. Compute.

$$\begin{bmatrix} i & 0 \\ 0 & -i \end{bmatrix}^3$$

12.5 DeMOIVRE'S THEOREM

An important theorem about powers and roots of complex numbers is named for the French mathematician DeMoivre (1667–1754). Let us consider a number $r \operatorname{cis} \theta$ and its square:

$$(r \operatorname{cis} \theta)^2 = (r \operatorname{cis} \theta)(r \operatorname{cis} \theta) = r \cdot r \operatorname{cis}(\theta + \theta) = r^2 \operatorname{cis} 2\theta.$$

Similarly, we see that

$$(r \operatorname{cis} \theta)^3 = r \cdot r \cdot r \operatorname{cis}(\theta + \theta + \theta) = r^3 \operatorname{cis} 3\theta.$$

The generalization of this is DeMoivre's theorem.

THEOREM 10

DeMoivre's theorem. **For any complex number $r \operatorname{cis} \theta$ and any natural number n, $(r \operatorname{cis} \theta)^n = r^n \operatorname{cis} n\theta$.**

Example 1 Find $(1 + i)^9$.

We first find polar notation: $1 + i = \sqrt{2} \operatorname{cis} 45°$. Then

$$\begin{aligned}
(1 + i)^9 &= (\sqrt{2} \operatorname{cis} 45°)^9 \\
&= \sqrt{2}^9 \operatorname{cis} 9 \cdot 45° \\
&= 2^{9/2} \operatorname{cis} 405° \\
&= 16\sqrt{2} \operatorname{cis} 45°. \qquad \text{405° has the same terminal side as 45°}
\end{aligned}$$

Roots of Complex Numbers

As we shall see, every nonzero complex number has two square roots. A number has three cube roots, four fourth roots, and so on. In general a nonzero complex number has n different nth roots. These can be found by the formula that we now state and prove.

THEOREM 11

The nth roots of a complex number $r \operatorname{cis} \theta$ are given by

$$r^{1/n} \operatorname{cis}\left(\frac{\theta}{n} + k \cdot \frac{360°}{n}\right),$$

where $k = 0, 1, 2, \ldots, n - 1$.

Proof. We show that this formula gives us n different roots, using DeMoivre's theorem. We take the expression for the nth roots and raise it to the nth power, to show that we get $r \operatorname{cis} \theta$:

$$\left[r^{1/n} \operatorname{cis}\left(\frac{\theta}{n} + k \cdot \frac{360°}{n}\right)\right]^n = (r^{1/n})^n \operatorname{cis}\left(\frac{\theta}{n} \cdot n + k \cdot n \cdot \frac{360°}{n}\right)$$
$$= r \operatorname{cis}(\theta + k \cdot 360°) = r \operatorname{cis} \theta.$$

Thus we know that the formula gives us nth roots for any natural number k. Next we show that there are at least n different roots. To see this, consider substituting 0, 1, 2, and so on, for k. When $k = n$ the cycle begins to repeat, but from 0 to $n - 1$ the angles obtained and their sines and cosines are all different. There cannot be more than n different nth roots. This fact follows from the *fundamental theorem of algebra,* considered in Chapter 13. □

Example 2 Find the square roots of $2 + 2\sqrt{3}i$.

We first find polar notation: $2 + 2\sqrt{3}i = 4 \operatorname{cis} 60°$. Then

$$(4 \operatorname{cis} 60°)^{1/2} = 4^{1/2} \operatorname{cis}\left(\frac{60°}{2} + k \cdot \frac{360°}{2}\right), \quad k = 0, 1,$$
$$= 2 \operatorname{cis}\left(30° + k \cdot \frac{360°}{2}\right), \quad k = 0, 1.$$

Thus the roots are $2 \operatorname{cis} 30°$ and $2 \operatorname{cis} 210°$, or
$$\sqrt{3} + i \quad \text{and} \quad -\sqrt{3} - i.$$

In Example 2 it should be noted that the two square roots of a number were additive inverses of each other. This is true of the square roots of any complex number. To see this, let us find the square roots of any complex number $r \operatorname{cis} \theta$:

$$(r \operatorname{cis} \theta)^{1/2} = r^{1/2} \operatorname{cis}\left(\frac{\theta}{2} + k \cdot \frac{360°}{2}\right), \quad k = 0, 1,$$
$$= r^{1/2} \operatorname{cis} \frac{\theta}{2} \quad \text{or} \quad r^{1/2} \operatorname{cis}\left(\frac{\theta}{2} + 180°\right).$$

Now let us look at these numbers on a graph (Fig. 5). They lie on a line, so if one number has binomial notation $a + bi$, the other has binomial notation $-a - bi$. Hence their sum is 0 and they are additive inverses of each other.

Example 3 Find the cube roots of 1. Locate them on a graph.

$$1 = 1 \operatorname{cis} 0°$$
$$(1 \operatorname{cis} 0°)^{1/3} = 1^{1/3} \operatorname{cis}\left(\frac{0°}{3} + k \cdot \frac{360°}{3}\right), \quad k = 0, 1, 2.$$

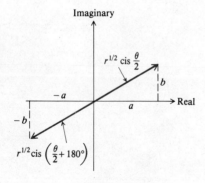

Figure 5

The roots are 1 cis 0°, 1 cis 120°, and 1 cis 240°, or

$$1, \; -\frac{1}{2} + \frac{\sqrt{3}}{2}i, \; \text{and} \; -\frac{1}{2} - \frac{\sqrt{3}}{2}i.$$

See Fig. 6.
Note in the example that the graphs of the cube roots lie equally spaced about a circle. This is true of the nth roots of any complex number.

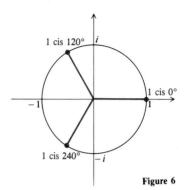

Figure 6

EXERCISE SET 12.5

Raise the number to the given power and write polar notation for the answer.

1. $\left(2 \text{ cis } \frac{\pi}{3}\right)^3$
2. $\left(3 \text{ cis } \frac{\pi}{2}\right)^4$
3. $\left(2 \text{ cis } \frac{\pi}{6}\right)^6$
4. $\left(2 \text{ cis } \frac{\pi}{5}\right)^5$
5. $(1 + i)^6$
6. $(1 - i)^6$

Raise the number to the given power and write rectangular notation for the answer.

7. $(2 \text{ cis } 240°)^4$
8. $(2 \text{ cis } 120°)^4$
9. $(1 + \sqrt{3}i)^4$
10. $(-\sqrt{3} + i)^6$
11. $\left(\frac{1}{\sqrt{2}} + \frac{1}{\sqrt{2}}i\right)^{10}$
12. $\left(\frac{1}{\sqrt{2}} - \frac{1}{\sqrt{2}}i\right)^{12}$
13. $\left(\frac{\sqrt{3}}{2} + \frac{1}{2}i\right)^{12}$
14. $\left(\frac{\sqrt{3}}{2} - \frac{1}{2}i\right)^{14}$
15. Solve $x^2 = -1 + \sqrt{3}i$. That is, find the square roots of $-1 + \sqrt{3}i$.
16. Solve $x^2 = -\sqrt{3} - i$. That is, find the square roots of $-\sqrt{3} - i$.
17. Solve: $x^3 = i$.
18. ▤ Solve: $x^3 = 68.4321$.
19. Find and graph the fourth roots of 16.
20. Find and graph the fourth roots of i.

☆

Solve. Evaluate square roots.

21. $x^2 + (1 - i)x + i = 0$
22. $3x^2 + (1 + 2i)x + 1 - i = 0$

CHAPTER 12 REVIEW

Simplify.

1. $(2 - 2i)(3 + 4i)$
2. $(3 - 5i) - (2 - i)$
3. $(6 + 2i) + (-4 - 3i)$
4. $\dfrac{2 - 3i}{1 - 3i}$

5. Solve for x and y: $4x + 2i = 8 - (2 + y)i$.

6. Find a polynomial in $\bar{z}$ that is the conjugate of $3z^3 + z - 7$.

7. Find an equation having the solutions $1 - 2i$, $1 + 2i$.

Solve.

8. $5x^2 - 4x + 1 = 0$

9. $x^2 + 3ix - 1 = 0$

10. Find the square roots of $4i$.

11. Graph the pair of complex numbers $-3 - 2i$, $4 + 7i$, and their sum.

12. Find rectangular notation for $2(\cos 135° + i \sin 135°)$.

13. Find polar notation for $1 + i$.

14. Find the product of $7 \operatorname{cis} 18°$ and $10 \operatorname{cis} 32°$.

15. Find the cube roots of $1 + i$.

16. Let $f(z) = 3z^2 - 2\bar{z} + 5i$. Find $f(2 - i)$.

17. Solve: $3z + 2\bar{z} = 5 + 2i$.

18. Solve: $2x - 3y = 7 + 7i$,
$\phantom{\text{Solve: }}3x + 2y = 4 - 9i$.

19. Solve: $x^3 - 27 = 0$.

20. Graph: $|z - (1 + i)| = 1$.

13

Polynomials and Rational Functions

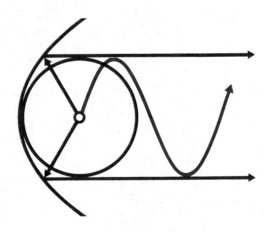

13.1 POLYNOMIALS AND POLYNOMIAL FUNCTIONS

We have already considered polynomials. We now make a formal definition.

DEFINITION

A *polynomial* is any expression of the type

$$a_n x^n + a_{n-1} x^{n-1} + \cdots + a_2 x^2 + a_1 x + a_0,$$

where n is a nonnegative integer and $a_0, \ldots, a_n$, are numbers.

The numbers mentioned in this definition are complex numbers, but in certain cases may be restricted to real numbers, or rational numbers or integers. Some or all of them may be 0.

TERMINOLOGY

In a polynomial:

1. The numbers $a_0, \ldots, a_n$ are called *coefficients*.

2. The first nonzero coefficient, a_n, is called the *leading coefficient*. The number n is called the *degree* of the polynomial, except the polynomial consisting only of the number 0 has no degree.

Example 1 We list some polynomials and their degrees below.

POLYNOMIAL	DEGREE
$5x^3 - 3x^2 + i$	3
$-4x + \sqrt{2}$	1
25	0
0	No degree

Zeros or Roots of Polynomials

When a number is substituted for the variable in a polynomial, the result is some unique number. Thus every polynomial defines a function. We often refer to polynomials, therefore, using function notation, such as $P(x)$. If a number a makes a polynomial 0 (that is, $P(a) = 0$), then a is said to be a *zero* of $P(x)$. Note that a is also a solution of the equation $P(x) = 0$. We often say that a is a *root* of the equation $P(x) = 0$

and also a *root* of the polynomial P(x).*

Example 2 If $P(x) = x^3 + 2x^2 - 5x - 6$, is 3 a zero of P(x)?

We substitute 3 into the polynomial:
$$P(3) = 3^3 + 2(3)^2 - 5 \cdot 3 - 6 = 24.$$

Since $P(3) \neq 0$, 3 is not a zero of the polynomial.

Example 3 Is -1 a root of P(x) of Example 2?
$$P(-1) = (-1)^3 + 2(-1)^2 - 5(-1) - 6 = 0.$$

Since $P(-1) = 0$, -1 is a root of the polynomial.

Division of Polynomials and Factors

When we divide one polynomial by another we obtain a quotient and a remainder. If the remainder is 0, then the divisor is a *factor* of the dividend.

Example 4 Divide, to find whether $x^2 + 9$ is a factor of $x^4 - 81$.

$$\begin{array}{r} x^2 - 9 \\ x^2 + 9 \overline{)x^4 - 81} \\ \underline{x^4 + 9x^2 } \\ -9x^2 - 81 \\ \underline{-9x^2 - 81} \\ 0 \end{array}$$

Note how spaces have been left for missing terms in the dividend.

Since the remainder is 0, we know that $x^2 + 9$ is a factor.

Example 5 Divide, to find whether $x^2 + 3x - 1$ is a factor of $x^4 - 81$.

$$\begin{array}{r} x^2 - 3x + 10 \\ x^2 + 3x - 1 \overline{)x^4 - 81} \\ \underline{x^4 + 3x^3 - x^2 } \\ -3x^3 + x^2 \\ \underline{-3x^3 - 9x^2 + 3x } \\ 10x^2 - 3x - 81 \\ \underline{10x^2 + 30x - 10} \\ -33x - 71 \end{array}$$

---

*Terminology varies from author to author. Traditionally, one referred to the *zeros* of a function and the *roots* of an equation. Subsequently it was found to be convenient in exposition to speak also of the *roots* of a function, especially a polynomial function.

In some books the traditional distinction is made. The authors of this book have found that the freer use of terminology facilitates simplification of the exposition.

Since the remainder is not 0, we know that $x^2 + 3x - 1$ is not a factor of $x^4 - 81$.

In general, when we divide a polynomial $P(x)$ by a divisor $d(x)$ we obtain some polynomial $Q(x)$ for a quotient and some polynomial $R(x)$ for a remainder. The remainder must either be 0 or have degree less than that of $d(x)$. To check a division we multiply the quotient by the divisor and add the remainder, to see if we get the dividend. Thus these polynomials are related as follows:

$$P(x) = d(x) \cdot Q(x) + R(x).$$

Example 6 If $P(x) = x^4 - 81$ and $d(x) = x^2 + 9$, then $Q(x) = x^2 - 9$ and $R(x) = 0$, and

$$\underbrace{x^4 - 81}_{P(x)} = \underbrace{(x^2 + 9)}_{d(x)} \cdot \underbrace{(x^2 - 9)}_{Q(x)} + \underbrace{0}_{R(x)}.$$

Example 7 If $P(x) = x^4 - 81$ and $d(x) = x^2 + 3x - 1$, then $Q(x) = x^2 - 3x + 10$ and $R(x) = -33x - 71$, and

$$\underbrace{x^4 - 81}_{P(x)} = \underbrace{(x^2 + 3x - 1)}_{d(x)} \cdot \underbrace{(x^2 - 3x + 10)}_{Q(x)} + \underbrace{(-33x - 71)}_{R(x)}.$$

EXERCISE SET 13.1

Determine the degree of each polynomial.

1. $x^4 - 3x^2 + 1$
2. $2x^5 - x^4 + \frac{1}{4}x - 7$
3. $-2x + 5$
4. $3x - \sqrt{\pi}$
5. $2x^2 - 3x + 4$
6. $\frac{1}{4}x^2 - 7$
7. 3
8. 0

9. Determine whether 2, 3, and -1 are roots, or zeros, of $P(x) = x^3 + 6x^2 - x - 30$.

10. Determine whether 2, 3, and -1 are roots, or zeros, of $P(x) = 2x^3 - 3x^2 + x - 1$.

11. For $P(x)$ in Exercise 9, which of the following are factors of $P(x)$?
    a) $x - 2$ b) $x - 3$ c) $x + 1$

12. For $P(x)$ in Exercise 10, which of the following are factors of $P(x)$?
    a) $x - 2$ b) $x - 3$ c) $x + 1$

In each of the following, a polynomial $P(x)$ and a divisor $d(x)$ are given. Find the quotient $Q(x)$ and the remainder $R(x)$ when $P(x)$ is divided by $d(x)$ and express $P(x)$ in the form $d(x) \cdot Q(x) + R(x)$.

13. $P(x) = x^3 + 6x^2 - x - 30$,
    $d(x) = x - 2$

14. $P(x) = 2x^3 - 3x^2 + x - 1$,
    $d(x) = x - 2$

15. $P(x)$ as in Exercise 13,
    $d(x) = x - 3$

16. $P(x)$ as in Exercise 14,
    $d(x) = x - 3$

17. $P(x) = x^3 - 8$,
    $d(x) = x + 2$
18. $P(x) = x^3 + 27$,
    $d(x) = x + 1$
19. $P(x) = x^4 + 9x^2 + 20$,
    $d(x) = x^2 + 4$
20. $P(x) = x^4 + x^2 + 2$,
    $d(x) = x^2 + x + 1$
21. $P(x) = 5x^7 - 3x^4 + 2x^2 - 3$,
    $d(x) = 2x^2 - x + 1$
22. $P(x) = 6x^5 + 4x^4 - 3x^2 + x - 2$,
    $d(x) = 3x^2 + 2x - 1$
23. For $P(x) = x^5 - 64$,
    a) find $P(2)$;
    b) find the remainder when $P(x)$ is divided by $x - 2$, and compare your answer to (a);
    c) find $P(-1)$;
    d) find the remainder when $P(x)$ is divided by $x + 1$, and compare your answer to (c).
24. For $P(x) = x^3 + x^2$,
    a) find $P(-1)$;
    b) find the remainder when $P(x)$ is divided by $x + 1$ and compare your answer to (a);
    c) find $P(2)$;
    d) find the remainder when $P(x)$ is divided by $x - 2$, and compare your answer to (c).

25. For $P(x) = 2x^2 - ix + 1$,
    a) find $P(-i)$;
    b) find the remainder when $P(x)$ is divided by $x + i$.
26. For $P(x) = 2x^2 + ix - 1$,
    a) find $P(i)$;
    b) find the remainder when $P(x)$ is divided by $x - i$.

27. Under what conditions is a polynomial function of real variables an even function, that is, $P(x) = P(-x)$ for all x? Does your answer still hold for complex variables? Explain.

28. Under what conditions is a polynomial function of real variables an odd function, that is, $-P(x) = P(-x)$ for all x? Does your answer still hold for complex variables?

29. Find a rule for determining the degree of the product of two polynomials with real coefficients.

30. Find a rule for determining the degree of the sum of two polynomials with real coefficients.

13.2 THE REMAINDER AND FACTOR THEOREMS

Some of the exercises in the previous exercise set illustrate the following theorem.

THEOREM 1

The remainder theorem. **If a number r is substituted for x in the polynomial $P(x)$, then the result $P(r)$ is the remainder that would be obtained by dividing $P(x)$ by $x - r$.**

Proof. The equation $P(x) = d(x) \cdot Q(x) + R(x)$ is the basis of this proof. If we divide $P(x)$ by $x - r$, we obtain a quotient $Q(x)$ and a remainder $R(x)$ related as follows:

$$P(x) = (x - r) \cdot Q(x) + R(x).$$

The remainder $R(x)$ must either be 0 or have degree less than $x - r$. Thus $R(x)$ must be a constant. Let us call this constant R. In the above expression we get a true sentence whenever we replace x by any number. Let us replace x by r. We get

$$P(r) = (r - r) \cdot Q(r) + R$$
$$P(r) = 0 \cdot Q(r) + R$$
$$P(r) = R.$$

This tells us that the function value $P(r)$ is the remainder obtained when we divide $P(x)$ by $x - r$. □

Theorem 1 motivates us to find a rapid way of dividing by $x - r$ in order to find function values.

Synthetic Division

To streamline division, we can arrange the work so that duplicate and unnecessary writing are avoided. Consider the following.

A.
$$\begin{array}{r} 4x^2 + 5x + 11 \\ x - 2 \overline{\smash{)}4x^3 - 3x^2 + x + 7} \\ \underline{4x^3 - 8x^2\phantom{{}+x+7}} \\ 5x^2 + x\phantom{{}+7} \\ \underline{5x^2 - 10x\phantom{{}+7}} \\ 11x + 7 \\ \underline{11x - 22} \\ 29 \end{array}$$

B.
$$\begin{array}{r} 4 \quad 5 \quad 11 \\ 1 - 2 \overline{\smash{)}4 - 3 + 1 + 7} \\ \underline{4 - 8\phantom{{}+1+7}} \\ 5 + 1\phantom{{}+7} \\ \underline{5 - 10\phantom{{}+7}} \\ 11 + 7 \\ \underline{11 - 22} \\ 29 \end{array}$$

The division in (B) is the same as that in (A), but we wrote only the coefficients. The colored numerals are duplicated, so we look for an arrangement in which they are not duplicated. We can also simplify things by using the additive inverse of -2 and then adding instead of subtracting. When things are thus "collapsed," we have the algorithm known as *synthetic division*.

13.2 THE REMAINDER AND FACTOR THEOREMS

C. Synthetic division

$$\begin{array}{c|cccc} 2 & 4 & -3 & 1 & 7 \\ & & 8 & 10 & 22 \\ \hline & 4 & 5 & 11 & 29 \end{array}$$

We "bring down" the 4. Then multiply it by the 2, to get 8. Add to get 5. Multiply 5 by 2 to get 10. We repeat this process. The last number, 29, is the remainder. The others are the coefficients of the quotient.

Example 1 Use synthetic division to find the quotient and remainder:

$$(2x^3 + 7x^2 - 5) \div (x + 3).$$

First note that $x + 3 = x - (-3)$.

$$\begin{array}{c|cccc} -3 & 2 & 7 & 0 & -5 \\ & & -6 & -3 & 9 \\ \hline & 2 & 1 & -3 & 4 \end{array}$$ Note: We must write 0's for missing terms.

The quotient is $2x^2 + x - 3$. The remainder is 4.

We can now apply synthetic division to the finding of function values for polynomials.

Example 2 $P(x) = 2x^5 - 3x^4 + x^3 - 2x^2 + x - 8$. Find $P(10)$.

By Theorem 1, $P(10)$ is the remainder when $P(x)$ is divided by $x - 10$. We use synthetic division to find that remainder.*

$$\begin{array}{c|cccccc} 10 & 2 & -3 & 1 & -2 & 1 & -8 \\ & & 20 & 170 & 1710 & 17{,}080 & 170{,}810 \\ \hline & 2 & 17 & 171 & 1708 & 17{,}081 & 170{,}802 \end{array}$$

Thus $P(10) = 170{,}802$.

Example 3 Determine whether -4 is a zero or root of $P(x)$, where $P(x) = x^3 + 8x^2 + 8x - 32$.

We use synthetic division and Theorem 1 to find $P(-4)$.

$$\begin{array}{c|cccc} -4 & 1 & 8 & 8 & -32 \\ & & -4 & -16 & 32 \\ \hline & 1 & 4 & -8 & 0 \end{array}$$

Since $P(-4) = 0$, the number -4 is a root of $P(x)$.

---

*Compare this with the work involved in a direct calculation!

We now consider the following useful corollary of the remainder theorem.

THEOREM 2

The factor theorem. **For a polynomial $P(x)$, if $P(r) = 0$, then the polynomial $x - r$ is a factor of $P(x)$.**

Proof. If we divide $P(x)$ by $x - r$, we obtain a quotient and remainder related as follows:
$$P(x) = (x - r) \cdot Q(x) + P(r).$$
Then if $P(r) = 0$, we have
$$P(x) = (x - r) \cdot Q(x),$$
so $x - r$ is a factor of $P(x)$.

This theorem is very useful in factoring polynomials, and hence in the solving of equations. □

Example 4 Let $P(x) = x^3 + 2x^2 - 5x - 6$. Factor $P(x)$ and thus solve the equation $P(x) = 0$.

We look for linear factors of the form $x - r$. Let us try $x - 1$. We use synthetic division to see whether $P(1) = 0$.

$$\underline{1|} \quad \begin{array}{cccc} 1 & 2 & -5 & -6 \\ & 1 & 3 & -2 \\ \hline 1 & 3 & -2 & -8 \end{array}$$

We know that $x - 1$ is not a factor of $P(x)$. We try $x + 1$.

$$\underline{-1|} \quad \begin{array}{cccc} 1 & 2 & -5 & -6 \\ & -1 & -1 & 6 \\ \hline 1 & 1 & -6 & 0 \end{array}$$

We know that $x + 1$ is one factor and the quotient, $x^2 + x - 6$, is another. Thus
$$P(x) = (x + 1)(x^2 + x - 6).$$
The trinomial is easily factored, so we have
$$P(x) = (x + 1)(x + 3)(x - 2).$$
To solve the equation $P(x) = 0$, we use the principle of zero products. The solutions are -1, -3, and 2.

EXERCISE SET 13.2

Use synthetic division to find the quotient and remainder.

1. $(2x^4 + 7x^3 + x - 12) \div (x + 3)$
2. $(x^3 - 7x^2 + 13x + 3) \div (x - 2)$
3. $(x^3 - 2x^2 - 8) \div (x + 2)$
4. $(x^3 - 3x + 10) \div (x - 2)$
5. $(x^4 - 1) \div (x - 1)$
6. $(x^5 + 32) \div (x + 2)$
7. $(2x^4 + 3x^2 - 1) \div \left(x - \frac{1}{2}\right)$
8. $(3x^4 - 2x^2 + 2) \div \left(x - \frac{1}{4}\right)$
9. $(x^4 - y^4) \div (x - y)$
10. $(x^3 + 3ix^2 - 4ix - 2) \div (x + i)$

Use synthetic division to find the function values.

11. $P(x) = x^3 - 6x^2 + 11x - 6$
Find $P(1)$, $P(-2)$, and $P(3)$.
12. $P(x) = x^3 + 7x^2 - 12x - 3$
Find $P(-3)$, $P(-2)$, and $P(1)$.
13. $P(x) = 2x^5 - 3x^4 + 2x^3 - x + 8$
Find $P(20)$ and $P(-3)$.
14. $P(x) = x^5 - 10x^4 + 20x^3 - 5x - 100$
Find $P(-10)$ and $P(5)$.

Using synthetic division, determine whether the numbers are roots of the polynomials.

15. $-3, 2;\quad P(x) = 3x^3 + 5x^2 - 6x + 18$
16. $-4, 2;\quad P(x) = 3x^3 + 11x^2 - 2x + 8$
17. $-3, \frac{1}{2};\quad P(x) = x^3 - \frac{7}{2}x^2 + x - \frac{3}{2}$
18. $i, -i, -2;\quad P(x) = x^3 + 2x^2 + x + 2$

Factor the polynomial $P(x)$. Then solve the equation $P(x) = 0$.

19. $P(x) = x^3 + 4x^2 + x - 6$
20. $P(x) = x^3 + 5x^2 - 2x - 24$
21. $P(x) = x^3 - 6x^2 + 3x + 10$
22. $P(x) = x^3 + 2x^2 - 13x + 10$
23. $P(x) = x^3 - x^2 - 14x + 24$
24. $P(x) = x^3 - 3x^2 - 10x + 24$
25. $P(x) = x^4 - x^3 - 19x^2 + 49x - 30$
26. $P(x) = x^4 + 11x^3 + 41x^2 + 61x + 30$

Solve.

27. $x^3 + 2x^2 - 13x + 10 > 0$
28. $x^4 - 22x^2 + 51x - 30 < 0$
29. Find k so that $x + 2$ is a factor of $x^3 - kx^2 + 3x + 7k$.
30. ▦ Given that
$$f(x) = 2.13x^5 - 42.1x^3 + 17.5x^2 + 0.953x - 1.98,$$
find $f(3.21)$ (a) by synthetic division; (b) by substitution.
31. For what values of k will the remainder be the same when $x^2 + kx + 4$ is divided by $x - 1$ or $x + 1$?
32. When $x^2 - 3x + 2k$ is divided by $x + 2$ the remainder is 7. Find the value of k.

33. Devise a way to use the method of synthetic division when the divisor is a polynomial such as $bx - r$.
34. Prove that $x - a$ is a factor of $x^n - a^n$, for any natural number n.

35. (*Nondecimal notation*). Ordinary notation for numbers is *decimal* (base ten). For example,

$$4325 \text{ means } 4 \times 10^3 + 3 \times 10^2 + 2 \times 10 + 5.$$

When the base is a number other than ten, the notation is called *nondecimal* and is defined in a similar way. For example,

$$4312_5 \text{ means } 4 \times 5^3 + 3 \times 5^2 + 1 \times 5 + 2.$$

Synthetic division can be used to convert from a nondecimal base to base ten. For example to convert 31214_5 to base ten we divide as follows.

```
5 | 3   1    2    1     4
  |     15   80   410   2055
  ---------------------------
    3   16   82   411 | 2059
```

Decimal notation for this number is 2059.
a) Convert 4352_6 to decimal notation.
b) Prove that, in general, the procedure gives correct results.

36. Consider $P(x) = x^4 - x^3 - 1$ and $Q(x) = x^4 - x^3 - 2x^2 + 1$. Show that if $P(x) = 0$, then $Q(x^2) = 0$.

13.3 THEOREMS ABOUT ROOTS

The Fundamental Theorem of Algebra

A linear, or first-degree, polynomial $ax + b$ (where $a \neq 0$, of course) has just one root, $-b/a$. From Chapter 12 we know that any quadratic polynomial, with complex numbers for coefficients, has at least one, and at most two roots. The following theorem is a generalization. A proof is beyond this text.

THEOREM 3

The fundamental theorem of algebra. **Every polynomial of degree greater than 0, with complex coefficients, has at least one root in the system of complex numbers.***

This is a very powerful theorem. Note that while it guarantees that a root exists, it does not tell how to find it. We now develop some theory that can help in finding roots. First, we prove a corollary of the fundamental theorem of algebra.

---

* It is well to keep in mind that when we speak of "roots," or "zeros," of a polynomial $P(x)$, we are also speaking about "roots" or "solutions" of the polynomial equation $P(x) = 0$.

THEOREM 4

Every polynomial of degree n, where $n > 0$, having complex coefficients, can be factored into n linear factors.

Proof. Let us consider any polynomial of degree n, say $P(x)$. By the fundamental theorem, it has a root r_1. By the factor theorem, $x - r_1$ is a factor of $P(x)$. Thus we know that
$$P(x) = (x - r_1) \cdot Q_1(x),$$
where $Q_1(x)$ is the quotient that would be obtained upon dividing $P(x)$ by $x - r_1$. Let the leading coefficient of $P(x)$ be a_n. By considering the actual division process we see that the leading coefficient of $Q_1(x)$ is also a_n and that the degree of $Q_1(x)$ is $n - 1$. Now if the degree of $Q_1(x)$ is greater than 0, then it has a root r_2, and we have
$$Q_1(x) = (x - r_2) \cdot Q_2(x),$$
where the degree of $Q_2(x)$ is $n - 2$ and the leading coefficient is a_n. Thus we have
$$P(x) = (x - r_1)(x - r_2) \cdot Q_2(x).$$
This process can be continued until a quotient $Q_n(x)$ is obtained having degree 0. The leading coefficient will be a_n, so $Q_n(x)$ is actually the constant a_n. We now have the following:
$$P(x) = a_n(x - r_1)(x - r_2)(x - r_3) \cdots (x - r_n).$$
This completes the proof. We have actually shown a little more than what the theorem states. We see that $P(x)$ has been factored with one constant factor a_n and n linear factors having leading coefficient 1. □

To find the roots of a polynomial, we can attempt to factor it.

Examples

1. $x^4 + x^3 - 13x^2 - x + 12$ can be factored as
$$(x - 3)(x + 4)(x + 1)(x - 1),$$
so the roots are 3, -4, -1, and 1.
2. $3x^4 - 15x^3 + 18x^2 + 12x - 24$ can be factored as
$$3(x - 2)(x - 2)(x - 2)(x + 1),$$
so the roots are 2 and -1.

In Example 2 the factor $x - 2$ occurs three times. In a case like this we sometimes say that the root we obtain, 2, has a *multiplicity* of three.

From the preceding theorem and the above examples, we have the following.

THEOREM 5

Every polynomial of degree n, where $n > 0$, has at least one root and at most n roots.

Given several numbers, we can find a polynomial having them as its roots.

Example 3 Find a polynomial of degree three, having the roots -2, 1, and $3i$.

By Theorem 2, such a polynomial has factors $x + 2$, $x - 1$, and $x - 3i$, so we have

$$P(x) = a_n(x + 2)(x - 1)(x - 3i).$$

The number a_n can be any nonzero number. The simplest polynomial will be obtained if we let it be 1. If we then multiply the factors we obtain

$$P(x) = x^3 + (1 - 3i)x^2 + (-2 - 3i)x + 6i.$$

Example 4 Find a polynomial of degree 5 with -1 as a root of multiplicity 3, 4 as a root of multiplicity 1, and 0 as a root of multiplicity 1.

Proceeding as in Example 3, letting $a_n = 1$, we obtain

$$P(x) = (x + 1)^3(x - 4)(x - 0),$$

or

$$P(x) = x^5 - x^4 - 9x^3 - 11x^2 - 4x.$$

Roots of Polynomials with Real Coefficients

Consider the quadratic equation $x^2 - 2x + 2 = 0$, with real coefficients. Its roots are $1 + i$ and $1 - i$. Note that they are complex conjugates. This generalizes to any polynomial with real coefficients.

THEOREM 6

If a complex number z is a root of a polynomial $P(x)$ of degree greater than or equal to 1 with real coefficients, then its conjugate $\bar{z}$ is also a root. (Nonreal roots occur in conjugate pairs.)

Proof. Let
$$P(x) = a_n x^n + a_{n-1} x^{n-1} + \cdots + a_1 x + a_0,$$
where the coefficients are real numbers. Suppose z is a complex root of $P(x)$. Then $P(z) = 0$, or
$$a_n z^n + a_{n-1} z^{n-1} + \cdots + a_1 z + a_0 = 0.$$
Now let us find the conjugate of each side of the equation. First note that $\overline{0} = 0$, since 0 is a real number. Then we have the following.

$$\begin{aligned}
0 = \overline{0} &= \overline{a_n z^n + a_{n-1} z^{n-1} + \cdots + a_1 z + a_0} \\
&= \overline{a_n z^n} + \overline{a_{n-1} z^{n-1}} + \cdots + \overline{a_1 z} + \overline{a_0} && \text{By Theorem 3, Ch. 12} \\
&= \overline{a_n} \cdot \overline{z^n} + \overline{a_{n-1}} \cdot \overline{z^{n-1}} + \cdots + \overline{a_1} \cdot \overline{z} + \overline{a_0} && \text{By Theorem 4, Ch. 12} \\
&= a_n \overline{z^n} + a_{n-1} \overline{z^{n-1}} + \cdots + a_1 \overline{z} + a_0 && \text{By Theorem 5, Ch. 12} \\
&= a_n \overline{z}^n + a_{n-1} \overline{z}^{n-1} + \cdots + a_1 \overline{z} + a_0. && \text{By Theorem 6, Ch. 12}
\end{aligned}$$

Thus $P(\overline{z}) = 0$, so $\overline{z}$ is a root of the polynomial. □

For Theorem 6, it is essential that the coefficients be real numbers. This can be seen by considering Example 3. In that polynomial the root $3i$ occurs, but its conjugate does not. This can happen because some of the coefficients of the polynomial are not real.

Rational Coefficients

When a polynomial has rational numbers for coefficients, certain irrational roots also occur in pairs, as described in the following theorem.

THEOREM 7

Suppose $P(x)$ is a polynomial with rational coefficients and of degree greater than 0. Then if either of the following is a root, so is the other: $a + c\sqrt{b}$, $a - c\sqrt{b}$, a and c rational, b not a square.

This theorem can be proved in a manner analogous to the one above, but we shall not do it here. The theorem can be used to help in finding roots.

Example 5 Suppose a polynomial of degree 6 with rational coefficients has $-2 + 5i$, $-2i$, and $1 - \sqrt{3}$ as some of its roots. Find the other roots.

The other roots are $-2 - 5i$, $2i$, and $1 + \sqrt{3}$. There are no other roots since the degree is 6.

Example 6 Find a polynomial of lowest degree with rational coefficients that has $1 - \sqrt{2}$ and $1 + 2i$ as some of its roots.

The polynomial must also have the roots $1 + \sqrt{2}$ and $1 - 2i$. Thus the polynomial is

$$[x - (1 - \sqrt{2})][x - (1 + \sqrt{2})][x - (1 + 2i)][x - (1 - 2i)],$$

or

$$(x^2 - 2x - 1)(x^2 - 2x + 5),$$

or

$$x^4 - 4x^3 + 8x^2 - 8x - 5.$$

Example 7 Let $P(x) = x^4 - 5x^3 + 10x^2 - 20x + 24$. Find the other roots of $P(x)$, given that $2i$ is a root.

Since $2i$ is a root, we know that $-2i$ is also a root. Thus

$$P(x) = (x - 2i)(x + 2i) \cdot Q(x)$$

for some $Q(x)$. Since $(x - 2i)(x + 2i) = x^2 + 4$, we know that

$$P(x) = (x^2 + 4) \cdot Q(x).$$

We find, using division, that $Q(x) = x^2 - 5x + 6$, and since we can factor $x^2 - 5x + 6$, we get

$$P(x) = (x^2 + 4)(x - 2)(x - 3).$$

Thus the other roots are $-2i$, 2, and 3.

EXERCISE SET 13.3

Find the roots of each polynomial, or polynomial equation, and state the multiplicity of each.

1. $(x + 3)^2(x - 1)$
2. $-8(x - 3)^2(x + 4)^3 x^4 = 0$
3. $x^3(x - 1)^2(x + 4)$
4. $(x^2 - 5x + 6)^2 = 0$

Find a polynomial of degree 3 with the given numbers as roots.

5. $-2, 3, 5$
6. $2, i, -i$
7. $-3, 2i, -2i$
8. $1 + 4i, 1 - 4i, -1$
9. $\sqrt{2}, -\sqrt{2}, \sqrt{3}$.   Are the coefficients rational?

10. Find a polynomial equation of degree 4 with -2 as a root of multiplicity 1, 3 as a root of multiplicity 2, and -1 as a root of multiplicity 1.

Suppose a polynomial or polynomial equation of degree 5 with rational coefficients has the given numbers as roots. Find the other roots.

11. $6, -3 + 4i, 4 - \sqrt{5}$

12. $-2, 3, 4, 1 - i$

Find a polynomial of lowest degree with rational coefficients that has the given numbers as some of its roots.

13. $1 + i, 2$

14. $2 - i, -1$

15. $-4i, 5$

16. $2 - \sqrt{3}, 1 + i$

17. $\sqrt{5}, -3i$

18. $-\sqrt{2}, 4i$

Given that the polynomial or polynomial equation has the given root, find the other roots.

19. $x^4 - 5x^3 + 7x^2 - 5x + 6; \quad -i$

20. $x^4 - 16 = 0; \quad 2i$

21. $x^3 - 6x^2 + 13x - 20 = 0; \quad 4$

22. $x^3 - 8; \quad 2$

Solve the following equations. Use synthetic division, the quadratic formula, the theorems of this section, or whatever else you think might help.

23. $x^3 - 4x^2 + x - 4 = 0$

24. $x^3 - x^2 - 7x + 15 = 0$

25. $x^4 - 2x^3 - 2x - 1 = 0$

27. What does the fundamental theorem of algebra tell you about this equation?

$$2\sin^5 x - 2\sin^3 x + \sin x = \frac{1}{8}$$

26. The equation $x^2 + 2ax + a^2 = 0$ has a double root. Find it.

28. Prove that a polynomial with positive coefficients cannot have a positive root.

29. Prove that every polynomial of odd degree, with real coefficients, has at least one real root.

30. Prove that there is no polynomial P that defines the cosine function, $P(x) \equiv \cos x$.

31. Prove that a polynomial of degree n cannot have more than $n/2$ roots of multiplicity 2.

32. Prove that every polynomial with real coefficients has a factorization into linear and quadratic factors (with real coefficients).

33. Prove that if $P(z) = u$, where P is any polynomial with real coefficients, then $P(\bar{z}) = \bar{u}$.

34. Prove that $P(x)$ is a factor of $Q(x)$, where P and Q are polynomial functions, if and only if every root of $P(x)$ is also a root of $Q(x)$.

13.4 RATIONAL ROOTS

Integer Coefficients

It is not always easy to find the roots of a polynomial. However, if a polynomial has integer coefficients, there is a procedure that will yield all of the rational roots.

THEOREM 8

The rational roots theorem. Let
$$P(x) = a_n x^n + a_{n-1} x^{n-1} + \cdots + a_1 x + a_0,$$
where all the coefficients are integers. Consider a rational number denoted by c/d, where c and d are relatively prime (having no common factor besides 1 and -1). If c/d is a root of $P(x)$, then c is a factor of a_0 and d is a factor of a_n.

Proof. Since c/d is a root of $P(x)$, we know that

$$a_n \left(\frac{c}{d}\right)^n + a_{n-1}\left(\frac{c}{d}\right)^{n-1} + \cdots + a_1\left(\frac{c}{d}\right) + a_0 = 0. \tag{1}$$

We multiply by d^n and get the equation

$$a_n c^n + a_{n-1} c^{n-1} d + \cdots + a_1 c d^{n-1} + a_0 d^n = 0. \tag{2}$$

Then we have

$$a_n c^n = (-a_{n-1} c^{n-1} - \cdots - a_1 c d^{n-2} - a_0 d^{n-1})d.$$

Note that d is a factor of $a_n c^n$. Now d is not a factor of c because c and d are relatively prime. Thus d is not a factor of c^n. So d is a factor of a_n.

In a similar way we can show from Eq. (2) that

$$a_0 d^n = (-a_n c^{n-1} - a_{n-1} c^{n-2} d - \cdots - a_1 d^{n-1})c.$$

Thus c is a factor of $a_0 d^n$. Again, c is not a factor of d^n, so it must be a factor of a_0. □

Example 1 Let $P(x) = 3x^4 - 11x^3 + 10x - 4$. Find the rational roots of $P(x)$. If possible, find the other roots.

By the rational roots theorem, if c/d is a root of $P(x)$, then c must be a factor of -4 and d must be a factor of 3. Thus the possibilities for c and d are

$$c: \ 1, -1, 4, -4, 2, -2; \qquad d: \ 1, -1, 3, -3.$$

Then the resulting possibilities for c/d are

$$\frac{c}{d}: \ 1, -1, 4, -4, \frac{1}{3}, -\frac{1}{3}, \frac{4}{3}, -\frac{4}{3}, \frac{2}{3}, -\frac{2}{3}, 2, -2.$$

Of these 12 possibilities, we know that at most 4 of them could be roots because $P(x)$ is of degree 4. To find which are roots we could use substitution, but synthetic division is usually more efficient.

We try 1:

$$\begin{array}{r|rrrrr} 1 & 3 & -11 & 0 & 10 & -4 \\ & & 3 & -8 & -8 & 2 \\ \hline & 3 & -8 & -8 & 2 & -2. \end{array}$$

We try -1:

$$\begin{array}{r|rrrrr} -1 & 3 & -11 & 0 & 10 & -4 \\ & & -3 & 14 & -14 & 4 \\ \hline & 3 & -14 & 14 & -4 & 0. \end{array}$$

Thus $P(1) = -2$, so 1 is not a root; and $P(-1) = 0$, so -1 is a root. We use synthetic division again, to see whether -1 is a double root:

$$\begin{array}{r|rrrr} -1 & 3 & -14 & 14 & -4 \\ & & -3 & 17 & -31 \\ \hline & 3 & -17 & 31 & -35. \end{array}$$

It is not. Using the results of the second synthetic division above, we can express $P(x)$ as follows:

$$P(x) = (x + 1)(3x^3 - 14x^2 + 14x - 4).$$

We now use $3x^3 - 14x^2 + 14x - 4$ and check the other possible roots. We try $\frac{2}{3}$:

$$\begin{array}{r|rrrr} \frac{2}{3} & 3 & -14 & 14 & -4 \\ & & 2 & -8 & 4 \\ \hline & 3 & -12 & 6 & 0. \end{array}$$

Thus $P(\frac{2}{3}) = 0$, so $\frac{2}{3}$ is a root. Again using the results of the synthetic division, we can express $P(x)$ as follows:

$$P(x) = (x + 1)\left(x - \frac{2}{3}\right)(3x^2 - 12x + 6).$$

Since the factor $3x^2 - 12x + 6$ is quadratic, we can use the quadratic formula to find that the other roots are $2 + \sqrt{2}$ and $2 - \sqrt{2}$. These are irrational numbers. Thus the rational roots are -1 and $\frac{2}{3}$.

Example 2 Let $P(x) = x^3 + 6x^2 + x + 6$. Find the rational roots of $P(x)$. If possible, find the other roots.

By the rational roots theorem, if c/d is a root of $P(x)$, then c must be a factor of 6 and d must be a factor of 1. Thus the possibilities for c and d are

$$c: \ 1, -1, 2, -2, 3, -3, 6, -6; \qquad d: \ 1, -1.$$

Then the resulting possibilities for c/d are

$$\frac{c}{d}: \quad 1, -1, 2, -2, 3, -3, 6, -6.$$

Note that these are the same as the possibilities for c. If the leading coefficient is 1, we need only check the factors of the last coefficient as possibilities for rational roots.

There is another aid in eliminating possibilities for rational roots. Note that all coefficients of $P(x)$ are positive. Thus when any positive number is substituted in $P(x)$, we get a positive value, never 0. Therefore no positive number can be a root. Thus the only possibilities for roots are

$$-1, -2, -3, -6.$$

We try -6:

$$\begin{array}{r|rrrr} -6 & 1 & 6 & 1 & 6 \\ & & -6 & 0 & -6 \\ \hline & 1 & 0 & 1 & 0. \end{array}$$

Thus $P(-6) = 0$, so -6 is a root. We could divide again to determine whether -6 is a double root, but since we can now factor $P(x)$ as a product of a linear polynomial and a quadratic polynomial, it is preferable to proceed that way. We have

$$P(x) = (x + 6)(x^2 + 1).$$

Now $x^2 + 1$ has the complex roots i and $-i$. Thus the only rational root of $P(x)$ is -6.

Example 3 Find the rational roots of $x^4 + 2x^3 + 2x^2 - 4x - 8$.

Since the leading coefficient is 1, the only possibilities for rational roots are the factors of the last coefficient, -8:

$$1, -1, 2, -2, 4, -4, 8, -8.$$

But, using substitution or synthetic division, we find that none of the possibilities is a root. We leave it to the student to verify this. Thus there are no rational roots.

The polynomial of Example 3 has no rational roots. We can approximate the irrational roots of such a polynomial by graphing it and determining the x-intercepts.

Rational Coefficients

Suppose some (or all) of the coefficients of a polynomial are rational, but not integers. After multiplying on both sides by the LCM of the denominators, we can then find the rational roots.

Example 4 Let $P(x) = \frac{1}{12}x^3 - \frac{1}{12}x^2 - \frac{2}{3}x + 1$. Find the rational roots of $P(x)$.

The LCM of the denominators is 12. When we multiply on both sides by 12, we get

$$12P(x) = x^3 - x^2 - 8x + 12.$$

This equation is equivalent to the first, and all coefficients on the right are integers. Thus any root of $12P(x)$ is a root of $P(x)$. We leave it to the student to verify that 2 and -3 are the rational roots, in fact the only roots, of $P(x)$.

EXERCISE SET 13.4

Use Theorem 8 to list all *possible* rational roots.

1. $x^5 - 3x^2 + 1$
2. $x^7 + 37x^5 - 6x^2 + 12$
3. $15x^6 + 47x^2 + 2$
4. $10x^{25} + 3x^{17} - 35x + 6$

Find the rational roots, if they exist, of each polynomial or equation. If possible, find the other roots. Then write the equation or polynomial in factored form.

5. $x^3 + 3x^2 - 2x - 6$
6. $x^3 - x^2 - 3x + 3 = 0$
7. $5x^4 - 4x^3 + 19x^2 - 16x - 4 = 0$
8. $3x^4 - 4x^3 + x^2 + 6x - 2$
9. $x^4 - 3x^3 - 20x^2 - 24x - 8$
10. $x^4 + 5x^3 - 27x^2 + 31x - 10$
11. $x^3 + 3x^2 - x - 3 = 0$
12. $x^3 + 5x^2 - x - 5$
13. $x^3 + 8$
14. $x^3 - 8 = 0$
15. $\frac{1}{3}x^3 - \frac{1}{2}x^2 - \frac{1}{6}x + \frac{1}{6}$
16. $\frac{2}{3}x^3 - \frac{1}{2}x^2 + \frac{2}{3}x - \frac{1}{2}$

Find only the rational roots.

17. $x^4 + 32$
18. $x^6 + 8 = 0$
19. $x^3 - x^2 - 4x + 3 = 0$
20. $2x^3 + 3x^2 + 2x + 3$
21. $x^4 + 2x^3 + 2x^2 - 4x - 8 = 0$
22. $x^4 + 6x^3 + 17x^2 + 36x + 66 = 0$
23. $x^5 - 5x^4 + 5x^3 + 15x^2 - 36x + 20$
24. $x^5 - 3x^4 - 3x^3 + 9x^2 - 4x + 12$

☆

25. The volume of a cube is 64 cm³. Find the length of a side. (*Hint:* Solve $x^3 - 64 = 0$.)

26. The volume of a cube is 125 cm³. Find the length of a side.

27. An open box of volume 48 cm³ can be made from a piece of tin 10 cm on a side by cutting a square from each corner and folding up the edges. What is the length of a side of the squares?

28. An open box of volume 500 cm³ can be made from a piece of tin 20 cm on a side by cutting a square from each corner and folding up the edges. What is the length of a side of the squares?

29. Prove that if n is an even positive integer and b is positive, then $x^n + b$ has no rational root.

30. Prove that if n is an even positive integer, then $-3x^n - 8$ has no rational root.

31. Show that $\sqrt{5}$ is irrational by considering the equation $x^2 - 5 = 0$.

32. Generalize the result of Exercise 31 to find which integers have rational square roots.

13.5 FURTHER HELPS IN FINDING ROOTS

Descartes' Rule of Signs

A rule that helps determine the number of positive real roots of a polynomial is due to Descartes.* To use the rule, we must have the polynomial arranged in descending or ascending order, with no zero terms written in. Then we determine the number of *variations of sign*, that is, the number of times, in going through the polynomial, that successive coefficients are of different sign.

Example 1 Determine the number of variations of sign in the polynomial $2x^6 - 3x^2 + x + 4$.

The number of variations of sign is two.

We now state Descartes' rule, without proof.

THEOREM 9

Descartes' rule of signs. The number of positive real roots of a polynomial with real coefficients is either

1. the same as the number of its variations of sign, or
2. less than the number of its variations of sign by a positive even integer.

A root of multiplicity m must be counted m times.

---

*Rene Descartes (1596-1650) was a French mathematician and philosopher.

Examples In each case, what does Descartes' rule of signs tell you about the number of positive real roots?

2. $2x^5 - 5x^2 + 3x + 6$

The number of variations of sign is two. Therefore the number of positive real roots is either 2 or is less than 2 by 2, 4, 6, etc. Thus the number of positive roots is either 2 or 0, a negative number of roots having no meaning.

3. $5x^4 - 3x^3 + 7x^2 - 12x + 4$

There are four variations of sign. Thus the number of positive real roots is either

$$4 \text{ or } 4 - 2 \text{ or } 4 - 4.$$

That is, the number of roots is 4, 2, or 0.

4. $6x^5 - 2x - 5 = 0$

The number of variations of sign is 1. Therefore there is exactly one positive real root.

Negative Roots

Descartes' rule can also be used to help determine the number of negative roots of a polynomial. Consider the graphs (see Figs. 1 and 2) of a polynomial equation $y = P(x)$ and its reflection across the y-axis, that is, $y = P(-x)$. The points at which the graphs cross the x-axis are the roots of the polynomials. From the graphs we see that the number of positive roots of $P(-x)$ is the same as the number of negative roots of $P(x)$.

Figure 1

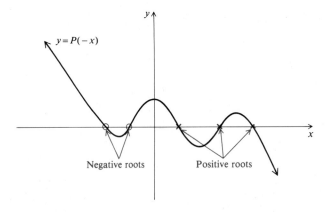

Figure 2

THEOREM 10

Corollary to Descartes' rule of signs. The number of negative real roots of a polynomial $P(x)$ with real coefficients is either

1. the number of variations of sign of $P(-x)$, or
2. less than the number of variations of sign of $P(-x)$ by a positive even integer.

To apply this rule, we construct $P(-x)$ by replacing x by $-x$ wherever it occurs and then count the variations of sign.

Examples In each case, what does Descartes' rule of signs tell you about the number of negative real roots?

5. $5x^4 + 3x^3 + 7x^2 + 12x + 4$

$$P(-x) = 5(-x)^4 + 3(-x)^3 + 7(-x)^2 + 12(-x) + 4 \quad \text{Replacing x by } -x$$
$$= 5x^4 - 3x^3 + 7x^2 - 12x + 4 \quad \text{Simplifying*}$$

There are four variations of sign, so the number of negative roots is either 4 or 2 or 0.

6. $2x^5 - 5x^2 + 3x + 6$

$$P(-x) = 2(-x)^5 - 5(-x)^2 + 3(-x) + 6 \quad \text{Replacing x by } -x$$
$$= -2x^5 - 5x^2 - 3x + 6 \quad \text{Simplifying}$$

There is one variation of sign, so there is exactly one negative root.

Upper Bounds on Roots

Recall that when a polynomial $P(x)$ is divided by $x - a$ we obtain a quotient $Q(x)$ and a remainder R, related as follows:

$$P(x) = (x - a) \cdot Q(x) + R.$$

---

*It is simple to construct $P(-x)$ mechanically. If $P(x)$ is in descending order with no missing terms, we change the sign of every second term beginning with the second term from the right.

Change signs.

Let us consider the case in which a is positive, R is positive or zero, and all the coefficients of $Q(x)$ are nonnegative. Then $P(b)$ will be positive for any positive number b that is greater than a. We show this as follows.

$$P(b) = \underbrace{(b-a)}_{\text{positive}} \cdot \underbrace{Q(b)}_{\text{positive}} + \underbrace{R}_{\text{nonnegative}}$$

$b - a > 0$ because we assumed that $b > a$;

$Q(b) > 0$ because all coefficients are nonnegative and the number b we are substituting is positive. (We know that $Q(x)$ is not the zero polynomial, so it has some nonzero coefficients.)

R is assumed to be nonnegative.

This shows that there can be no root of $P(x)$ greater than the positive number a. If there were such a root r it would make $P(r) = 0$ and this cannot occur, because all values of $P(x)$ are positive when x is greater than a. In terms of synthetic division, this tells us the following.

If this number is positive $a \mid$ □ □ □ □ □

□ □ □ □

□ □ □ □ R

and all of the numbers in the bottom row are nonnegative,

then there is no root greater than a. In other words, the number a is an *upper bound* to all the roots of $P(x)$. Of course, if the number R is 0, we know that a is actually a root, as well as an upper bound.

We state this result as a theorem.

THEOREM 11

If when a polynomial is divided by $x - a$, where a is positive, the remainder and all coefficients of the quotient are nonnegative, the number a is an upper bound to the roots of the polynomial. (All of the roots are less than or equal to a.)

Example 7 Determine an upper bound to the roots of $3x^4 - 11x^3 + 10x - 4$.

We choose some *positive* number a and divide by $x - a$, using synthetic division. Let's try 1 for a.

$$\begin{array}{r|rrrrr} 1 & 3 & -11 & 0 & 10 & -4 \\ & & 3 & -8 & -8 & 2 \\ \hline & 3 & -8 & -8 & 2 & -2 \end{array}$$

Since some of the coefficients are negative the theorem does not guarantee that 1 is an upper bound.

We choose a larger number, this time 4.

$$\begin{array}{r|rrr|rr} 4 & 3 & -11 & 0 & 10 & -4 \\ & & 12 \leftarrow & 4 & 16 & 104 \\ \hline & 3 & 1 & 4 & 26 & 100 \end{array}$$ One reason for choosing 4 is that we know *this* number must be at least 11 in order for the addition to give a nonnegative result.

Since there are no negative numbers in the bottom row, 4 is an upper bound.

CAUTION:

Remember, in applying Theorem 11 the number a must be a *positive* number.

Lower Bounds

A corollary of Theorem 11 can be used to find lower bounds of roots of a polynomial. Consider the graph shown in Fig. 3. Here we have an upper bound, a, to the roots of the polynomial $P(-x)$. Now let us consider the reflection across the y-axis, $y = P(x)$, shown in Fig. 4. It is easy to see that the negative number $-a$ is a lower bound to the roots of $P(x)$.

Figure 3

Figure 4

THEOREM 12

The number $-a$, where a is positive, is a lower bound to the roots of the polynomial $P(x)$, if when $P(-x)$ is divided by $x - a$ the remainder and all coefficients of the quotient are nonnegative.

Example 8 Determine a lower bound to the roots of $P(x)$, where $P(x) = 3x^4 + 11x^3 + 10x - 4$.

We first construct $P(-x)$ by replacing x by $-x$:

$$3(-x)^4 + 11(-x)^3 + 10(-x) - 4$$
$$3x^4 - 11x^3 - 10x - 4. \quad \text{Simplifying}$$

Next, we use synthetic division, dividing by $x - 4$:

$$\begin{array}{r|rrrrr} 4 & 3 & -11 & 0 & -10 & -4 \\ & & 12 & 4 & 16 & 24 \\ \hline & 3 & 1 & 4 & 6 & 20. \end{array}$$

Since none of the numbers in the bottom row is negative, 4 is an upper bound to the roots of $P(-x)$. Thus -4 is a lower bound to the roots of $3x^4 + 11x^3 + 10x - 4$, which is $P(x)$.

Example 9 Determine a lower bound to the roots of $P(x)$, where $P(x) = 5x^3 - 12x^2 + 2x + 3$.

We construct $P(-x)$ by replacing x by $-x$ (or by simply changing the sign of every other term, beginning at the second term from the right):

$$P(-x) = 5(-x)^3 - 12(-x)^2 + 2(-x) + 3$$
$$= -5x^3 - 12x^2 - 2x + 3.$$

We try synthetic division.

$$\begin{array}{r|rrrr} a & -5 & -12 & -2 & 3 \\ & & b & & \\ \hline & -5 & c & & \end{array}$$

Recall that the number a must be positive in this procedure. It is clear that no positive number a will give us a positive number in position b. Therefore it will be impossible to get a nonnegative number in position c.

In a case like this we will use $-P(-x)$ instead of $P(-x)$, since the two polynomials have the same roots. To change to $-P(-x)$ all we need do is change the sign of every coefficient in $P(-x)$. Then we proceed as before.

$$
\begin{array}{r|rrrr}
1 & 5 & 12 & 2 & -3 \\
  &   & 5 & 17 & 19 \\
\hline
  & 5 & 17 & 19 & 16
\end{array}
$$

The number -1 is a lower bound.

In determining bounds for roots, if the leading coefficient of $P(x)$ is negative, we simply convert to $-P(x)$ by changing the sign of every term. Then we proceed as before.

A Shortcut

On p. 480 you were cautioned to use only positive numbers when seeking upper bounds. Likewise, when looking for lower bounds we have used only positive numbers, in accordance with Theorem 12. In looking for lower bounds there is a way to use negative numbers in the synthetic division that may save a small amount of time. We illustrate, using the polynomial of Example 8. The following is what we have already done.

$$
\begin{array}{r|rrrrr}
4 & 3 & -11 & 0 & -10 & -4 \\
  &   & 12 & 4 & 16 & 24 \\
\hline
  & 3 & 1 & 4 & 6 & 20
\end{array}
\leftarrow \text{This is } P(-x).
$$

Now let us use -4 with $P(x)$ and compare.

$$
\begin{array}{r|rrrrr}
-4 & 3 & 11 & 0 & 10 & -4 \\
   &   & -12 & 4 & -16 & 24 \\
\hline
   & 3 & -1 & 4 & -6 & 20
\end{array}
\leftarrow \text{This is } P(x).
$$

Note that the only difference is a change of sign in the indicated columns. This illustrates the following corollary to Theorem 12.

THEOREM 13

> If when a polynomial is divided by $x - a$, where a is negative, a will be a lower bound to the roots if in the result of the synthetic division the first number is nonnegative, the second nonpositive, the third nonnegative, the fourth nonpositive, and so on.

Example 10 Determine a lower bound to the roots of $P(x)$, where $P(x) = 5x^3 - 12x^2 + 2x + 3$.

We try -2:

$$
\begin{array}{r|rrrr}
-2 & 5 & -12 & 2 & 3 \\
   &   & -10 & 44 & -92 \\
\hline
   & 5 & -22 & 46 & -89.
\end{array}
$$

Since the odd-numbered coefficients (from left to right) are nonnegative and the even-numbered ones are nonpositive, -2 is a lower bound to the roots.

Example 11 What does Descartes' rule of signs tell you about the roots of $P(x)$? Find upper and lower bounds to the real roots.

$$P(x) = 4x^4 + 3x^3 + x - 1$$

a) There is one variation of sign, so there is just one real positive root.
b) $P(-x) = 4(-x)^4 + 3(-x)^3 + (-x) - 1$
   $ = 4x^4 \phantom{{}+{}} - 3x^3 \phantom{{}+{}} - x \phantom{{}+{}} - 1$

There is just one variation of sign, so there is just one negative root.

c) Since the equation is of degree four it has four roots. Therefore there are two nonreal roots.
d) We look for an upper bound.

$$
\begin{array}{r|rrrrr}
1 & 4 & 3 & 0 & 1 & -1 \\
  &   & 4 & 7 & 7 & 8 \\
\hline
  & 4 & 7 & 7 & 8 & 7
\end{array}
$$

The number 1 is an upper bound. Let's try $\frac{1}{4}$.

$$
\begin{array}{r|rrrrr}
\frac{1}{4} & 4 & 3 & 0 & 1 & -1 \\
            &   & 1 & 1 & \frac{1}{4} & \frac{5}{16} \\
\hline
            & 4 & 4 & 1 & \frac{5}{4} & -\frac{11}{16}
\end{array}
$$

We do not know whether or not $\frac{1}{4}$ is an upper bound,* but we look no further.

e) We look for a lower bound, using Theorem 13.

$$
\begin{array}{r|rrrrr}
-1 & 4 & 3 & 0 & 1 & -1 \\
   &   & -4 & 1 & -1 & 0 \\
\hline
   & 4 & -1 & 1 & 0 & -1
\end{array}
$$

---

*The converse of Theorem 11 is not true. That is, if the positive number a is an upper bound to the roots, it is not necessarily true that when we divide by $x - a$ all the coefficients of the quotient and the remainder will be nonnegative. Thus when we divide and find negative numbers in the bottom row, this does not mean that the number a is *not* an upper bound. We simply do not know whether it is or not.

We do not know whether or not -1 is a lower bound.

$$
\begin{array}{r|rrrrr}
-2 & 4 & 3 & 0 & 1 & -1 \\
   &   & -8 & 10 & -20 & 38 \\
\hline
   & 4 & -5 & 10 & -19 & 37
\end{array}
$$

The number -2 is a lower bound and not a root (because $R \neq 0$).

To summarize, we have learned that $P(x)$ has two real roots, one between 0 and 1 and one between -2 and 0. It also has two nonreal roots.

From the information obtained in Example 11, we can actually say a little more. In (d) we found that $P(1) = 7$ (a positive number) and $P(\frac{1}{4}) = -\frac{11}{16}$ (a negative number). Thus $P(x)$ must be 0 for some number between $\frac{1}{4}$ and 1. One of the roots is therefore between $\frac{1}{4}$ and 1.

EXERCISE SET 13.5

What does Descartes' rule of signs tell you about the number of positive real roots?

1. $3x^5 - 2x^2 + x - 1$ **2.** $5x^6 - 3x^3 + x^2 - x$
3. $6x^7 + 2x^2 + 5x + 4 = 0$ **4.** $-3x^5 - 7x^3 - 4x - 5 = 0$
5. $3p^{18} + 2p^4 - 5p^2 + p + 3$ **6.** $5t^{12} - 7t^4 + 3t^2 + t + 1$

What does Descartes' rule of signs tell you about the number of negative real roots?

7. $3x^5 - 2x^2 + x - 1$ **8.** $5x^6 - 3x^3 + x^2 - x$
9. $6x^7 + 2x^2 + 5x + 4 = 0$ **10.** $-3x^5 - 7x^3 - 4x - 5 = 0$
11. $3p^{18} + 2p^3 - 5p^2 + p + 3$ **12.** $5t^{11} - 7t^4 + 3t^2 + t + 1$

Determine an upper bound to the roots.

13. $3x^4 - 15x^2 + 2x - 3$ **14.** $4x^4 - 14x^2 + 4x - 2$
15. $6x^3 - 17x^2 - 3x - 1$ **16.** $5x^3 - 15x^2 + 5x - 4$

Determine a lower bound to the roots.

17. $3x^4 - 15x^3 + 2x - 3$ **18.** $4x^4 - 17x^3 + 3x - 2$
19. $6x^3 + 15x^2 + 3x - 1$ **20.** $6x^3 + 12x^2 + 5x - 3$

What does Descartes' rule of signs tell you about the roots of the polynomial or equation? Find upper and lower bounds to the real roots.

21. $x^4 - 2x^2 + 12x - 8$ **22.** $x^4 - 6x^2 + 20x - 24$
23. $x^4 - 2x^2 - 8 = 0$ **24.** $3x^4 - 5x^2 - 4 = 0$
25. $x^4 - 9x^2 - 6x + 4$ **26.** $x^4 - 21x^2 + 4x + 6$
27. $x^4 + 3x^2 + 2 = 0$ **28.** $x^4 + 5x^2 + 6 = 0$

29. Prove that for n a positive even integer, $x^n - 1$ has only two real roots.

31. Show that $x^4 + ax^2 + bx - c$, where a, b, and c are positive, has just two nonreal roots.

30. Prove that for n an odd positive integer, $x^n - 1$ has only one real root.

32. In the three-body problem that arises in astronomy the following equation occurs:

$r^5 + (3 - k)r^4 + (3 - 2k)r^3 - kr^2 - 2kr - k = 0$,

where $0 < k < 1$. Show that this equation has just one positive real solution.

13.6 GRAPHS OF POLYNOMIAL FUNCTIONS

Graphs of first-degree polynomial functions are lines; graphs of second-degree, or quadratic, functions are parabolas. We now consider polynomials of higher degree. We begin with some general principles to keep in mind while graphing. We consider only polynomials with real coefficients.

1. Every polynomial function is a continuous function, whose domain is the set of all real numbers. The graph of any function must pass the vertical line test, so no polynomial function can have a graph that looks like the one in Fig. 5.

2. Unless a polynomial function is linear, no part of its graph is straight. A common error is to make a graph look like the black one in Fig. 6, rather than the colored one.

3. A polynomial of degree n cannot have more than n real roots. This means that the graph cannot cross the x-axis more than n times, so we

Figure 5

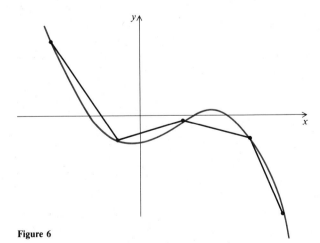

Figure 6

know something about how the curve "wiggles." Third-degree, or *cubic,* functions have graphs like those shown in Fig. 7.

Figure 7

In Fig. 7(a) the graph crosses the axis three times, so there are three real roots. In (b) and (c) there is only one x-intercept, so there is only one real root in each case. The graph of a cubic cannot look like the one in Fig. 8 because there would be a possibility that it might cross the x-axis more than three times.

Graphs of fourth-degree, or *quartic,* polynomials look like those shown in Fig. 9. In (a) there are four real roots, in (b) there is one, and in (c) there are two.

Figure 8

Figure 9

Multiple roots occur at points like those shown in Fig. 10.

Figure 10

When a graph looks like the one in Fig. 11, missing the x-axis at one of its opportunities, a pair of nonreal roots occurs. There is no easy way to find them from the graph.

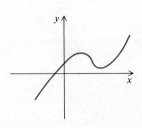

Figure 11

4. The leading term of a polynomial tells a lot about how the graph looks for values of x with large absolute value. This is so because for such values of x the contributions of the other terms are relatively minor. First, let us suppose that the coefficient is positive. Then for large positive values of x, the function value will be positive and increasing. Thus we know that as we move far to the right the graph looks like the one in Fig. 12(a).

If the exponent of the leading term is even, the same thing happens as we move far to the left, as in Fig. 12(b). If the exponent is odd, then the function values are negative and they decrease as we move far to the left, as in Fig. 12(c). If the leading coefficient is negative, the graphs will of course be reflected across the x-axis.

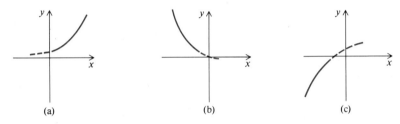

(a)   (b)   (c)

Figure 12

To graph polynomials, keep in mind previously established results and proceed as follows.

TO GRAPH POLYNOMIALS

a) Look at the degree of the polynomial and its leading coefficient. This gives a lot of information about the general shape of the graph.

b) Look for symmetries, as covered in Chapter 3.

c) Make a table of values using synthetic division.

d) Find the y-intercept and as many x-intercepts as possible (the latter are roots of the polynomial). In doing this, recall the theorems about roots, including Descartes' rule of signs.

e) Plot the points and connect them appropriately.

Example 1 Graph: $P(x) = 2x^3 - x + 2$.

a) This polynomial is of degree 3 with leading coefficient positive. Thus as we move far to the right, function values will increase beyond bound. As we move far to the left, they will be negative and decrease beyond bound. The curve will have the general shape of a cubic, one of those shown in Fig. 7.

b) The function is not odd or even. However, $2x^3 - x$ is an odd function, with the origin as a point of symmetry. The function $P(x)$ is a translation of this upward two units, hence the point $(0, 2)$ is a point of symmetry.

c) We make a table of values. The first entry is merely the y-intercept. The other rows contain the bottom line in synthetic division (which is really all we need to write ordinarily).

	2	0	−1	2	(x, y)	
0				2	(0, 2)	The y-intercept
1	2	2	1	3	(1, 3)	Since all numbers are positive, 1 is an upper bound to the roots.
2	2	4	7	16	(2, 16)	
−1	2	−2	1	1	(−1, 1)	
−2	2	−4	7	−12	(−2, −12)	Since signs alternate, −2 is a lower bound to the roots.

d) Because we know the curve is symmetric with respect to the point $(0, 2)$, we do not really need to calculate $P(x)$ for negative values of x. We can also see from this table that the graph will cross the x-axis between -1 and -2. This is because $P(-1)$ is positive (and the graph there is above the axis), whereas $P(-2)$ is negative (and the graph there is below the axis). Somewhere between -1 and -2, then, $P(x)$ must be 0. Thus there is a real root of $P(x)$ between -1 and -2.

By Descartes' rule of signs, we know that there are either 2 or 0 positive real roots. From our table of values, we see that $P(x)$ is positive and increasing rapidly. These facts would lead us to *suspect* that there are no positive roots. If there are any, they must be between 0 and 1 because 1 is an upper bound.

Now $P(-x) = -2x^3 + x + 2$, so Descartes' rule tells us there is just one negative root. Thus the one we have already found is the only one. We also know that if there are no positive roots, there will be two nonreal ones, and they will be conjugates of each other.

e) We now plot points and connect them appropriately.* See Fig. 13. In this example, it is not easy to see how to draw the curve. It might

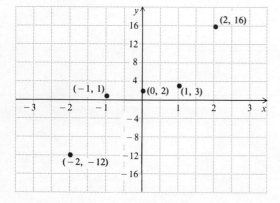

Figure 13

---

*In this graph we are using different scales on the x-axis and the y-axis. We are reluctant to do so, but it is sometimes necessary.

look like any of those shown in Fig. 14 so far as we know from the points plotted. We therefore need to plot more points. Because of symmetry, it will be sufficient to consider positive values of x.

	2	0	−1	2	(x, y)
0.1	2	0.2	0.98	1.9	(0.1, 1.9)
0.3	2	0.6	−0.82	1.75	(0.3, 1.75)
0.5	2	1.0	−0.5	1.75	(0.5, 1.75)
0.7	2	1.4	−0.02	1.99	(0.7, 1.99)
0.9	2	1.8	0.62	2.56	(0.9, 2.56)

Figure 14

We now have enough points to determine the shape of the graph (see Fig. 15). We can now draw the graph confidently (see Fig. 16).

Figure 15

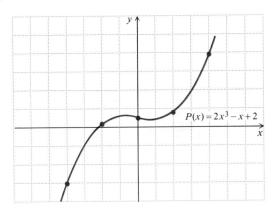

$P(x) = 2x^3 - x + 2$

Figure 16

Solving Equations

Whenever we find the roots, or zeros, of a function $P(x)$ we have solved the equation $P(x) = 0$. One means of finding approximate solutions of polynomial equations is by graphing. We graph the function $y = P(x)$ and note where the graph crosses the x-axis. In Example 1, there is a real solution of the equation $2x^3 - x + 2 = 0$ at about -1.1.

We can approximate roots more precisely by doing further calculation.

Example 2 Find a better approximation of the solution of $2x^3 - x + 2 = 0$, which is near -1.1.

We know that $P(-1) = 1$ and $P(-2) = -12$, so there must be a root between -1 and -2. To find a better approximation we use synthetic division.

	2	0	-1	2	
-1.1	2	-2.2	1.42	0.44	Since $P(-1.1) > 0$, we have not established a root between -1 and -1.1.
-1.2	2	-2.4	1.88	-0.26	Since $P(-1.2) < 0$ and $P(-1.1) > 0$, there is a root between -1.1 and -1.2. We'll try -1.15, halfway between.
-1.15	2	-2.30	1.65	0.11	
-1.16	2	-2.30	1.69	0.04	
-1.17	2	-2.34	1.74	-0.03	We have a root between -1.16 and -1.17.

In this manner the approximation can be made as close as we please.

EXERCISE SET 13.6

Graph.

1. $P(x) = x^3 - 3x^2 - 2x - 6$
2. $P(x) = x^3 + 4x^2 - 3x - 12$
3. $P(x) = 2x^4 + x^3 - 7x^2 - x + 6$
4. $P(x) = 3x^4 + 5x^3 + 5x^2 - 5x - 6$
5. $P(x) = x^5 - 2x^4 - x^3 + 2x^2$
6. $P(x) = x^5 + 4x^4 - 5x^3 - 14x^2 - 8x$

Graph the corresponding polynomial functions in order to find approximate solutions of the equations.

7. $x^3 - 3x - 2 = 0$
8. $x^3 - 3x^2 + 3 = 0$
9. $x^3 - 3x - 4 = 0$
10. $x^3 - 3x^2 + 5 = 0$

11. $x^4 + x^2 + 1 = 0$

12. $x^4 + 2x^2 + 2 = 0$

13. $x^4 - 6x^2 + 8 = 0$

14. $x^4 - 4x^2 + 2 = 0$

15. $x^5 + x^4 - x^3 - x^2 - 2x - 2 = 0$

16. $x^5 - 2x^4 - 2x^3 + 4x^2 - 3x + 6 = 0$

17. ▦ The following equation has a solution between 0 and 1. Approximate it to hundredths.

$$2x^5 + 2x^3 - x^2 - 1 = 0$$

18. ▦ The following equation has a solution between 1 and 2. Approximate it to hundredths.

$$x^4 - 2x^3 - 3x^2 + 4x + 2$$

In each of the following, graph and then approximate the irrational roots to hundredths.

19. ▦ $P(x) = x^3 - 2x^2 - x + 4$

20. ▦ $P(x) = x^3 - 4x^2 + x + 3$

21. The equation $\log x = \sin x$ has a solution between 2 and 3. Find it, to the nearest tenth.

22. ▦ Graph: $P(x) = 5.8x^4 - 2.3x^2 - 6.1$.

23. *Nested evaluation.* A procedure for evaluating a polynomial is as follows. Given a polynomial, such as $3x^4 - 5x^3 + 4x^2 - 5$, successively factor out x, as shown:

$$x(x(x(3x - 5) + 4)) - 5$$

Given a value for x, substitute it in the innermost parentheses and work your way out, at each step multiplying, then adding or subtracting. Show that this process is identical to synthetic division.

13.7 RATIONAL FUNCTIONS

A *rational function* is a function definable as the quotient of two polynomials. Here are some examples:

$$y = \frac{x^2 + 3x - 5}{x + 4}, \quad y = \frac{5}{x^2 + 3}, \quad y = \frac{3x^5 - 5x + 2}{4}.$$

DEFINITION

A *rational function* is a function that can be defined as $y = P(x)/Q(x)$, where $P(x)$ and $Q(x)$ are polynomials having no common factors other than 1 and -1, and with $Q(x)$ not the zero polynomial.

Polynomial functions are themselves special kinds of rational functions, since $Q(x)$ can be the polynomial 1. Here we are interested in rational functions in which the denominator is not a constant. We begin with the simplest such function.

Figure 17

Figure 18

Figure 19

Example 1 Graph the function $y = 1/x$.

a) Note that the domain of this function consists of all real numbers except 0.

b) For nonzero x or y, the above equation is equivalent to $xy = 1$. Now it is easy to see that the graph is symmetric with respect to the line $y = x$ (because interchanging x and y produces an equivalent equation).

c) The graph is also symmetric with respect to the origin (because replacing x with $-x$ and y with $-y$ produces an equivalent equation).

d) Now we find some values, keeping in mind the two symmetries.

x	1	2	3	4	5
y	1	$\frac{1}{2}$	$\frac{1}{3}$	$\frac{1}{4}$	$\frac{1}{5}$

e) We plot these points. We use one symmetry to get other points in the first quadrant. We use the other symmetry to get the points in the third quadrant. (Fig. 17). The points indicated by • have been obtained from the table. Those marked • have been obtained by reflection across the line $y = x$. Following this, the points marked ∘ have been obtained by reflection across the origin.

Asymptotes

Note that this curve does not touch either axis, but comes very close. As $|x|$ becomes very large the curve comes very near to the x-axis. In fact, we can find points as close to the x-axis as we please by choosing x large enough. We say that the curve approaches the x-axis *asymptotically,* and we say that the x-axis is an *asymptote* to the curve. The y-axis is also an asymptote to this curve.

Using the ideas of transformations we can easily graph certain variations of the above function.

Example 2 $y = -1/x$ is a reflection across the x-axis (or the y-axis; the result is the same). See Fig. 18.

Example 3 $y = 1/(x - 2)$ is a translation two units to the right (Fig. 19). Note that the line $x = 2$ is a vertical asymptote.

Example 4 Graph the function $y = 1/x^2$.

a) Note that this function is defined for all x except 0. Therefore the line $x = 0$ is an asymptote.

b) As $|x|$ gets very large, y approaches 0. Therefore the x-axis is also an asymptote for both positive and negative values of x.

c) This function is even. Therefore it is symmetric with respect to the y-axis.

d) All function values are positive. Therefore the entire graph is above the x-axis.

e) With this much information, we can already sketch a rough graph of the function. However, a table of values will help.

x	1	2	3	4	$\frac{1}{2}$	$\frac{1}{3}$
y	1	$\frac{1}{4}$	$\frac{1}{9}$	$\frac{1}{16}$	4	9

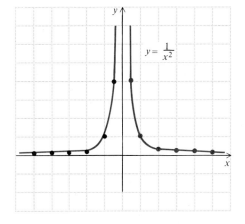

Figure 20

The graph is shown in Fig. 20. Points marked • have been obtained from the table. Points marked • have been obtained by reflection across the y-axis.

Occurrence of Asymptotes

It is important in graphing rational functions to determine where the asymptotes, if any, occur. Vertical asymptotes are easy to locate when a denominator is factored. The x-values that make a denominator 0 but do not make the numerator 0 are those of the vertical asymptotes.

Examples Determine the vertical asymptotes.

5. $y = \dfrac{3x - 2}{x(x - 5)(x + 3)}$.

The vertical asymptotes are the lines $x = 0$, $x = 5$, and $x = -3$.

6. $y = \dfrac{x - 2}{x^3 - x}$.

We factor the denominator:
$$x^3 - x = x(x + 1)(x - 1).$$

The vertical asymptotes are $x = 0$, $x = -1$, and $x = 1$.

Horizontal asymptotes occur when the degree of the numerator is the same as or less than that of the denominator. Let us first consider a function for which the degree of the numerator is less than that of the denominator:

$$y = \frac{2x + 3}{x^3 - 2x^2 + 4}.$$

We shall multiply by $\dfrac{1/x^3}{1/x^3}$, to obtain

$$y = \frac{\dfrac{2}{x^2} + \dfrac{3}{x^3}}{1 - \dfrac{2}{x} + \dfrac{4}{x^3}}.$$

Let us now consider what happens to the function values as $|x|$ becomes very large. Each expression with x in its denominator takes on smaller and smaller values, approaching 0. Thus the numerator approaches 0 and the denominator approaches 1; hence the entire expression takes on values closer and closer to 0. Therefore the x-axis is an asymptote. Whenever the degree of a numerator is less than that of the denominator, the x-axis will be an asymptote.

Next, we consider a function for which the numerator and denominator have the same degree:

$$y = \frac{3x^2 + 2x - 4}{2x^2 - x + 1} = \frac{3x^2 + 2x - 4}{2x^2 - x + 1} \cdot \frac{\dfrac{1}{x^2}}{\dfrac{1}{x^2}}$$

$$= \frac{3 + \dfrac{2}{x} - \dfrac{4}{x^2}}{2 - \dfrac{1}{x} + \dfrac{1}{x^2}}.$$

As $|x|$ gets very large the numerator approaches 3 and the denominator approaches 2. Therefore the function values get very close to $\frac{3}{2}$, and thus the line $y = \frac{3}{2}$ is an asymptote. From this example, we can see that the asymptote in such cases can be determined by dividing the leading coefficients of the two polynomials.

Examples Determine the horizontal asymptotes.

7. $y = \dfrac{5x^3 - x^2 + 7}{3x^3 + x - 10}.$

The line $y = \frac{5}{3}$ is an asymptote.

8. $y = \dfrac{-7x^4 - 10x^2 + 1}{11x^4 + x - 2}.$

The line $y = -\frac{7}{11}$ is an asymptote.

There are asymptotes that are neither vertical nor horizontal. They are called *oblique*, and they occur when the degree of the numerator is greater than that of the denominator by 1. To find the asymptote we divide the numerator by the denominator. Consider

$$y = \frac{2x^2 - 3x - 1}{x - 2}.$$

When we divide the numerator by the denominator we obtain a quotient of $2x + 1$ and a remainder of 1. Thus

$$y = 2x + 1 + \frac{1}{x - 2}.$$

Now we can see that when $|x|$ becomes very large, $1/(x - 2)$ approaches 0, and the y-values thus approach $2x + 1$. This means that the graph comes closer and closer to the straight line $y = 2x + 1$.

Example 9 Find and draw the asymptotes of the function

$$y = \frac{2x^2 - 11x - 10}{x - 4}.$$

a) We first note that $x = 4$ is a vertical asymptote.

b) We divide, to obtain $y = 2x - 3 - \frac{22}{x - 4}$.

Thus the line $y = 2x - 3$ is an asymptote (Fig. 21).

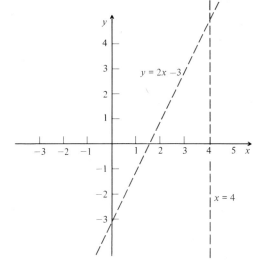

Figure 21

Zeros or Roots

Zeros, or roots, of a rational function occur when the numerator is 0 but the denominator is not 0. The zeros of a function occur at points where the graph crosses the x-axis. Therefore knowing the zeros helps in making a graph. If the numerator can be factored, the zeros are easy to determine.

Example 10 Find the zeros of the function $y = \frac{x^3 - x^2 - 6x}{x^2 - 3x + 2}$.

We factor numerator and denominator:

$$y = \frac{x(x + 2)(x - 3)}{(x - 1)(x - 2)}.$$

The values making the numerator zero are 0, -2, and 3. Since none of these makes the denominator zero, they are the zeros of the function.

The following is an outline of a procedure to be followed in graphing rational functions.

TO GRAPH A RATIONAL FUNCTION

a) Determine any symmetries.

b) Determine any horizontal or oblique asymptotes and sketch them.

c) Factor the denominator and the numerator.
   i) Determine any vertical asymptotes and sketch them.
   ii) Determine the zeros, if possible, and plot them.

d) Make a table, showing where the function values are positive and where negative.

e) Make a table of values and plot them.

f) Sketch the curve.

Example 11 Graph: $y = \dfrac{x^3 - x^2 - 6x}{x^2 - 3x + 2}$.

We follow the outline above.

a) The function is neither even nor odd and no symmetries are apparent.

b) The degree of the numerator is one greater than that of the denominator. Dividing numerator by denominator will show that the line $y = x + 2$ is an oblique asymptote.

c) The numerator and denominator are easily factorable. We have

$$y = \frac{x(x + 2)(x - 3)}{(x - 1)(x - 2)}.$$

The zeros are $0, -2$, and 3 and there are vertical asymptotes at $x = 1$ and $x = 2$.

d) The zeros and asymptotes divide the x-axis into intervals in a natural way, pictured as follows. We tabulate signs in these intervals.

INTERVAL	x + 2	x	x − 1	x − 2	x − 3	y
$x < -2$	−	−	−	−	−	−
$-2 < x < 0$	+	−	−	−	−	+
$0 < x < 1$	+	+	−	−	−	−
$1 < x < 2$	+	+	+	−	−	+
$2 < x < 3$	+	+	+	+	−	−
$3 < x$	+	+	+	+	+	+

e) Next we make a table of function values and plot them.

x	$\frac{1}{2}$	$\frac{3}{2}$	$\frac{5}{2}$	4	5	-1	-3	-5
y	$-\frac{25}{6}$	$\frac{63}{2}$	$-\frac{15}{2}$	4	$\frac{35}{6}$	$\frac{2}{3}$	$-\frac{9}{10}$	$-\frac{20}{7}$

f) Using all available information, we draw the graph. See Fig. 22. The lower left-hand part of the graph is shown in Fig. 23. The curve crosses the oblique asymptote at $(-2, 0)$ and then, moving to the left, comes back, close to the asymptote from above.*

Figure 22

Figure 23

Example 12 Graph: $y = \dfrac{1}{x^2 + 1}$.

We follow the outline above.

a) The function is even, so the graph is symmetric with respect to the y-axis.

---

*Mathematical folklore often has it that no curve ever crosses its asymptote. This example shows the folklore to be false.

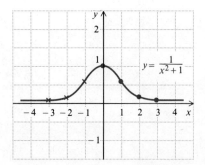

Figure 24

b) The degree of the denominator is greater than that of the numerator. Thus the x-axis is an asymptote.

c) The denominator is not factorable. Neither is the numerator. Therefore there are no vertical asymptotes and there are no zeros.

d) The numerator is a positive constant and the denominator can never be negative, so this function has only positive values. Thus no table of signs is necessary.

e) We tabulate some values.

x	0	1	2	3
y	1	$\frac{1}{2}$	$\frac{1}{5}$	$\frac{1}{10}$

f) We sketch the graph (Fig. 24).

EXERCISE SET 13.7

Graph these functions.

1. $y = \dfrac{1}{x-3}$
2. $y = \dfrac{1}{x-5}$
3. $y = \dfrac{-2}{x-5}$
4. $y = \dfrac{-3}{x-3}$
5. $y = \dfrac{2x+1}{x}$
6. $y = \dfrac{3x-1}{x}$
7. $y = \dfrac{1}{(x-2)^2}$
8. $y = \dfrac{-2}{(x-3)^2}$
9. $y = \dfrac{2}{x^2}$
10. $y = \dfrac{1}{3x^2}$
11. $y = \dfrac{1}{x^2+3}$
12. $y = \dfrac{-1}{x^2+2}$
13. $y = \dfrac{x-1}{x+2}$
14. $y = \dfrac{x-2}{x+1}$
15. $y = \dfrac{3x}{x^2+5x+4}$
16. $y = \dfrac{x+3}{2x^2-5x-3}$
17. $y = \dfrac{x^2-4}{x-1}$
18. $y = \dfrac{x^2-9}{x+1}$
19. $y = \dfrac{x^2+x-2}{2x^2+1}$
20. $y = \dfrac{x^2-2x-3}{3x^2+2}$
21. $y = \dfrac{x-1}{x^2-2x-3}$
22. $y = \dfrac{x+2}{x^2+2x-15}$
23. $y = \dfrac{x+2}{(x-1)^3}$
24. $y = \dfrac{x-3}{(x+1)^3}$
25. $y = \dfrac{x^3+1}{x}$
26. $y = \dfrac{x^3-1}{x}$
27. $y = \dfrac{x^3+2x^2-15x}{x^2-5x-14}$
28. $y = \dfrac{x^3+2x^2-3x}{x^2-25}$

☆

29. $y = \dfrac{1}{x^2+3} - 5$
30. $y = \dfrac{-1}{x^2+2} + 4$
31. $y = \dfrac{1}{|x+2|}$
32. $y = \left|\dfrac{1}{x} - 2\right|$
33. $y = \left|\dfrac{1}{x-2} - 3\right|$
34. $y = \dfrac{3x}{|x^2+5x+4|}$

13.8 PARTIAL FRACTIONS

There are situations in which it is helpful to do the reverse of adding fractional expressions, that is, to *decompose* a fractional expression into a sum of several fractional expressions. The following illustrates:

$$\frac{3}{x+2} - \frac{2}{2x-3} = \frac{4x-13}{2x^2+x-6}.$$

Adding is straightforward. The present problem is to take the sum and find the fractions which were added to get it. That procedure turns out to be rather straightforward also, and it depends upon the following theorem, whose proof we omit.

THEOREM 14

> Any rational expression $P(x)/Q(x)$, where the degree of the numerator is less than that of the denominator, can be decomposed into partial fractions. To each linear factor $ax + b$ of $Q(x)$ corresponds a fraction $A/(ax + b)$. To each linear factor $ax + b$ occurring twice in $Q(x)$ there correspond two fractions: $B_1/(ax + b)$ and $B_2/(ax + b)^2$. If $ax + b$ occurs three times there corresponds an additional fraction, $B_3/(ax + b)^3$, and so on. To each quadratic factor $ax^2 + bx + c$ of $Q(x)$ corresponds a fraction $(Ax + B)/(ax^2 + bx + c)$. If the quadratic factor occurs twice there is an additional fraction, $(Cx + D)/(ax^2 + bx + c)^2$, and so on.* The expressions in the numerators $A, B, \ldots$ are constants, hence they do not depend on x.

For our first example, we use the fractional expression in the opening paragraph.

Example 1 Decompose into partial fractions: $\dfrac{4x-13}{2x^2+x-6}$.

We begin by factoring the denominator: $(x + 2)(2x - 3)$. By Theorem 14 we know that there are constants A and B such that

$$\frac{4x-13}{(x+2)(2x-3)} = \frac{A}{x+2} + \frac{B}{2x-3}.$$

To determine A and B, we add on the right:

$$\frac{4x-13}{(x+2)(2x-3)} = \frac{A(2x-3) + B(x+2)}{(x+2)(2x-3)}.$$

---

*This theorem covers all situations, because any polynomial $Q(x)$ with real coefficients has a factorization into linear and quadratic factors.

Next, we equate the numerators:

$$4x - 13 = A(2x - 3) + B(x + 2) = 2Ax - 3A + Bx + 2B \quad (1)$$
$$4x - 13 = (2A + B)x + (2B - 3A). \quad (2)$$

Next, we equate corresponding coefficients of Eq. 2:*

$$4 = 2A + B \quad \text{The coefficients of the x-terms}$$
$$-13 = 2B - 3A. \quad \text{The constant terms}$$

We have two equations in A and B. We solve, to obtain

$$A = 3 \quad \text{and} \quad B = -2.$$

The decomposition is as follows:

$$\frac{3}{x + 2} - \frac{2}{2x - 3}.$$

To check, we would add, to see if we get the original expression.

Example 2 Decompose into partial fractions: $\dfrac{7x^2 - 29x + 24}{(2x - 1)(x - 2)^2}.$

By Theorem 14, the decomposition looks like the following:

$$\frac{A}{2x - 1} + \frac{B}{x - 2} + \frac{C}{(x - 2)^2}.$$

As in Example 1, we add and equate numerators. This gives us

$$7x^2 - 29x + 24 = A(x - 2)^2 + B(2x - 1)(x - 2) + C(2x - 1).$$

This time we use a timesaving approach to determine A, B, and C. Since the equation containing A, B, and C is true for all x, we can substitute any value of x whatever and still have a true equation. We let $2x - 1 = 0$, or $x = \frac{1}{2}$. Then, we get

$$7\left(\frac{1}{2}\right)^2 - 29 \cdot \frac{1}{2} + 24 = A\left(\frac{1}{2} - 2\right)^2 + 0.$$

Solving, we obtain $A = 5$. Next, we let $x - 2 = 0$, or $x = 2$. Substituting, we get

$$7(2)^2 - 29(2) + 24 = 0 + C(2 \cdot 2 - 1).$$

---

*To see that this is valid, note that Eq. 2 is equivalent to $(2A + B - 4)x + (2B - 3A + 13) = 0$. By Theorem 14, A and B do not depend on x. Thus we have a polynomial function $P(x)$ that has the value 0 for all x. Since a polynomial $P(x)$ has no more than n real roots, where n is the degree of $P(x)$, it follows that $P(x)$ is the zero polynomial. This means that every coefficient must be 0.

Solving, we obtain $C = -2$. To find B we now equate the coefficients of x^2:

$$7 = A + 2B.$$

Substituting for A and solving gives us $B = 1$. The decomposition is as follows:

$$\frac{5}{(2x-1)} + \frac{1}{(x-2)} - \frac{2}{(x-2)^2}.$$

Example 3 Decompose into partial fractions: $\dfrac{x^2 - 17x + 35}{(x^2+1)(x-4)}$.

By Theorem 14, the decomposition looks like the following:

$$\frac{Ax+B}{x^2+1} + \frac{C}{x-4}.$$

Adding and equating numerators, we get

$$x^2 - 17x + 35 = (Ax+B)(x-4) + C(x^2+1).$$

Letting $x = 4$, we get

$$4^2 - 17 \cdot 4 + 35 = 0 + C(4^2 + 1).$$

Solving, we obtain $C = -1$. Equating the coefficients of x^2, we get $1 = A + C$. Thus we know that $A = 2$. Equating the constant terms, we get $35 = -4B + C$. This gives us $B = -9$. The decomposition is as follows:

$$\frac{2x-9}{x^2+1} - \frac{1}{x-4}.$$

Theorem 14 refers only to rational expressions $P(x)/Q(x)$ in which the degree of the numerator is less than the degree of the denominator. As we show in the following example, this is not much of a restriction. For an expression not satisfying the condition, we can divide numerator by denominator. The result will be some polynomial plus a fraction $R(x)/Q(x)$, where $R(x)$ is the remainder. The latter fraction meets the conditions of Theorem 14, hence it can be decomposed.

Example 4 Decompose into partial fractions: $\dfrac{6x^3 + 5x^2 - 7}{3x^2 - 2x - 1}$.

Since the degree of the numerator is not less than that of the denominator, we divide:

$$\begin{array}{r} 2x + 3 \\ 3x^2 - 2x - 1 \overline{\smash{\big)}\, 6x^3 + 5x^2 - 7} \\ \underline{6x^3 - 4x^2 - 2x } \\ 9x^2 + 2x \\ \underline{9x^2 - 6x - 3} \\ 8x - 4 \end{array}$$

The original expression is thus equivalent to the following:

$$2x + 3 + \frac{8x - 4}{3x^2 - 2x - 1}.$$

We decompose the fraction. The decomposition is as follows:

$$\frac{8x - 4}{(3x + 1)(x - 1)} = \frac{5}{3x + 1} + \frac{1}{x - 1}.$$

The final result is the following:

$$2x + 3 + \frac{5}{3x + 1} + \frac{1}{x - 1}.$$

EXERCISE SET 13.8

Decompose into partial fractions.

1. $\dfrac{x + 7}{(x - 3)(x + 2)}$

2. $\dfrac{2x}{(x + 1)(x - 1)}$

3. $\dfrac{7x - 1}{6x^2 - 5x + 1}$

4. $\dfrac{13x + 46}{12x^2 - 11x - 15}$

5. $\dfrac{3x^2 - 11x - 26}{(x^2 - 4)(x + 1)}$

6. $\dfrac{5x^2 + 9x - 56}{(x - 4)(x - 2)(x + 1)}$

7. $\dfrac{9}{(x + 2)^2(x - 1)}$

8. $\dfrac{x^2 - x - 4}{(x - 2)^3}$

9. $\dfrac{2x^2 + 3x + 1}{(x^2 - 1)(2x - 1)}$

10. $\dfrac{x^2 - 10x + 13}{(x^2 - 5x + 6)(x - 1)}$

11. $\dfrac{x^4 - 3x^3 - 3x^2 + 10}{(x + 1)^2(x - 3)}$

12. $\dfrac{10x^3 - 15x^2 - 35x}{x^2 - x - 6}$

13. $\dfrac{2x^2 - 11x + 5}{(x - 3)(x^2 + 2x - 5)}$

14. $\dfrac{26x^2 + 208x}{(x^2 + 1)(x + 5)}$

15. $\dfrac{6 + 26x - x^2}{(2x - 1)(x + 2)^2}$

16. $\dfrac{5x^3 + 6x^2 + 5x}{(x^2 - 1)(x + 1)^3}$

17. Decompose into partial fractions.

$$\frac{x}{x^4 - a^4}$$

18. Decompose into partial fractions.

$$\frac{9x^3 - 24x^2 + 48x}{(x - 2)^4(x + 1)}$$

[*Hint:* Let the expression equal

$$\frac{A}{x + 1} + \frac{P(x)}{(x - 2)^4}$$

and find $P(x)$.]

19. Decompose into partial fractions and then graph by addition of ordinates.

$$y = \frac{3x}{x^2 + 5x + 4}$$

20. Decompose into partial fractions and then graph by addition of ordinates.

$$y = \frac{x - 1}{x^2 - 2x - 3}$$

Decompose into partial fractions.

21. $\dfrac{1 + \ln x^2}{(\ln x + 2)(\ln x - 3)^2}$

22. $\dfrac{1}{e^{-x} + 3 + 2e^x}$

23. $\dfrac{\cos^2 \theta - 1 + \sin \theta}{\sin^3 \theta - \sin^2 \theta - 2 \sin \theta}$

24. $\dfrac{\csc^2 \theta}{2 - 4 \cot \theta + \csc^2 \theta}$

CHAPTER 13 REVIEW

1. Find the remainder when $x^4 + 3x^3 + 3x^2 + 3x + 2$ is divided by $x + 2$.

2. Use synthetic division to find the quotient and remainder.

$$(2x^4 - 6x^3 + 7x^2 - 5x + 1) \div (x + 2)$$

3. Find a polynomial of degree 3 with roots 0, 1, and 2.

4. Find a polynomial of lowest degree having roots 1 and -1, and having 2 as a root of multiplicity 2 and -3 as a root of multiplicity 3.

5. Find a complete factorization of $x^3 - 1$.

6. Use synthetic division to find $P(3)$.

$$P(x) = 2x^4 - 3x^3 + x^2 - 3x + 7$$

7. Factor the polynomial $P(x)$. Then solve the equation $P(x) = 0$.

$$P(x) = x^3 + 7x^2 + 7x - 15$$

8. Determine whether $x + 1$ is a factor of

$$x^3 + 6x^2 + x + 30.$$

9. Find k such that $x + 3$ is a factor of $x^3 + kx^2 + kx - 15$.

10. The equation $x^4 - 81 = 0$ has $3i$ for a root. Find the other roots.

11. The equation $x^2 - 8x + c = 0$ has a double root. Find it.

12. A polynomial of degree 4 with rational coefficients has roots $-8 - 7i$ and $10 + \sqrt{5}$. Find the other roots.

13. List all possible rational roots of $2x^6 - 12x^4 + 17x^2 + 12$.

14. What does Descartes' rule of signs tell you about the number of positive real roots of $3x^{12} + 3x^4 - 7x^2 + x + 5$?

15. What does Descartes' rule of signs tell you about the number of negative real roots of $6x^8 - 12x^4 + 5x^2 + x + 2$?

16. Find an upper bound to the roots of $2x^4 - 7x^2 + 2x - 1$.

17. Find a lower bound to the roots of $12x^3 + 24x^2 + 10x - 6$.

18. The equation $x^5 + x^4 - x^3 - x^2 - 2x - 2$ has a root between 1 and 2. Approximate this root to hundredths.

19. Graph $P(x) = x^3 - 3x^2 + 3$. Use the graph to approximate the irrational roots.

20. Graph $y = \dfrac{x^2 + x - 6}{x^2 - x - 20}$.

21. Decompose into partial fractions.

$$\frac{5}{(x+2)^2(x+1)}$$

Equations of Second Degree and Their Graphs

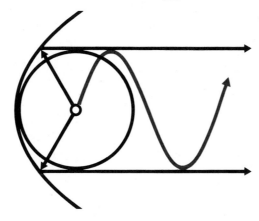

EQUATIONS OF SECOND DEGREE AND THEIR GRAPHS

14.1 CONIC SECTIONS

Cones

Suppose C is a circle with center O and P is a point not in the same plane as C such that the line $\overleftrightarrow{OP}$ is perpendicular to the plane of the circle C (see Fig. 1). The set of points on all lines through P and a point of the circle form a *right circular cone* (or *conical surface*). Any line contained in the surface is called a *surface element*. Note that there are two parts or *nappes* of a cone. Point P is called the *vertex* and line $\overleftrightarrow{OP}$ is called the *axis*.

Figure 1

Conic Sections

The intersection of any plane with a cone is a *conic section*. Some conic sections are shown in Figs. 2 and 3.

In this chapter we study equations of second degree and their graphs. Graphs of most such equations are conic sections.

(a) Ellipse (b) Parabola (c) Hyperbola

Figure 2

Lines

Some second-degree equations have graphs consisting of a line. Some have graphs consisting of two lines. The graphs are conic sections except when the two lines are parallel.

(a) Line

(b) Intersecting lines

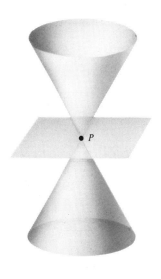

(c) Single point

Figure 3

Example 1 Graph: $3x^2 + 2xy - y^2 = 0$.

We factor and use the principle of zero products:

$$(3x - y)(x + y) = 0$$
$$3x - y = 0 \quad \text{or} \quad x + y = 0$$
$$y = 3x \quad \text{or} \quad y = -x.$$

The graph consists of two intersecting lines (Fig. 4).

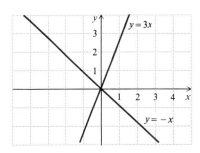

Figure 4

Example 2 Graph: $(y - x)(y - x - 1) = 0$.

$$y - x = 0 \quad \text{or} \quad y - x - 1 = 0 \qquad \text{Using the principle of}$$
$$y = x \quad \text{or} \quad y = x + 1 \qquad \text{zero products}$$

The graph consists of two parallel lines (Fig. 5).

Figure 5

Example 3 Graph: $y^2 + x^2 = 2xy$.

$$y^2 - 2xy + x^2 = 0$$
$$(y - x)(y - x) = 0$$
$$y - x = 0 \quad \text{or} \quad y - x = 0$$
$$y = x \quad \text{or} \quad y = x$$

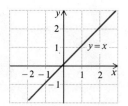

Figure 6

The graph consists of a single line (Fig. 6).

Generally a second-degree equation with 0 on one side and a factorable expression on the other has a graph consisting of one or two lines. A special case is $xy = 0$, in which the graph consists of the coordinate axes.

Single Points or No Points

When a plane intersects only the vertex of a cone, the result is a single point. The following is an equation for such a conic section.

Example 4 Graph: $x^2 + 4y^2 = 0$.

The expression $x^2 + 4y^2$ is not factorable in the real-number system. The only real-number solution of the equation is $(0, 0)$.

Example 5 Graph: $3x^2 + 7y^2 = -2$.

Since squares of numbers are never negative, the left side of the equation can never be negative. The equation has no real-number solutions, hence has no graph.

Figure 7

Circles

Some equations of second degree have graphs that are circles. Circles are defined as follows.

DEFINITION

A *circle* is the locus or set of all points in a plane that are at a fixed distance from a fixed point in that plane.

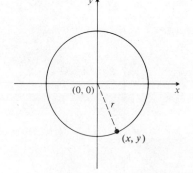

Figure 8

When a plane intersects a cone as in Fig. 7, perpendicular to the axis of the cone, a circle is formed. We shall prove this later.

We first obtain an equation for a circle centered at the origin (Fig. 8). We let r represent the radius. If a point (x, y) is on the circle, then by

the definition of a circle and the distance formula we have
$$r^2 = x^2 + y^2.$$
We have shown that if a point (x, y) is on the circle, then $x^2 + y^2 = r^2$. We now prove the converse, that is, if a point (x, y) satisfies the equation $x^2 + y^2 = r^2$, then it is on the circle. We assume that (x, y) satisfies $x^2 + y^2 = r^2$. Then we have
$$(x - 0)^2 + (y - 0)^2 = r^2.$$
Taking the principal square root, we obtain
$$\sqrt{(x - 0)^2 + (y - 0)^2} = r.$$
Thus the distance from (x, y) to $(0, 0)$ is r, and the point (x, y) is on the circle. The two parts of this proof together show that the equation $x^2 + y^2 = r^2$ gives *all* the points of the circle *and no others*.

When a circle is translated so that its center is the point (h, k), an equation for it can be found by replacing x by $x - h$ and y by $y - k$ (Fig. 9).

The equation in standard form of a circle with center (h, k) and radius r is
$$(x - h)^2 + (y - k)^2 = r^2.$$

We now prove that a conic section obtained when the plane is perpendicular to the axis of a cone is actually a circle. Consider a plane α that cuts a right circular cone perpendicular to its axis, and of course parallel to its base (Fig. 10). Then look at the cross sections on any two

Figure 9

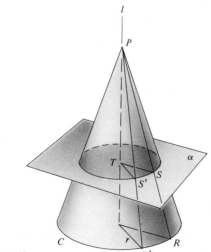

Figure 10

EQUATIONS OF SECOND DEGREE AND THEIR GRAPHS

(a)   (b)

Figure 11

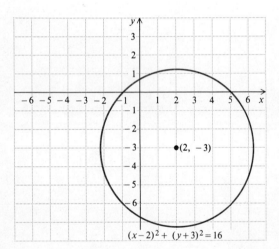

Figure 12

planes containing the axis of the cone l (Fig. 11). Angles PTS and PTS' are right angles because α is perpendicular to l. Thus $\triangle PTS \simeq \triangle PTS'$. Then $\overline{TS} \simeq \overline{TS'}$, for any two points S and S' on the intersection of the cone and the plane α. Thus the intersection consists of all points that are a distance TS from T and is thus a circle, centered at T.

Example 6 Find the center and radius of $(x - 2)^2 + (y + 3)^2 = 16$. Then graph the circle.

We may first rewrite the equation as $(x - 2)^2 + [y - (-3)]^2 = 4^2$. Then the *center* is $(2, -3)$ and the *radius* is 4. The graph is then easy to draw, as shown, using a compass (Fig. 12).

Completing the square allows us to find the standard form for the equation of a circle.

Example 7 Find the center and radius of the circle

$$x^2 + y^2 + 8x - 2y + 15 = 0.$$

We complete the square twice to get the standard form:

$$(x^2 + 8x +) + (y^2 - 2y +) = -15$$
$$(x^2 + 8x + 16) + (y^2 - 2y + 1) = -15 + 16 + 1$$
$$(x + 4)^2 + (y - 1)^2 = 2$$
$$[x - (-4)]^2 + (y - 1)^2 = (\sqrt{2})^2.$$

The *center* is $(-4, 1)$ and the *radius* is $\sqrt{2}$.

Example 8 Find an equation of a circle with center $(-2, -3)$ that passes through the point $(1, 1)$.

Since $(-2, -3)$ is the center, we have

$$(x + 2)^2 + (y + 3)^2 = r^2.$$

The circle passes through $(1, 1)$. We find r by substituting in the above equation:

$$(1 + 2)^2 + (1 + 3)^2 = r^2$$
$$9 + 16 = r^2$$
$$25 = r^2$$
$$5 = r.$$

Then $(x + 2)^2 + (y + 3)^2 = 25$ is an equation of the circle.

EXERCISE SET 14.1

Graph.

1. $x^2 - y^2 = 0$ **2.** $x^2 - 9y^2 = 0$ **3.** $3x^2 + xy - 2y^2 = 0$

4. $x^2 - xy - 2y^2 = 0$ **5.** $2x^2 + y^2 = 0$ **6.** $5x^2 + y^2 = -3$

Find an equation of a circle with center and radius as given.

7. Center: $(0, 0)$ Radius: 7

8. Center: $(-2, 7)$ Radius: $\sqrt{5}$

Find the center and radius of each circle. Then graph the circle.

9. $(x + 1)^2 + (y + 3)^2 = 4$

10. $(x - 2)^2 + (y + 3)^2 = 1$

Find the center and radius of each circle.

11. $(x - 8)^2 + (y + 3)^2 = 40$

12. $(x + 5)^2 + (y - 1)^2 = 75$

13. $x^2 + y^2 + 8x - 6y - 15 = 0$

14. $x^2 + y^2 + 25x + 10y + 12 = 0$

15. $x^2 + y^2 - 4x = 0$

16. $x^2 + y^2 + 10y - 75 = 0$

17. $x^2 + y^2 + 8.246x - 6.348y - 74.35 = 0$

18. $x^2 + y^2 + 25.074x + 10.004y + 12.054 = 0$

19. $9x^2 + 9y^2 = 1$

20. $16x^2 + 16y^2 = 1$

Find an equation of a circle satisfying the given conditions.

21. Center $(0, 0)$, passing through $(-3, 4)$

22. Center $(3, -2)$, passing through $(11, -2)$

☆ ──────────────────────────────────

Find an equation of a circle satisfying the given conditions.

23. Center $(2, 4)$ and tangent (touching at one point) to the x-axis

24. Center $(-3, -2)$ and tangent to the y-axis

25. Find an equation of a circle such that the endpoints of a diameter are $(5, -3)$ and $(-3, 7)$.

26. a) Graph $x^2 + y^2 = 4$. Is this relation a function?
b) Solve $x^2 + y^2 = 4$ for y.
c) Graph $y = \sqrt{4 - x^2}$ and determine whether it is a function. Find the domain and range.
d) Graph $y = -\sqrt{4 - x^2}$ and determine whether it is a function. Find the domain and range.

27. Show that the following equation is an equation of a circle with center (h, k) and radius r.

$$\begin{vmatrix} x - h & -(y - k) \\ y - k & x - h \end{vmatrix} = r^2$$

Determine whether each of the following points lies on the unit circle $x^2 + y^2 = 1$.

28. $(0, -1)$

29. $\left(\dfrac{\sqrt{3}}{2}, -\dfrac{1}{2}\right)$

30. $(\sqrt{2} + \sqrt{3}, 0)$

31. $\left(\dfrac{\pi}{4}, \dfrac{4}{\pi}\right)$

32. $(\cos \theta, \sin \theta)$

33. $(\sin \phi, \cos \phi)$

34. $(\tan \theta, \sec \theta)$

35. $(\sec \theta, \tan \theta)$

★ ──────────────────────────────────

36. Prove that $\angle ABC$ is a right angle. Assume point B is on the circle whose radius is a and whose center is at the origin. (*Hint:* Use slopes and an equation of the circle.)

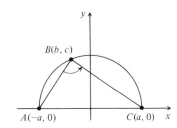

14.2 ELLIPSES

Some equations of second degree have graphs that are ellipses. Ellipses are defined as follows.

DEFINITION

An *ellipse* is the locus or set of all points P in a plane such that the sum of the distances from P to two fixed points F_1 and F_2 in the plane is constant. The points F_1 and F_2 are called *foci* (singular, *focus*) of the ellipse.

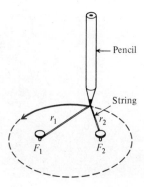

Figure 13

When a plane intersects a cone as shown in Fig. 2(a) on p. 506, not perpendicular to the axis of the cone, an ellipse is formed. We shall prove this later.

Here is a way to draw an ellipse. Stick two tacks in a piece of cardboard (see Fig. 13). These will be the foci F_1 and F_2. Attach a piece of string to the tacks. The length of the string will be the constant sum of the distances from the foci to points on the ellipse. Take a pencil and pull the string tight. Now swing the pencil around, keeping the string tight.

Now let us find equations for ellipses. We first consider an equation of an ellipse whose center is at the origin and whose foci lie on one of the coordinate axes (Fig. 14). Suppose we have an ellipse with foci $F_1(-c, 0)$ and $F_2(c, 0)$. If $P(x, y)$ is a point on the ellipse, then $F_1P + F_2P$ is the given constant distance. We will call it $2a$:

$$F_1P + F_2P = 2a.$$

By the distance formula,

$$\sqrt{(x + c)^2 + y^2} + \sqrt{(x - c)^2 + y^2} = 2a,$$

or

$$\sqrt{(x + c)^2 + y^2} = 2a - \sqrt{(x - c)^2 + y^2}.$$

Figure 14

Squaring, we get

$$x^2 + 2cx + c^2 + y^2 = 4a^2 - 4a\sqrt{(x - c)^2 + y^2} + x^2 - 2cx + c^2 + y^2,$$

or

$$-4a^2 + 4cx = -4a\sqrt{(x - c)^2 + y^2},$$

then

$$-a^2 + cx = -a\sqrt{(x - c)^2 + y^2}$$

Squaring again, we get

$$a^4 - 2a^2cx + c^2x^2 = a^2x^2 - 2cxa^2 + a^2c^2 + a^2y^2,$$

or

$$x^2(a^2 - c^2) + a^2y^2 = a^2(a^2 - c^2).$$

It follows from Fig. 15 that when P is at $(0, b)$, $b^2 = a^2 - c^2$. Substituting b^2 for $a^2 - c^2$ in the last equation, we have the equation of the ellipse $b^2x^2 + a^2y^2 = a^2b^2$, or the following:

$$\frac{x^2}{a^2} + \frac{y^2}{b^2} = 1. \quad \text{Standard form of an equation of an ellipse}$$

We have proved that if a point is on the ellipse, then its coordinates satisfy this equation. We also need to know the converse, that is, if the coordinates of a point satisfy this equation, then the point is on the ellipse. The proof of the latter will be omitted here. In the above, the longer axis of symmetry $\overline{A'A}$ is called the *major axis*. The shorter axis of symmetry $\overline{B'B}$ is called the *minor axis*. The intersection of these axes is called the *center*. The points A, A', B, and B' are called *vertices*. If the center of an ellipse is at the origin, the vertices are also the intercepts.

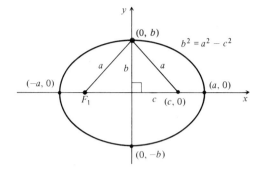

Figure 15

Ellipses as Stretched Circles

Let us consider a unit circle centered at the origin:

$$x^2 + y^2 = 1.$$

If we replace x by x/a and y by y/b we get an equation of an ellipse:

$$\left(\frac{x}{a}\right)^2 + \left(\frac{y}{b}\right)^2 = 1 \quad \text{or} \quad \frac{x^2}{a^2} + \frac{y^2}{b^2} = 1.$$

It follows that an ellipse is a circle transformed by a stretch or a shrink in the x-direction and also in the y-direction. If $a = 2$, for example, the unit circle is stretched in the x-direction by a factor of 2. In any case, the x-intercepts become a and $-a$ and the y-intercepts become b and $-b$.

Example 1 For the ellipse $x^2 + 16y^2 = 16$, find the vertices and the foci. Then graph the ellipse.

a) We first multiply by $\frac{1}{16}$:

$$\frac{x^2}{16} + \frac{y^2}{1} = 1, \quad \text{or} \quad \frac{x^2}{4^2} + \frac{y^2}{1^2} = 1.$$

Thus $a = 4$ and $b = 1$. Then two of the vertices are $(-4, 0)$ and $(4, 0)$. These are also x-intercepts. The other vertices are $(0, 1)$ and $(0, -1)$. These are also y-intercepts. Since $c^2 = a^2 - b^2$, we have $c^2 = 16 - 1$, so $c = \sqrt{15}$ and the foci are $(-\sqrt{15}, 0)$ and $(\sqrt{15}, 0)$.

b) The graph is shown in Fig. 16.

Figure 16

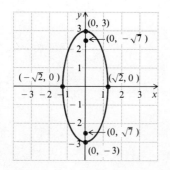

Figure 17

Example 2 Graph this ellipse and its foci: $9x^2 + 2y^2 = 18$.

a) We first multiply by $\frac{1}{18}$:

$$\frac{x^2}{2} + \frac{y^2}{9} = 1 \quad \text{or} \quad \frac{x^2}{(\sqrt{2})^2} + \frac{y^2}{3^2} = 1.$$

Thus $a = \sqrt{2}$ and $b = 3$.

b) Since $b > a$ the foci are on the y-axis and the major axis lies along the y-axis. To find c in this case we proceed as follows:

$$c^2 = b^2 - a^2 = 9 - 2 = 7$$
$$c = \sqrt{7}.$$

c) The graph is shown in Fig. 17.

If the center of an ellipse is not at the origin, but at some point (h, k), then the standard equation is as follows:

$$\frac{(x-h)^2}{a^2} + \frac{(y-k)^2}{b^2} = 1. \quad \text{Ellipse, center at } (h, k)$$

Example 3 For the ellipse

$$16x^2 + 4y^2 + 96x - 8y + 84 = 0,$$

find the center, vertices, foci. Then graph the ellipse.

a) We first complete the square to get standard form:

$$16(x^2 + 6x +) + 4(y^2 - 2y +) = -84$$
$$16(x^2 + 6x + 9) + 4(y^2 - 2y + 1) = -84 + 144 + 4$$
$$16(x+3)^2 + 4(y-1)^2 = 64$$
$$\frac{(x+3)^2}{2^2} + \frac{(y-1)^2}{4^2} = 1.$$

Thus the center is $(-3, 1)$, $a = 2$, and $b = 4$.

b) The vertices of $x^2/2^2 + y^2/4^2 = 1$ are $(2, 0)$, $(-2, 0)$, $(0, 4)$, and $(0, -4)$; and since $c^2 = 16 - 4 = 12$, $c = 2\sqrt{3}$, and its foci are $(0, 2\sqrt{3})$ and $(0, -2\sqrt{3})$.

c) Then the vertices and foci of the translated ellipse are found by translation in the same way in which the center has been translated. Thus the vertices are

$$(-3+2, 1), \quad (-3-2, 1), \quad (-3, 1+4), \quad (-3, 1-4),$$
or
$$(-1, 1), \quad (-5, 1), \quad (-3, 5), \quad (-3, -3).$$

The foci are $(-3, 1 + 2\sqrt{3})$ and $(-3, 1 - 2\sqrt{3})$.

d) The graph is shown in Fig. 18.

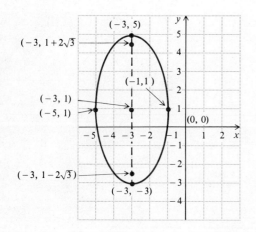

Figure 18

Ellipses have many applications. Earth satellites travel in elliptical orbits. The planets travel around the sun in elliptical orbits with the sun at one focus (Fig. 19). An interesting attraction found in museums is the *whispering gallery*. It is elliptical (Fig. 20). Persons with their heads at the foci can whisper and hear each other clearly while persons at other positions cannot hear them. This happens because sound waves emanating from one focus are reflected to the other focus, being concentrated there.

Planetary orbit

Figure 19

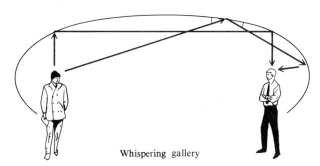

Whispering gallery

Figure 20

We now prove that a conic section obtained when the plane is not perpendicular to the axis of a cone is actually an ellipse. Consider a cone cut by a plane α not perpendicular to the axis (Fig. 21). Now consider two spheres tangent to the cone at circles C_1 and C_2 and to plane α at points F_1 and F_2, respectively. Then consider any point P on the intersection of α with the cone. Then $\overline{PF_2}$ is tangent to the large sphere at F_2, and $\overline{PT_2}$, a segment on an element of the cone, is tangent to the large sphere at T_2. Hence $PF_2 = PT_2$.

Similarly, $\overline{PF_1}$ and $\overline{PT_1}$ are tangent to the small sphere, hence $PF_1 = PT_1$. Thus $PF_1 + PF_2 = PT_1 + PT_2$. But $PT_1 + PT_2$ is constant, being the distance between the circles C_1 and C_2. Therefore for all points P in the intersection of α with the cone, $PF_1 + PF_2$ is constant. The curve is thus an ellipse.

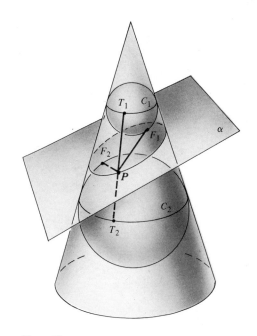

Figure 21

EXERCISE SET 14.2

For each ellipse find the center, the vertices, and the foci, and draw a graph.

1. $\dfrac{x^2}{4} + \dfrac{y^2}{1} = 1$
2. $\dfrac{x^2}{1} + \dfrac{y^2}{4} = 1$
3. $\dfrac{(x-1)^2}{4} + \dfrac{(y-2)^2}{1} = 1$
4. $\dfrac{(x-1)^2}{1} + \dfrac{(y-2)^2}{4} = 1$
5. $\dfrac{(x+3)^2}{25} + \dfrac{(y-2)^2}{16} = 1$
6. $\dfrac{(x-2)^2}{25} + \dfrac{(y+3)^2}{16} = 1$
7. $16x^2 + 9y^2 = 144$
8. $9x^2 + 16y^2 = 144$
9. $3(x+2)^2 + 4(y-1)^2 = 192$
10. $4(x-5)^2 + 3(y-5)^2 = 192$
11. $2x^2 + 3y^2 = 6$
12. $5x^2 + 7y^2 = 35$
13. $4x^2 + 9y^2 = 1$
14. $25x^2 + 16y^2 = 1$
15. $4x^2 + 9y^2 - 16x + 18y - 11 = 0$
16. $x^2 + 2y^2 - 10x + 8y + 29 = 0$
17. $4x^2 + y^2 - 8x - 2y + 1 = 0$
18. $9x^2 + 4y^2 + 54x - 8y + 49 = 0$

For each ellipse find the center and vertices.

19. $4x^2 + 9y^2 - 16.025x + 18.0927y - 11.346 = 0$
20. $9x^2 + 4y^2 + 54.063x - 8.016y + 49.872 = 0$

☆

Find equations of the ellipses with the following vertices. (*Hint:* Graph the vertices.)

21. $(2, 0), (-2, 0), (0, 3), (0, -3)$
22. $(1, 0), (-1, 0), (0, 4), (0, -4)$
23. $(1, 1), (5, 1), (3, 6), (3, -4)$
24. $(-1, -1), (-1, 5), (-3, 2), (1, 2)$

Find equations of the ellipses satisfying the given conditions.

25. Center at $(-2, 3)$ with major axis of length 4 and parallel to the y-axis, minor axis of length 1.

26. Vertices $(3, 0)$ and $(-3, 0)$ and containing the point $\left(2, \dfrac{22}{3}\right)$.

27. a) Graph $9x^2 + y^2 = 9$. Is this relation a function?
    b) Solve $9x^2 + y^2 = 9$ for y.
    c) Graph $y = 3\sqrt{1 - x^2}$ and determine whether it is a function. Find the domain and range.
    d) Graph $y = -3\sqrt{1 - x^2}$ and determine whether it is a function. Find the domain and range.

28. Describe the graph of $\dfrac{x^2}{a^2} + \dfrac{y^2}{b^2} = 1$ when $a^2 = b^2$.

★

29. The unit square on the left in the figure is transformed to the rectangle on the right by a stretch or a shrink in the x-direction and a stretch or a shrink in the y-direction.

a) Use the above result to develop a formula for the area of the ellipse $\dfrac{x^2}{a^2} + \dfrac{y^2}{b^2} = 1$. (*Hint:* The area of the circle $x^2 + y^2 = r^2$ is $\pi \cdot r \cdot r$.)

b) Use the result of (a) to find the area of $\dfrac{x^2}{16} + \dfrac{y^2}{25} = 1$.

c) Use the result of (a) to find the area of $\dfrac{x^2}{4} + \dfrac{y^2}{3} = 1$.

30. The maximum distance of the earth from the sun is 9.3×10^7 miles. The minimum distance is 9.1×10^7 miles. The sun is at one focus of the elliptical orbit. Find the distance from the sun to the other focus.

31. A point moves so that its x-coordinate is given by $x = a \cos t$ and its y-coordinate is given by $y = b \sin t$, where a and b are constants and t is time. Show that the path of the point is an ellipse.

32. The toy shown in the figure is called a *vacuum grinder*. It consists of a rod hinged to two blocks A and B that slide in perpendicular grooves. One grasps the knob at C and grinds. Determine (and prove) whether or not the path of the handle C is an ellipse.

14.3 HYPERBOLAS

Some equations of second degree have graphs that are hyperbolas. Hyperbolas are defined as follows.

DEFINITION

A *hyperbola* is a locus or set of all points P in a plane such that the absolute value of the difference of the distances from P to two fixed points F_1 and F_2 in the plane is constant. The points F_1 and F_2 are called *foci* (singular, *focus*) and the midpoint of the segment joining them is called the *center*.

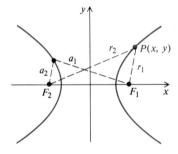

$|r_2 - r_1| = \text{constant} = |a_2 - a_1|$

Figure 22

When a plane intersects a cone as shown in Fig. 2(c) on p. 506, parallel to the axis of the cone, a hyperbola is formed. Note that a hyperbola has two parts. These are called *branches* (Fig. 22).

Now let us find equations for hyperbolas. We first consider an equation of a hyperbola whose center is at the origin and whose foci lie on one of the coordinate axes. Suppose we have a hyperbola as shown in Fig. 23, with foci $F_1(c, 0)$ and $F_2(-c, 0)$ on the x-axis. We consider a point $P(x, y)$ in the first quadrant. The proof for the other quadrants is similar to the proof that follows.

We know that $PF_2 > PF_1$, so $PF_2 - PF_1 > 0$ and $|PF_2 - PF_1| = PF_2 - PF_1$. Let the constant difference be $2a$. Then

$$|PF_2 - PF_1| = PF_2 - PF_1 = 2a.$$

In the triangle F_2PF_1, $PF_2 - PF_1 < F_1F_2$, or $2a < 2c$; therefore $a < c$.

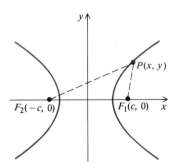

Figure 23

Using the distance formula, we have
$$\sqrt{(x+c)^2 + y^2} - \sqrt{(x-c)^2 + y^2} = 2a,$$
or
$$\sqrt{(x+c)^2 + y^2} = 2a + \sqrt{(x-c)^2 + y^2}.$$
Squaring, we get
$$x^2 + 2xc + c^2 + y^2 = 4a^2 + 4a\sqrt{(x-c)^2 + y^2} + x^2 - 2xc + c^2 + y^2,$$
which simplifies to
$$4cx - 4a^2 = 4a\sqrt{(x-c)^2 + y^2},$$
or
$$cx - a^2 = a\sqrt{(x-c)^2 + y^2}.$$
Squaring again, we get
$$c^2x^2 - 2a^2cx + a^4 = a^2x^2 - 2a^2cx + a^2c^2 + a^2y^2,$$
or
$$x^2(c^2 - a^2) - a^2y^2 = a^2(c^2 - a^2).$$

Since $c > a$, $c^2 > a^2$, so $c^2 - a^2$ is positive. We represent $c^2 - a^2$ by b^2. The previous equation then becomes
$$x^2b^2 - a^2y^2 = a^2b^2,$$
or the following:

$$\frac{x^2}{a^2} - \frac{y^2}{b^2} = 1. \qquad \text{Standard equation of a hyperbola, foci on } x\text{-axis}$$

We have shown that if a point is on the hyperbola it satisfies this equation. We also need to know the converse: If a point satisfies the equation, then it is on the hyperbola. We omit the proof.

Figure 24 is a hyperbola. Points $V_1\,(a, 0)$ and $V_2\,(-a, 0)$ are called the *vertices,* and the line segment $\overline{V_1V_2}$ is called the *transverse axis.* The line segment from $(0, b)$ to $(0, -b)$ is called the *conjugate axis.* Note that when we replace either or both of x by $-x$ and/or y by $-y$, we get an equivalent equation. Thus the hyperbola is symmetric with respect to the origin, and the x- and y-axes are lines of symmetry.

The lines $y = (b/a)x$ and $y = -(b/a)x$ are *asymptotes.* They have slopes b/a and $-b/a$.

Example 1 For the hyperbola $9x^2 - 16y^2 = 144$, find the vertices, the foci, and the asymptotes. Then graph the hyperbola.

a) We first multiply by $\frac{1}{144}$ to find the standard form:
$$\frac{x^2}{16} - \frac{y^2}{9} = 1.$$

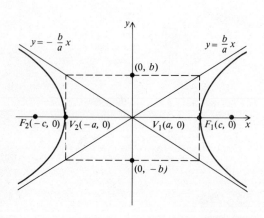

Figure 24

Thus $a = 4$ and $b = 3$. The vertices are $(4, 0)$ and $(-4, 0)$. Since $b^2 = c^2 - a^2$, $c = \sqrt{a^2 + b^2} = \sqrt{4^2 + 3^2} = 5$. Thus the foci are $(5, 0)$ and $(-5, 0)$. The asymptotes are $y = \frac{3}{4}x$ and $y = -\frac{3}{4}x$.

b) To graph the hyperbola it is helpful to first graph the asymptotes. An easy way to do this is to draw the rectangle shown in Fig. 25. Then draw the branches of the hyperbola outward from the vertices toward the asymptotes.

Why are $y = (b/a)x$ and $y = -(b/a)x$ asymptotes? To answer this we solve $x^2/16 - y^2/9 = 1$ for y^2:

$$-16y^2 = 144 - 9x^2$$

$$16y^2 = 9x^2 - 144$$

$$y^2 = \frac{1}{16}(9x^2 - 144)$$

$$y^2 = \frac{9x^2 - 144}{16}.$$

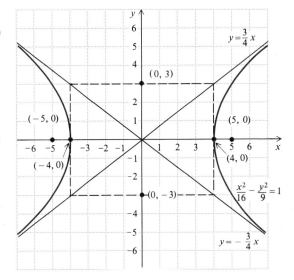

Figure 25

From this last equation, we see that as $|x|$ gets larger the term -144 is very small compared to $9x^2$, so y^2 gets close to $9x^2/16$. That is, when $|x|$ is large,

$$y^2 \approx \frac{9x^2}{16},$$

so

$$y \approx \left|\frac{3}{4}x\right|$$

or

$$y \approx \pm\frac{3}{4}x.$$

Thus the lines $y = \frac{3}{4}x$ and $y = -\frac{3}{4}x$ are asymptotes.

The foci of a hyperbola can be on the y-axis. In this case, the equation is as follows:

$$\frac{y^2}{b^2} - \frac{x^2}{a^2} = 1. \quad \text{Standard equation of a hyperbola, foci on } y\text{-axis}$$

In this case the slopes of the asymptotes are still $\pm b/a$ and it is still true that $c^2 = a^2 + b^2$. There are now y-intercepts and they are $\pm b$.

Example 2 For the hyperbola $25y^2 - 16x^2 = 400$, find the vertices, the foci, and the asymptotes. Then draw a graph.

a) We first multiply by $\frac{1}{400}$ to find the standard form:

$$\frac{y^2}{16} - \frac{x^2}{25} = 1.$$

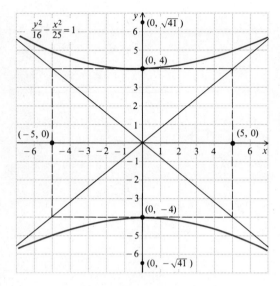

Thus $a = 5$ and $b = 4$. The vertices are $(0, 4)$ and $(0, -4)$. Since $c = \sqrt{4^2 + 5^2} = \sqrt{41}$, the foci are $(0, \sqrt{41})$ and $(0, -\sqrt{41})$. The asymptotes are $y = \frac{4}{5}x$ and $y = -\frac{4}{5}x$.

b) The graph is shown in Fig. 26.

If the center of a hyperbola is not at the origin, but at some point (h, k), then the standard equation is one of the following:

$$\frac{(x - h)^2}{a^2} - \frac{(y - k)^2}{b^2} = 1, \quad \text{Hyperbola, transverse axis parallel to } x\text{-axis}$$

$$\frac{(y - k)^2}{b^2} - \frac{(x - h)^2}{a^2} = 1. \quad \text{Hyperbola, transverse axis parallel to } y\text{-axis}$$

Example 3 For the hyperbola

$$4x^2 - y^2 + 24x + 4y + 28 = 0,$$

find the center, the vertices, the foci, and the asymptotes. Then graph the hyperbola.

Figure 26

a) We complete the square to find standard form:

$$4(x^2 + 6x +) - (y^2 - 4y +) = -28$$
$$4(x^2 + 6x + 9) - (y^2 - 4y + 4) = -28 + 36 - 4$$
$$4(x + 3)^2 - (y - 2)^2 = 4$$
$$\frac{(x + 3)^2}{1} - \frac{(y - 2)^2}{4} = 1.$$

The center is $(-3, 2)$.

b) Consider $x^2/1 - y^2/4 = 1$. We have $a = 1$ and $b = 2$. The vertices of this hyperbola are $(1, 0)$ and $(-1, 0)$. Also, $c = \sqrt{1^2 + 2^2} = \sqrt{5}$, so the foci are $(\sqrt{5}, 0)$ and $(-\sqrt{5}, 0)$. The asymptotes are $y = 2x$ and $y = -2x$.

c) The vertices, foci, and asymptotes of the translated hyperbola are found in the same way in which the center has been translated. The vertices are $(-3 + 1, 2)$, $(-3 - 1, 2)$, or $(-2, 2)$, $(-4, 2)$. The foci are $(-3 + \sqrt{5}, 2)$ and $(-3 - \sqrt{5}, 2)$. The asymptotes are

$$y - 2 = 2(x + 3) \quad \text{and} \quad y - 2 = -2(x + 3).$$

d) The graph is shown in Fig. 27.

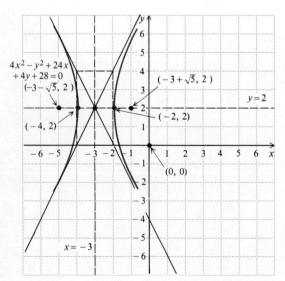

Figure 27

If a hyperbola has its center at the origin, but its axes at 45° to the coordinate axes, it has a simple equation:

$$xy = c, \quad c \text{ a nonzero constant.} \quad \text{Hyperbola, asymptotes the coordinate axes}$$

If c is positive, the branches of the hyperbola lie in the first and third quadrants (Fig. 28a). If c is negative, the branches lie in the second and fourth quadrants (Fig. 28b). In either case the asymptotes are the x-axis and the y-axis.

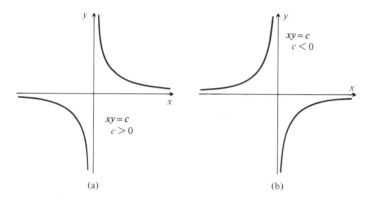

Figure 28

Hyperbolas also have many applications. A jet breaking the sound barrier creates a sonic boom whose wave front has the shape of a cone. This cone intersects the ground in one branch of a hyperbola (Fig. 29). Some comets travel in hyperbolic orbits. A cross section of an amphitheater may be half of one branch of a hyperbola.

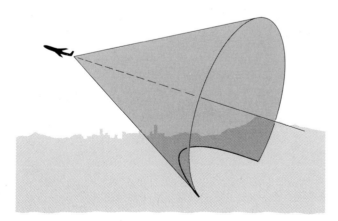

Figure 29

EXERCISE SET 14.3

For each hyperbola find the center, the vertices, the foci, and the asymptotes, and graph the hyperbola.

1. $\dfrac{x^2}{9} - \dfrac{y^2}{1} = 1$
2. $\dfrac{x^2}{1} - \dfrac{y^2}{9} = 1$
3. $\dfrac{(x-2)^2}{9} - \dfrac{(y+5)^2}{1} = 1$
4. $\dfrac{(x-2)^2}{1} - \dfrac{(y+5)^2}{9} = 1$
5. $\dfrac{(y+3)^2}{4} - \dfrac{(x+1)^2}{16} = 1$
6. $\dfrac{(y+3)^2}{25} - \dfrac{(x+1)^2}{16} = 1$
7. $x^2 - 4y^2 = 4$
8. $4x^2 - y^2 = 4$
9. $4y^2 - x^2 = 4$
10. $y^2 - 4x^2 = 4$
11. $x^2 - y^2 = 2$
12. $x^2 - y^2 = 3$
13. $x^2 - y^2 = \dfrac{1}{4}$
14. $x^2 - y^2 = \dfrac{1}{9}$
15. $x^2 - y^2 - 2x - 4y - 4 = 0$
16. $4x^2 - y^2 + 8x - 4y - 4 = 0$
17. $36x^2 - y^2 - 24x + 6y - 41 = 0$
18. $9x^2 - 4y^2 + 54x + 8y + 45 = 0$

Graph.

19. $xy = 1$
20. $xy = -4$
21. $xy = -8$
22. $xy = 3$

23. Find the center, the vertices, and the asymptotes.

$$x^2 - y^2 - 2.046x - 4.088y - 4.228 = 0$$

Find an equation of a hyperbola having:

24. Vertices at $(1, 0)$ and $(-1, 0)$; and foci at $(2, 0)$ and $(-2, 0)$.

25. Asymptotes $y = \dfrac{3}{2}x$ and $y = -\dfrac{3}{2}x$ and one vertex $(2, 0)$.

26. a) Graph $x^2 - 4y^2 = 4$. Is this relation a function?
    b) Solve $x^2 - 4y^2 = 4$ for y.
    c) Graph $y = \dfrac{1}{2}\sqrt{x^2 - 4}$ and determine whether it is a function. Find the domain and range.
    d) Graph $y = -\dfrac{1}{2}\sqrt{x^2 - 4}$ and determine whether it is a function. Find the domain and range.

27. Show that the equation

$$\begin{vmatrix} \dfrac{x-h}{a} & \dfrac{y-k}{b} \\ \dfrac{y-k}{b} & \dfrac{x-h}{a} \end{vmatrix} = 1$$

is an equation of a hyperbola with center (h, k).

★

28. A rifle at point A fires a bullet which hits a target at B (see the figure). A person at C hears the sound of the rifle shot and the sound of the bullet hitting the target simultaneously. Describe the set of all such points C.

29. In a navigation system called *Loran*, a radio transmitter at M (the *master* station) sends out pulses (see the figure). Each pulse triggers another transmitter at S (the *slave* station), which then also transmits a pulse. A ship or airplane at A receives pulses from both M and S and a device measures the difference in the time at which they arrive at A. Knowing this difference in time, the navigator can locate the vessel as being somewhere along a curve predrawn on a chart. What is the shape of that curve?

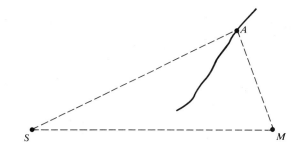

14.4 PARABOLAS

Some equations of second degree have graphs that are parabolas. Parabolas are defined as follows.

DEFINITION

A *parabola* is a locus or set of all points P in a plane equidistant from a fixed line and a fixed point in the plane. The fixed line is called the *directrix* and the fixed point is called the *focus*.

When a plane intersects a cone parallel to an element of the cone, as shown in Fig. 2(b) on p. 506, a parabola is formed.

Now let us find equations for parabolas. Given the focus F and the directrix l, we place coordinate axes as in Fig. 30. The y-axis contains F and is perpendicular to l. The x-axis is halfway between F and l. We shall call the distance from F to the x-axis p. Then F has coordinates $(0, p)$ and l has the equation $y = -p$.

Let $P(x, y)$ be any point of the parabola and consider $\overline{PG}$ perpendicular to the line $y = -p$. The coordinates of G are $(x, -p)$. By definition of a parabola,

$$PF = PG.$$

Then using the distance formula, we have

$$\sqrt{(x-0)^2 + (y-p)^2} = \sqrt{(x-x)^2 + (y+p)^2}.$$

Squaring, we get

$$x^2 + y^2 - 2py + p^2 = y^2 + 2py + p^2$$
$$x^2 = 4py.$$

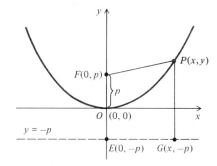

Figure 30

Thus we have the following:
$$x^2 = 4py.$$

Standard equation of a parabola with focus at $(0, p)$ and directrix $y = -p$. The vertex is $(0, 0)$ and the y-axis is the only line of symmetry.

We have shown that if $p(x, y)$ is on the parabola, then its coordinates satisfy this equation. The converse is also true, but we omit the proof.

Note that if $p > 0$, as above, the graph opens upward. If $p < 0$, the graph opens downward and the focus and directrix exchange sides of the x-axis.

The inverse of the above parabola is described as follows:
$$y^2 = 4px.$$

Standard equation of a parabola with focus at $(p, 0)$ and directrix $x = -p$. The vertex is $(0, 0)$ and the x-axis is the only line of symmetry.

Example 1 For the parabola $y = x^2$, find the vertex, the focus, and the directrix, and draw a graph.

We first write $x^2 = 4py$:
$$x^2 = 4\left(\frac{1}{4}\right)y.$$

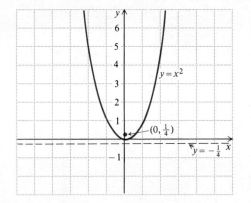

Figure 31

See the graph in Fig. 31.
Vertex: $(0, 0)$
Focus: $(0, \frac{1}{4})$
Directrix: $y = -\frac{1}{4}$

Example 2 For the parabola $y^2 = -12x$, find the vertex, the focus, and the directrix, and draw a graph.

We first write $y^2 = 4px$:
$$y^2 = 4(-3)x.$$

See the graph in Fig. 32.
Vertex: $(0, 0)$
Focus: $(-3, 0)$
Directrix: $x = -(-3) = 3$

Example 3 Find an equation of a parabola with focus $(5, 0)$ and directrix $x = -5$.

The focus is on the x-axis and $x = -5$ is the directrix, so the line of symmetry is the x-axis. Thus the equation is of the type
$$y^2 = 4px.$$

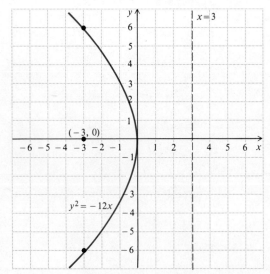

Figure 32

Since $p = 5$, the equation is $y^2 = 20x$.

Example 4 Find an equation of a parabola with focus $(0, -7)$ and directrix $y = 7$.

The focus is on the y-axis and $y = 7$ is the directrix, so the line of symmetry is the y-axis. Thus the equation is of the type
$$x^2 = 4py.$$
Since $p = -7$, we obtain $x^2 = -28y$.

If a parabola is translated so that its vertex is (h, k) and its axis of symmetry is parallel to the y-axis, it has an equation as follows:
$$(x - h)^2 = 4p(y - k),$$
where the vertex is (h, k), the focus is $(h, k + p)$, and the directrix is $y = k - p$.

If a parabola is translated so that its vertex is (h, k) and its axis of symmetry is parallel to the x-axis, it has an equation as follows:
$$(y - k)^2 = 4p(x - h),$$
where the vertex is (h, k), the focus is $(h + p, k)$, and the directrix is $x = h - p$.

Example 5 For the parabola
$$x^2 + 6x + 4y + 5 = 0,$$
find the vertex, the focus, and the directrix, and graph the parabola.

We complete the square:
$$\begin{aligned} x^2 + 6x &= -4y - 5 \\ x^2 + 6x + 9 &= -4y - 5 + 9 \\ (x + 3)^2 &= -4y + 4 = 4(-1)(y - 1). \end{aligned}$$

See the graph in Fig. 33.
Vertex: $(-3, 1)$
Focus: $(-3, 1 + (-1))$ or $(-3, 0)$
Directrix: $y = 1 - (-1) = 2$

Example 6 For the parabola
$$y^2 + 6y - 8x - 31 = 0,$$
find the vertex, the focus, and the directrix, and draw a graph.

We complete the square:
$$\begin{aligned} y^2 + 6y &= 8x + 31 \\ y^2 + 6y + 9 &= 8x + 31 + 9 \\ (y + 3)^2 &= 8x + 40 = 8(x + 5) = 4(2)(x + 5). \end{aligned}$$

Figure 33

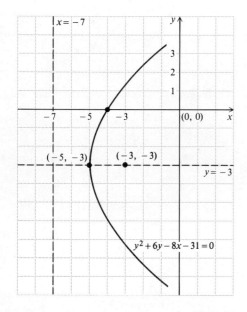

Figure 34

See the graph in Fig. 34.
Vertex: $(-5, -3)$
Focus: $(-5 + 2, -3)$ or $(-3, -3)$
Directrix: $x = -5 - 2 = -7$

Parabolas have many applications. Cross sections of headlights are parabolas. The bulb is located at the focus. All light from this point is reflected outward, parallel to the axis of symmetry (Fig. 35a). Radar and radio antennas may have cross sections that are parabolas. Incoming radio waves are reflected and concentrated at the focus (Fig. 35b). Cables hung between structures to form suspension bridges form parabolas. When a cable supports only its own weight it does not form a parabola, but rather a curve called a *catenary*.

(a)   (b)

Figure 35

EXERCISE SET 14.4

For each parabola find the vertex, the focus, and the directrix, and graph the parabola.

1. $x^2 = 8y$
2. $x^2 = 16y$
3. $y^2 = -6x$
4. $y^2 = -2x$
5. $x^2 - 4y = 0$
6. $y^2 + 4x = 0$
7. $y = 2x^2$
8. $y = \frac{1}{2}x^2$
9. $(x + 2)^2 = -6(y - 1)$
10. $(y - 3)^2 = -20(x + 2)$
11. $x^2 + 2x + 2y + 7 = 0$
12. $y^2 + 6y - x + 16 = 0$
13. $x^2 - y - 2 = 0$
14. $x^2 - 4x - 2y = 0$
15. $y = x^2 + 4x + 3$
16. $y = x^2 + 6x + 10$
17. $4y^2 - 4y - 4x + 24 = 0$
18. $4y^2 + 4y - 4x - 16 = 0$

Find an equation of a parabola satisfying the given conditions.

19. Focus $(4, 0)$, directrix $x = -4$
20. Focus $\left(0, \frac{1}{4}\right)$, directrix $y = -\frac{1}{4}$
21. Focus $(-\sqrt{2}, 0)$, directrix $x = \sqrt{2}$
22. Focus $(0, -\pi)$, directrix $y = \pi$
23. Focus $(3, 2)$, directrix $x = -4$
24. Focus $(-2, 3)$, directrix $y = -3$
25. Graph each of the following, using the same set of axes.

$$x^2 - y^2 = 0, \quad x^2 - y^2 = 1,$$
$$x^2 + y^2 = 1, \quad y = x^2.$$

26. Graph each of the following using the same set of axes.

$$x^2 - 4y^2 = 0, \quad x^2 - 4y^2 = 1,$$
$$x^2 + 4y^2 = 1, \quad x = 4y^2.$$

For each parabola find the vertex, the focus, and the directrix.

27. 🔲 $x^2 = 8056.25y$

28. 🔲 $y^2 = -7645.88x$

29. Find an equation of the following parabola: Line of symmetry parallel to the y-axis, vertex $(-1, 2)$, and passing through $(-3, 1)$.

30. a) Graph $(y - 3)^2 = -20(x + 1)$. Is this relation a function?

b) In general, is $(y - k)^2 = 4p(x - h)$ a function?

31. Show that the equation

$$\begin{vmatrix} y - k & x - h \\ 4p & y - k \end{vmatrix} = 0$$

is an equation of a parabola with vertex (h, k).

★

32. A microphone with 3-inch diameter is placed at the focus of a parabolic reflector. The microphone is 5 inches from the vertex of the reflector and is 5 inches from the outer plane of the reflector, as shown in the figure. If we assume that the reflecting surface is 90% efficient, how much does the reflector magnify the incoming sound?

33. The cables of a suspension bridge are 50 ft above the roadbed at the ends of the bridge and 10 ft above it in the center of the bridge (see the figure). The roadbed is 200 ft long. Vertical cables are to be spaced every 20 ft along the bridge. Calculate the lengths of these vertical cables.

34. Prove that when a cable supports a load distributed uniformly horizontally, it hangs in the shape of a parabola. *Hint:* Proceed as follows.

a) Place a coordinate system as shown in the figure with the origin at the lowest point of the cable.

b) For a point $P(x, y)$ on the cable, write an equation of rotation equilibrium. The forces involved are the tensions in the cable, F_1 and F_2, and the weight supported, W (which is a function of x). The weight of the cable is essentially neglected. Use point P as the center of rotation.

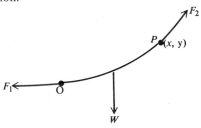

c) Solve for y.

14.5 SYSTEMS OF FIRST-DEGREE AND SECOND-DEGREE EQUATIONS

Figure 36

When we studied systems of linear equations, we solved them both graphically and algebraically. Here we study systems in which one equation is of first degree and one is of second degree. We will use graphical and then algebraic methods of solving.

We consider a system of equations, an equation of a circle and an equation of a line. Let us think about the possible ways in which a circle and a line can intersect. The three possibilities are shown in Fig. 36. For L_1 there is no point of intersection, hence the system of equations has no real solution. For L_2 there is one point of intersection, hence one real solution. For L_3 there are two points of intersection, hence two real solutions.

Example 1 Solve this system graphically:

$$x^2 + y^2 = 25,$$
$$3x - 4y = 0.$$

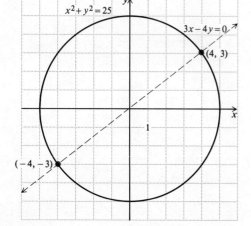

Figure 37

We graph the two equations, using the same axes (Fig. 37). The points of intersection have coordinates that must satisfy both equations. The solutions seem to be $(4, 3)$ and $(-4, -3)$. We check.

Check: For $(-4, -3)$. You should do the check for $(4, 3)$.

$x^2 + y^2 = 25$		$3x - 4y = 0$	
$(-4)^2 + (-3)^2$	25	$3(-4) - 4(-3)$	0
$16 + 9$		$-12 + 12$	
25		0	

Remember that we used the *addition* and *substitution* methods to solve systems of linear equations. In solving systems where one equation is of first degree and one is of second degree, it is preferable to use the *substitution* method.

Example 2 Solve this system algebraically:

$$x^2 + y^2 = 25, \qquad (1)$$
$$3x - 4y = 0. \qquad (2)$$

First solve the linear equation (2) for x:

$$x = \frac{4}{3}y.$$

14.5 SYSTEMS OF FIRST-DEGREE AND SECOND-DEGREE EQUATIONS

Then substitute $\frac{4}{3}y$ for x in Eq. (1) and solve for y:

$$\left(\frac{4}{3}y\right)^2 + y^2 = 25$$

$$\frac{16}{9}y^2 + y^2 = 25$$

$$\frac{16}{9}y^2 + \frac{9}{9}y^2 = 25$$

$$\frac{25}{9}y^2 = 25$$

$$\frac{9}{25} \cdot \frac{25}{9}y^2 = \frac{9}{25} \cdot 25$$

$$y^2 = 9$$

$$y = \pm 3.$$

Now substitute these numbers into the linear equation and solve for x:

$$x = \frac{4}{3}(3), \quad \text{or} \quad 4;$$

$$x = \frac{4}{3}(-3), \quad \text{or} \quad -4.$$

The pairs (4, 3) and (−4, −3) check, hence are solutions.

Sometimes an equation may not take on a familiar form like those studied previously in this chapter. We can still rely on the substitution method for solution.

Example 3 Solve the system

$$y + 3 = 2x, \qquad (1)$$
$$x^2 + 2xy = -1 \qquad (2)$$

First solve the linear equation (1) for y:

$$y = 2x - 3.$$

Then substitute $2x - 3$ for y in Eq. (2) and solve for x:

$$x^2 + 2x(2x - 3) = -1$$
$$x^2 + 4x^2 - 6x = -1$$
$$5x^2 - 6x + 1 = 0$$
$$(5x - 1)(x - 1) = 0$$
$$5x - 1 = 0 \quad \text{or} \quad x - 1 = 0$$
$$x = \frac{1}{5} \quad \text{or} \quad x = 1.$$

Now substitute these numbers into the linear equation and solve for y:

$$y = 2\left(\frac{1}{5}\right) - 3, \quad \text{or} \quad -\frac{13}{5};$$

$$y = 2(1) - 3, \quad \text{or} \quad -1.$$

The pairs $\left(\frac{1}{5}, -\frac{13}{5}\right)$ and $(1, -1)$ check and are solutions.

Applied Problems

Example 4 The perimeter of a rectangular field is 204 m and the area is 2565 m². Find the dimensions of the field.

We first translate the conditions of the problem to equations, using w for the width and l for the length (see Fig. 38):

$$\text{Perimeter: } 2w + 2l = 204,$$
$$\text{Area: } lw = 2565.$$

Figure 38

Now we solve the system

$$2w + 2l = 204,$$
$$lw = 2565,$$

and get the solution (45, 57). Then we check in the original problem: The perimeter is $2 \cdot 45 + 2 \cdot 57$, or 204. The area is $45 \cdot 57$, or 2565. The numbers check, so the answer is $l = 57$ m, $w = 45$ m.

EXERCISE SET 14.5

Solve each system graphically. Then solve algebraically.

1. $x^2 + y^2 = 25,$
   $y - x = 1$
2. $x^2 + y^2 = 100,$
   $y - x = 2$
3. $y^2 - x^2 = 9,$
   $2x - 3 = y$
4. $x + y = -6,$
   $xy = -7$
5. $4x^2 + 9y^2 = 36,$
   $3y + 2x = 6$
6. $9x^2 + 4y^2 = 36,$
   $3x + 2y = 6$
7. $y^2 = x + 3,$
   $2y = x + 4$
8. $y = x^2,$
   $3x = y + 2$

Solve.

9. $x^2 + 4y^2 = 25,$
   $x + 2y = 7$
10. $y^2 - x^2 = 16,$
    $2x - y = 1$
11. $x^2 - xy + 3y^2 = 27,$
    $x - y = 2$
12. $2y^2 + xy + x^2 = 7,$
    $x - 2y = 5$
13. $3x + y = 7,$
    $4x^2 + 5y = 56$
14. $2y^2 + xy = 5,$
    $4y + x = 7$

Applied problems

15. The sum of two numbers is 14 and the sum of their squares is 106. What are the numbers?

16. The sum of two numbers is 15 and the difference of their squares is also 15. What are the numbers?

17. A rectangle has perimeter 28 cm and the length of a diagonal is 10 cm. What are its dimensions?

18. A rectangle has perimeter 6 m and the length of a diagonal $\sqrt{5}$ m. What are its dimensions?

19. A rectangle has area 20 in^2 and perimeter 18 in. Find its dimensions.

20. A rectangle has area 2 yd^2 and perimeter 6 yd. Find its dimensions.

Solve.

21. $x^2 + y^2 = 19{,}380{,}510.36,$
$\quad 27{,}942.25x - 6.125y = 0$

22. $2x + 2y = 1660,$
$\quad xy = 35{,}325$

---

23. Given the area A and the perimeter P of a rectangle, show that the length L and the width W are given by the formulas

$$L = \frac{1}{4}(P + \sqrt{P^2 - 16A}),$$

$$W = \frac{1}{4}(P - \sqrt{P^2 - 16A}).$$

24. Show that a hyperbola does not intersect its asymptotes. That is, solve the system

$$\frac{x^2}{a^2} - \frac{y^2}{b^2} = 1,$$

$$y = \frac{b}{a}x \quad \left(\text{or, } y = -\frac{b}{a}x\right).$$

25. Find an equation of a circle that passes through the points $(2, 4)$ and $(3, 3)$ and whose center is on the line $3x - y = 3$.

26. Find an equation of a circle that passes through the points $(7, 3)$ and $(5, 5)$ and whose center is on the line $y - 4x = 1$.

14.6 SYSTEMS OF SECOND-DEGREE EQUATIONS

We now consider systems of two second-degree equations. Figure 39 shows ways in which a circle and a hyperbola can intersect.

4 real solutions

3 real solutions

2 real solutions

1 real solution

0 real solutions

Figure 39

Example 1 Solve this system graphically:

$$x^2 + y^2 = 25,$$

$$\frac{x^2}{25} - \frac{y^2}{25} = 1.$$

We graph the two equations using the same axes (Fig. 40). The points of intersection have coordinates that must satisfy both equations; the solutions seem to be $(5, 0)$ and $(-5, 0)$.

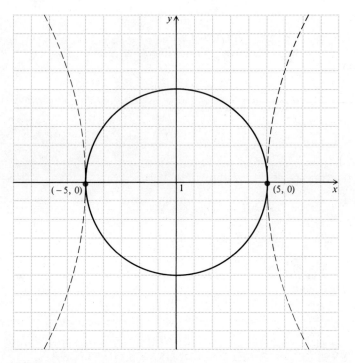

Figure 40

Check: Since $(5)^2 = 25$ and $(-5)^2 = 25$, we can do both checks at once.

$$
\begin{array}{c|c}
x^2 + y^2 = 25 & \dfrac{x^2}{25} - \dfrac{y^2}{25} = 1 \\
\hline
(\pm 5)^2 + 0^2 \quad \bigg| \quad 25 & \dfrac{(\pm 5)^2}{25} - \dfrac{0^2}{25} \quad \bigg| \quad 1 \\
25 + 0 & \dfrac{25}{25} - 0 \\
25 & 1
\end{array}
$$

To solve systems of two second-degree equations we can use either the substitution or the addition method.

Example 2 Solve this system:

$$2x^2 + 5y^2 = 22 \quad (1)$$
$$3x^2 - y^2 = -1. \quad (2)$$

Here we use the addition method.

$$2x^2 + 5y^2 = 22$$

From the second equation, we get

$$\begin{aligned}\underline{15x^2 - 5y^2 = -5} &\quad \text{Multiplying by 5}\\ 17x^2 = 17 &\quad \text{Adding}\\ x^2 = 1&\\ x = \pm 1.&\end{aligned}$$

If $x = 1$, $x^2 = 1$, and if $x = -1$, $x^2 = 1$, so substituting 1 or -1 for x in Eq. (2) we have

$$\begin{aligned}3 \cdot 1^2 - y^2 &= -1\\ y^2 &= 4\\ y &= \pm 2.\end{aligned}$$

Thus if $x = 1$, $y = 2$ or $y = -2$, and if $x = -1$, $y = 2$ or $y = -2$. The possible solutions are $(1, 2)$, $(1, -2)$, $(-1, 2)$, and $(-1, -2)$.

Check: Since $(2)^2 = 4$, $(-2)^2 = 4$, $(1)^2 = 1$, and $(-1)^2 = 1$, we can check all four pairs at one time.

$$\begin{array}{c|c}2x^2 + 5y^2 = 22 & \\ \hline 2(\pm 1)^2 + 5(\pm 2)^2 & 22\\ 2 + 20 & \\ 22 & \end{array} \qquad \begin{array}{c|c}3x^2 - y^2 = -1 & \\ \hline 3(\pm 1)^2 - (\pm 2)^2 & -1\\ 3 - 4 & \\ -1 & \end{array}$$

Example 3 Solve the system

$$\begin{aligned}x^2 + 4y^2 &= 20, & (1)\\ xy &= 4. & (2)\end{aligned}$$

Here we use the substitution method. First we solve Eq. (2) for y:

$$y = \frac{4}{x}.$$

Then we substitute $4/x$ for y in Eq. (1) and solve for x:

$$x^2 + 4\left(\frac{4}{x}\right)^2 = 20$$

$$x^2 + \frac{64}{x^2} = 20$$

$$\begin{aligned}x^4 + 64 &= 20x^2 &&\text{Multiplying by } x^2\\ x^4 - 20x^2 + 64 &= 0&&\\ u^2 - 20u + 64 &= 0 &&\text{Letting } u = x^2\\ (u - 16)(u - 4) &= 0.&&\end{aligned}$$

Then $x^2 = 4$ or $x^2 = -4$ so $x = 2$ or $x = -2$. Since $y = 4/x$, if $x = 4$, $y = 1$; if $x = -4$, $y = -1$; if $x = 2$, $y = 2$; if $x = -2$, $y = -2$. The solutions are $(4, 1)$, $(-4, -1)$, $(2, 2)$, and $(-2, -2)$.

An Applied Problem

Example 4 The area of a rectangle is 300 yd² and the length of a diagonal is 25 yd. Find the dimensions.

First we make a drawing (see Fig. 41). We use l for the length and w for the width and translate to equations.

From the Pythagorean theorem: $l^2 + w^2 = 25^2$
Area: $lw = 300$

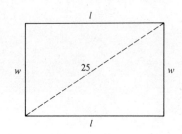

Figure 41

Now we *solve* the system

$$l^2 + w^2 = 625,$$
$$lw = 300,$$

and get the solutions $(15, 20)$ and $(-15, -20)$. Now we check in the original problem: $15^2 + 20^2 = 25^2$ and $15 \cdot 20 = 300$, so $(15, 20)$ is a solution of the problem. Lengths of sides cannot be negative so $(-15, -20)$ is not a solution. The answer is $l = 20$ yd and $w = 15$ yd.

EXERCISE SET 14.6

Solve each system graphically. Then solve algebraically:

1. $x^2 + y^2 = 25,$
   $y^2 = x + 5$
2. $y = x^2,$
   $x = y^2$
3. $x^2 + y^2 = 9,$
   $x^2 - y^2 = 9$
4. $y^2 - 4x^2 = 4,$
   $4x^2 + y^2 = 4$
5. $x^2 + y^2 = 25,$
   $xy = 12$
6. $x^2 - y^2 = 16,$
   $x + y^2 = 4$
7. $x^2 + y^2 = 4,$
   $16x^2 + 9y^2 = 144$
8. $x^2 + y^2 = 25,$
   $25x^2 + 16y^2 = 400$

Solve.

9. $x^2 + y^2 = 16,$
   $y^2 - 2x^2 = 10$
10. $x^2 + y^2 = 14,$
    $x^2 - y^2 = 4$
11. $x^2 + y^2 = 5,$
    $xy = 2$
12. $x^2 + y^2 = 20,$
    $xy = 8$
13. $x^2 + y^2 = 13,$
    $xy = 6$
14. $x^2 + 4y^2 = 20,$
    $xy = 4$
15. $x^2 + y^2 + 6y + 5 = 0,$
    $x^2 + y^2 - 2x - 8 = 0$
16. $2xy + 3y^2 = 7,$
    $3xy - 2y^2 = 4$
17. ▦ $18.465x^2 + 788.723y^2 = 6408,$
    $106.535x^2 - 788.723y^2 = 2692$
18. ▦ $0.319x^2 + 2688.7y^2 = 56,548,$
    $0.306x^2 - 2688.7y^2 = 43,452$

19. Find two numbers whose product is 156 if the sum of their squares is 313.

20. Find two numbers whose product is 60 if the sum of their squares is 136.

21. The area of a rectangle is $\sqrt{3}$ m² and the length of a diagonal is 2 m. Find the dimensions.

22. The area of a rectangle is $\sqrt{2}$ m² and the length of a diagonal is $\sqrt{3}$ m. Find the dimensions.

23. A garden contains two square peanut beds. Find the length of each bed if the sum of their areas is 832 ft², and the difference of their areas is 320 ft².

24. A certain amount of money saved for 1 yr at a certain interest rate yielded $7.50. If the principal had been $25 more and the interest rate 1% less, the interest would have been the same. Find the principal and the rate.

25. Find an equation of the circle that passes through the points (4, 6), (−6, 2), and (1, −3).

26. Find an equation of the circle that passes through the points (2, 3), (4, 5), and (0, −3).

27. Find k such that the following equations have two roots in common.

$$(x - 2)^4 - (x - 2) = 0,$$
$$x^2 - kx + k = 0$$

CHAPTER 14 REVIEW

1. Graph $2x^2 - 3xy - 2y^2 = 0$.

2. Find an equation of the circle with center $(-2, 6)$ and radius $\sqrt{13}$.

3. Find an equation of the circle having its center at (3, 4) and passing through the origin.

4. Find the center and radius of the circle
$$2x^2 + 2y^2 - 3x - 5y + 3 = 0.$$

5. Find an equation of the circle having a diameter with endpoints $(-3, 5)$ and $(7, 3)$.

6. Find an equation of the parabola with directrix $y = \frac{3}{2}$ and focus $(0, -\frac{3}{2})$.

7. Find the focus, vertex, and directrix of the parabola $y^2 = -12x$.

8. Find the center, vertices, and foci of the ellipse
$$16x^2 + 25y^2 - 64x + 50y - 311 = 0.$$
Then graph the ellipse.

9. Find an equation of the ellipse having vertices (3, 0) and (0, 4) and centered at the origin.

10. Find the center, vertices, foci, and asymptotes of the hyperbola
$$x^2 - 2y^2 + 4x + y - \tfrac{1}{8} = 0.$$

11. Solve.
$$x^2 - 16y = 0,$$
$$x^2 - y^2 = 64$$

12. Solve.
$$4x^2 + 4y^2 = 65,$$
$$6x^2 - 4y^2 = 25$$

13. The sum of two numbers is 11 and the sum of their squares is 65. Find the numbers.

14. The sides of a triangle are 8, 10, and 14. Find the altitude to the longest side.

15

Sequences, Series, and Mathematical Induction

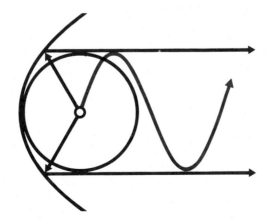

15.1 SEQUENCES

A sequence is a set in which the order of the elements is specified. It can be defined precisely as a certain kind of function.

DEFINITION

A *sequence* is a function having for its domain some set of natural numbers.

As an example, consider the sequence given by
$$a(n) = 2^n, \quad \text{or} \quad a_n = 2^n.*$$
Some of the function values (also known as *terms* of the sequence) are as follows:
$$a_1 = 2^1 = 2,$$
$$a_2 = 2^2 = 4,$$
$$a_3 = 2^3 = 8,$$
$$a_6 = 2^6 = 64.$$

The first term of the sequence is a_1, the fifth term is a_5, and the nth term or *general* term is a_n. This sequence can also be denoted in the following ways:

a) $2, 4, 8, \ldots$;

b) $2, 4, 8, \ldots, 2^n$.

Example 1 Find the first four terms and the 57th term of the sequence whose general term is given by $a_n = (-1)^n/(n+1)$.

$$a_1 = \frac{(-1)^1}{1+1} = -\frac{1}{2}, \quad a_2 = \frac{(-1)^2}{2+1} = \frac{1}{3},$$

$$a_3 = \frac{(-1)^3}{3+1} = -\frac{1}{4}, \quad a_4 = \frac{(-1)^4}{4+1} = \frac{1}{5},$$

$$a_{57} = \frac{(-1)^{57}}{57+1} = \frac{-1}{58}.$$

When a sequence is described merely by naming the first few terms, we do not know for sure what the general term is, but the reader is expected to make a guess by looking for a pattern.

---

*The notation a_n means the same as $a(n)$, but is more commonly used with sequences.

Examples For each sequence, make a guess at the general term.

2. 1, 4, 9, 16, 25, ...
   These are squares of numbers, so the general term may be n^2.

3. $\sqrt{1}, \sqrt{2}, \sqrt{3}, \sqrt{4}, \ldots$
   These are square roots of numbers, so the general term may be $\sqrt{n}$.

4. $-1, 2, -4, 8, -16, \ldots$
   These are powers of 2 with alternating signs, so the general term may be $(-1)^n 2^{n-1}$.

5. 2, 4, 8, ...
   If we see the pattern of powers of 2, we will see 16 as the next term and guess 2^n for the general term. If we see that we can get the second term by adding 2, the third term by adding 4, and the next term by adding 6, and so on, we will see 14 as the next term. A general term for this sequence is $n^2 - n + 2$.

A sequence may be defined by what is called a *recursive definition*. Such a definition lists the first term and tells how to get the $(k + 1)$st term from the kth term.

Example 6 Find the first five terms of the sequence defined by

$$a_1 = 5,$$
$$a_{k+1} = 2a_k - 3, \quad \text{for } k \geq 1.$$

$a_1 = 5,$
$a_2 = 2a_1 - 3 = 2 \cdot 5 - 3 = 7,$
$a_3 = 2a_2 - 3 = 2 \cdot 7 - 3 = 11,$
$a_4 = 2a_3 - 3 = 2 \cdot 11 - 3 = 19,$
$a_5 = 2a_4 - 3 = 2 \cdot 19 - 3 = 35.$

Sums

With any sequence there is associated another sequence, obtained by taking sums of terms of the given sequence. The general term of this sequence of "partial sums" is denoted S_n. Consider the sequence

$$3, 5, 7, 9, \ldots, 2n + 1.$$

We construct some terms of the sequence of partial sums.

$S_1 = 3,$ This is the first term of the given sequence.
$S_2 = 3 + 5 = 8,$ This is the sum of the first two terms.
$S_3 = 3 + 5 + 7 = 15,$ The sum of the first three terms
$S_4 = 3 + 5 + 7 + 9 = 24.$ The sum of the first four terms

A sum like any of the above is also known as a *series*.

Sums, or series, like those above can be handily denoted when the general term is known, using the Greek letter Σ (sigma).* The sum of the first four terms (S_4) of the above sequence can be named as follows:

$$\sum_{k=1}^{4} (2k + 1).$$

This is read "the sum as k goes from 1 to 4 of $(2k + 1)$."

Examples Rename the following sums without using sigma notation.

7. $\displaystyle\sum_{k=1}^{5} k^2 = 1^2 + 2^2 + 3^2 + 4^2 + 5^2$

8. $\displaystyle\sum_{k=1}^{4} (-1)^k(2k) = (-1)^1(2 \cdot 1) + (-1)^2(2 \cdot 2) + (-1)^3(2 \cdot 3) + (-1)^4(2 \cdot 4)$
$= -2 + 4 - 6 + 8$

9. $\displaystyle\sum_{k=10}^{12} \log k = \log 10 + \log 11 + \log 12$

To find sigma notation for a sum we must find a formula for the general term.

Examples Write sigma notation for these sums.

10. $x + x^2 + x^3 + x^4 + x^5 + x^6 = \displaystyle\sum_{k=1}^{6} x^k$

11. $\sin \dfrac{\pi}{2} - \sin \dfrac{3\pi}{2} + \sin \dfrac{5\pi}{2} - \sin \dfrac{7\pi}{2} = \displaystyle\sum_{k=1}^{4} (-1)^{k+1} \sin \left(\dfrac{2k-1}{2}\right) \pi$

EXERCISE SET 15.1

In each of the following the general, or nth, term of a sequence is given. In each case find the first four terms, the 10th term, and the 15th term.

1. $a_n = 3n + 1$
2. $a_n = (n - 1)(n - 2)(n - 3)$
3. $a_n = \dfrac{1}{n}$
4. $a_n = \dfrac{(-1)^n}{n}$
5. $a_n = \dfrac{n}{n + 1}$
6. $a_n = (-\tfrac{1}{2})^{n-1}$

---

*The letter Σ corresponds to S, for "sum."

For each of the following sequences find the general term or nth term. Answers may vary.

7. 1, 3, 5, 7, 9, ...

8. 3, 9, 27, 81, 243, ...

9. $\dfrac{2}{3}, \dfrac{3}{4}, \dfrac{4}{5}, \dfrac{5}{6}, \dfrac{6}{7}, \ldots$

10. $\sqrt{2}, \sqrt{4}, \sqrt{6}, \sqrt{8}, \sqrt{10}, \ldots$

11. $\sqrt{3}, 3, 3\sqrt{3}, 9, 9\sqrt{3}, \ldots$

12. $1 \cdot 2, 2 \cdot 3, 3 \cdot 4, 4 \cdot 5, \ldots$

13. log 1, log 10, log 100, log 1000, ...

14. 2, 4, 6, 8, 10, 12, 14, 16, ...

Find the first four terms of each recursively defined sequence.

15. $a_1 = 4, \; a_{k+1} = 1 + \dfrac{1}{a_k}$

16. $a_1 = 16, \; a_{k+1} = \sqrt{a_k}$

17. $a_1 = 64, \; a_{k+1} = \sqrt{a_k}$

18. $a_1 = e^9, \; a_{k+1} = \ln a_k$

For each sequence find the associated sum.

19. 1, 2, 3, 4, 5, 6, 7, ... Find S_7.

20. 1, −3, 5, −7, 9, −11, ... Find S_8.

21. 2, 4, 6, 8, ... Find S_5.

22. $1, \dfrac{1}{4}, \dfrac{1}{9}, \dfrac{1}{16}, \dfrac{1}{25}, \ldots$. Find S_5.

Rename without using Σ notation.

23. $\displaystyle\sum_{k=1}^{5} \dfrac{1}{2k}$

24. $\displaystyle\sum_{k=1}^{6} \dfrac{1}{2k+1}$

25. $\displaystyle\sum_{k=0}^{5} 2^k$

26. $\displaystyle\sum_{k=4}^{7} \sqrt{2k-1}$

27. $\displaystyle\sum_{k=7}^{10} \log k$

28. $\displaystyle\sum_{k=0}^{4} \pi k$

Write sigma notation for each sum.

29. $\dfrac{1}{2} + \dfrac{2}{3} + \dfrac{3}{4} + \dfrac{4}{5} + \dfrac{5}{6} + \dfrac{6}{7}$

30. $-2 + 4 - 8 + 16 - 32 + 64$

31. $\dfrac{1}{1^2} + \dfrac{1}{2^2} + \dfrac{1}{3^2} + \dfrac{1}{4^2}$

32. $\dfrac{1}{2} + \dfrac{2}{4} + \dfrac{3}{8} + \dfrac{4}{16}$

Find the first five terms of each sequence.

33. $a_n = \dfrac{1}{2n} \log 1000^n$

34. $a_n = i^n, \; i = \sqrt{-1}$

35. $a_n = \sin \dfrac{n\pi}{2}$

36. $a_n = \ln (1 \cdot 2 \cdot 3 \cdots n)$

37. $a_n = \text{Cos}^{-1}(-1)^n$

38. $a_n = |\cos (x + \pi n)|$

39. *Fibonacci sequence.* Find the first six terms of this recursively defined sequence.

$a_1 = 1, \quad a_2 = 1, \quad a_{k+1} = a_k + a_{k-1}$

40. a) Find the first few terms of the sequence

$a_n = n^2 - n + 41.$

b) What pattern do you observe?

c) Find the 41st term. Does the pattern you found in (b) still hold?

Find decimal notation, rounded to six decimal places, for the first six terms of each sequence.

41. $a_n = \left(1 + \frac{1}{n}\right)^n$

42. $a_n = \sqrt{n+1} - \sqrt{n}$

43. $a_1 = 2$, $a_{k+1} = \sqrt{1 + \sqrt{a_k}}$

44. $a_1 = 2$, $a_{k+1} = \frac{1}{2}\left(a_k + \frac{2}{a_k}\right)$

★

For each sequence find a formula for S_n.

45. $a_n = 3$

46. $a_n = \ln n$

47. $a_n = \frac{1}{n} - \frac{1}{n+1}$

48. $a_n = i^{n+1} - i^n$, $i = \sqrt{-1}$

15.2 ARITHMETIC SEQUENCES AND SERIES

Arithmetic Sequences

Consider this sequence:

$$2, 5, 8, 11, 14, 17, \ldots.$$

Note that adding 3 to any term produces the following term. Or, the difference between any term and the preceding one is 3. Sequences in which the difference between successive terms is constant are called *arithmetic sequences*.*

We now find a formula for the general, or nth, term of any arithmetic sequence. Let us denote the difference between successive terms (called the *common difference*) by d, and write out the first few terms.

$$\begin{aligned}
a_1 & \\
a_2 &= a_1 + d \\
a_3 &= a_2 + d = (a_1 + d) + d = a_1 + 2d \\
a_4 &= a_3 + d = (a_1 + 2d) + d = a_1 + 3d
\end{aligned}$$

Generalizing, we obtain the following.

THEOREM 1

The nth term of an arithmetic sequence is given by

$$a_n = a_1 + (n-1)d,$$

where d is the common difference between successive terms.

Example 1 Find the 14th term of the arithmetic sequence 4, 7, 10, 13, ...

---

*Sometimes called arithmetic *progressions*.

First note that $a_1 = 4$, $d = 3$, and $n = 14$. Using the formula of Theorem 1, we obtain
$$a_{14} = 4 + (14 - 1) \cdot 3$$
$$= 4 + 39$$
$$= 43.$$

Example 2 In the sequence of Example 1, which term is 301? That is, what is n if $a_n = 301$?

We substitute into the formula of Theorem 1 and solve for n:
$$a_n = a_1 + (n - 1)d$$
$$301 = 4 + (n - 1) \cdot 3$$
$$301 = 4 + 3n - 3$$
$$300 = 3n$$
$$100 = n.$$

The 100th term is 301.

Given two terms and their places in an arithmetic sequence, we can construct the sequence.

Example 3 The third term of an arithmetic sequence is 8 and the sixteenth term is 47. Find a_1 and d and construct the sequence.

We know that $a_3 = 8$ and $a_{16} = 47$. Thus we would have to add d thirteen times to get from 8 to 47. That is,
$$13d = 47 - 8 = 39.$$

Solving, we obtain
$$d = 3.$$

Since $a_3 = 8$, we subtract d twice to get to a_1. Thus
$$a_1 = 2.$$

The sequence is 2, 5, 8, 11, ...

Sums

We now look for a formula for the sum of the first n terms of an arithmetic sequence. Let us denote the terms as follows:
$$a_1, \ (a_1 + d), \ (a_1 + 2d), \ldots, (a_n - 2d), \ (a_n - d), \ a_n.$$
Then the sum of the first n terms is given by
$$S_n = a_1 + (a_1 + d) + (a_1 + 2d) + \cdots + (a_n - 2d) + (a_n - d) + a_n.$$
Reversing the right side, we obtain
$$S_n = a_n + (a_n - d) + (a_n - 2d) + \cdots + (a_1 + 2d) + (a_1 + d) + a_1.$$

Adding these two equations, we obtain
$$2S_n = (a_1 + a_n) + (a_1 + a_n) + \cdots + (a_1 + a_n),$$
a total of n terms on the right. Thus $2S_n = n(a_1 + a_n)$. Dividing by 2 gives us the formula we seek.

THEOREM 2

The sum of the first n terms of an arithmetic sequence is given by
$$S_n = \frac{n}{2}(a_1 + a_n).$$

We now derive a variant of this formula, which we can use in case a_n is not known. We substitute the expression for a_n given in Theorem 1, $a_n = a_1 + (n-1)d$, into the formula of Theorem 2:
$$S_n = \frac{n}{2}(a_1 + [a_1 + (n-1)d])$$
We now have the following.

THEOREM 3

The sum of the first n terms of an arithmetic sequence is given by
$$S_n = \frac{n}{2}[2a_1 + (n-1)d].$$

Example 4 Find the sum of the first 100 natural numbers.

The series is $1 + 2 + 3 + \cdots + 100$, so $a_1 = 1$, $a_n = 100$, and $n = 100$. We substitute into the formula of Theorem 2:
$$S_n = \frac{n}{2}(a_1 + a_n)$$
$$S_{100} = \frac{100}{2}(1 + 100) = 5050.$$

Example 5 Find the sum of the first 14 terms of the arithmetic sequence 2, 5, 8, 11, 14, ...

We note that $a_1 = 2$, $d = 3$, and $n = 14$. We use the formula of Theorem 3:
$$S_n = \frac{n}{2}[2a_1 + (n-1)d],$$
$$S_{14} = \frac{14}{2} \cdot [2 \cdot + (14-1)3] = 7 \cdot [4 + 13 \cdot 3] = 7 \cdot 43$$
$$S_{14} = 301.$$

Example 6 Find the sum $\sum_{k=1}^{13} (4k + 5)$.

It is helpful to write out a few terms:
$$9 + 13 + 17 + \cdots.$$

We see that the sequence is arithmetic, with $a_1 = 9$, $d = 4$, and $n = 13$. We use the formula of Theorem 3:

$$S_n = \frac{n}{2}[2a_1 + (n-1)d],$$

$$S_{13} = \frac{13}{2}[2 \cdot 9 + (13 - 1)4] = \frac{13}{2}[18 + 12 \cdot 4] = \frac{13}{2} \cdot 66$$

$$S_{13} = 429.$$

Example 7 A family saves money in an arithmetic sequence. They save $500 one year, $600 the next, $700 the next and so on, for 10 years. How much do they save in all, exclusive of interest?

The amount saved is given by the series
$$\$500 + \$600 + \$700 + \cdots$$

We have $a_1 = 500$, $n = 10$, and $d = 100$. We use the formula of Theorem 3:

$$S_n = \frac{n}{2}[2a_1 + (n-1)d],$$

$$S_{10} = \frac{10}{2}[2 \cdot 500 + (10 - 1)100]$$

$$S_{10} = 9500.$$

They save $9500.

If p, m, and q form an arithmetic sequence, it can be shown (see Exercise 35) that $m = (p + q)/2$. We call m the *arithmetic mean* of p and q. Given two numbers p and q, if we find k other numbers $m_1, m_2, m_3, \ldots, m_k$ such that the following is an arithmetic sequence,

$$p, \quad m_1, \quad m_2, \ldots, m_k, \quad q,$$

we say that we have "inserted k arithmetic means between p and q."

Example 8 Insert three arithmetic means between 4 and 13.

We look for m_1, m_2, and m_3 such that $4, m_1, m_2, m_3, 13$ is an arithmetic sequence. In this case, $a_1 = 4$, $n = 5$, and $a_5 = 13$. We use the formula of Theorem 1:

$$a_n = a_1 + (n - 1)d$$
$$13 = 4 + (5 - 1)d.$$

Then $d = 2\frac{1}{4}$, so we have

$$m_1 = a_1 + d = 4 + 2\frac{1}{4} = 6\frac{1}{4},$$

$$m_2 = m_1 + d = 6\frac{1}{4} + 2\frac{1}{4} = 8\frac{1}{2},$$

$$m_3 = m_2 + d = 8\frac{1}{2} + 2\frac{1}{4} = 10\frac{3}{4}.$$

EXERCISE SET 15.2

Exercises 1–10 refer to arithmetic sequences.

1. Find the 12th term of $2, 6, 10, \ldots$.

2. Find the 11th term of $0.07, 0.12, 0.17, \ldots$.

3. In the sequence in Exercise 1, which term is 106?

4. In the sequence in Exercise 2, which term is 1.67?

5. If $a_1 = 5$ and $d = 6$, what is a_{17}?

6. If $a_1 = 14$ and $d = -3$, what is a_{20}?

7. If $d = 4$ and $a_8 = 33$, what is a_1?

8. If $a_1 = 8$ and $a_{11} = 26$, what is d?

9. If $a_1 = 5$, $d = -3$ and $a_n = -76$, what is n?

10. If $a_1 = 25$, $d = -14$, and $a_n = -507$, what is n?

11. In an arithmetic sequence $a_{17} = -40$ and $a_{28} = -73$. Find a_1 and d. Find the first 5 terms of the sequence.

12. A bomb drops from an airplane and falls 16 ft the first sec, 48 ft the second sec, and so on, forming an arithmetic sequence. How many feet will the bomb fall during the 20th sec?

13. Find the sum of the first 20 terms of the sequence $5, 8, 11, 14, \ldots$.

14. Find the sum of the first 15 terms of the sequence $5, \frac{55}{7}, \frac{75}{7}, \frac{95}{7}, \ldots$.

15. Find the sum of the odd numbers from 1 to 99, inclusive.

16. Find the sum of the multiples of 7, from 7 to 98 inclusive.

17. An arithmetic sequence has $a_1 = 2$, $d = 5$, and $n = 20$. What is S_n?

18. Find the sum of all multiples of 4 that are between 14 and 523.

19. Find the sum of $\sum_{k=1}^{16} (7k - 76)$.

20. Find the sum of $\sum_{k=1}^{12} (6k - 3)$.

21. How many poles will be in a pile of telephone poles if there are 30 in the first layer, 29 in the second, and so on until there is one in the last layer?

22. If a student saved 10¢ on October 1, 20¢ on October 2, 30¢ on October 3, etc., how much would be saved during October? (October has 31 days.)

23. Insert four arithmetic means between 4 and 13.

24. Insert three arithmetic means between 5 and -3.

☆

25. Find a formula for the sum of the first n odd natural numbers.

26. Find three numbers in an arithmetic sequence such that the sum of the first and third is 10 and the product of the first and second is 15.

27. The zeros of this polynomial form an arithmetic sequence. Find them.

$$x^4 + 4x^3 - 84x^2 - 176x + 640$$

28. Insert enough arithmetic means between 1 and 50 so the sum of the resulting arithmetic series will be 459.

29. Suppose that the lengths of the sides of a right triangle form an arithmetic sequence. Prove that the triangle is similar to a right triangle whose sides have lengths 3, 4, and 5.

Straight-line depreciation. A company buys an office machine for $5200 on January 1 of a given year. It is expected to last for eight years at the end of which time its *trade-in,* or *salvage, value* will be $1100. If the company figures the decline in value to be the same each year, then the *book value* or *salvage value* after n years, $0 \leq n \leq 8$, is given by an arithmetic sequence,

$$V_n = C - n\left(\frac{C - S}{L}\right),$$

where C = the original cost of the item ($5200), L = the years of expected life (8), and S = the salvage value ($1100).

30. Find the general term for the straight-line depreciation of the office machine.

31. Find the salvage value after 0 years, 1 year, 2 years, 3 years, 4 years, 7 years, 8 years.

32. ▦ Find the first 10 terms of the arithmetic sequence for which

$$a_1 = \$8760 \quad \text{and} \quad d = -\$798.23.$$

33. ▦ Find the sum of the first ten terms of the sequence in Exercise 31.

34. Prove that an expression for the nth term of an arithmetic sequence defines a linear function.

35. Prove that if p, m, and q form an arithmetic sequence, then

$$m = \frac{p + q}{2}.$$

15.3 GEOMETRIC SEQUENCES AND SERIES

Geometric Sequences

Consider this sequence:

$$3, 6, 12, 24, 48, 96, \ldots.$$

Note that multiplying any term by 2 produces the following term. That is, the ratio of any term and the preceding one is 2. Sequences in which the ratio of successive terms is constant (but different from 1) are called *geometric sequences.**

We now find a formula for the general, or nth, term of any geometric sequence. Let us denote the ratio of successive terms (called the *common ratio*) by r, and write out the first few terms.

$$a_1$$
$$a_2 = a_1 r$$
$$a_3 = a_2 r = (a_1 r)r = a_1 r^2$$
$$a_4 = a_3 r = (a_1 r^2)r = a_1 r^3$$

Generalizing, we obtain the following.

---
*Sometimes called geometric *progressions.*

THEOREM 4

The nth term of a geometric sequence is given by
$$a_n = a_1 r^{n-1},$$
where r is the common ratio of successive terms.

Example 1 Find the 6th term of the geometric sequence 4, 20, 100, ...

First note that $a_1 = 4$, $n = 6$, and $r = 5$. We use the formula of Theorem 4:

$$a_n = a_1 r^{n-1}$$
$$a_6 = 4 \cdot 5^{6-1} = 4 \cdot 5^5$$
$$= 12{,}500.$$

Example 2 Find the 11th term of the geometric sequence 64, -32, 16, -8, ...

Note that $a_1 = 64$, $n = 11$, and $r = -\tfrac{1}{2}$. We use the formula of Theorem 4:

$$a_n = a_1 r^{n-1}$$
$$a_{11} = 64 \cdot \left(-\frac{1}{2}\right)^{11-1}$$
$$= 64 \cdot \left(-\frac{1}{2}\right)^{10}$$
$$= 2^6 \cdot \frac{1}{2^{10}}$$
$$= \frac{1}{2^4} \quad \text{or} \quad \frac{1}{16}.$$

Example 3 A college student borrows \$600 at 12% interest compounded annually. The loan is paid off at the end of 3 years. How much is paid back?

For principal P, at 12% interest, $P + 0.12P$ will be owed at the end of a year. This is also $1.12P$, and is the principal for the second year, so at the end of that year $1.12(1.12P)$ is owed. Thus the principal at the beginnings of successive years is

$$P, \quad 1.12P, \quad 1.12^2 P, \quad 1.12^3 P, \quad \text{and so on.}$$

We now have a geometric sequence with $a_1 = 600$, $n = 4$, and $r = 1.12$.

15.3 GEOMETRIC SEQUENCES AND SERIES

We use the formula of Theorem 4:

$$a_n = a_1 r^{n-1}$$
$$a_4 = 600 \cdot (1.12)^{4-1}$$
$$= 600(1.12)^3$$
$$= 600 \cdot 1.404928$$
$$= 842.96.$$

The amount paid back is $842.96.

Note that the equation $a_4 = \$600(1.12)^3$, used in Example 3, conforms to the compound interest formula $A = P(1 + i)^t$, developed in Chapter 2. If we consult that development we see that the numbers $A_1, A_2, A_3,$ and so on, form a geometric sequence, where $A_1 = P$, $A_2 = P(1 + i)$, $r = 1 + i$, and $n - 1 = t$. (Note that "after t years" and "at the beginning of the nth year" have the same meaning here.)

Sums

We now look for a formula for the sum of the first n terms of a geometric sequence.* Let us denote the terms as follows:

$$a_1, \quad a_1 r, \quad a_1 r^2, \quad a_1 r^3, \ldots, \quad a_1 r^{n-1}.$$

Then the sum of the first n terms is given by

$$S_n = a_1 + a_1 r + a_1 r^2 + \cdots + a_1 r^{n-2} + a_1 r^{n-1}.$$

We next multiply on both sides of this equation by $-r$, obtaining

$$-rS_n = -a_1 r - a_1 r^2 - \cdots - a_1 r^{n-1} - a_1 r^n.$$

We now add the two equations, obtaining

$$S_n - rS_n = a_1 - a_1 r^n.$$

Solving for S_n gives us the formula we seek.

THEOREM 5

The sum of the first n terms of a geometric sequence is given by

$$S_n = \frac{a_1 - a_1 r^n}{1 - r}, \quad r \neq 1.$$

We now derive a variant of this formula, which we can use when a_n is known. Using Theorem 4 we substitute a_n for $a_1 r^{n-1}$ into the formula of Theorem 5. We then have the following.

---

*It is not uncommon to see or hear the redundant terminology "sum of a geometric series."

THEOREM 6

The sum of the first n terms of a geometric sequence is given by

$$S_n = \frac{a_1 - ra_n}{1 - r}, \quad r \neq 1.$$

Example 4 Find the sum of the first 6 terms of the geometric sequence 3, 6, 12, 24, ...

Note that $a_1 = 3$, $n = 6$, and $r = 2$. We use the formula of Theorem 5:

$$S_n = \frac{a_1 - a_1 r^n}{1 - r}$$

$$S_6 = \frac{3 - 3 \cdot 2^6}{1 - 2} = \frac{3 - 192}{-1}, \quad \text{or} \quad 189.$$

Example 5 Find the sum $\sum_{k=1}^{11} (0.3)^k$.

This is a geometric series. The first term is 0.3, $r = 0.3$, and $n = 11$. We use the formula of Theorem 5:

$$S_n = \frac{a_1 - a_1 r^n}{1 - r}$$

$$S_{11} = \frac{0.3 - 0.3 \cdot (0.3)^{11}}{1 - 0.3} = \frac{0.3 - (0.3)^{12}}{0.7}$$

$$= 0.42857 +.$$

Example 6 Someone offers you a job for 30 days, the pay being 1¢ the first day and 2¢ the second day, and doubled every succeeding day. Would you take the job? Calculate your pay for the 30 days.

Your pay is the sum of the geometric series (in dollars)

$$0.01, \quad 2(0.01), \quad 2^2(0.01), \quad 2^3(0.01), \ldots, \quad 2^{29}(0.01).$$

We note that $a_1 = 0.01$, $n = 30$, and $r = 2$. We use the formula of Theorem 5:

$$S_n = \frac{a_1 - a_1 r^n}{1 - r}$$

$$S_{30} = \frac{0.01 - 0.01(2)^{30}}{1 - 2} = \frac{0.01(1 - 2^{30})}{-1} = \frac{0.01(1 - 1073741824)}{-1}$$

$$= 10{,}737{,}418.23.$$

Your pay: $10,737,418.23

EXERCISE SET 15.3

1. Find the 10th term of the geometric sequence $\frac{8}{243}, \frac{4}{81}, \frac{2}{27}, \ldots$.

2. Find the 5th term of the geometric sequence $2, -10, 50, \ldots$.

3. Find the sum of the first 8 terms of the geometric sequence $5, 10, 20, \ldots$.

4. Find the sum of the first 6 terms of the geometric sequence $16, -8, 4, \ldots$.

5. Find the sum of the first 7 terms of the geometric sequence $\frac{1}{18}, -\frac{1}{6}, \frac{1}{2}, \ldots$.

6. Find the sum of the following geometric sequence: $-8, 4, -2, \ldots, -\frac{1}{32}$.

7. Find the sum

$$\sum_{k=1}^{6} \left(\frac{1}{2}\right)^{k-1}.$$

8. Find the sum

$$\sum_{k=1}^{10} 2^k.$$

9. A college student borrows $800 at 8% interest compounded annually. She pays off the loan in full at the end of 2 yr. How much does she pay?

10. A college student borrows $1000 at 8% interest compounded annually. He pays off the loan in full at the end of 4 yr. How much does he pay?

11. A ping-pong ball is dropped from a height of 16 ft and always rebounds $\frac{1}{4}$ of the distance of the previous fall. What distance does it rebound the 6th time?

12. Gaintown has a population of 100,000 now and the population is increasing 10% every year. What will be the population in 5 yr?

13. ▦ Find the sum of the first 5 terms of the geometric sequence

$$\$1000, \$1000(1.08), \$1000(1.08)^2, \ldots.$$

Round to the nearest cent.

14. ▦ Find the sum of the first 6 terms of the geometric sequence

$$\$200, \$200(1.13), \$200(1.13)^2, \ldots.$$

Round to the nearest cent.

☆ ─────────────────────────────

For Exercises 15–17, assume that $a_1, a_2, a_3, \ldots,$ is a geometric sequence.

15. Show that

$$a_1^2, a_2^2, a_3^2, \ldots,$$

is a geometric sequence.

16. Show that

$$a_1^{-3}, a_2^{-3}, a_3^{-3}, \ldots,$$

is a geometric sequence.

17. Show that

$$\ln a_1, \ln a_2, \ln a_3, \ldots,$$

is an arithmetic sequence.

18. Show that

$$5^{a_1}, 5^{a_2}, 5^{a_3}, \ldots,$$

is a geometric sequence, if $a_1, a_2, a_3, \ldots,$ is an arithmetic sequence.

19. If the positive numbers p, G, and q form a geometric sequence, show that

$$G = \sqrt{pq}.$$

G is called the *geometric mean* of the numbers p and q.

20. Find the geometric mean between each pair of numbers.
 a) 4 and 9 b) 2 and 6
 c) $\frac{1}{2}$ and $\frac{1}{3}$ d) $\sqrt{5} + \sqrt{2}$ and $\sqrt{5} - \sqrt{2}$

21. ▦ A piece of paper is 0.01 in. thick. It is folded in such a way that its thickness is doubled each time for 20 times. How thick is the result?

22. ▦ A superball dropped from the top of the Washington Monument (556 ft high) always rebounds $\frac{3}{4}$ of the distance of the previous fall. How far up and down has it traveled when it hits the ground for the sixth time?

23. (*An annuity*). A person decides to save money in a savings account for retirement. At the beginning of each year $1000 is invested at 8% compounded annually. How much is in the retirement fund at the end of 40 years?

25. Prove that the geometric mean of two positive numbers p and q is less than or equal to the arithmetic mean.

24. (*Yearly annuity payment*). A person wants to have $388,585 accumulated in a retirement fund over a period of 40 years. How much would he have to deposit at the beginning of each year for 40 years at 8% compounded annually?

26. A square has sides one meter in length. Another square is formed whose vertices are the midpoints of the original square. Inside this square another is formed whose vertices are the midpoints of the second square, and so on.

  a) Find the first five terms of the sequence of areas of the squares.

  b) Find the nth term.

  c) Find a formula for the sum of the areas of the first n squares.

27. The zeros of the polynomial

$$x^3 + 7x^2 - 21x - 27$$

form a geometric sequence. Find them.

15.4 INFINITE GEOMETRIC SEQUENCES AND SERIES

Recall the definition of a *sequence* as a function having for its domain some set of natural numbers. If that domain is an infinite (unending) set of natural numbers, then the sequence is called an *infinite sequence*. In the examples that follow, the domain will be the set of *all* natural numbers, even though this is not a requirement of the definition.

Let us look at two examples of infinite geometric sequences and their sequences of partial sums.

A. $2, 4, 8, 16, \ldots, 2^n, \ldots$

$$S_1 = 2$$
$$S_2 = 6$$
$$S_3 = 14$$
$$S_4 = 30$$
$$S_5 = 62$$

The terms of the sequence S_n get larger and larger, without bound. Now let us look at the next sequence.

B. $\frac{1}{2}, \frac{1}{4}, \frac{1}{8}, \frac{1}{16}, \ldots, \frac{1}{2^n}, \ldots$

$$S_1 = \frac{1}{2}$$

$$S_2 = \frac{1}{2} + \frac{1}{4} = \frac{3}{4}$$

$$S_3 = S_2 + \frac{1}{8} = \frac{3}{4} + \frac{1}{8} = \frac{7}{8}$$

$$S_4 = S_3 + \frac{1}{16} = \frac{15}{16}$$

$$S_5 = S_4 + \frac{1}{32} = \frac{31}{32}$$

It appears that we can write a formula for S_n as follows:

$$S_n = \frac{2^n - 1}{2^n}.$$

Later (see Example 3) we shall prove that this is correct. In this formula the numerator is less than the denominator for all values of n. Yet as n increases the numerator gets very close to the denominator. Thus the values of S_n approach the number 1 more and more closely. We say that S_n "approaches 1 as a limit."

Sums

Sequence B in the above example is infinite. For any finite number of terms, the sum is given by

$$S_n = \sum_{k=1}^{n} \frac{1}{2^k} = \frac{2^n - 1}{2^n}.$$

Since S_n approaches a limit, we *define* that limit to be the "sum" of the infinite sequence of terms, and use the following notation.*

$$S_\infty = \sum_{k=1}^{\infty} \frac{1}{2^k} = 1$$

---

* $\sum_{k=1}^{\infty} \frac{1}{2^k}$ is read "the sum, as k goes from 1 to infinity, of $\frac{1}{2^k}$."

Not all infinite sequences have sums, for example, sequence A above. In those cases the notation $\sum_{k=1}^{\infty} a_k$ is meaningless. An infinite sum is also called an *infinite series*.

We now find a formula for the "sum" of an infinite geometric sequence by looking at values of S_n as n becomes very large. Recall that

$$S_n = \frac{a_1 - a_1 r^n}{1 - r} = \frac{a_1}{1 - r} - \frac{a_1 r^n}{1 - r}.$$

As n becomes large, r^n will also become large if $r > 1$. If r is negative and $|r| > 1$, then the values of r^n will alternate between positive and negative numbers having large absolute value. In either case S_n will not approach a limit. If $r = 1$, the formula does not hold, and the sequence will not be geometric. However, if $|r| < 1$, then the values of r^n approach 0 as n becomes large, and we have the following.

THEOREM 7

An infinite geometric sequence has a sum if and only if $|r| < 1$. That sum is given by

$$S_\infty = \frac{a_1}{1 - r}.$$

Example 1 Find the sum of the infinite geometric sequence $5, \frac{5}{2}, \frac{5}{4}, \frac{5}{8}, \ldots,$ if it exists. That is, find

$$\sum_{k=1}^{\infty} \frac{5}{2^{k-1}}.$$

Note that $a_1 = 5$. The sum exists because $r = \frac{1}{2}$, which is less than 1. We use the formula of Theorem 7:

$$S_\infty = \frac{a_1}{1 - r}$$

$$= \frac{5}{1 - \frac{1}{2}} = 10.$$

Repeating Decimals

Any repeating decimal represents a geometric series. For example,

$$0.33333\ldots = 0.3 + 0.03 + 0.003 + 0.0003 + \cdots.$$

15.4 INFINITE GEOMETRIC SEQUENCES AND SERIES

In this series the common ratio is 0.1. It is a common series and the sum is known to be $\frac{1}{3}$. Consider the following.

$$0.35\overline{3535}{}^* = 0.35 + 0.0035 + 0.000035 + \cdots$$

In this case the common ratio is 0.01. In any repeating decimal the common ratio will be 0.1, 0.01, 0.001, or 10^{-n} for some natural number n. Thus in any repeating decimal the common ratio is less than 1, so the infinite sum exists.

Example 2 Find fractional notation for $0.27\overline{2727}$.

Note that $a_1 = 0.27$ and that $r = 0.01$. We use the formula of Theorem 7:

$$S_\infty = \frac{a_1}{1-r}$$

$$S_\infty = \frac{0.27}{1-0.01}$$

$$= \frac{0.27}{0.99}$$

$$= \frac{3}{11}.$$

Fractional notation for $0.27\overline{27}$ is $\frac{3}{11}$. This can be checked by dividing 3 by 11.

Example 3 Find fractional notation for $0.416\overline{66}$.

Note that $0.416\overline{6} = 0.41 + 0.00666\overline{6}$, and that $0.00\overline{666}$ is a geometric series, with $a_1 = 0.006$ and $r = 0.1$. We find fractional notation for this sum first:

$$S_\infty = \frac{a_1}{1-r}$$

$$= \frac{0.006}{1-0.1}$$

$$= \frac{0.006}{0.9}$$

$$= \frac{6}{900}.$$

The number in question is then $0.41 + \frac{6}{900}$ or $\frac{41}{100} + \frac{6}{900}$. Adding and simplifying, we obtain $\frac{5}{12}$. The reader should check this by dividing.

---
*The bar indicates the repeating cycle.

Example 4 *The economic multiplier.* Recently, Congress passed a law increasing the income tax exemption from $750 to $1000. For a family of four this meant that an additional $250 per person, or $1000, was available for spending. Let us suppose that they spent 90% of this amount, that the people who received it spent 90% of it, the people who received that amount spent 90% of it, and so on.

According to certain economic theory, the money that this effectively put into the economy can be calculated as the sum of an infinite geometric sequence, as follows:

$$\$1000 + \$1000(0.90) + \$1000(0.90)^2 + \$1000(0.90)^3 + \cdots.$$

Using the formula of Theorem 7 we find this amount to be

$$\frac{\$1000}{1 - 0.90}, \quad \text{or} \quad \$10{,}000.$$

EXERCISE SET 15.4

Determine which of the following infinite geometric sequences have sums.

1. $5, 10, 20, 40, \ldots$
2. $16, 8, 4, 2, \ldots$
3. $6, 2, \dfrac{2}{3}, \dfrac{2}{9}, \ldots$
4. $2, -4, 8, -16, 32, \ldots$
5. $1, 0.1, 0.01, 0.001, 0.0001, \ldots$
6. $-\dfrac{5}{3}, -\dfrac{10}{9}, -\dfrac{20}{27}, \ldots$
7. $1, -\dfrac{1}{5}, \dfrac{1}{25}, -\dfrac{1}{125}, \ldots$
8. $6, \dfrac{42}{5}, \dfrac{294}{25}, \ldots$

Find these infinite sums. (The series are geometric.)

9. $4 + 2 + 1 + \cdots$
10. $7 + 3 + \dfrac{9}{7} + \cdots$
11. $25 + 20 + 16 + \cdots$
12. $12 + 9 + \dfrac{27}{4} + \cdots$
13. $\sum\limits_{k=1}^{\infty} \dfrac{1}{2^{k-1}}$
14. $\sum\limits_{k=1}^{\infty} \dfrac{8}{3}\left(\dfrac{1}{2}\right)^{k-1}$
15. $\sum\limits_{k=1}^{\infty} 16(0.1)^{k-1}$
16. $\sum\limits_{k=1}^{\infty} 4(0.6)^{k-1}$
17. ▨ $\$1000(1.08)^{-1} + \$1000(1.08)^{-2} + \$1000(1.08)^{-3} + \cdots$
18. ▨ $\$500(1.12)^{-1} + \$500(1.12)^{-2} + \$500(1.12)^{-3} + \cdots$

Find fractional notation for these numbers.

19. $0.7\overline{777}$
20. $0.5\overline{333}$
21. $0.2\overline{12121}$
22. $0.6\overline{4444}$
23. $5.1\overline{5151}5$
24. $0.4\overline{125125}$

25. The government makes an $8,000,000,000 expenditure for a new type of aircraft. If 85% of this gets spent again, and 85% of that gets spent again, and so on, what is the total effect on the economy?

26. Repeat Exercise 25 for $9,400,000,000 and 99%.

27. (*Advertising effect*). A company is marketing a new product in a city of 5,000,000 people. They plan an advertising campaign that they think will induce 40% of the people to buy the product. They estimate that if those people like the product, they will induce 40% (of the 40% of 5,000,000) more to buy the product, and those will induce 40%, and so on. In all, how many people will buy the product as a result of the advertising campaign? What percentage of the population is this?

28. How far (up and down) will a ball travel before stopping if it is dropped from a height of 12 ft, and each rebound is $\frac{1}{3}$ of the previous distance? (*Hint:* Use an infinite geometric series.)

29. The function e^x is given by the infinite series
$$e^x = 1 + x + \frac{x^2}{2!} + \frac{x^3}{3!} + \cdots.$$
($n!$ is read "n factorial," and is defined to be the product $1 \cdot 2 \cdot 3 \cdot \cdots \cdot n$). Approximate e using the first six terms of the series.

30. The function $\cos x$ is given by the infinite series
$$\cos x = 1 - \frac{x^2}{2!} + \frac{x^4}{4!} - \frac{x^6}{6!} + \cdots.$$
Approximate $\cos(\pi/10)$ using the first four terms of the series.

31. The infinite sequence
$$2, \frac{1}{2}, \frac{1}{2 \cdot 3}, \frac{1}{2 \cdot 3 \cdot 4}, \frac{1}{2 \cdot 3 \cdot 4 \cdot 5}, \frac{1}{2 \cdot 3 \cdot 4 \cdot 5 \cdot 6}, \cdots$$
is not geometric, but does have a sum. Find $S_1, S_2, S_3, S_4, S_5,$ and S_6. Make a conjecture about the value of S_∞.

32. The infinite sequence
$$\frac{1}{1 \cdot 3}, \frac{1}{3 \cdot 5}, \frac{1}{5 \cdot 7}, \frac{1}{7 \cdot 9}, \frac{1}{9 \cdot 11}, \frac{1}{11 \cdot 13}, \cdots$$
is not geometric, but does have a sum. Find $S_1, S_2, S_3, S_4, S_5,$ and S_6. Make a conjecture about the value of S_∞.

33. Referring to Exercise 26 in Exercise Set 15.3, assume the procedure is continued infinitely. What is the sum of the areas?

34. An infinite sequence is defined recursively by
$$a_1 = 2, \quad a_{k+1} = \frac{1}{2}\left(a_k + \frac{2}{a_k}\right).$$
Find decimal notation, rounded to six decimal places, for $S_1, S_2, S_3, S_4, S_5,$ and S_6. Make a conjecture about the value of S_∞.

35. By calculating values and graphing, show that the following infinite series of functions approximates a sawtooth wave (commonly used in electronics).
$$\frac{1}{2} - \frac{1}{\pi}\sin x - \frac{1}{2\pi}\sin 2x - \frac{1}{3\pi}\sin 3x - \cdots$$

15.5 MATHEMATICAL INDUCTION

Sequences of Statements

Infinite sequences of statements occur often in mathematics. In an infinite sequence of statements there is of course a statement for each natural number. For example, consider the sentence

For x between 0 and 1, $a < x^n < 1$.

Let us think of this as $S(n)$ or S_n. Substituting natural numbers for n gives a sequence of statements. We list a few of them.

Statement 1	(S_1):*	For x between 0 and 1, $a < x^1 < 1$.
Statement 2	(S_2):	For x between 0 and 1, $a < x^2 < 1$.
	S_3:	For x between 0 and 1, $a < x^3 < 1$.
	S_4:	For x between 0 and 1, $a < x^4 < 1$.

Example 1 List the first few statements in the sequence obtainable from $\log n < n$.

This time S_n is "$\log n < n$."

S_1: $\log 1 < 1$
S_2: $\log 2 < 2$
S_3: $\log 3 < 3$

Many sequences of statements concern sums.

Example 2 List the first few statements in the sequence obtainable from $1 + 3 + 5 + \cdots + (2n - 1) = n^2$.

This time the entire equation is S_n.

S_1: $1 = 1^2$
S_2: $1 + 3 = 2^2$
S_3: $1 + 3 + 5 = 3^2$
S_4: $1 + 3 + 5 + 7 = 4^2$

Proving Infinite Sequences of Statements

The method of proof of this section, called *mathematical induction*, allows us to prove infinite sequences of statements. They are usually verbalized somewhat as follows:

For all natural numbers n, S_n,

where S_n is some sentence such as those in the preceding examples. Of course we cannot prove each statement of an infinite sequence individually. Instead, we try to show that whenever S_k holds, then S_{k+1} must also hold. We abbreviate this as $S_k \to S_{k+1}$. (This is read "S_k implies S_{k+1}.") If we can establish that this holds for all natural numbers k, we then have the following.

$S_1 \to S_2$ meaning whenever S_1 holds, S_2 must hold;
$S_2 \to S_3$ meaning whenever S_2 holds, S_3 must hold;
$S_3 \to S_4$ meaning whenever S_3 holds, S_4 must hold;
and so on, indefinitely.

---

*Note that S_1, S_2, and so on do *not* represent sums in this context.

At this stage we do not yet know whether there is any k for which S_k holds. All we know is that *if* S_k holds, then S_{k+1} must hold. So we now show that S_k holds for some k, usually $k = 1$. We are then in a good position because we have the following.

S_1 is true.   We have verified, or proved this.

$S_1 \to S_2$ This means that whenever S_1 holds, S_2 must hold.

Therefore S_2 is true.

$S_2 \to S_3$ This means that whenever S_2 holds, S_3 must hold.

Therefore S_3 is true,

and so on.

We conclude that S_n is true for all natural numbers n.* We now state the principle of mathematical induction.

THE PRINCIPLE OF MATHEMATICAL INDUCTION

We can prove an infinite sequence of statements S_n by showing that

1. S_1 is true (this is called the *basis* step);
2. for all natural numbers k, $S_k \to S_{k+1}$ (this is called the *induction* step).

When learning to do proofs by mathematical induction, it is helpful to first write out S_n, S_1, S_k, and S_{k+1}. This helps to identify what is to be assumed and what is to be deduced.

Example 3 Prove: For every natural number n, $n < 2^n$.

We first list S_n, S_1, S_k, and S_{k+1}.

$$\begin{aligned} S_n: & \quad n < 2^n \\ S_1: & \quad 1 < 2^1 \\ S_k: & \quad k < 2^k \\ S_{k+1}: & \quad k+1 < 2^{k+1} \end{aligned}$$

1. Basis step. S_1 as listed is obviously true.
2. Induction step. We assume S_k as hypothesis and try to show that it implies S_{k+1}.

   For any k,

   $k < 2^k$ By hypothesis (this is S_k)

   $2k < 2 \cdot 2^k$ Multiplying on both sides by 2

   $2k < 2^{k+1}$.   Adding exponents on the right

---

*This is reminiscent of knocking over dominoes. The first one hits the second one, the second one hits the third, and so on.

Now, since k is any natural number,

$$1 \leq k$$
$$k + 1 \leq k + k \quad \text{Adding } k \text{ on both sides}$$
$$k + 1 \leq 2k.$$

Thus we have

$$k + 1 \leq 2k < 2^{k+1},$$
and $\quad k + 1 \leq 2^{k+1}.\quad$ This is S_{k+1}.

We have now shown that $S_k \to S_{k+1}$ for any natural number k. Thus the induction step and the proof are complete. We have proved that for every natural number n, $n < 2^n$.

Example 4 Prove: For every natural number n,

$$1 + 3 + 5 + 7 + \cdots + (2n - 1) = n^2.$$

We first list S_n, S_1, S_k, and S_{k+1}.

S_n: $\quad 1 + 3 + 5 + \cdots + (2n - 1) = n^2$
S_1: $\quad 1 = 1^2$
S_k: $\quad 1 + 3 + 5 + \cdots + (2k - 1) = k^2$
S_{k+1}: $\quad 1 + 3 + 5 + \cdots + (2k - 1) + [2(k + 1) - 1] = (k + 1)^2$

1. *Basis step.* S_1 as listed is obviously true.

2. *Induction step.* We assume S_k as hypothesis and try to show that it implies S_{k+1}.

$$1 + 3 + 5 + \cdots + (2k - 1) = k^2 \quad \text{for any natural number } k.$$

This is the hypothesis (it is S_k). Let us add $[2(k + 1) - 1]$ on both sides. This gives us

$$1 + 3 + \cdots + (2k - 1) + [2(k + 1) - 1] = k^2 + [2(k + 1) - 1]$$
$$= k^2 + 2k + 1$$
$$= (k + 1)^2.$$

We have arrived at S_{k+1}. Thus we have shown that for all natural numbers k, $S_k \to S_{k+1}$. This completes the induction step. The proof is complete. We have proved that the equation is true for all natural numbers n.

Example 5 Prove:

$$\sum_{p=1}^{n} \frac{1}{2^p} = \frac{2^n - 1}{2^n}, \quad \text{for all natural numbers } n.$$

In other words, prove that, for all natural numbers n,

$$\frac{1}{2} + \frac{1}{4} + \frac{1}{8} + \cdots + \frac{1}{2^n} = \frac{2^n - 1}{2^n}.$$

We first list S_n, S_1, S_k, and S_{k+1}.

S_n: $\quad \dfrac{1}{2} + \dfrac{1}{4} + \dfrac{1}{8} + \cdots + \dfrac{1}{2^n} = \dfrac{2^n - 1}{2^n}$

S_1: $\quad \dfrac{1}{2} = \dfrac{2^1 - 1}{2^1}$

S_k: $\quad \dfrac{1}{2} + \dfrac{1}{4} + \cdots + \dfrac{1}{2^k} = \dfrac{2^k - 1}{2^k}$

S_{k+1}: $\quad \dfrac{1}{2} + \dfrac{1}{4} + \dfrac{1}{8} + \cdots + \dfrac{1}{2^k} + \dfrac{1}{2^{k+1}} = \dfrac{2^{k+1} - 1}{2^{k+1}}$

1. *Basis step.* $\dfrac{2^1 - 1}{2^1} = \dfrac{2 - 1}{2} = \dfrac{1}{2}$, so S_1 is true.

2. *Induction step.* Using S_k as hypothesis, we have, for all natural numbers k,

$$\frac{1}{2} + \frac{1}{4} + \frac{1}{8} + \cdots + \frac{1}{2^k} = \frac{2^k - 1}{2^k}.$$

Let us add $1/2^{k+1}$ on both sides. Then we have

$$\frac{1}{2} + \frac{1}{4} + \cdots + \frac{1}{2^k} + \frac{1}{2^{k+1}} = \frac{2^k - 1}{2^k} + \frac{1}{2^{k+1}}$$

$$= \frac{2^k - 1}{2^k} \cdot \frac{2}{2} + \frac{1}{2^{k+1}}$$

$$= \frac{(2^k - 1) \cdot 2 + 1}{2^{k+1}} = \frac{2^{k+1} - 1}{2^{k+1}}.$$

We have arrived at S_{k+1}. Thus we have shown that for all natural numbers k, $S_k \to S_{k+1}$. This completes the induction and the proof is complete. We have proved that the equation is true for all natural numbers n.

EXERCISE SET 15.5

Use mathematical induction. In Exercises 1–18 prove for every natural number n.

1. $1 + 2 + 3 + \cdots + n = \dfrac{n(n + 1)}{2}$

2. $4 + 8 + 12 + \cdots + 4n = 2n(n + 1)$

3. $1 + 5 + 9 + \cdots + (4n - 3) = n(2n - 1)$

4. $3 + 6 + 9 + \cdots + 3n = \dfrac{3n(n + 1)}{2}$

5. $\dfrac{1}{1 \cdot 2} + \dfrac{1}{2 \cdot 3} + \cdots + \dfrac{1}{n(n + 1)} = \dfrac{n}{n + 1}$

6. $2 + 4 + 8 + \cdots + 2^n = 2(2^n - 1)$

7. $n < n + 1$

8. $2 \leq 2^n$

9. $3^n < 3^{n+1}$

10. $2n \leq 2^n$

11. $1^3 + 2^3 + 3^3 + \cdots + n^3 = \dfrac{n^2(n + 1)^2}{4}$

12. $\dfrac{1}{1 \cdot 2 \cdot 3} + \dfrac{1}{2 \cdot 3 \cdot 4} + \dfrac{1}{3 \cdot 4 \cdot 5} + \cdots + \dfrac{1}{n(n + 1)(n + 2)} = \dfrac{n(n + 3)}{4(n + 1)(n + 2)}$

☆

13. $\left(1 + \dfrac{1}{1}\right)\left(1 + \dfrac{1}{2}\right)\left(1 + \dfrac{1}{3}\right) \cdots \left(1 + \dfrac{1}{n}\right) = n + 1$

14. $a_1 + (a_1 + d) + (a_2 + d) + \cdots + [a_1 + (n - 1)d] = \dfrac{n}{2}[2a_1 + (n - 1)d]$ [This is a formula for the sum of the first n terms of an *arithmetic* sequence (Theorem 3).]

15. $a_1 + a_1 r + a_1 r^2 + \cdots + a_1 r^{n-1} = \dfrac{a_1 - a_1 r^n}{1 - r}$ [This is a formula for the sum of the first n terms of a *geometric* sequence (Theorem 5).]

16. $\cos n\pi = (-1)^n$
[Hint: Use the identity for $\cos(\alpha + \beta)$.]

17. DeMoivre's theorem. For any angle θ and any nonnegative number r, and $i^2 = -1$,

$$[r(\cos\theta + i\sin\theta)]^n = r^n(\cos n\theta + i\sin n\theta).$$

18. $|\sin(nx)| \leq n|\sin x|$

19. For every natural number $n \geq 2$,

$$\cos x \cdot \cos 2x \cdot \cos 4x \cdot \ldots \cdot \cos 2^{n-1}x = \dfrac{\sin 2^n x}{2^n \sin x}.$$

20. For every natural number $n \geq 2$,

$$\log_a(b_1 b_2 \cdots b_n) = \log_a b_1 + \log_a b_2 + \cdots + \log_a b_n.$$

21. For every natural number $n \geq 2$,

$$\left(1 - \dfrac{1}{2^2}\right)\left(1 - \dfrac{1}{3^2}\right) \cdots \left(1 - \dfrac{1}{n^2}\right) = \dfrac{n + 1}{2n}.$$

Prove the following for any complex numbers $z_1, \ldots, z_n$, where $i^2 = -1$ and $\bar{z}$ is the conjugate of z (see Section 12.2).

22. $\overline{z^n} = \bar{z}^n$

23. $\overline{z_1 + z_2 + \cdots + z_n} = \bar{z}_1 + \bar{z}_2 + \cdots + \bar{z}_n$

24. $\overline{z_1 \cdot z_2 \cdot \ldots \cdot z_n} = \bar{z}_1 \cdot \bar{z}_2 \cdot \ldots \cdot \bar{z}_n$

25. i^n is either $1, -1, i,$ or $-i$.

For any integers a and b, b is a factor of a if there exists an integer c such that $a = bc$. Prove the following for any natural number n.

26. 3 is a factor of $n^3 + 2n$

27. 2 is a factor of $n^2 + n$

28. 5 is a factor of $n^5 - n$

29. 3 is a factor of $n(n + 1)(n + 2)$

30. For every natural number $n \geq 2$,

$$\dfrac{1}{\sqrt{1}} + \dfrac{1}{\sqrt{2}} + \dfrac{1}{\sqrt{3}} + \cdots + \dfrac{1}{\sqrt{n}} > \sqrt{n}.$$

31. Bernoulli's inequality. For any real number $a > 1$ and any natural number $n \geq 2$,

$$(1 + a)^n > 1 + na.$$

32. Consider the statement: For every natural number n, $n = n + 1$.

   a) Can you prove the basis step? If so, do it.
   b) Can you prove the induction step? If so, do it.
   c) Is the statement true?

33. Find the error in this proof.
Statement. Everyone is of the same sex.
Proof. Let S_n be the statement: If A is a set of n people, then all the people are of the same sex.
Clearly, S_1 is true. Assume S_k. Let A be a set of $k + 1$ people. Then A is the union of two overlapping sets B and C, each containing k people. (Consider the illustration for $k = 5$.) By S_k all the people in B are of the same sex. Since B and C overlap, all the people in A are of the same sex, and $S_k \to S_{k+1}$.

34. *The Tower of Hanoi problem.* There are three pegs on a board. On one peg are n disks, each smaller than the one on which it rests. The problem is to move this pile of disks to another peg. The final order must be the same, but you can move only one disk at a time and you can never place a larger disk on a smaller one.

   a) What is the *least* number of moves it takes to move three disks?
   b) What is the *least* number of moves it takes to move four disks?
   c) What is the *least* number of moves it takes to move two disks?
   d) What is the *least* number of moves it takes to move one disk?
   e) Conjecture a formula for the *least* number of moves it takes to move n disks. Prove it by mathematical induction.

CHAPTER 15 REVIEW

1. Find the 10th term in the arithmetic sequence $\frac{3}{4}, \frac{13}{12}, \frac{17}{12}, \ldots$.

2. Find the 6th term in the arithmetic sequence $a - b$, $a, a + b, \ldots$.

3. Find the sum of the first 18 terms of the arithmetic sequence $4, 7, 10, \ldots$.

4. Find the sum of the first 30 positive integers.

5. The first term of an arithmetic sequence is 5. The 17th term is 53. Find the 3rd term.

6. The common difference in an arithmetic sequence is 3. The 10th term is 23. Find the first term.

7. For a geometric sequence, $a_1 = -2, r = 2$, and $a_n = -64$. Find n and S_n.

8. For a geometric sequence, $r = \frac{1}{2}, n = 5$, and $S_n = \frac{31}{2}$. Find a_1 and a_n.

9. Write the first three terms of the infinite geometric sequence with $r = 0.01$ and $S_\infty = \frac{3}{11}$.

10. Write the first three terms of the infinite geometric sequence with $r = -\frac{1}{3}$ and $S_\infty = \frac{3}{8}$.

11. Find fractional notation for $2.\overline{13}$.

12. Insert four arithmetic means between 5 and 9.

13. Find the geometric mean between 5 and 9.

14. A golf ball is dropped from a height of 30 ft to the pavement, and the rebound is $\frac{1}{4}$ the distance it drops. If, after each descent, it continues to rebound $\frac{1}{4}$ the distance dropped, what is the total distance the ball has traveled when it reaches the pavement on its 10th descent?

15. You receive 10¢ on the first day of the year, 12¢ on the 2nd day, 14¢ on the 3rd day, and so on. How much will you receive on the 365th day? What is the sum of all these 365 gifts?

16. The present population of a city is 30,000. Its population is supposed to double every 10 yr. What will its population be at the end of 80 yr?

17. The sides of a square are each 16 in. long. A second square is inscribed by joining the midpoints of the sides, successively. In the second square we repeat the process, inscribing a third square. If this process is continued indefinitely, what is the sum of the perimeters of all of the squares? (*Hint:* Use an infinite geometric series.)

18. A pendulum is moving back and forth in such a way that it traverses an arc 10 cm in length and thereafter arcs that are $\frac{4}{7}$ the length of the previous arc. What is the sum of the arc lengths that the pendulum traverses?

Use mathematical induction.

19. Prove: For all natural numbers n,
$$1 + 4 + 7 + \cdots + (3n - 2) = \frac{n(3n - 1)}{2}.$$

20. Prove: For all natural numbers n,
$$1 + 3 + 3^2 + \cdots + 3^{n-1} = \frac{3^n - 1}{2}.$$

21. Prove: For every natural number $n \geq 2$,
$$\left(1 - \frac{1}{2}\right) \cdot \left(1 - \frac{1}{3}\right) \cdot \cdots \cdot \left(1 - \frac{1}{n}\right) = \frac{1}{n}.$$

22. Explain why the following cannot be proved by mathematical induction. For every natural number n,
   a) $3 + 5 + \cdots + (2n + 1) = (n + 1)^2$;
   b) $1 + 3 + \cdots + (2n - 1) = n^2 + 3$.

23. Suppose a and b are geometric sequences. Prove that the sequence c is a geometric sequence where $c_n = a_n b_n$.

24. Suppose a is an arithmetic sequence. Prove that c is a geometric sequence, where $c_n = b^{a_n}$, for some positive number b.

25. Suppose a is an arithmetic sequence. Under what conditions is b an arithmetic sequence where:
   a) $b_n = |a_n|$?   b) $b_n = a_n + 8$?   c) $b_n = 7a_n$?
   d) $b_n = \frac{1}{a_n}$?   e) $b_n = \log a_n$?   f) $b_n = a_n^3$?

26. The zeros of this polynomial form an arithmetic sequence. Find them.
$$x^4 - 4x^3 - 4x^2 + 16x$$

16

Combinatorics and Probability

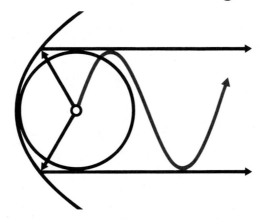

16.1 PERMUTATIONS

Combinatorics

We shall develop means of determining the number of ways a set of objects can be arranged or combined, certain objects can be chosen, or a succession of events can occur. The study of such things is called *combinatorics*.

Example 1 How many 3-letter code symbols can be formed with the letters A, B, and C *without* repetition?

Consider placing letters in these frames.

We can select any of the 3 letters for the first letter in the symbol. Once this letter has been selected, the second is selected from the remaining 2 letters. Then the third is already determined since there is only 1 letter left. The possibilities can be arrived at with a *tree diagram* (see Fig. 1). There are $3 \cdot 2 \cdot 1$ possibilities. The set of all of them is

$$\{ABC, ACB, BAC, BCA, CAB, CBA\}.$$

Suppose we perform an experiment such as selecting letters (as in the preceding example), flipping a coin, or drawing a card. The results are called *outcomes*. An *event* is a set of outcomes. When several events occur together, we say that the event is *compound*. The following principle is concerned with compound events.

Figure 1

FUNDAMENTAL COUNTING PRINCIPLE

In a compound event in which the first event can occur independently in n_1 ways, the second in n_2 ways, and the kth event in n_k ways, the total number of ways the compound event can occur is

$$n_1 \cdot n_2 \cdot n_3 \cdot \cdots \cdot n_k.$$

Example 2 How many 3-letter code symbols can be formed with the letters A, B, and C *with* repetition?

There are 3 choices for the first letter, and since we allow repetition, 3 choices for the second and 3 for the third. Thus by the fundamental counting principle there are $3 \cdot 3 \cdot 3$, or 27, choices.

Permutations

A *permutation* of a set of n objects is an ordered arrangement of the objects. Consider, for example, a set of 4 objects

$$\{A, B, C, D\}.$$

To find the number of ordered arrangements of the set, we select a first letter; there are 4 choices. Then we select a second letter; there are 3 choices. Then we select a third letter; there are 2 choices. Finally there is 1 choice for the last selection. Thus by the fundamental counting principle, there are $4 \cdot 3 \cdot 2 \cdot 1$, or 24 permutations of a set of 4 objects.

We can find a formula for the number of permutations of a set of n objects. We have n choices for the first selection, $n - 1$ for the second, $n - 2$ for the third, and so on. For the nth selection there is only one choice.

THEOREM 1

The total number of permutations of a set of n objects, denoted $_nP_n$, is given by

$$_nP_n = n(n - 1)(n - 2) \cdots 3 \cdot 2 \cdot 1.$$

Example 3 Find (a) $_4P_4$ and (b) $_7P_7$.

a) $_4P_4 = 4 \cdot 3 \cdot 2 \cdot 1 = 24$ b) $_7P_7 = 7 \cdot 6 \cdot 5 \cdot 4 \cdot 3 \cdot 2 \cdot 1 = 5040$

Factorial Notation

Products of successive natural numbers, such as $7 \cdot 6 \cdot 5 \cdot 4 \cdot 3 \cdot 2 \cdot 1$ and $9 \cdot 8 \cdot 7 \cdot 6 \cdot 5 \cdot 4 \cdot 3 \cdot 2 \cdot 1$, are used so often that it is convenient to adopt a notation for them.

For the product $7 \cdot 6 \cdot 5 \cdot 4 \cdot 3 \cdot 2 \cdot 1$ we write 7!, read "7 factorial."

DEFINITION

$$n! = n(n - 1)(n - 2) \cdots 3 \cdot 2 \cdot 1.$$

Here are some examples.

$$
\begin{aligned}
7! &= 7 \cdot 6 \cdot 5 \cdot 4 \cdot 3 \cdot 2 \cdot 1 = 5040 \\
6! &= 6 \cdot 5 \cdot 4 \cdot 3 \cdot 2 \cdot 1 = 720 \\
5! &= 5 \cdot 4 \cdot 3 \cdot 2 \cdot 1 = 120 \\
4! &= 4 \cdot 3 \cdot 2 \cdot 1 = 24 \\
3! &= 3 \cdot 2 \cdot 1 = 6 \\
2! &= 2 \cdot 1 = 2 \\
1! &= 1 = 1
\end{aligned}
$$

We also define 0! to be 1. We do this so that certain formulas and theorems can be stated concisely.

We can now simplify the formula of Theorem 1 as:

$$_nP_n = n!$$

Note that $8! = 8 \cdot 7!$ We can see this as follows.

By the definition of factorial notation,

$$8! = 8 \cdot 7 \cdot 6 \cdot 5 \cdot 4 \cdot 3 \cdot 2 \cdot 1$$
$$= 8 \cdot (7 \cdot 6 \cdot 5 \cdot 4 \cdot 3 \cdot 2 \cdot 1)$$
$$= 8 \cdot 7!.$$

Generalizing, we get:

For any natural number n, $n! = n(n-1)!$.

By using this result repeatedly, we can further manipulate factorial notation.

Example 4 Rewrite 7! with a factor of 5!.

$$7! = 7 \cdot 6 \cdot 5!$$

Permutations of n Objects Taken r at a Time

Consider a set of 6 objects. In how many ways can we construct an ordered subset having three members? We can select the first object in 6 ways. There are then 5 choices for the second and then 4 choices for the third. By the fundamental counting principle, there are then $6 \cdot 5 \cdot 4$ ways to construct the subset. In other words, there are $6 \cdot 5 \cdot 4$ permutations of a set of 6 objects taken three at a time. Note that

$$6 \cdot 5 \cdot 4 = \frac{6 \cdot 5 \cdot 4 \cdot 3 \cdot 2 \cdot 1}{3 \cdot 2 \cdot 1} \quad \text{or} \quad \frac{6!}{3!}.$$

Consider a set of n objects and the selecting of an ordered subset of r objects. The first object can be selected in n ways. The second can be selected in $n-1$ ways, and so on. The rth one can be selected in $n-(r-1)$ ways. By the fundamental counting principle, the total number of permutations is

$$n(n-1)(n-2) \cdots [n-(r-1)].$$

We now multiply by 1:

$$n(n-1)(n-2) \cdots [n-(r-1)] \frac{(n-r)!}{(n-r)!}$$

$$= \frac{n(n-1)(n-2)(n-3) \cdots [n-(r-1)](n-r)!}{(n-r)!}$$

16.1 PERMUTATIONS 569

The numerator is now the product of all natural numbers from n to 1, hence is $n!$. Thus the total number of permutations is

$$\frac{n!}{(n-r)!}.$$

This gives us the following theorem.

THEOREM 2

The number of permutations of a set of n objects taken r at a time, denoted $_nP_r$, is given by

$$_nP_r = n(n-1)(n-2) \cdots [n-(r-1)] \qquad (1)$$

$$= \frac{n!}{(n-r)!}. \qquad (2)$$

Formula (1) is most useful in application, but formula (2) will be important in a development given in Section 16.2.

Example 5 Compute $_6P_4$ using both formulas in Theorem 2.

Using formula (1), we have

$_6P_4 = 6 \cdot 5 \cdot 4 \cdot 3$ Note that the 6 in $_6P_4$ shows where to start and the 4 shows how many factors there are.

$= 360.$

Using formula (2) in Theorem 2, we have

$$_6P_4 = \frac{6!}{(6-4)!} = \frac{6!}{2!} = \frac{6 \cdot 5 \cdot 4 \cdot 3 \cdot 2 \cdot 1}{2 \cdot 1} = 6 \cdot 5 \cdot 4 \cdot 3 = 360.$$

Example 6 How many ways can letters of the set $\{A, B, C, D, E, F, G\}$ be arranged to form code words of (a) 7 letters? (b) 5 letters? (c) 4 letters? (d) 2 letters?

a) $_7P_7 = 7 \cdot 6 \cdot 5 \cdot 4 \cdot 3 \cdot 2 \cdot 1 = 5040$
b) $_7P_5 = 7 \cdot 6 \cdot 5 \cdot 4 \cdot 3 \qquad = 2520$
c) $_7P_4 = 7 \cdot 6 \cdot 5 \cdot 4 \qquad = 840$
d) $_7P_2 = 7 \cdot 6 \qquad = 42$

Example 7 A baseball manager arranges his batting order as follows: The 4 infielders will bat first, and then the outfielders, catcher, and pitcher will follow, not necessarily in that order. How many batting orders are possible?

The infielders can bat in 4! different ways; the rest in 5! different ways. Then by the fundamental counting principle we have $_4P_4 \cdot {_5P_5} = 4! \cdot 5!$, or 2880, possible batting orders.

Permutations of Sets with Nondistinguishable Objects

Consider a set of 7 marbles, four of which are red and three of which are black. Although the marbles are all different, when they are lined up, one black marble will look just like any other black marble. In this sense we say that the marbles are nondistinguishable.

Ⓡ Ⓡ Ⓡ Ⓡ Ⓑ Ⓑ Ⓑ

We know that there are 7! permutations of this set. Many of them will look alike, however. We develop a formula for finding the number of distinguishable permutations.

Consider a set of n objects in which n_1 are of one kind, n_2 are of a second kind, ..., n_k are of the kth kind. By Theorem 1 we know that the total number of permutations of the set is $n!$. Let P be the number of distinguishable permutations. For each of these P permutations there are $n_1!$ actual permutations, obtained by permuting the objects of the first kind. For each of these $P \cdot n_1!$ permutations there are $n_2!$ actual permutations, obtained by permuting the objects of the second kind. And so on. By the fundamental counting principle, the total number of actual permutations is

$$P \cdot n_1! \cdot n_2! \cdots n_k!$$

Then we have $P \cdot n_1! \cdot n_2! \cdots n_k! = n!$

Solving for P, we obtain

$$P = \frac{n!}{n_1! n_2! \cdots n_k!}$$

This proves the following theorem.

THEOREM 3

For a set of n objects in which n_1 are of one kind, n_2 are of another kind, ..., n_k are of a kth kind, the number of distinguishable permutations is
$$\frac{n!}{n_1! \cdot n_2! \cdots n_k!}.$$

Example 8 How many distinguishable ways can the letters of the word CINCINNATI be arranged?

Note: There are 2C's, 3I's, 3N's, 1A, and 1T, for a total of 10. Thus

$$P = \frac{10!}{2! \cdot 3! \cdot 3! \cdot 1! \cdot 1!}, \quad \text{or} \quad 50{,}400.$$

Circular Arrangements

Suppose we arrange the 4 letters A, B, C, and D in a circular arrangement as shown in Fig. 2. Note that the arrangements ABCD, BCDA, CDAB, and DABC are not distinguishable. For each circular arrangement there are 4 distinguishable arrangements on a line. Thus if there are P circular arrangements, these yield $4 \cdot P$ arrangements on a line, which we know is 4!. Thus $4 \cdot P = 4!$, so $P = 4!/4$, or 3!. To generalize:

Figure 2

THEOREM 4

The number of distinct (different) circular arrangements of n objects is $(n-1)!$.

Example 9 How many ways can nine different foods be arranged around a lazy Susan?

$$(9-1)! = 8! = 8 \cdot 7 \cdot 6 \cdot 5 \cdot 4 \cdot 3 \cdot 2 \cdot 1 = 40{,}320$$

Repeated Use of the Same Object

Example 10 How many 5-letter code symbols can be formed with the letters A, B, C, and D if we allow a letter to occur more than once?

We have five spaces:

We can select the first letter in 4 ways, the second in 4 ways, and so on. Thus there are 4^5, or 1024, arrangements. Generalizing we have the following.

THEOREM 5

The number of distinct arrangements of n objects taken r at a time, allowing for repetition, is n^r.

EXERCISE SET 16.1

Compute.

1. $_4P_3$
2. $_7P_5$
3. $_{10}P_7$
4. $_{10}P_3$

Table 5 at the back of the book can help with many of these problems.

5. How many 5-digit numbers can be named using the digits 5, 6, 7, 8, and 9 without repetition? with repetition?

6. How many 4-digit numbers can be named using the digits 2, 3, 4, and 5 without repetition? with repetition?

7. How many ways can 5 students be arranged in a straight line? in a circle?

8. How many ways can 7 athletes be arranged in a straight line? in a circle?

9. How many distinguishable ways can the letters of the word PEACE be arranged?

10. How many distinguishable ways can the letters of the word GROOVY be arranged?

11. How many 7-digit phone numbers can be formed with the digits 0, 1, 2, 3, 4, 5, 6, 7, 8, and 9, assuming that no digit is used more than once and that the first digit is not 0?

12. A program is planned and is to have five rock numbers and 4 speeches. In how many ways can this be done if a rock number and a speech are to alternate and the rock numbers come first?

13. Suppose the expression $a^2b^3c^4$ is expressed without exponents. In how many ways can this be done?

14. Suppose the expression a^3bc^2 is rewritten without exponents. In how many ways can this be done?

15. In how many ways could King Arthur and his 12 knights sit at his Round Table?

16. How many ways can 4 people be seated at a bridge table?

17. A penny, nickel, dime, quarter, and half dollar (if you have one) are arranged in a straight line. (a) Considering just the coins, in how many ways can they be lined up? (b) Considering the coins and heads and tails, in how many ways can they be lined up?

18. A penny, nickel, dime, and quarter are arranged in a straight line. (a) Considering just the coins, in how many ways can they be lined up? (b) Considering the coins and heads and tails, in how many ways can they be lined up?

19. ▪ Compute $_{52}P_4$.

20. ▪ Compute $_{50}P_5$.

21. A professor is going to grade her 24 students on a curve. She will give 3 A's, 5 B's, 9 C's, 4 D's, and 3 F's. In how many ways can she do this?

22. A professor is planning to grade his 20 students on a curve. He will give 2 A's, 5 B's, 8 C's, 3 D's, and 2 F's. In how many ways can he do this?

23. ▪ How many distinguishable code symbols can be formed from the letters of the word MATH? BUSINESS? PHILOSOPHICAL?

24. ▪ How many distinguishable code symbols can be formed from the letters of the word ORANGE? BIOLOGY? MATHEMATICS?

25. ▪ A state forms its license plates by first listing a number that corresponds to the county in which the car owner lives (the names of the counties are alphabetized and the number is its location in the order). Then the plate lists a letter of the alphabet and this is followed by a number from 1 to 9999. How many such plates are possible when there are 80 counties?

26. How many code symbols can be formed using 4 out of 5 letters of A, B, C, D, and E if the letters

   a) are not repeated?
   b) can be repeated?
   c) are not repeated but must begin with D?
   d) are not repeated but must end with DE?

Solve for n.

27. $_nP_5 = 7 \cdot {_nP_4}$
28. $_nP_4 = 8 \cdot {_{n-1}P_3}$
29. $_nP_5 = 9 \cdot {_{n-1}P_4}$
30. $_nP_4 = 8 \cdot {_nP_3}$

31. In a single-elimination sports tournament consisting of n teams, a team is eliminated when it loses one game. How many games are required to complete the tournament?

32. In a double-elimination softball tournament consisting of n teams, a team is eliminated when it loses two games. At most how many games are required to complete the tournament?

16.2 COMBINATIONS

Permutations of a set are subsets arranged in order. *Combinations* of a set are simply subsets and are not considered with regard to order.

Example 1 How many combinations are there of the set $\{A, B, C, D\}$ taken 3 at a time?

The combinations are the subsets $\{A, B, C\}$, $\{A, C, D\}$, $\{B, C, D\}$, $\{A, B, D\}$. There are four of them.

DEFINITION

The number of combinations taken r at a time from a set of n objects, denoted $_nC_r$, is the number of subsets containing exactly r objects.

We shall develop a formula for finding $_nC_r$, but we first define some convenient symbolism.

DEFINITION

$\binom{n}{r}$ is defined to mean $\dfrac{n!}{r!(n-r)!}$.

The notation $\binom{n}{r}$ is read "n over r," or (for reasons we will see later) "binomial coefficient n over r." It does *not* mean $n \div r$.

Example 2 Evaluate $\binom{5}{2}$.

$$\binom{5}{2} = \frac{5!}{2!(5-2)!} = \frac{5!}{2!3!}$$
$$= \frac{5 \cdot 4 \cdot 3!}{2!3!} = \frac{5 \cdot 4}{2}$$
$$= 10$$

We now find a formula for $_nC_r$. Consider a set of n objects. The number of subsets with r members is $_nC_r$. Each subset with r members can be arranged in $r!$ ways. By the fundamental counting principle we have $_nC_r \cdot r!$ ordered subsets with r members. This gives us

$$_nP_r = {_nC_r} \cdot r!.$$

Then

$$_nC_r = \frac{_nP_r}{r!}.$$

Since

$$_nP_r = \frac{n!}{(n-r)!}, \quad \text{Theorem 2}$$

we get

$$_nC_r = \frac{n!}{r!(n-r)!},$$

or

$$_nC_r = \binom{n}{r}.$$

This proves the following.

THEOREM 6

The number of combinations of a set of n objects taken r at a time is given by

$$_nC_r = \binom{n}{r} = \frac{n!}{r!(n-r)!}.$$

Example 3 Compute $_7C_4$.

By Theorem 6 we have $_7C_4 = \binom{7}{4}$.

$$\binom{7}{4} = \frac{7!}{4!3!} = \frac{7 \cdot 6 \cdot 5 \cdot 4!}{3!4!} = \frac{7 \cdot 6 \cdot 5}{3 \cdot 2} = 35$$

Example 4 How many committees can be formed from a set of five governors and seven senators if each committee contains three governors and four senators?

The three governors can be selected in $_5C_3$ ways and the four senators can be selected in $_7C_4$ ways. Using the fundamental counting principle, it follows that the number of possible committees is

$$_5C_3 \cdot {_7C_4} = 10 \cdot 35 = 350.$$

EXERCISE SET 16.2

Compute. Table 5 at the back of the book or a calculator can be of help with these problems.

1. $_9C_5$

2. $_{14}C_2$

3. $\binom{50}{2}$

4. $\binom{40}{3}$

5. $_nC_3$

6. $_nC_2$

7. There are 23 students in a fraternity. How many sets of 4 officers can be selected?

8. How many basketball games can be played in a 9-team league if each team plays all other teams once? twice?

9. On a test a student is to select 6 out of 10 questions. In how many ways can he do this?

10. On a test a student is to select 7 out of 11 questions. In how many ways can she do this?

11. How many lines are determined by 8 points, no 3 of which are collinear? How many triangles are determined by the same points if no 4 are coplanar?

12. How many lines are determined by 7 points, no 3 of which are collinear? How many triangles are determined by the same points if no 4 are coplanar?

13. Of the first 10 questions on a test, a student must answer 7. On the second 5 questions, she must answer 3. In how many ways can this be done?

14. Of the first 8 questions on a test, a student must answer 6. On the second 4 questions, he must answer 3. In how many ways can this be done?

15. Suppose the Senate of the United States consists of 58 Democrats and 42 Republicans. How many committees consisting of 6 Democrats and 4 Republicans could be formed? You need not simplify the expression.

16. Suppose the Senate of the United States consists of 63 Republicans and 37 Democrats. How many committees consisting of 8 Republicans and 12 Democrats could be formed? You need not simplify the expression.

17. How many 5-card poker hands consisting of 3 aces and 2 cards that are not aces are possible with a 52-card deck? See Fig. 5 on p. 582 for a description of a 52-card deck.

18. How many 5-card poker hands consisting of 2 kings and 3 cards that are not kings are possible with a 52-card deck?

19. How many 5-card poker hands are possible with a 52-card deck?

20. How many 13-card bridge hands are possible with a 52-card deck?

21. There are 8 points on a circle. How many triangles can be inscribed with these points as vertices?

22. There are n points on a circle. How many quadrilaterals can be inscribed with these points as vertices?

23. A set of 5 parallel lines crosses another set of 8 parallel lines at angles that are not right angles. How many parallelograms are formed?

24. Prove: For any natural numbers n and $r \leq n$,
$$\binom{n}{r} = \binom{n}{n-r}.$$

Solve for n.

25. $\binom{n+1}{3} = 2 \cdot \binom{n}{2}$

26. $\binom{n}{n-2} = 6$

27. $\binom{n+2}{4} = 6 \cdot \binom{n}{2}$

28. $\binom{n}{3} = 2 \cdot \binom{n-1}{2}$

29. How many line segments are determined by the 5 vertices of a pentagon? Of these, how many are diagonals?

30. How many line segments are determined by the 6 vertices of a hexagon? Of these, how many are diagonals?

31. How many line segments are determined by the n vertices of an n-gon? Of these, how many are diagonals? Use mathematical induction to prove the result for the diagonals.

32. Prove: For any natural numbers n and $r \leq n$,
$$\binom{n}{r-1} + \binom{n}{r} = \binom{n+1}{r}.$$

16.3 BINOMIAL SERIES

Consider the following expanded powers of $(a + b)^n$, where $a + b$ is any binomial. Look for patterns.

$$(a + b)^0 = 1$$
$$(a + b)^1 = a + b$$
$$(a + b)^2 = a^2 + 2ab + b^2$$
$$(a + b)^3 = a^3 + 3a^2b + 3ab^2 + b^3$$
$$(a + b)^4 = a^4 + 4a^3b + 6a^2b^2 + 4ab^3 + b^4$$
$$(a + b)^5 = a^5 + 5a^4b + 10a^3b^2 + 10a^2b^3 + 5ab^4 + b^5$$

Each expansion is a polynomial that can be regarded as a series. There are some patterns to be noted in the expansions.

1. In each term, the sum of the exponents is n.

2. The exponents of a start with n and decrease to 0. The last term has no factor of a. The first term has no factor of b. The exponents of b start in the second term with 1 and increase to n.

3. There is one more term than the degree of the polynomial. The expansion of $(a + b)^n$ has $n + 1$ terms.

We now find a way to determine the coefficients. Let us consider the nth power of a binomial $(a + b)$:

$$(a + b)^n = \underbrace{(a + b)(a + b)(a + b) \cdots (a + b)}_{n \text{ factors}}.$$

When we multiply, we will find all possible products of a's and b's. For example, when we multiply all the first terms we will get n factors of a, or a^n. Thus the first term in the expansion is a^n, and the first coefficient is 1. Similarly, the coefficient of the last term is 1.

To get a term such as the $a^{n-r}b^r$ term, we will take a's from $n - r$ factors and b's from r factors. Thus we take n objects, $n - r$ of them a's and r of them b's. The number of ways we can do this is

$$\frac{n!}{(n - r)!r!}.$$

This is $\binom{n}{r}$. Thus the $(r + 1)$st term in the expansion is $\binom{n}{r}a^{n-r}b^r$. We now have a theorem.

THEOREM 7

The binomial theorem. For any binomial $(a + b)$ and any natural number n,

$$(a + b)^n = \binom{n}{0}a^n + \binom{n}{1}a^{n-1}b + \binom{n}{2}a^{n-2}b^2 + \cdots + \binom{n}{n}b^n.$$

Sigma notation for a binomial series is as follows:

$$(a + b)^n = \sum_{r=0}^{n} \binom{n}{r} a^{n-r} b^r.$$

Because of Theorem 7, $\binom{n}{r}$ is called a *binomial coefficient*. It should now be apparent why 0! is defined to be 1. In the binomial expansion we want $\binom{n}{0}$ to equal 1 and we also want the definition

$$\binom{n}{r} = \frac{n!}{(n-r)!r!}$$

to hold for all whole numbers n and r. Thus we must have

$$\binom{n}{0} = \frac{n!}{(n-0)!0!} = \frac{n!}{n!0!} = 1.$$

This will be satisfied if 0! is defined to be 1.

Example 1 Find the 7th term of $(4x - y^3)^9$.

We let $r = 6$, $n = 9$, $a = 4x$, and $b = -y^3$ in the formula $\binom{n}{r} a^{n-r} b^r$. Then

$$\binom{9}{6}(4x)^3(-y^3)^6 = \frac{9!}{6!3!}(4x)^3(-y^3)^6 = \frac{9 \cdot 8 \cdot 7 \cdot 6!}{3! \cdot 6!} 64x^3 y^{18}$$

$$= 5376x^3 y^{18}.$$

Example 2 Expand: $(x^2 - 2y)^5$.

Note that $a = x^2$, $b = -2y$, and $n = 5$. Then using the binomial theorem we have

$$(x^2 - 2y)^5 = \binom{5}{0}(x^2)^5 + \binom{5}{1}(x^2)^4(-2y) + \binom{5}{2}(x^2)^3(-2y)^2$$

$$+ \binom{5}{3}(x^2)^2(-2y)^3 + \binom{5}{4}x^2(-2y)^4 + \binom{5}{5}(-2y)^5$$

$$= \frac{5!}{0!5!}x^{10} + \frac{5!}{1!4!}x^8(-2y) + \frac{5!}{2!3!}x^6(-2y)^2$$

$$+ \frac{5!}{3!2!}x^4(-2y)^3 + \frac{5!}{4!1!}x^2(-2y)^4 + \frac{5!}{5!0!}(-2y)^5$$

$$= x^{10} - 10x^8 y + 40x^6 y^2 - 80x^4 y^3 + 80x^2 y^4 - 32y^5.$$

Example 3 Expand: $\left(\dfrac{2}{x} + 3\sqrt{x}\right)^4$.

Note that $a = 2/x$, $b = 3\sqrt{x}$, and $n = 4$. Then using the binomial theorem we have

$$\left(\dfrac{2}{x} + 3\sqrt{x}\right)^4 = \binom{4}{0} \cdot \left(\dfrac{2}{x}\right)^4 + \binom{4}{1} \cdot \left(\dfrac{2}{x}\right)^3 (3\sqrt{x})$$

$$+ \binom{4}{2} \cdot \left(\dfrac{2}{x}\right)^2 (3\sqrt{x})^2$$

$$+ \binom{4}{3}\left(\dfrac{2}{x}\right)(3\sqrt{x})^3 + \binom{4}{4}(3\sqrt{x})^4$$

$$= \dfrac{4!}{0!4!} \cdot \dfrac{16}{x^4} + \dfrac{4!}{1!3!} \cdot \dfrac{8}{x^3} 3\sqrt{x} + \dfrac{4!}{2!2!} \cdot \dfrac{4}{x^2} \cdot 9x$$

$$+ \dfrac{4!}{3!1!} \cdot \dfrac{2}{x} \cdot 27x^{3/2} + \dfrac{4!}{4!0!} \cdot 81x^2$$

$$= \dfrac{16}{x^4} + \dfrac{96}{x^{5/2}} + \dfrac{216}{x} + 216\sqrt{x} + 81x^2.$$

Subsets

Suppose a set has n objects. The number of subsets containing r members is $\binom{n}{r}$, by Theorem 6. The total number of subsets of a set is the number with 0 elements, plus the number with 1 element, plus the number with 2 elements, and so on. The total number of subsets of a set with n members is

$$\binom{n}{0} + \binom{n}{1} + \binom{n}{2} + \cdots + \binom{n}{n}.$$

Now let us expand $(1 + 1)^n$:

$$(1 + 1)^n = \binom{n}{0} + \binom{n}{1} + \binom{n}{2} + \cdots + \binom{n}{n}.$$

Thus the total number of subsets is $(1 + 1)^n$ or 2^n. We have proved the following theorem.

THEOREM 8

The total number of subsets of a set with n members is 2^n.

Example 4 The set $\{A, B, C, D, E\}$ has how many subsets?

The set has 5 members, so the number of subsets is 2^5, or 32.

Example 5 A fast-food restaurant makes hamburgers 256 ways using combinations of 8 seasonings. Show why.

The total number of combinations is

$$\binom{8}{0} + \binom{8}{1} + \cdots + \binom{8}{8} = 2^8$$
$$= 256.$$

EXERCISE SET 16.3

Find the indicated term of the binomial expression.

1. 3rd, $(a + b)^6$
2. 6th, $(x + y)^7$
3. 12th, $(a - 2)^{14}$
4. 11th, $(x - 3)^{12}$
5. 5th, $(2x^3 - \sqrt{y})^8$
6. 4th, $\left(\dfrac{1}{b^2} + \dfrac{b}{3}\right)^7$
7. Middle, $(2u - 3v^2)^{10}$
8. Middle two, $(\sqrt{x} + \sqrt{3})^5$

Expand.

9. $(m + n)^5$
10. $(a - b)^4$
11. $(x^2 - 3y)^5$
12. $(3c - d)^6$
13. $(x^{-2} + x^2)^4$
14. $\left(\dfrac{1}{\sqrt{x}} - \sqrt{x}\right)^6$
15. $(1 - 1)^n$
16. $(1 + 3)^n$
17. $(\sqrt{2} + 1)^6 - (\sqrt{2} - 1)^6$
18. $(1 - \sqrt{2})^4 + (1 + \sqrt{2})^4$
19. $(\sqrt{3} - t)^4$
20. $(\sqrt{5} + t)^6$

Determine the number of subsets:

21. Of a set of 7 members
22. Of a set of 6 members
23. Of the set of letters of the English alphabet, which contains 26 letters
24. Of the set of letters of the Greek alphabet, which contains 24 letters

☆

Expand.

25. $(\sqrt{2} - i)^4$, where $i^2 = -1$
26. $(1 + i)^6$, where $i^2 = -1$
27. $(\sin t - \csc t)^7$
28. $(\tan \theta + \cot \theta)^{11}$
29. Find a formula for $(a - b)^n$. Use sigma notation.
30. Expand and simplify $\dfrac{(x + h)^n - x^n}{h}$. Use sigma notation.

Solve for x.

31. $\sum_{r=0}^{8} \binom{8}{r} x^{8-r} 3^r = 0$

32. $\sum_{r=0}^{4} \binom{4}{r} 5^{4-r} x^r = 64$

33. $\sum_{r=0}^{5} \binom{5}{r} (-1)^r x^{5-r} 3^r = 32$

34. $\sum_{r=0}^{9} \binom{9}{r} \sin^r x = 0$

35. A money clip contains one each of the following bills: $1, $2, $5, $10, $20, $50, and $100. How many different sums of money can be formed using the bills?

Simplify.

36. $\sum_{r=0}^{23} \binom{23}{r} (\log_a x)^{23-r} (\log_a t)^r$

37. $\ln \left[\sum_{r=0}^{15} (\cos t)^{30-2r} (\sin t)^{2r} \right]$

38. Use mathematical induction and the property $\binom{k}{r-1} + \binom{k}{r} = \binom{k+1}{r}$ to prove the binomial theorem.

16.4 PROBABILITY

We say that when a coin is tossed the chances that it will fall heads are 1 out of 2, or that the *probability* that it will fall heads is $\frac{1}{2}$. Of course this does not mean that if a coin is tossed 10 times it will necessarily fall heads exactly 5 times. If the coin is tossed a great number of times, however, it will fall heads very nearly half of them.

Experimental and Theoretical Probability

If we toss a coin a large number of times, say 1000, and count the number of heads, we can determine the probability. If there are 503 heads, we would calculate the probability to be

$$\frac{503}{1000} \quad \text{or} \quad 0.503.$$

This is an *experimental* determination of probability.

If we consider a coin and reason that it is just as likely to fall heads as tails we would calculate the probability to be $\frac{1}{2}$. This is a *theoretical* determination of probability. You might ask "What is the *true* probability?" In fact there is none. Experimentally we can determine probabilities within certain limits. These may or may not agree with what we obtain theoretically.

We need some terminology before we can continue. Suppose we perform an experiment such as flipping a coin, throwing a dart, drawing a card from a deck, or checking an item off an assembly line for quality. The results of an experiment are called the *outcomes*. The set of all possible outcomes is called the *sample space*. An *event* is a set of outcomes; that is, a subset of the sample space. For example, for the experiment "throwing a dart," suppose the board was A in Fig. 3. Then one event is "hitting red," which is a subset of the sample space "red, white, gray," assuming the dart hits the target somewhere.

We denote the probability that an event E occurs as P(E). For example, "getting a head" may be denoted by H. Then P(H) represents the probability of getting a head. When the outcomes of an experiment all have the same probability of occurring, we say that they are *equally likely*. To see the distinction between events that are equally likely and those that are not, consider the dartboards shown in Fig. 3. For dartboard A, the events, *hitting red, hitting white,* and *hitting gray* are equally likely, but for board B they are not. A sample space that can be expressed as a union of equally likely events allows us to calculate probabilities of other events.

Figure 3

Principle P. **If an event E can occur m ways out of n possible equally likely outcomes of a sample space S, the probability of that event is given by**

$$P(E) = \frac{m}{n}.$$

A die (pl., dice) is a cube, with six faces, each containing a number of dots from 1 to 6, as shown in Fig. 4.

Figure 4

Example 1 What is the probability of rolling a 3 on a die?

On a fair die there are 6 equally likely outcomes and there is 1 way to get a 3. By Principle P, $P(3) = \frac{1}{6}$.

Example 2 What is the probability of rolling an even number on a die?

The event is getting an *even* number. It can occur in 3 ways (getting 2, 4, or 6). The number of equally likely outcomes is 6. By Principle P, $P(\text{even}) = \frac{3}{6}$ or $\frac{1}{2}$.

We use a number of examples related to a standard bridge deck of 52 cards. Such a deck is made up as shown in Fig. 5.

Figure 5

Example 3 What is the probability of drawing an ace from a well-shuffled deck of 52 cards?

Since there are 52 outcomes (cards in the deck) and they are equally likely (from a well-shuffled deck) and there are 4 ways to obtain an ace, by Principle P we have

$$P(\text{drawing an ace}) = \frac{4}{52}, \quad \text{or} \quad \frac{1}{13}.$$

Example 4 Suppose we select, without looking, one marble from a bag containing 3 red marbles and 4 green marbles. What is the probability of selecting a red marble?

There are 7 equally likely ways of selecting any marble, and since the number of ways of getting a red marble is 3,

$$P(\text{selecting a red marble}) = \frac{3}{7}.$$

16.4 PROBABILITY

The following are some results that follow from Principle P.

THEOREM 9

If an event E cannot occur, then $P(E) = 0$.

For example in coin tossing, the event that a coin would land on its edge has probability 0.

THEOREM 10

If an event E is certain to occur (every trial is a success), then $P(E) = 1$.

For example in coin tossing, the event that a coin falls either heads or tails has probability 1.

In general,

THEOREM 11

The probability that an event E will occur is a number from 0 to 1.

$$0 \leq P(E) \leq 1.$$

Example 5 Suppose 2 cards are drawn from a well-shuffled deck of 52 cards. What is the probability that both of them are spades?

The number n of ways of drawing 2 cards from a deck of 52 is $_{52}C_2$. Now 13 of the 52 cards are spades, so the number m of ways of drawing 2 spades is $_{13}C_2$. Thus,

$$P(\text{getting 2 spades}) = \frac{m}{n} = \frac{_{13}C_2}{_{52}C_2} = \frac{78}{1326} = \frac{1}{17}.$$

Example 6 Suppose 2 people are selected at random from a group that consists of 6 men and 4 women. What is the probability that both of them are women?

The number of ways of selecting 2 people from a group of 10 is $_{10}C_2$. The number of ways of selecting 2 women from a group of 4 is $_4C_2$. Thus the probability of selecting 2 women from the group of 10 is P, where

$$P = \frac{_4C_2}{_{10}C_2} = \frac{6}{45} = \frac{2}{15}.$$

Example 7 Suppose 3 people are selected at random from a group that consists of 6 men and 4 women. What is the probability that 1 man and 2 women are selected?

The number of ways of selecting 3 people from a group of 10 is $_{10}C_3$. One man can be selected in $_6C_1$ ways, and 2 women can be selected in $_4C_2$ ways. By the fundamental counting principle we know that the number of ways of selecting 1 man and 2 women is $_6C_1 \cdot {_4C_2}$. Thus the probability is

$$\frac{_6C_1 \cdot {_4C_2}}{_{10}C_3}, \quad \text{or} \quad \frac{3}{10}.$$

Example 8 What is the probability of getting a total of 8 on a roll of a pair of dice? (Assume the dice are different, say one red and one black.)

On each die there are 6 possible outcomes. The outcomes are paired so there are $6 \cdot 6$, or 36, possible ways in which the two can fall. The pairs that total 8 are as shown in Fig. 6. Thus there are 5 possible ways of getting a total of 8, so the probability is $\frac{5}{36}$.

Red die

6	(1, 6)	(2, 6)	(3, 6)	(4, 6)	(5, 6)	(6, 6)
5	(1, 5)	(2, 5)	(3, 5)	(4, 5)	(5, 5)	(6, 5)
4	(1, 4)	(2, 4)	(3, 4)	(4, 4)	(5, 4)	(6, 4)
3	(1, 3)	(2, 3)	(3, 3)	(4, 3)	(5, 3)	(6, 3)
2	(1, 2)	(2, 2)	(3, 2)	(4, 2)	(5, 2)	(6, 2)
1	(1, 1)	(2, 1)	(3, 1)	(4, 1)	(5, 1)	(6, 1)
	1	2	3	4	5	6	Black die

Figure 6

Origin and Use of Probability

A desire to calculate odds in games of chance gave rise to the theory of probability. Today the theory of probability and its closely related field, mathematical statistics, have many applications, most of them not related to games of chance. Opinion polls, with such uses as predicting elections, are a familiar example. Quality control, in which a prediction about the percentage of faulty items manufactured is made without testing them all, is an important application, among many, in business. Still other applications are in the areas of genetics, medicine, and the kinetic theory of gases.

EXERCISE SET 16.4

Suppose we draw a card from a well-shuffled deck of 52 cards.

1. How many equally likely outcomes are there?
2. What is the probability of drawing a queen?
3. What is the probability of drawing a heart?
4. What is the probability of drawing a club?
5. What is the probability of drawing a 4?
6. What is the probability of drawing a red card?
7. What is the probability of drawing a black card?
8. What is the probability of drawing an ace or a deuce?
9. What is the probability of drawing a 9 or a king?

Suppose we select, without looking, one marble from a bag containing 4 red marbles and 10 green marbles.

10. What is the probability of selecting a red marble?
11. What is the probability of selecting a green marble?
12. What is the probability of selecting a purple marble?
13. What is the probability of selecting a white marble?

Suppose 4 cards are drawn from a well-shuffled deck of 52 cards.

14. What is the probability that all 4 are spades?
15. What is the probability that all 4 are hearts?
16. If 4 marbles are drawn at random all at once from a bag containing 8 white marbles and 6 black marbles, what is the probability that 2 will be white and 2 will be black?
17. From a group of 8 men and 7 women, a committee of 4 is chosen. What is the probability that 2 men and 2 women will be chosen?
18. What is the probability of getting a total of 6 on a roll of a pair of dice?
19. What is the probability of getting a total of 3 on a roll of a pair of dice?
20. What is the probability of getting snake eyes (a total of 2) on a roll of a pair of dice?
21. What is the probability of getting box-cars (a total of 12) on a roll of a pair of dice?
22. From a bag containing 5 nickels, 8 dimes, and 7 quarters, 5 coins are drawn at random all at once. What is the probability of getting 2 nickels, 2 dimes, and 1 quarter?
23. From a bag containing 6 nickels, 10 dimes, and 4 quarters, 6 coins are drawn at random all at once. What is the probability of getting 3 nickels, 2 dimes, and 1 quarter?

Roulette. A roulette wheel contains slots numbered 00, 0, 1, 2, 3, ..., 35, 36. Eighteen of the slots numbered 1 through 36 are colored red and eighteen are colored black. The 00 and 0 slots are uncolored. The wheel is spun, and a ball is rolled around the rim until it falls into a slot. What is the probability that the ball falls in:

24. a black slot?
25. a red slot?
26. a red or black slot?
27. the 00 slot?
28. the 0 slot?
29. either the 00 or 0 slot? (Here the house always wins.)
30. an odd-numbered slot?

Five-card poker hands and probabilities. In part (a) of each problem, give a reasoned expression as well as the answer. Read all the problems before beginning.

31. ▨ How many 5-card poker hands can be dealt from a standard 52-card deck?
32. ▨ A *royal flush* consists of a 5-card hand with A-K-Q-J-10 of the same suit.
    a) How many royal flushes are there?
    b) What is the probability of getting a royal flush?

33. A *straight flush* consists of five cards in sequence in the same suit, but excludes royal flushes.

    a) How many straight flushes are there?

    b) What is the probability of getting a straight flush?

34. *Four of a kind* is a 5-card hand in which 4 of the cards are of the same denomination, such as J-J-J-J-6, 7-7-7-7-A, or 2-2-2-2-5.

    a) How may are there?

    b) What is the probability of getting four of a kind?

★

35. A *full house* consists of a pair and three of a kind, such as Q-Q-Q-4-4.

    a) How many are there?

    b) What is the probability of getting a full house?

36. A *pair* is a 5-card hand in which just 2 of the cards are the same denomination, such as Q-Q-8-A-3.

    a) How many are there?

    b) What is the probability of getting a pair?

37. *Three of a kind* consists of a 5-card hand in which 3 of the cards are the same denomination, such as Q-Q-Q-10-7.

    a) How many are there?

    b) What is the probability of getting three of a kind?

38. A *flush* is a 5-card hand in which all the cards are the same suit, but not all in sequence (not a straight flush or royal flush).

    a) How many are there?

    b) What is the probability of getting a flush?

39. *Two pairs* is a hand such as Q-Q-3-3-A.

    a) How many are there?

    b) What is the probability of getting two pairs?

40. A *straight* is any five cards in sequence, but not in the same suit. For example, 4 of spades, 5 of spades, 6 of diamonds, 7 of hearts, and 8 of clubs.

    a) How many are there?

    b) What is the probability of getting a straight?

CHAPTER 16 REVIEW

1. In how many ways can 6 books be arranged on a shelf?

2. If 9 different signal flags are available, how many different displays are possible using 4 flags in a row?

3. The winner of a contest can choose any 8 of 15 prizes. How many different selections can be made?

4. The Greek alphabet contains 24 letters. How many fraternity or sorority names can be formed using 3 different letters?

5. In how many distinguishable ways can the letters of the word TENNESSEE be arranged?

6. A manufacturer of houses has one floor plan but achieves variety by having 3 different colored roofs, 4 different ways of attaching the garage, and 3 different types of entrance. Find the number of different houses that can be produced.

7. How many code symbols can be formed using 5 out of 6 of the letters of G, H, I, J, K, L if the letters:

    a) are not repeated?

    b) can be repeated?

    c) are not repeated but must begin with K?

    d) are not repeated but must end with IGH?

8. In how many different ways can 7 people be seated at a round table?

Solve for n.

9. $\binom{n}{n-1} = 36$

10. $4 \cdot \binom{n}{1} = \binom{n}{3}$

11. Find the 4th term of $(a + x)^{12}$.

12. Find the 12th term of $(a + x)^{18}$. Do not multiply out the factorials.

Expand.

13. $(m + n)^7$

14. $(x^2 + 3y)^4$

15. $(5i + 1)^6$, where $i^2 = -1$

16. $(\cos t + \sec t)^8$

Simplify.

17. $\sum_{r=0}^{10} (-1)^r (\log x)^{10-r} (\log y)^r$

18. What is the probability of rolling a 10 on a roll of a pair of dice? on a roll of one die?

19. From a deck of 52 cards 1 card is drawn. What is the probability that it is a club?

20. From a deck of 52 cards 3 are drawn at random without replacement. What is the probability that 2 are aces and 1 is a king.

Tables

Table 1 Powers, Roots, and Reciprocals

n	n^2	n^3	$\sqrt{n}$	$\sqrt[3]{n}$	$\sqrt{10n}$	$\frac{1}{n}$	n	n^2	n^3	$\sqrt{n}$	$\sqrt[3]{n}$	$\sqrt{10n}$	$\frac{1}{n}$
1	1	1	1.000	1.000	3.162	1.0000	51	2,601	132,651	7.141	3.708	22.583	.0196
2	4	8	1.414	1.260	4.472	.5000	52	2,704	140,608	7.211	3.733	22.804	.0192
3	9	27	1.732	1.442	5.477	.3333	53	2,809	148,877	7.280	3.756	23.022	.0189
4	16	64	2.000	1.587	6.325	.2500	54	2,916	157,464	7.348	3.780	23.238	.0185
5	25	125	2.236	1.710	7.071	.2000	55	3,025	166,375	7.416	3.803	23.452	.0182
6	36	216	2.449	1.817	7.746	.1667	56	3,136	175,616	7.483	3.826	23.664	.0179
7	49	343	2.646	1.913	8.367	.1429	57	3,249	185,193	7.550	3.849	23.875	.0175
8	64	512	2.828	2.000	8.944	.1250	58	3,364	195,112	7.616	3.871	24.083	.0172
9	81	729	3.000	2.080	9.487	.1111	59	3,481	205,379	7.681	3.893	24.290	.0169
10	100	1,000	3.162	2.154	10.000	.1000	60	3,600	216,000	7.746	3.915	24.495	.0167
11	121	1,331	3.317	2.224	10.488	.0909	61	3,721	226,981	7.810	3.936	24.698	.0164
12	144	1,728	3.464	2.289	10.954	.0833	62	3,844	238,328	7.874	3.958	24.900	.0161
13	169	2,197	3.606	2.351	11.402	.0769	63	3,969	250,047	7.937	3.979	25.100	.0159
14	196	2,744	3.742	2.410	11.832	.0714	64	4,096	262,144	8.000	4.000	25.298	.0156
15	225	3,375	3.873	2.466	12.247	.0667	65	4,225	274,625	8.062	4.021	25.495	.0154
16	256	4,096	4.000	2.520	12.648	.0625	66	4,356	287,496	8.124	4.041	25.690	.0152
17	289	4,913	4.123	2.571	13.038	.0588	67	4,489	300,763	8.185	4.062	25.884	.0149
18	324	5,832	4.243	2.621	13.416	.0556	68	4,624	314,432	8.246	4.082	26.077	.0147
19	361	6,859	4.359	2.668	13.784	.0526	69	4,761	328,509	8.307	4.102	26.268	.0145
20	400	8,000	4.472	2.714	14.142	.0500	70	4,900	343,000	8.367	4.121	26.458	.0143
21	441	9,261	4.583	2.759	14.491	.0476	71	5,041	357,911	8.426	4.141	26.646	.0141
22	484	10,648	4.690	2.802	14.832	.0455	72	5,184	373,248	8.485	4.160	26.833	.0139
23	529	12,167	4.796	2.844	15.166	.0435	73	5,329	389,017	8.544	4.179	27.019	.0137
24	576	13,824	4.899	2.884	15.492	.0417	74	5,476	405,224	8.602	4.198	27.203	.0135
25	625	15,625	5.000	2.924	15.811	.0400	75	5,625	421,875	8.660	4.217	27.386	.0133
26	676	17,576	5.099	2.962	16.125	.0385	76	5,776	438,976	8.718	4.236	27.568	.0132
27	729	19,683	5.196	3.000	16.432	.0370	77	5,929	456,533	8.775	4.254	27.749	.0130
28	784	21,952	5.292	3.037	16.733	.0357	78	6,084	474,552	8.832	4.273	27.928	.0128
29	841	24,389	5.385	3.072	17.029	.0345	79	6,241	493,039	8.888	4.291	28.107	.0127
30	900	27,000	5.477	3.107	17.321	.0333	80	6,400	512,000	8.944	4.309	28.284	.0125
31	961	29,791	5.568	3.141	17.607	.0323	81	6,561	531,441	9.000	4.327	28.460	.0123
32	1,024	32,768	5.657	3.175	17.889	.0312	82	6,724	551,368	9.055	4.344	28.636	.0122
33	1,089	35,937	5.745	3.208	18.166	.0303	83	6,889	571,787	9.110	4.362	28.810	.0120
34	1,156	39,304	5.831	3.240	18.439	.0294	84	7.056	592,704	9.165	4.380	28.983	.0119
35	1,225	42,875	5.916	3.271	18.708	.0286	85	7,225	614,125	9.220	4.397	29.155	.0118
36	1,296	46,656	6.000	3.302	18.974	.0278	86	7,396	636,056	9.274	4.414	29.326	.0116
37	1,369	50,653	6.083	3.332	19.235	.0270	87	7,569	658,503	9.327	4.431	29.496	.0115
38	1,444	54,872	6.164	3.362	19.494	.0263	88	7,744	681,472	9.381	4.448	29.665	.0114
39	1,521	59,319	6.245	3.391	19.748	.0256	89	7,921	704,969	9.434	4.465	29.833	.0112
40	1,600	64,000	6.325	3.420	20.000	.0250	90	8,100	729,000	9.487	4.481	30.000	.0111
41	1,681	68,921	6.403	3.448	20.248	.0244	91	8,281	753,571	9.539	4.498	30.166	.0110
42	1,764	74,088	6.481	3.476	20.494	.0238	92	8,464	778,688	9.592	4.514	30.332	.0109
43	1,849	79,507	6.557	3.503	20.736	.0233	93	8,649	804,357	9.644	4.531	30.496	.0108
44	1,936	85,184	6.633	3.530	20.976	.0227	94	8,836	830,584	9.695	4.547	30.659	.0106
45	2,025	91,125	6.708	3.557	21.213	.0222	95	9,025	857,375	9.747	4.563	30.822	.0105
46	2,116	97,336	6.782	3.583	21.448	.0217	96	9,216	884,736	9.798	4.579	30.984	.0104
47	2,209	103,823	6.856	3.609	21.679	.0213	97	9,409	912,673	9.849	4.595	31.145	.0103
48	2,304	110,592	6.928	3.634	21.909	.0208	98	9,604	941,192	9.899	4.610	31.305	.0102
49	2,401	117,649	7.000	3.659	22.136	.0204	99	9,801	970,299	9.950	4.626	31.464	.0101
50	2,500	125,000	7.071	3.684	22.361	.0200	100	10,000	1,000,000	10.000	4.642	31.623	.0100

Table 2 Common Logarithms

x	0	1	2	3	4	5	6	7	8	9
1.0	.0000	.0043	.0086	.0128	.0170	.0212	.0253	.0294	.0334	.0374
1.1	.0414	.0453	.0492	.0531	.0569	.0607	.0645	.0682	.0719	.0755
1.2	.0792	.0828	.0864	.0899	.0934	.0969	.1004	.1038	.1072	.1106
1.3	.1139	.1173	.1206	.1239	.1271	.1303	.1335	.1367	.1399	.1430
1.4	.1461	.1492	.1523	.1553	.1584	.1614	.1644	.1673	.1703	.1732
1.5	.1761	.1790	.1818	.1847	.1875	.1903	.1931	.1959	.1987	.2014
1.6	.2041	.2068	.2095	.2122	.2148	.2175	.2201	.2227	.2253	.2279
1.7	.2304	.2330	.2355	.2380	.2405	.2430	.2455	.2480	.2504	.2529
1.8	.2553	.2577	.2601	.2625	.2648	.2672	.2695	.2718	.2742	.2765
1.9	.2788	.2810	.2833	.2856	.2878	.2900	.2923	.2945	.2967	.2989
2.0	.3010	.3032	.3054	.3075	.3096	.3118	.3139	.3160	.3181	.3201
2.1	.3222	.3243	.3263	.3284	.3304	.3324	.3345	.3365	.3385	.3404
2.2	.3424	.3444	.3464	.3483	.3502	.3522	.3541	.3560	.3579	.3598
2.3	.3617	.3636	.3655	.3674	.3692	.3711	.3729	.3747	.3766	.3784
2.4	.3802	.3820	.3838	.3856	.3874	.3892	.3909	.3927	.3945	.3962
2.5	.3979	.3997	.4014	.4031	.4048	.4065	.4082	.4099	.4116	.4133
2.6	.4150	.4166	.4183	.4200	.4216	.4232	.4249	.4265	.4281	.4298
2.7	.4314	.4330	.4346	.4362	.4378	.4393	.4409	.4425	.4440	.4456
2.8	.4472	.4487	.4502	.4518	.4533	.4548	.4564	.4579	.4594	.4609
2.9	.4624	.4639	.4654	.4669	.4683	.4698	.4713	.4728	.4742	.4757
3.0	.4771	.4786	.4800	.4814	.4829	.4843	.4857	.4871	.4886	.4900
3.1	.4914	.4928	.4942	.4955	.4969	.4983	.4997	.5011	.5024	.5038
3.2	.5051	.5065	.5079	.5092	.5105	.5119	.5132	.5145	.5159	.5172
3.3	.5185	.5198	.5211	.5224	.5237	.5250	.5263	.5276	.5289	.5302
3.4	.5315	.5328	.5340	.5353	.5366	.5378	.5391	.5403	.5416	.5428
3.5	.5441	.5453	.5465	.5478	.5490	.5502	.5514	.5527	.5539	.5551
3.6	.5563	.5575	.5587	.5599	.5611	.5623	.5635	.5647	.5658	.5670
3.7	.5682	.5694	.5705	.5717	.5729	.5740	.5752	.5763	.5775	.5786
3.8	.5798	.5809	.5821	.5832	.5843	.5855	.5866	.5877	.5888	.5899
3.9	.5911	.5922	.5933	.5944	.5955	.5966	.5977	.5988	.5999	.6010
4.0	.6021	.6031	.6042	.6053	.6064	.6075	.6085	.6096	.6107	.6117
4.1	.6128	.6138	.6149	.6160	.6170	.6180	.6191	.6201	.6212	.6222
4.2	.6232	.6243	.6253	.6263	.6274	.6284	.6294	.6304	.6314	.6325
4.3	.6335	.6345	.6355	.6365	.6375	.6385	.6395	.6405	.6415	.6425
4.4	.6435	.6444	.6454	.6464	.6474	.6484	.6493	.6503	.6513	.6522
4.5	.6532	.6542	.6551	.6561	.6571	.6580	.6590	.6599	.6609	.6618
4.6	.6628	.6637	.6646	.6656	.6665	.6675	.6684	.6693	.6702	.6712
4.7	.6721	.6730	.6739	.6749	.6758	.6767	.6776	.6785	.6794	.6803
4.8	.6812	.6821	.6830	.6839	.6848	.6857	.6866	.6875	.6884	.6893
4.9	.6902	.6911	.6920	.6928	.6937	.6946	.6955	.6964	.6972	.6981
5.0	.6990	.6998	.7007	.7016	.7024	.7033	.7042	.7050	.7059	.7067
5.1	.7076	.7084	.7093	.7101	.7110	.7118	.7126	.7135	.7143	.7152
5.2	.7160	.7168	.7177	.7185	.7193	.7202	.7210	.7218	.7226	.7235
5.3	.7243	.7251	.7259	.7267	.7275	.7284	.7292	.7300	.7308	.7316
5.4	.7324	.7332	.7340	.7348	.7356	.7364	.7372	.7380	.7388	.7396
x	0	1	2	3	4	5	6	7	8	9

Table 2 (continued)

x	0	1	2	3	4	5	6	7	8	9
5.5	.7404	.7412	.7419	.7427	.7435	.7443	.7451	.7459	.7466	.7474
5.6	.7482	.7490	.7497	.7505	.7513	.7520	.7528	.7536	.7543	.7551
5.7	.7559	.7566	.7574	.7582	.7589	.7597	.7604	.7612	.7619	.7627
5.8	.7634	.7642	.7649	.7657	.7664	.7672	.7679	.7686	.7694	.7701
5.9	.7709	.7716	.7723	.7731	.7738	.7745	.7752	.7760	.7767	.7774
6.0	.7782	.7789	.7796	.7803	.7810	.7818	.7825	.7832	.7839	.7846
6.1	.7853	.7860	.7868	.7875	.7882	.7889	.7896	.7903	.7910	.7917
6.2	.7924	.7931	.7938	.7945	.7952	.7959	.7966	.7973	.7980	.7987
6.3	.7993	.8000	.8007	.8014	.8021	.8028	.8035	.8041	.8048	.8055
6.4	.8062	.8069	.8075	.8082	.8089	.8096	.8102	.8109	.8116	.8122
6.5	.8129	.8136	.8142	.8149	.8156	.8162	.8169	.8176	.8182	.8189
6.6	.8195	.8202	.8209	.8215	.8222	.8228	.8235	.8241	.8248	.8254
6.7	.8261	.8267	.8274	.8280	.8287	.8293	.8299	.8306	.8312	.8319
6.8	.8325	.8331	.8338	.8344	.8351	.8357	.8363	.8370	.8376	.8382
6.9	.8388	.8395	.8401	.8407	.8414	.8420	.8426	.8432	.8439	.8445
7.0	.8451	.8457	.8463	.8470	.8476	.8482	.8488	.8494	.8500	.8506
7.1	.8513	.8519	.8525	.8531	.8537	.8543	.8549	.8555	.8561	.8567
7.2	.8573	.8579	.8585	.8591	.8597	.8603	.8609	.8615	.8621	.8627
7.3	.8633	.8639	.8645	.8651	.8657	.8663	.8669	.8675	.8681	.8686
7.4	.8692	.8698	.8704	.8710	.8716	.8722	.8727	.8733	.8739	.8745
7.5	.8751	.8756	.8762	.8768	.8774	.8779	.8785	.8791	.8797	.8802
7.6	.8808	.8814	.8820	.8825	.8831	.8837	.8842	.8848	.8854	.8859
7.7	.8865	.8871	.8876	.8882	.8887	.8893	.8899	.8904	.8910	.8915
7.8	.8921	.8927	.8932	.8938	.8943	.8949	.8954	.8960	.8965	.8971
7.9	.8976	.8982	.8987	.8993	.8998	.9004	.9009	.9015	.9020	.9025
8.0	.9031	.9036	.9042	.9047	.9053	.9058	.9063	.9069	.9074	.9079
8.1	.9085	.9090	.9096	.9101	.9106	.9112	.9117	.9122	.9128	.9133
8.2	.9138	.9143	.9149	.9154	.9159	.9165	.9170	.9175	.9180	.9186
8.3	.9191	.9196	.9201	.9206	.9212	.9217	.9222	.9227	.9232	.9238
8.4	.9243	.9248	.9253	.9258	.9263	.9269	.9274	.9279	.9284	.9289
8.5	.9294	.9299	.9304	.9309	.9315	.9320	.9325	.9330	.9335	.9340
8.6	.9345	.9350	.9555	.9360	.9365	.9370	.9375	.9380	.9385	.9390
8.7	.9395	.9400	.9405	.9410	.9415	.9420	.9425	.9430	.9435	.9440
8.8	.9445	.9450	.9455	.9460	.9465	.9469	.9474	.9479	.9484	.9489
8.9	.9494	.9499	.9504	.9509	.9513	.9518	.9523	.9528	.9533	.9538
9.0	.9542	.9547	.9552	.9557	.9562	.9566	.9571	.9576	.9581	.9586
9.1	.9590	.9595	.9600	.9605	.9609	.9614	.9619	.9624	.9628	.9633
9.2	.9638	.9643	.9647	.9652	.9657	.9661	.9666	.9671	.9675	.9680
9.3	.9685	.9689	.9694	.9699	.9703	.9708	.9713	.9717	.9722	.9727
9.4	.9731	.9736	.9741	.9745	.9750	.9754	.9759	.9763	.9768	.9773
9.5	.9777	.9782	.9786	.9791	.9795	.9800	.9805	.9809	.9814	.9818
9.6	.9823	.9827	.9832	.9836	.9841	.9845	.9850	.9854	.9859	.9863
9.7	.9868	.9872	.9877	.9881	.9886	.9890	.9894	.9899	.9903	.9908
9.8	.9912	.9917	.9921	.9926	.9930	.9934	.9939	.9943	.9948	.9952
9.9	.9956	.9961	.9965	.9969	.9974	.9978	.9983	.9987	.9991	.9996
x	0	1	2	3	4	5	6	7	8	9

Table 3 Values of Trigonometric Functions

Degrees	Radians	Sin	Cos	Tan	Cot	Sec	Csc		
0° 00'	.0000	.0000	1.0000	.0000	——	1.000	——	1.5708	90° 00'
10	029	029	000	029	343.8	000	343.8	679	50
20	058	058	000	058	171.9	000	171.9	650	40
30	.0087	.0087	1.0000	.0087	114.6	1.000	114.6	1.5621	30
40	116	116	.9999	116	85.94	000	85.95	592	20
50	145	145	999	145	68.75	000	68.76	563	10
1° 00'	.0175	.0175	.9998	.0175	57.29	1.000	57.30	1.5533	89° 00'
10	204	204	998	204	49.10	000	49.11	504	50
20	233	233	997	233	42.96	000	42.98	475	40
30	.0262	.0262	.9997	.0262	38.19	1.000	38.20	1.5446	30
40	291	291	996	291	34.37	000	34.38	417	20
50	320	320	995	320	31.24	001	31.26	388	10
2° 00'	.0349	.0349	.9994	.0349	28.64	1.001	28.65	1.5359	88° 00'
10	378	378	993	378	26.43	001	26.45	330	50
20	407	407	992	407	24.54	001	24.56	301	40
30	.0436	.0436	.9990	.0437	22.90	1.001	22.93	1.5272	30
40	465	465	989	466	21.47	001	21.49	243	20
50	495	494	988	495	20.21	001	20.23	213	10
3° 00'	.0524	.0523	.9986	.0524	19.08	1.001	19.11	1.5184	87° 00'
10	553	552	985	553	18.07	002	18.10	155	50
20	582	581	983	582	17.17	002	17.20	126	40
30	.0611	.0610	.9981	.0612	16.35	1.002	16.38	1.5097	30
40	640	640	980	641	15.60	002	15.64	068	20
50	669	669	978	670	14.92	002	14.96	039	10
4° 00'	.0698	.0698	.9976	.0699	14.30	1.002	14.34	1.5010	86° 00'
10	727	727	974	729	13.73	003	13.76	981	50
20	756	756	971	758	13.20	003	13.23	952	40
30	.0785	.0785	.9969	.0787	12.71	1.003	12.75	1.4923	30
40	814	814	967	816	12.25	003	12.29	893	20
50	844	843	964	846	11.83	004	11.87	864	10
5° 00'	.0873	.0872	.9962	.0875	11.43	1.004	11.47	1.4835	85° 00'
10	902	901	959	904	11.06	004	11.10	806	50
20	931	929	957	934	10.71	004	10.76	777	40
30	.0960	.0958	.9954	.0963	10.39	1.005	10.43	1.4748	30
40	989	987	951	992	10.08	005	10.13	719	20
50	.1018	.1016	948	.1022	9.788	005	9.839	690	10
6° 00'	.1047	.1045	.9945	.1051	9.514	1.006	9.567	1.4661	84° 00'
10	076	074	942	080	9.255	006	9.309	632	50
20	105	103	939	110	9.010	006	9.065	603	40
30	.1134	.1132	.9936	.1139	8.777	1.006	8.834	1.4573	30
40	164	161	932	169	8.556	007	8.614	544	20
50	193	190	929	198	8.345	007	8.405	515	10
7° 00'	.1222	.1219	.9925	.1228	8.144	1.008	8.206	1.4486	83° 00'
10	251	248	922	257	7.953	008	8.016	457	50
20	280	276	918	287	7.770	008	7.834	428	40
30	.1309	.1305	.9914	.1317	7.596	1.009	7.661	1.4399	30
40	338	334	911	346	7.429	009	7.496	370	20
50	367	363	907	376	7.269	009	7.337	341	10
8° 00'	.1396	.1392	.9903	.1405	7.115	1.010	7.185	1.4312	82° 00'
10	425	421	899	435	6.968	010	7.040	283	50
20	454	449	894	465	6.827	011	6.900	254	40
30	.1484	.1478	.9890	.1495	6.691	1.011	6.765	1.4224	30
40	513	507	886	524	6.561	012	6.636	195	20
50	542	536	881	554	6.435	012	6.512	166	10
9° 00'	.1571	.1564	.9877	.1584	6.314	1.012	6.392	1.4137	81° 00'
		Cos	Sin	Cot	Tan	Csc	Sec	Radians	Degrees

Table 3 (continued)

Degrees	Radians	Sin	Cos	Tan	Cot	Sec	Csc		
9° 00'	.1571	.1564	.9877	.1584	6.314	1.012	6.392	1.4137	81° 00'
10	600	593	872	614	197	013	277	108	50
20	629	622	868	644	084	013	166	079	40
30	.1658	.1650	.9863	.1673	5.976	1.014	6.059	1.4050	30
40	687	679	858	703	871	014	5.955	1.4021	20
50	716	708	853	733	769	015	855	992	10
10° 00'	.1745	.1736	.9848	.1763	5.671	1.015	5.759	1.3963	80° 00'
10	774	765	843	793	576	016	665	934	50
20	804	794	838	823	485	016	575	904	40
30	.1833	.1822	.9833	.1853	5.396	1.017	5.487	1.3875	30
40	862	851	827	883	309	018	403	846	20
50	891	880	822	914	226	018	320	817	10
11° 00'	.1920	.1908	.9816	.1944	5.145	1.019	5.241	1.3788	79° 00'
10	949	937	811	974	066	019	164	759	50
20	978	965	805	.2004	4.989	020	089	730	40
30	.2007	.1994	.9799	.2035	4.915	1.020	5.016	1.3701	30
40	036	.2022	793	065	843	021	4.945	672	20
50	065	051	787	095	773	022	876	643	10
12° 00'	.2094	.2079	.9781	.2126	4.705	1.022	4.810	1.3614	78° 00'
10	123	108	775	156	638	023	745	584	50
20	153	136	769	186	574	024	682	555	40
30	.2182	.2164	.9763	.2217	4.511	1.024	4.620	1.3526	30
40	211	193	757	247	449	025	560	497	20
50	240	221	750	278	390	026	502	468	10
13° 00'	.2269	.2250	.9744	.2309	4.331	1.026	4.445	1.3439	77° 00'
10	298	278	737	339	275	027	390	410	50
20	327	306	730	370	219	028	336	381	40
30	.2356	.2334	.9724	.2401	4.165	1.028	4.284	1.3352	30
40	385	363	717	432	113	029	232	323	20
50	414	391	710	462	061	030	182	294	10
14° 00'	.2443	.2419	.9703	.2493	4.011	1.031	4.134	1.3265	76° 00'
10	473	447	696	524	3.962	·031	086	235	50
20	502	476	689	555	914	032	039	206	40
30	.2531	.2504	.9681	.2586	3.867	1.033	3.994	1.3177	30
40	560	532	674	617	821	034	950	148	20
50	589	560	667	648	776	034	906	119	10
15° 00'	.2618	.2588	.9659	.2679	3.732	1.035	3.864	1.3090	75° 00'
10	647	616	652	711	689	036	822	061	50
20	676	644	644	742	647	037	782	032	40
30	.2705	.2672	.9636	.2773	3.606	1.038	3.742	1.3003	30
40	734	700	628	805	566	039	703	974	20
50	763	728	621	836	526	039	665	945	10
16° 00'	.2793	.2756	.9613	.2867	3.487	1.040	3.628	1.2915	74° 00'
10	822	784	605	899	450	041	592	886	50
20	851	812	596	931	412	042	556	857	40
30	.2880	.2840	.9588	.2962	3.376	1.043	3.521	1.2828	30
40	909	868	580	994	340	044	487	799	20
50	938	896	572	.3026	305	045	453	770	10
17° 00'	.2967	.2924	.9563	.3057	3.271	1.046	3.420	1.2741	73° 00'
10	996	952	555	089	237	047	388	712	50
20	.3025	979	546	121	204	048	356	683	40
30	.3054	.3007	.9537	.3153	3.172	1.049	3.326	1.2654	30
40	083	035	528	185	140	049	295	625	20
50	113	062	520	217	108	050	265	595	10
18° 00'	.3142	.3090	.9511	.3249	3.078	1.051	3.236	1.2566	72° 00'
		Cos	Sin	Cot	Tan	Csc	Sec	Radians	Degrees

(Continued)

Table 3 (continued)

Degrees	Radians	Sin	Cos	Tan	Cot	Sec	Csc		
18° 00'	.3142	.3090	.9511	.3249	3.078	1.051	3.236	1.2566	72° 00'
10	171	118	502	281	047	052	207	537	50
20	200	145	492	314	018	053	179	508	40
30	.3229	.3173	.9483	.3346	2.989	1.054	3.152	1.2479	30
40	258	201	474	378	960	056	124	450	20
50	287	228	465	411	932	057	098	421	10
19° 00'	.3316	.3256	.9455	.3443	2.904	1.058	3.072	1.2392	71° 00'
10	345	283	446	476	877	059	046	363	50
20	374	311	436	508	850	060	021	334	40
30	.3403	.3338	.9426	.3541	2.824	1.061	2.996	1.2305	30
40	432	365	417	574	798	062	971	275	20
50	462	393	407	607	773	063	947	246	10
20° 00'	.3491	.3420	.9397	.3640	2.747	1.064	2.924	1.2217	70° 00'
10	520	448	387	673	723	065	901	188	50
20	549	475	377	706	699	066	878	159	40
30	.3578	.3502	.9367	.3739	2.675	1.068	2.855	1.2130	30
40	607	529	356	772	651	069	833	101	20
50	636	557	346	805	628	070	812	072	10
21° 00'	.3665	.3584	.9336	.3839	2.605	1.071	2.790	1.2043	69° 00'
10	694	611	325	872	583	072	769	1.2014	50
20	723	638	315	906	560	074	749	985	40
30	.3752	.3665	.9304	.3939	2.539	1.075	2.729	1.1956	30
40	782	692	293	973	517	076	709	926	20
50	811	719	283	.4006	496	077	689	897	10
22° 00'	.3840	.3746	.9272	.4040	2.475	1.079	2.669	1.1868	68° 00'
10	869	773	261	074	455	080	650	839	50
20	898	800	250	108	434	081	632	810	40
30	.3927	.3827	.9239	.4142	2.414	1.082	2.613	1.1781	30
40	956	854	228	176	394	084	595	752	20
50	985	881	216	210	375	085	577	723	10
23° 00'	.4014	.3907	.9205	.4245	2.356	1.086	2.559	1.1694	67° 00'
10	043	934	194	279	337	088	542	665	50
20	072	961	182	314	318	089	525	636	40
30	.4102	.3987	.9171	.4348	2.300	1.090	2.508	1.1606	30
40	131	.4014	159	383	282	092	491	577	20
50	160	041	147	417	264	093	475	548	10
24° 00'	.4189	.4067	.9135	.4452	2.246	1.095	2.459	1.1519	66° 00'
10	218	094	124	487	229	096	443	490	50
20	247	120	112	522	211	097	427	461	40
30	.4276	.4147	.9100	.4557	2.194	1.099	2.411	1.1432	30
40	305	173	088	592	177	100	396	403	20
50	334	200	075	628	161	102	381	374	10
25° 00'	.4363	.4226	.9063	.4663	2.145	1.103	2.366	1.1345	65° 00'
10	392	253	051	699	128	105	352	316	50
20	422	279	038	734	112	106	337	286	40
30	.4451	.4305	.9026	.4770	2.097	1.108	2.323	1.1257	30
40	480	331	013	806	081	109	309	228	20
50	509	358	001	841	066	111	295	199	10
26° 00'	.4538	.4384	.8988	.4877	2.050	1.113	2.281	1.1170	64° 00'
10	567	410	975	913	035	114	268	141	50
20	596	436	962	950	020	116	254	112	40
30	.4625	.4462	.8949	.4986	2.006	1.117	2.241	1.1083	30
40	654	488	936	.5022	1.991	119	228	054	20
50	683	514	923	059	977	121	215	1.1025	10
27° 00'	.4712	.4540	.8910	.5095	1.963	1.122	2.203	1.0996	63° 00'
		Cos	Sin	Cot	Tan	Csc	Sec	Radians	Degrees

Table 3 (continued)

Degrees	Radians	Sin	Cos	Tan	Cot	Sec	Csc		
27° 00'	.4712	.4540	.8910	.5095	1.963	1.122	2.203	1.0996	**63° 00'**
10	741	566	897	132	949	124	190	966	50
20	771	592	884	169	935	126	178	937	40
30	.4800	.4617	.8870	.5206	1.921	1.127	2.166	1.0908	30
40	829	643	857	243	907	129	154	879	20
50	858	669	843	280	894	131	142	850	10
28° 00'	.4887	.4695	.8829	.5317	1.881	1.133	2.130	1.0821	**62° 00'**
10	916	720	816	354	868	134	118	792	50
20	945	746	802	392	855	136	107	763	40
30	.4974	.4772	.8788	.5430	1.842	1.138	2.096	1.0734	30
40	.5003	797	774	467	829	140	085	705	20
50	032	823	760	505	816	142	074	676	10
29° 00'	.5061	.4848	.8746	.5543	1.804	1.143	2.063	1.0647	**61° 00'**
10	091	874	732	581	792	145	052	617	50
20	120	899	718	619	780	147	041	588	40
30	.5149	.4924	.8704	.5658	1.767	1.149	2.031	1.0559	30
40	178	950	689	696	756	151	020	530	20
50	207	975	675	735	744	153	010	501	10
30° 00'	.5236	.5000	.8660	.5774	1.732	1.155	2.000	1.0472	**60° 00'**
10	265	025	646	812	720	157	1.990	443	50
20	294	050	631	851	709	159	980	414	40
30	.5323	.5075	.8616	.5890	1.698	1.161	1.970	1.0385	30
40	352	100	601	930	686	163	961	356	20
50	381	125	587	969	675	165	951	327	10
31° 00'	.5411	.5150	.8572	.6009	1.664	1.167	1.942	1.0297	**59° 00'**
10	440	175	557	048	653	169	932	268	50
20	469	200	542	088	643	171	923	239	40
30	.5498	.5225	.8526	.6128	1.632	1.173	1.914	1.0210	30
40	527	250	511	168	621	175	905	181	20
50	556	275	496	208	611	177	896	152	10
32° 00'	.5585	.5299	.8480	.6249	1.600	1.179	1.887	1.0123	**58° 00'**
10	614	324	465	289	590	181	878	094	50
20	643	348	450	330	580	184	870	065	40
30	.5672	.5373	.8434	.6371	1.570	1.186	1.861	1.0036	30
40	701	398	418	412	560	188	853	1.0007	20
50	730	422	403	453	550	190	844	977	10
33° 00'	.5760	.5446	.8387	.6494	1.540	1.192	1.836	.9948	**57° 00'**
10	789	471	371	536	530	195	828	919	50
20	818	495	355	577	520	197	820	890	40
30	.5847	.5519	.8339	.6619	1.511	1.199	1.812	.9861	30
40	876	544	323	661	501	202	804	832	20
50	905	568	307	703	1.492	204	796	803	10
34° 00'	.5934	.5592	.8290	.6745	1.483	1.206	1.788	.9774	**56° 00'**
10	963	616	274	787	473	209	781	745	50
20	992	640	258	830	464	211	773	716	40
30	.6021	.5664	.8241	.6873	1.455	1.213	1.766	.9687	30
40	050	688	225	916	446	216	758	657	20
50	080	712	208	959	437	218	751	628	10
35° 00'	.6109	.5736	.8192	.7002	1.428	1.221	1.743	.9599	**55° 00'**
10	138	760	175	046	419	223	736	570	50
20	167	783	158	089	411	226	729	541	40
30	.6196	.5807	.8141	.7133	1.402	1.228	1.722	.9512	30
40	225	831	124	177	393	231	715	483	20
50	254	854	107	221	385	233	708	454	10
36° 00'	.6283	.5878	.8090	.7265	1.376	1.236	1.701	.9425	**54° 00'**
		Cos	Sin	Cot	Tan	Csc	Sec	Radians	Degrees

(Continued)

Table 3 (continued)

Degrees	Radians	Sin	Cos	Tan	Cot	Sec	Csc		
36° 00'	.6283	.5878	.8090	.7265	1.376	1.236	1.701	.9425	54° 00'
10	312	901	073	310	368	239	695	396	50
20	341	925	056	355	360	241	688	367	40
30	.6370	.5948	.8039	.7400	1.351	1.244	1.681	.9338	30
40	400	972	021	445	343	247	675	308	20
50	429	995	004	490	335	249	668	279	10
37° 00'	.6458	.6018	.7986	.7536	1.327	1.252	1.662	.9250	53° 00'
10	487	041	969	581	319	255	655	221	50
20	516	065	951	627	311	258	649	192	40
30	.6545	.6088	.7934	.7673	1.303	1.260	1.643	.9163	30
40	574	111	916	720	295	263	636	134	20
50	603	134	898	766	288	266	630	105	10
38° 00'	.6632	.6157	.7880	.7813	1.280	1.269	1.624	.9076	52° 00'
10	661	180	862	860	272	272	618	047	50
20	690	202	844	907	265	275	612	.9018	40
30	.6720	.6225	.7826	.7954	1.257	1.278	1.606	.8988	30
40	749	248	808	.8002	250	281	601	959	20
50	778	271	790	050	242	284	595	930	10
39° 00'	.6807	.6293	.7771	.8098	1.235	1.287	1.589	.8901	51° 00'
10	836	316	753	146	228	290	583	872	50
20	865	338	735	195	220	293	578	843	40
30	.6894	.6361	.7716	.8243	1.213	1.296	1.572	.8814	30
40	923	383	698	292	206	299	567	785	20
50	952	406	679	342	199	302	561	756	10
40° 00'	.6981	.6428	.7660	.8391	1.192	1.305	1.556	.8727	50° 00'
10	.7010	450	642	441	185	309	550	698	50
20	039	472	623	491	178	312	545	668	40
30	.7069	.6494	.7604	.8541	1.171	1.315	1.540	.8639	30
40	098	517	585	591	164	318	535	610	20
50	127	539	566	642	157	322	529	581	10
41° 00'	.7156	.6561	.7547	.8693	1.150	1.325	1.524	.8552	49° 00'
10	185	583	528	744	144	328	519	523	50
20	214	604	509	796	137	332	514	494	40
30	.7243	.6626	.7490	.8847	1.130	1.335	1.509	.8465	30
40	272	648	470	899	124	339	504	436	20
50	301	670	451	952	117	342	499	407	10
42° 00'	.7330	.6691	.7431	.9004	1.111	1.346	1.494	.8378	48° 00'
10	359	713	412	057	104	349	490	348	50
20	389	734	392	110	098	353	485	319	40
30	.7418	.6756	.7373	.9163	1.091	1.356	1.480	.8290	30
40	447	777	353	217	085	360	476	261	20
50	476	799	333	271	079	364	471	232	10
43° 00'	.7505	.6820	.7314	.9325	1.072	1.367	1.466	.8203	47° 00'
10	534	841	294	380	066	371	462	174	50
20	563	862	274	435	060	375	457	145	40
30	.7592	.6884	.7254	.9490	1.054	1.379	1.453	.8116	30
40	621	905	234	545	048	382	448	087	20
50	650	926	214	601	042	386	444	058	10
44° 00'	.7679	.6947	.7193	.9657	1.036	1.390	1.440	.8029	46° 00'
10	709	967	173	713	030	394	435	999	50
20	738	988	153	770	024	398	431	970	40
30	.7767	.7009	.7133	.9827	1.018	1.402	1.427	.7941	30
40	796	030	112	884	012	406	423	912	20
50	825	050	092	942	006	410	418	883	10
45° 00'	.7854	.7071	.7071	1.000	1.000	1.414	1.414	.7854	45° 00'
		Cos	Sin	Cot	Tan	Csc	Sec	Radians	Degrees

Table 4 Exponential Functions

x	e^x	e^{-x}	x	e^x	e^{-x}	x	e^x	e^{-x}
0.00	1.0000	1.0000	0.55	1.7333	0.5769	3.6	36.598	0.0273
0.01	1.0101	0.9900	0.60	1.8221	0.5488	3.7	40.447	0.0247
0.02	1.0202	0.9802	0.65	1.9155	0.5220	3.8	44.701	0.0224
0.03	1.0305	0.9704	0.70	2.0138	0.4966	3.9	49.402	0.0202
0.04	1.0408	0.9608	0.75	2.1170	0.4724	4.0	54.598	0.0183
0.05	1.0513	0.9512	0.80	2.2255	0.4493	4.1	60.340	0.0166
0.06	1.0618	0.9418	0.85	2.3396	0.4274	4.2	66.686	0.0150
0.07	1.0725	0.9324	0.90	2.4596	0.4066	4.3	73.700	0.0136
0.08	1.0833	0.9231	0.95	2.5857	0.3867	4.4	81.451	0.0123
0.09	1.0942	0.9139	1.0	2.7183	0.3679	4.5	90.017	0.0111
0.10	1.1052	0.9048	1.1	3.0042	0.3329	4.6	99.484	0.0101
0.11	1.1163	0.8958	1.2	3.3201	0.3012	4.7	109.95	0.0091
0.12	1.1275	0.8869	1.3	3.6693	0.2725	4.8	121.51	0.0082
0.13	1.1388	0.8781	1.4	4.0552	0.2466	4.9	134.29	0.0074
0.14	1.1503	0.8694	1.5	4.4817	0.2231	5	148.41	0.0067
0.15	1.1618	0.8607	1.6	4.9530	0.2019	6	403.43	0.0025
0.16	1.1735	0.8521	1.7	5.4739	0.1827	7	1096.6	0.0009
0.17	1.1853	0.8437	1.8	6.0496	0.1653	8	2981.0	0.0003
0.18	1.1972	0.8353	1.9	6.6859	0.1496	9	8103.1	0.0001
0.19	1.2092	0.8270	2.0	7.3891	0.1353	10	22026	0.00005
0.20	1.2214	0.8187	2.1	8.1662	0.1225	11	59874	0.00002
0.21	1.2337	0.8106	2.2	9.0250	0.1108	12	162,754	0.000006
0.22	1.2461	0.8025	2.3	9.9742	0.1003	13	442,413	0.000002
0.23	1.2586	0.7945	2.4	11.023	0.0907	14	1,202,604	0.0000008
0.24	1.2712	0.7866	2.5	12.182	0.0821	15	3,269,017	0.0000003
0.25	1.2840	0.7788	2.6	13.464	0.0743			
0.26	1.2969	0.7711	2.7	14.880	0.0672			
0.27	1.3100	0.7634	2.8	16.445	0.0608			
0.28	1.3231	0.7558	2.9	18.174	0.0550			
0.29	1.3364	0.7483	3.0	20.086	0.0498			
0.30	1.3499	0.7408	3.1	22.198	0.0450			
0.35	1.4191	0.7047	3.2	24.533	0.0408			
0.40	1.4918	0.6703	3.3	27.113	0.0369			
0.45	1.5683	0.6376	3.4	29.964	0.0334			
0.50	1.6487	0.6065	3.5	33.115	0.0302			

Table 5 Factorials and Large Powers of 2

n	n!	2^n
0	1	1
1	1	2
2	2	4
3	6	8
4	24	16
5	120	32
6	720	64
7	5040	128
8	40,320	256
9	362,880	512
10	3,628,800	1024
11	39,916,800	2048
12	479,001,600	4096
13	6,227,020,800	8192
14	87,178,291,200	16,384
15	1,307,674,368,000	32,768
16	20,922,789,888,000	65,536
17	355,687,428,096,000	131,072
18	6,402,373,705,728,000	262,144
19	121,645,100,408,832,000	524,288
20	2,432,902,008,176,640,000	1,048,576

Table 6 Tables of Measures

LENGTH

1 *kilometer* (km) = 1000 meters (m)
1 *hectometer* (hm) = 100 meters
1 *dekameter* (dam) = 10 meters
1 *decimeter* (dm) = 0.1 meter
1 *centimeter* (cm) = 0.01 meter
1 *millimeter* (mm) = 0.001 meter

MASS OR WEIGHT

1 *kilogram* (kg) = 1000 grams (g)
1 *hectogram* (hg) = 100 grams
1 *dekagram* (dag) = 10 grams
1 *decigram* (dg) = 0.1 gram
1 *centigram* (dg) = 0.01 gram
1 metric ton (MT or t) = 1000 kilograms

AREA

1 *hectare* (ha) = 100 are (a), or 10,000 sq m (m^2)
1 *are* (a) = 100 sq m (m^2)
1 *centare* (ca) = 0.01 are, or 1 m^2

The word "are" is pronounced "AIR."

VOLUME

1000 cubic centimeters (cc or cm^3) = 1 liter (L)
1 cubic centimeter (cc) = 1 milliliter (mL)
1 mL of water weighs 1 g
1 stere = 1 cubic meter

Metric-American Conversions (Approximate)

LENGTH

1 m = 39.37 in. = 3.3 ft
1 in. = 2.54 cm
1 km = 0.62 mi
1 mi = 1.6 km
1 cm = 3/8 in.

MASS OR WEIGHT

1 kg = 2.2 lb
1 MT = 1.1 tons
1 lb = 454 g
1 oz = 28 g

AREA

1 hectare = 2.47 acres
1 are = 120 sq yd

VOLUME

1 liter = 1.057 qt = 2.1 pt
1 cup = 240 mL
1 ounce (liquid) = 30 mL
1 gallon = 3.78 liters
1 tablespoon = 15 mL
1 teaspoon = 5 mL

Table 7 Natural Logarithms (ln x)

x	0.00	0.01	0.02	0.03	0.04	0.05	0.06	0.07	0.08	0.09
1.0	0.0000	0.0100	0.0198	0.0296	0.0392	0.0488	0.0583	0.0677	0.0770	0.0862
1.1	0.0953	0.1044	0.1133	0.1222	0.1310	0.1398	0.1484	0.1570	0.1655	0.1740
1.2	0.1823	0.1906	0.1989	0.2070	0.2151	0.2231	0.2311	0.2390	0.2469	0.2546
1.3	0.2624	0.2700	0.2776	0.2852	0.2927	0.3001	0.3075	0.3148	0.3221	0.3293
1.4	0.3365	0.3436	0.3507	0.3577	0.3646	0.3716	0.3784	0.3853	0.3920	0.3988
1.5	0.4055	0.4121	0.4187	0.4253	0.4318	0.4383	0.4447	0.4511	0.4574	0.4637
1.6	0.4700	0.4762	0.4824	0.4886	0.4947	0.5008	0.5068	0.5128	0.5188	0.5247
1.7	0.5306	0.5365	0.5423	0.5481	0.5539	0.5596	0.5653	0.5710	0.5766	0.5822
1.8	0.5878	0.5933	0.5988	0.6043	0.6098	0.6152	0.6206	0.6259	0.6313	0.6366
1.9	0.6419	0.6471	0.6523	0.6575	0.6627	0.6678	0.6729	0.6780	0.6831	0.6881
2.0	0.6931	0.6981	0.7031	0.7080	0.7130	0.7178	0.7227	0.7275	0.7324	0.7372
2.1	0.7419	0.7467	0.7514	0.7561	0.7608	0.7655	0.7701	0.7747	0.7793	0.7839
2.2	0.7885	0.7930	0.7975	0.8020	0.8065	0.8109	0.8154	0.8198	0.8242	0.8286
2.3	0.8329	0.8372	0.8416	0.8459	0.8502	0.8544	0.8587	0.8629	0.8671	0.8713
2.4	0.8755	0.8796	0.8838	0.8879	0.8920	0.8961	0.9002	0.9042	0.9083	0.9123
2.5	0.9163	0.9203	0.9243	0.9282	0.9322	0.9361	0.9400	0.9439	0.9478	0.9517
2.6	0.9555	0.9594	0.9632	0.9670	0.9708	0.9746	0.9783	0.9821	0.9858	0.9895
2.7	0.9933	0.9969	1.0006	1.0043	1.0080	1.0116	1.0152	0.0188	1.0225	1.0260
2.8	1.0296	1.0332	1.0367	1.0403	1.0438	1.0473	1.0508	1.0543	1.0578	1.0613
2.9	1.0647	1.0682	1.0716	1.0750	1.0784	1.0818	1.0852	1.0886	1.0919	1.0953
3.0	1.0986	1.1019	1.1053	1.1086	1.1119	1.1151	1.1184	1.1217	1.1249	1.1282
3.1	1.1314	1.1346	1.1378	1.1410	1.1442	1.1474	1.1506	1.1537	1.1569	1.1600
3.2	1.1632	1.1663	1.1694	1.1725	1.1756	1.1787	1.1817	1.1848	1.1878	1.1909
3.3	1.1939	1.1970	1.2000	1.2030	1.2060	1.2090	1.2119	1.2149	1.2179	1.2208
3.4	1.2238	1.2267	1.2296	1.2326	1.2355	1.2384	1.2413	1.2442	1.2470	1.2499
3.5	1.2528	1.2556	1.2585	1.2613	1.2641	1.2669	1.2698	1.2726	1.2754	1.2782
3.6	1.2809	1.2837	1.2865	1.2892	1.2920	1.2947	1.2975	1.3002	1.3029	1.3056
3.7	1.3083	1.3110	1.3137	1.3164	1.3191	1.3218	1.3244	1.3271	1.3297	1.3324
3.8	1.3350	1.3376	1.3403	1.3429	1.3455	1.3481	1.3507	1.3533	1.3558	1.3584
3.9	1.3610	1.3635	1.3661	1.3686	1.3712	1.3737	1.3762	1.3788	1.3813	1.3838
4.0	1.3863	1.3888	1.3913	1.3938	1.3962	1.3987	1.4012	1.4036	1.4061	1.4085
4.1	1.4110	1.4134	1.4159	1.4183	1.4207	1.4231	1.4255	1.4279	1.4303	1.4327
4.2	1.4351	1.4375	1.4398	1.4422	1.4446	1.4469	1.4493	1.4516	1.4540	1.4563
4.3	1.4586	1.4609	1.4633	1.4656	1.4679	1.4702	1.4725	1.4748	1.4770	1.4793
4.4	1.4816	1.4839	1.4861	1.4884	1.4907	1.4929	1.4952	1.4974	1.4996	1.5019
4.5	1.5041	1.5063	1.5085	1.5107	1.5129	1.5151	1.5173	1.5195	1.5217	1.5239
4.6	1.5261	1.5282	1.5304	1.5326	1.5347	1.5369	1.5390	1.5412	1.5433	1.5454
4.7	1.5476	1.5497	1.5518	1.5539	1.5560	1.5581	1.5602	1.5623	1.5644	1.5665
4.8	1.5686	1.5707	1.5728	1.5748	1.5769	1.5790	1.5810	1.5831	1.5851	1.5872
4.9	1.5892	1.5913	1.5933	1.5953	1.5974	1.5994	1.6014	1.6034	1.6054	1.6074
5.0	1.6094	1.6114	1.6134	1.6154	1.6174	1.6194	1.6214	1.6233	1.6253	1.6273
5.1	1.6292	1.6312	1.6332	1.6351	1.6371	1.6390	1.6409	1.6429	1.6448	1.6467
5.2	1.6487	1.6506	1.6525	1.6544	1.6563	1.6582	1.6601	1.6620	1.6639	1.6658
5.3	1.6677	1.6696	1.6715	1.6734	1.6752	1.6771	1.6790	1.6808	1.6827	1.6845
5.4	1.6864	1.6882	1.6901	1.6919	1.6938	1.6956	1.6974	1.6993	1.7011	1.7029
5.5	1.7047	1.7066	1.7084	1.7102	1.7120	1.7138	1.7156	1.7174	1.7192	1.7210
5.6	1.7228	1.7246	1.7263	1.7281	1.7299	1.7317	1.7334	1.7352	1.7370	1.7387
5.7	1.7405	1.7422	1.7440	1.7457	1.7475	1.7492	1.7509	1.7527	1.7544	1.7561
5.8	1.7579	1.7596	1.7613	1.7630	1.7647	1.7664	1.7682	1.7699	1.7716	1.7733
5.9	1.7750	1.7766	1.7783	1.7800	1.7817	1.7834	1.7851	1.7867	1.7884	1.7901

Table 7 (continued)

x	0.00	0.01	0.02	0.03	0.04	0.05	0.06	0.07	0.08	0.09
6.0	1.7918	1.7934	1.7951	1.7967	1.7984	1.8001	1.8017	1.8034	1.8050	1.8066
6.1	1.8083	1.8099	1.8116	1.8132	1.8148	1.8165	1.8181	1.8197	1.8213	1.8229
6.2	1.8245	1.8262	1.8278	1.8294	1.8310	1.8326	1.8342	1.8358	1.8374	1.8390
6.3	1.8406	1.8421	1.8437	1.8453	1.8469	1.8485	1.8500	1.8516	1.8532	1.8547
6.4	1.8563	1.8579	1.8594	1.8610	1.8625	1.8641	1.8656	1.8672	1.8687	1.8703
6.5	1.8718	1.8733	1.8749	1.8764	1.8779	1.8795	1.8810	1.8825	1.8840	1.8856
6.6	1.8871	1.8886	1.8901	1.8916	1.8931	1.8946	1.8961	1.8976	1.8991	1.9006
6.7	1.9021	1.9036	1.9051	1.9066	1.9081	1.9095	1.9110	1.9125	1.9140	1.9155
6.8	1.9169	1.9184	1.9199	1.9213	1.9228	1.9242	1.9257	1.9272	1.9286	1.9301
6.9	1.9315	1.9330	1.9344	1.9359	1.9373	1.9387	1.9402	1.9416	1.9430	1.9445
7.0	1.9459	1.9473	1.9488	1.9502	1.9516	1.9530	1.9544	1.9559	1.9573	1.9587
7.1	1.9601	1.9615	1.9629	1.9643	1.9657	1.9671	1.9685	1.9699	1.9713	1.9727
7.2	1.9741	1.9755	1.9769	1.9782	1.9796	1.9810	1.9824	1.9838	1.9851	1.9865
7.3	1.9879	1.9892	1.9906	1.9920	1.9933	1.9947	1.9961	1.9974	1.9988	2.0001
7.4	2.0015	2.0028	2.0042	2.0055	2.0069	2.0082	2.0096	2.0109	2.0122	2.0136
7.5	2.0149	2.0162	2.0176	2.0189	2.0202	2.0215	2.0229	2.0242	2.0255	2.0268
7.6	2.0282	2.0295	2.0308	2.0321	2.0334	2.0347	2.0360	2.0373	2.0386	2.0399
7.7	2.0412	2.0425	2.0438	2.0451	2.0464	2.0477	2.0490	2.0503	2.0516	2.0528
7.8	2.0541	2.0554	2.0567	2.0580	2.0592	2.0605	2.0618	2.0631	2.0643	2.0665
7.9	2.0669	2.0681	2.0694	2.0707	2.0719	2.0732	2.0744	2.0757	2.0769	2.0782
8.0	2.0794	2.0807	2.0819	2.0832	2.0844	2.0857	2.0869	2.0882	2.0894	2.0906
8.1	2.0919	2.0931	2.0943	2.0956	2.0968	2.0980	2.0992	2.1005	2.1017	2.1029
8.2	2.1041	2.1054	2.1066	2.1078	2.1090	2.1102	2.1114	2.1126	2.1133	2.1150
8.3	2.1163	2.1175	2.1187	2.1199	2.1211	2.1223	2.1235	2.1247	2.1258	2.1270
8.4	2.1282	2.1294	2.1306	2.1318	2.1330	2.1342	2.1353	2.1365	2.1377	2.1389
8.5	2.1401	2.1412	2.1424	2.1436	2.1448	2.1459	2.1471	2.1483	2.1494	2.1506
8.6	2.1518	2.1529	2.1541	2.1552	2.1564	2.1576	2.1587	2.1599	2.1610	2.1622
8.7	2.1633	2.1645	2.1656	2.1668	2.1679	2.1691	2.1702	2.1713	2.1725	2.1736
8.8	2.1748	2.1759	2.1770	2.1782	2.1793	2.1804	2.1815	2.1827	2.1838	2.1849
8.9	2.1861	2.1872	2.1883	2.1894	2.1905	2.1917	2.1928	2.1939	2.1950	2.1961
9.0	2.1972	2.1983	2.1994	2.2006	2.2017	2.2028	2.2039	2.2050	2.2061	2.2072
9.1	2.2083	2.2094	2.2105	2.2116	2.2127	2.2138	2.2148	2.2159	2.2170	2.2181
9.2	2.2192	2.2203	2.2214	2.2225	2.2235	2.2246	2.2257	2.2268	2.2279	2.2289
9.3	2.2300	2.2311	2.2322	2.2332	2.2343	2.2354	2.2364	2.2375	2.2386	2.2396
9.4	2.2407	2.2418	2.2428	2.2439	2.2450	2.2460	2.2471	2.2481	2.2492	2.2502
9.5	2.2513	2.2523	2.2534	2.2544	2.2555	2.2565	2.2576	2.2586	2.2597	2.2607
9.6	2.2618	2.2628	2.2638	2.2649	2.2659	2.2670	2.2680	2.2690	2.2701	2.2711
9.7	2.2721	2.2732	2.2742	2.2752	2.2762	2.2773	2.2783	2.2793	2.2803	2.2814
9.8	2.2824	2.2834	2.2844	2.2854	2.2865	2.2875	2.2885	2.2895	2.2905	2.2915
9.9	2.2925	2.2935	2.2946	2.2956	2.2966	2.2976	2.2986	2.2996	2.3006	2.3016

Examples.

$$\begin{aligned} \ln 96{,}700 &= \ln 9.67 + 4 \ln 10 \\ &= 2.2690 + 9.2103 \\ &= 11.4793. \end{aligned}$$

$$\begin{aligned} \ln 0.00967 &= \ln 9.67 - 3 \ln 10 \\ &= 2.2690 - 6.9078 \\ &= -4.6388. \end{aligned}$$

$\ln 10$	= 2.3026	$7 \ln 10$	= 16.1181
$2 \ln 10$	= 4.6052	$8 \ln 10$	= 18.4207
$3 \ln 10$	= 6.9078	$9 \ln 10$	= 20.7233
$4 \ln 10$	= 9.2103	$10 \ln 10$	= 23.0259
$5 \ln 10$	= 11.5129	$11 \ln 10$	= 25.3284
$6 \ln 10$	= 13.8155	$12 \ln 10$	= 27.6310

Note: Adapted from *Functional Approach to Precalculus*, 2nd ed., Mustafa A. Munem and James P. Yizze (New York, NY: Worth Publishers, Inc., © 1974), pp. 500–501. Reproduced by permission of the publisher.

Answers to Selected Odd-numbered Exercises

CHAPTER 1
Exercise Set 1.1 pp. 9–11

1. 3, 14 **3.** $\sqrt{3}, -\sqrt{7}, \sqrt[3]{2}$ **5.** $-6, 0, 3, -2, 14$ **7.** Rational **9.** Rational **11.** Rational **13.** Irrational **15.** Irrational
17. Irrational **19.** Rational **21.** Irrational **23.** 12 **25.** 47 **27.** 7, 7 **29.** $-57, -57$ **31.** -87 **33.** -16 **35.** -10.3
37. $\frac{39}{10}$ **39.** 28 **41.** -49.2 **43.** 210 **45.** $-\frac{833}{5}$ **47.** 5 **49.** $-\frac{1}{7}$ **51.** $-\frac{3}{49}$ **53.** 25 **55.** -4 **57.** 18 **59.** -11.6
61. $-\frac{7}{2}$ **63.** (a) 1.96, 1.9881, 1.999396, 1.999962, 1.999990; (b) $\sqrt{2}$ **65.** Identity $(+)$ **67.** Distributive
69. Commutativity $(+)$ **71.** Associativity $(\times)$ **73.** Inverse $(\times)$ **75.** $7 - 5 \neq 5 - 7$; $7 - 5 = 2, 5 - 7 = -2$
77. $16 \div (4 \div 2) \neq (16 \div 4) \div 2$; $16 \div (4 \div 2) = 8, (16 \div 4) \div 2 = 2$

Exercise Set 1.2 pp. 14–15

1. 2^{-1} **3.** 1 **5.** 4^3 **7.** $6x^5$ **9.** $15a^{-1}b^5$ **11.** $72x^5$ **13.** $-18x^7yz$ **15.** b^3 **17.** x^3y^{-3} **19.** 1 **21.** $3ab^2$ **23.** $\frac{4}{7}xyz^{-5}$
25. $8a^3b^6$ **27.** $16x^{12}$ **29.** $-16x^{12}$ **31.** $36a^4b^6c^2$ **33.** $\frac{1}{25}c^2d^4$ **35.** 1 **37.** 32 **39.** $\frac{27}{4}a^8b^{-10}c^{18}$ **41.** $\frac{3}{4}xy$ **43.** 5.8×10^7
45. 3.65×10^5 **47.** 2.7×10^{-6} **49.** 2.7×10^{-2} **51.** 9.1×10^{11} **53.** 400,000 **55.** 0.0062 **57.** 7,690,000,000,000
59. 9,460,000,000,000 **61.** $-25, 25$ **63.** $-1.1664, 1.1664$ **65.** $1044.65

Exercise Set 1.3 pp. 18–19

1. 4, 3, 2, 1, 0; 4 **3.** 3, 6, 6, 0; 6 **5.** 5, 6, 2, 1, 0; 6 **7.** $3x^2y - 5xy^2 + 7xy + 2$ **9.** $-10pq^2 - 5p^2q + 7pq - 4p + 2q + 3$
11. $3x + 2y - 2z - 3$ **13.** $5x\sqrt{y} - 4y\sqrt{x} - \frac{2}{5}$ **15.** $-5x^3 + 7x^2 - 3x + 6$ **17.** $-2x^2 + 6x - 2$ **19.** $6a - 5b - 2c + 4d$
21. $x^4 - 3x^3 - 4x^2 + 9x - 3$ **23.** $9x\sqrt{y} - 3y\sqrt{x} + 9.1$ **25.** $-1.047p^2q - 2.479pq^2 + 8.879pq - 104.144$

Exercise Set 1.4 p. 21

1. $6x^3 + 4x^2 + 32x - 64$ **3.** $4a^3b^2 - 10a^2b^2 + 3ab^3 + 4ab^2 - 6b^3 + 4a^2b - 2ab + 3b^2$ **5.** $a^3 - b^3$ **7.** $4x^2 + 8xy + 3y^2$
9. $12x^3 + x^2y - \frac{3}{2}xy - \frac{1}{8}y^2$ **11.** $2x^3 - 2\sqrt{2}x^2y - \sqrt{2}xy^2 + 2y^3$ **13.** $4x^2 + 12xy + 9y^2$ **15.** $4x^4 - 12x^2y + 9y^2$
17. $4x^6 + 12x^3y^2 + 9y^4$ **19.** $\frac{1}{4}x^4 - \frac{3}{5}x^2y + \frac{9}{25}y^2$ **21.** $0.25x^2 + 0.70xy^2 + 0.49y^4$ **23.** $9x^2 - 4y^2$ **25.** $x^4 - y^2z^2$
27. $9x^4 - 2$ **29.** $4x^2 + 12xy + 9y^2 - 16$ **31.** $x^4 + 6x^2y + 9y^2 - y^4$ **33.** $x^4 - 1$ **35.** $16x^4 - y^4$
37. $0.002601x^2 + 0.00408xy + 0.0016y^2$ **39.** $2462.0358x^2 - 945.0214x - 38.908$ **41.** $a^3 + 3a^2b + 3ab^2 + b^3$
43. $P + 3Pi + 3Pi^2 + Pi^3$

Exercise Set 1.5 p. 24

1. $3ab(6a - 5b)$ **3.** $(a + c)(b - 2)$ **5.** $(x + 6)(x + 3)$ **7.** $(3x - 5)(3x + 5)$ **9.** $4x(y^2 - z)(y^2 + z)$ **11.** $(y - 3)^2$
13. $(1 - 4x)^2$ **15.** $(2x - \sqrt{5})(2x + \sqrt{5})$ **17.** $(xy - 7)^2$ **19.** $4a(x + 7)(x - 2)$ **21.** $(a + b + c)(a + b - c)$
23. $(x + y - a - b)(x + y + a + b)$ **25.** $5(y^2 + 4x^2)(y - 2x)(y + 2x)$ **27.** $(x + 2)(x^2 - 2x + 4)$
29. $3(x - \frac{1}{2})(x^2 + \frac{1}{2}x + \frac{1}{4})$ **31.** $(x + 0.1)(x^2 - 0.1x + 0.01)$ **33.** $3(z - 2)(z^2 + 2z + 4)$
35. $(a - t)(a + t)(a^2 - at + t^2)(a^2 + at + t^2)$ **37.** $2ab(2a^2 + 3b^2)(4a^4 - 6a^2b^2 + 9b^4)$ **39.** $(x + 4.19524)(x - 4.19524)$
41. $37(x + 0.626y)(x - 0.626y)$ **43.** $h(3x^2 + 3xh + h^2)$

Exercise Set 1.6 p. 28

1. 12 **3.** -6 **5.** 8 **7.** $\frac{4}{5}$ **9.** 2 **11.** $-\frac{3}{2}$ **13.** -2 **15.** $\frac{3}{2}, \frac{2}{3}$ **17.** $0, 1, -2$ **19.** $\frac{2}{3}, -1$ **21.** 4, 1 **23.** $-1, -2$
25. $-\frac{5}{3}, 4, \frac{5}{2}$ **27.** $0, \frac{1}{4}, -\frac{1}{4}$ **29.** $x > 3$ **31.** $x \geq -\frac{5}{12}$ **33.** $y \geq \frac{22}{13}$ **35.** $x \leq \frac{15}{34}$ **37.** $x < 1$ **39.** $\{x \mid x > 2.5\}$
41. $\{t \mid t^2 = 5\}$ **43.** 0.7892 **45.** $x < -0.7848$

Exercise Set 1.7 pp. 34–36

1. All numbers except 0, 1 **3.** All numbers except 0, 3, −2 **5.** $\dfrac{x-2}{x+3}$; all numbers except −3 **7.** $\dfrac{1}{x-y}$
9. $\dfrac{(x+5)(2x+3)}{7x}$ **11.** $\dfrac{a+2}{a-5}$ **13.** $m+n$ **15.** $\dfrac{3(x-4)}{2(x+4)}$ **17.** $\dfrac{1}{x+y}$ **19.** $\dfrac{x-y-z}{x+y+z}$ **21.** 1 **23.** $\dfrac{y-2}{y-1}$ **25.** $\dfrac{x+y}{2x-3y}$
27. $\dfrac{3x-4}{x^2-4}$ **29.** $\dfrac{3y-10}{(y-5)(y+4)}$ **31.** $\dfrac{4x-8y}{x^2-y^2}$ **33.** $\dfrac{3x-4}{(x-2)(x-1)}$ **35.** $\dfrac{5a^2+10ab-4b^2}{(a-b)(a+b)}$ **37.** $\dfrac{11x^2-18x+8}{(x+2)(x-2)^2}$ **39.** 0
41. $\dfrac{x+y}{x}$ **43.** $\dfrac{a^2-1}{a^2+1}$ **45.** $\dfrac{c^2-2c+4}{c}$ **47.** $\dfrac{xy}{x-y}$ **49.** $x-y$ **51.** $\dfrac{x^2-y^2}{xy}$ **53.** $\dfrac{1+a}{1-a}$ **55.** $\dfrac{b+a}{b-a}$ **57.** $2x+h$
59. $3x^2+3xh+h^2$ **61.** x^5

Exercise Set 1.8 p. 42

1. $\{x\,|\,x\geq 3\}$ **3.** $\{x\,|\,x\leq \tfrac{3}{4}\}$ **5.** $9|xy|$ **7.** $3a^2|b|$ **9.** 11 **11.** $4|x|$ **13.** $|b+1|$ **15.** $-3x$ **17.** $|x-2|$ **19.** 2
21. $6\sqrt{5}$ **23.** $3\sqrt[3]{2}$ **25.** $\dfrac{8\sqrt{2}|c|}{d^2}$ **27.** $3\sqrt{2}$ **29.** $2x^2y\sqrt{6}$ **31.** $3x\sqrt[3]{4y}$ **33.** $2(x+4)\sqrt[3]{(x+4)^2}$ **35.** $\sqrt{7b}$ **37.** 2 **39.** $\dfrac{1}{2x}$
41. $\sqrt{a+b}$ **43.** $\dfrac{3a\sqrt{2b}}{4b}$ **45.** $\dfrac{y}{5z^2}\sqrt[3]{20x^2z^2}$ **47.** $8x^2\sqrt[3]{2}$ **49.** $ab^2x^2y\sqrt{a}$ **51.** $10.124x^2y$ **53.** $\dfrac{0.5933a\sqrt{b}}{b}$

Exercise Set 1.9 p. 45

1. $-12\sqrt{5}-2\sqrt{2}$ **3.** $19\sqrt[3]{x^2}-3x$ **5.** $4y\sqrt{3}-2y\sqrt{6}$ **7.** 1 **9.** $t-2x\sqrt{t}+x^2$ **11.** $10\sqrt{7}$ **13.** x **15.** $\dfrac{3(3-\sqrt{5})}{2}$
17. $\dfrac{2\sqrt[3]{6}}{3}$ **19.** $\dfrac{8x-20\sqrt{xy}-6x\sqrt{y}+15y\sqrt{x}}{4x-25y}$ **21.** $\dfrac{2-5a}{6(\sqrt{2}-\sqrt{5a})}$ **23.** $\dfrac{x}{x+2-2\sqrt{x+1}}$ **25.** $\dfrac{a}{3(\sqrt{a+3}+\sqrt{3})}$
27. $\dfrac{(2+x^2)\sqrt{1+x^2}}{1+x^2}$ **29.** Let $a=16$ and $b=9$. Then $\sqrt{a+b}=25$ and $\sqrt{a}+\sqrt{b}=7$.

Exercise Set 1.10 pp. 47–48

1. $\sqrt[4]{x^3}$ **3.** $(\sqrt[4]{16})^3$ or $\sqrt[4]{(2^4)^3}$ or 8 **5.** $\tfrac{1}{5}$ **7.** $\dfrac{a}{b}\sqrt[4]{ab}$ **9.** $20^{2/3}$ **11.** $13^{5/4}$ **13.** $11^{1/6}$ **15.** $5^{5/6}$ **17.** 4 **19.** $2y^2$
21. $(a^2+b^2)^{1/3}$ **23.** $3ab^3$ **25.** $\dfrac{m^2n^4}{2}$ **27.** $8a^{4/2}$ or $8a^2$ **29.** $\dfrac{x^{-3}}{3^{-1}b^2}$ or $\dfrac{3}{x^3b^2}$ **31.** $xy^{1/3}$ or $x\sqrt[3]{y}$ **33.** $\sqrt[6]{288}$ **35.** $\sqrt[12]{x^{11}y^7}$
37. $a\sqrt[6]{a^5}$ **39.** $(a+x)\sqrt[12]{(a+x)^{11}}$ **41.** 24.685 **43.** 43.138 **45.** 2510.472 **47.** 5.56 ft **49.** 7.07 ft **51.** $a^{a/2}$

Exercise Set 1.11 pp. 50–51

1. 12 yd **3.** 48 hr **5.** 3 g **7.** 8 m **9.** 12 ft³ **11.** $\dfrac{7}{10}\dfrac{\text{kg}^2}{\text{m}^2}$ **13.** $720\dfrac{\text{lb-mi}^2}{\text{hr}^2\text{-ft}}$ **15.** $\dfrac{15}{2}\dfrac{\text{cm}^5\text{-kg}}{\text{sec}^3}$ **17.** 6 ft **19.** 172,800 sec
21. 600 g/cm **23.** 2,160,000 cm² **25.** 150 ¢/hr **27.** 5,865,696,000,000 mi/yr **29.** 1621.8 m/min **31.** 1638.4 km²

Chapter 1 Review pp. 51–53

1. $12, -3, -1, -19, 31, 0$ **3.** All except $\sqrt{7}, \sqrt[3]{10}$ **5.** $\sqrt{7}, \sqrt[3]{10}$ **7.** -4 **9.** -5 **11.** -6 **13.** -3000 **15.** $\frac{31}{24}$
17. Distributive **19.** Commutativity ($\times$) **21.** 0.00041 **23.** 4.321×10^4 **25.** $6x^9y^{-6}z^6$ **27.** -2 **29.** $\frac{x+y}{xy}$
31. $25x^4 - 10x^2\sqrt{2} + 2$ **33.** $x^3 + t^3$ **35.** $4y^4 + 24x^2y^3 + 36x^4y^2$ **37.** $3a(2a - 3b^2)(2a + 3b^2)$ **39.** $x(9x - 1)(x + 4)$
41. $(3x^2 + 5y^2)(9x^4 - 15x^2y^2 + 25y^4)$ **43.** $\sqrt[3]{(a+b)^2}$ **45.** $\frac{m^4n^2}{3}$ **47.** -20 **49.** $-2, 1$ **51.** $x \geq 5$ **53.** $\frac{x-5}{(x+3)(x+5)}$
55. $166\frac{2}{3} \frac{\text{mi}}{\text{min}}$

CHAPTER 2

Exercise Set 2.1 pp. 60–61

1. Yes **3.** No **5.** No **7.** $\{0, 3\}$ **9.** $\{0, 2, -\frac{1}{3}\}$ **11.** $\{\frac{3}{2}, -\frac{2}{3}, 1\}$ **13.** $\{\frac{1}{2}, 0, -3\}$ **15.** $\{-2\}$ **17.** $\{6\}$ **19.** $\emptyset$ **21.** $\emptyset$
23. $\{8, -5\}$ **25.** $\{\frac{5}{3}\}$ **27.** $\{0, 2.1522\}$ **29.** $\{0.94656\}$ **31.** $\{-1, 1, -\frac{1}{5}\}$ **33.** $\{-2, 1, -1\}$ **35.** (a) Equivalent to (b); (b) not equivalent to (c); (c) equivalent to (d) **37.** Identity **39.** Identity **41.** Not an identity

Exercise Set 2.2 pp. 70–71

1. $w = \frac{P - 2l}{2}$ **3.** $I = \frac{E}{R}$ **5.** $T_1 = \frac{T_2 P_1 V_1}{P_2 V_2}$ **7.** $v_1 = \frac{H}{Sm} + v_2$ **9.** $p = \frac{Fm}{m - F}$ **11.** $x = \frac{5 + ab}{a - b}$ **13.** $x = -\frac{a}{9}$
15. 44% **17.** 6% **19.** \$14,500; \$16,095 **21.** \$650 **23.** 26°, 130°, 24° **25.** 68 m, 93 m **27.** 91% **29.** 2 cm **31.** 810,000
33. 32 mph **35.** 12 km/h **37.** $1\frac{34}{71}$ hr **39.** 6.21 hr **41.** (a) \$1137.50; (b) \$1142.23; (c) \$1144.75; (d) \$1147.37; (e) \$1147.40

Exercise Set 2.3 pp. 76–77

1. $\pm\sqrt{7}$ **3.** $\pm\frac{\sqrt{5}}{3}$ **5.** $\pm\sqrt{\frac{b}{a}}$ **7.** $7 \pm \sqrt{5}$ **9.** $h \pm \sqrt{a}$ **11.** $-3 \pm \sqrt{5}$ **13.** $\{3, -10\}$ **15.** $\frac{2 \pm \sqrt{14}}{5}$ **17.** $-5, \frac{3}{2}$
19. $1, -5$ **21.** $2, -\frac{1}{2}$ **23.** $-1, -\frac{5}{3}$ **25.** $6 \pm \sqrt{33}$ **27.** No real solution **29.** Two real solutions **31.** $x^2 + 2x - 99 = 0$
33. $x^2 - 14x + 49 = 0$ **35.** $x^2 - \frac{4}{5}x - \frac{12}{25} = 0$ **37.** $x^2 - \left(\frac{c+d}{2}\right)x + \frac{cd}{4} = 0$ **39.** $x^2 - 4\sqrt{2}x + 6 = 0$ **41.** $1.1754,$
-0.4254 **43.** $2, -\frac{3}{2}$ **45.** $\frac{-3 \pm \sqrt{41}}{2}$ **47.** $\frac{3}{2}, \frac{2}{3}$ **49.** $-0.1 \pm \sqrt{0.31}$ **51.** $\frac{-1 \pm \sqrt{1 + 4\sqrt{2}}}{2}$ **53.** $\frac{-\sqrt{5} \pm \sqrt{5 + 4\sqrt{3}}}{2}$
55. $\frac{\sqrt{6} \pm \sqrt{6 + 8\sqrt{10}}}{4}$ **57.** $-2, \frac{3}{4}$ **59.** $\frac{1 \pm \sqrt{113}}{2}$ **61.** $3 \pm \sqrt{5}$

Exercise Set 2.4 pp. 80–81

1. $d = \sqrt{\frac{kM_1M_2}{F}}$ **3.** $t = \sqrt{\frac{2S}{a}}$ **5.** $t = \frac{-v_0 \pm \sqrt{v_0^2 - 64s}}{-32}$ **7.** $n = \frac{3 + \sqrt{9 + 8d}}{2}$ **9.** $i = -1 + \sqrt{\frac{A}{P}}$ **11.** 18.75%

13. 11%   **15.** 9   **17.** 2 ft   **19.** 4.685 cm   **21.** A: 15 mph; B: 20 mph   **23.** (a) 3.91 sec; (b) 1.906 sec; (c) 79.6 m   **25.** 7
27. 12   **29.** 3.237 cm   **31.** $2, -\frac{3}{k}$ **33.** $\frac{1}{m+n}, \frac{-2}{m+n}$

Exercise Set 2.5   pp. 83–84

1. $\frac{5}{3}$ **3.** $\pm\sqrt{2}$ **5.** ∅   **7.** 4   **9.** ∅   **11.** −6   **13.** 3, −1   **15.** $\frac{80}{9}$ **17.** 62.4459   **19.** $L = \frac{gT^2}{4\pi^2}$, $g = \frac{4L\pi^2}{T^2}$ **21.** 208 mi
23. 14,400 ft   **25.** $5 \pm 2\sqrt{2}$ **27.** $-\frac{8}{9}$ **29.** 2   **31.** $\frac{-5 + \sqrt{61}}{18}$

Exercise Set 2.6   pp. 87–88

1. 1, 81   **3.** $\pm\sqrt{5}$ **5.** −27, 8   **7.** 16   **9.** 7, 5, −1, 1   **11.** $1, 4, \frac{5 \pm \sqrt{37}}{2}$ **13.** $\pm\sqrt{2 + \sqrt{6}}$ **15.** $-\frac{1}{2}, \frac{1}{3}$ **17.** −1, 2
19. $-1 \pm \sqrt{3}, \frac{9 \pm \sqrt{89}}{2}$ **21.** $\frac{100}{99}$ **23.** $-\frac{6}{7}$ **25.** 132.66 ft   **27.** 2.0486   **29.** 1, 4

Exercise Set 2.7   pp. 92–93

1. $y = \frac{3}{2}x$ **3.** $y = \frac{0.0015}{x^2}$ **5.** $y = \frac{xz}{w}$ **7.** $y = \frac{5}{4}\frac{xz}{w^2}$ **9.** y is doubled   **11.** y is multiplied by $\frac{1}{n^2}$ **13.** 532,500 tons
15. L is multiplied by 16   **17.** 68.56 m   **19.** If p varies directly as q, then $p = kq$. Thus, $q = \frac{1}{k}p$, so q varies directly as p.
21. $\frac{\pi}{4}$

Chapter 2 Review   pp. 93–94

1. −1   **3.** $\frac{4}{3}, -2$ **5.** $\frac{3 \pm \sqrt{57}}{6}$ **7.** $\pm\sqrt{\frac{3 \pm \sqrt{5}}{2}}$ **9.** $\pm\sqrt{3}, 0$ **11.** 5   **13.** 8, −2   **15.** Two real-number solutions
17. $h = \frac{v^2}{2g}$ **19.** 94%   **21.** $1\frac{1}{2}$ hr   **23.** 4.5   **25.** 8, 15, 17   **27.** $y = \frac{0.5}{x^2}$ **29.** 50 volts   **31.** $-(a + c)$

CHAPTER 3

Exercise Set 3.1   p. 98

1. {(0, a), (0, b), (0, c), (2, a), (2, b), (2, c), (4, a), (4, b), (4, c), (5, a), (5, b), (5, c)}   **3.** {(−1, 0), (−1, 1), (−1, 2), (0, 1), (0, 2), (1, 2)}
5. {(−1, −1), (−1, 0), (−1, 1), (−1, 2), (0, 0), (0, 1), (0, 2), (1, 1), (1, 2), (2, 2)}   **7.** {(−1, −1), (0, 0), (1, 1), (2, 2)}   **9.** Domain, {0, 1};
range, {0, 1, 2}

Exercise Set 3.2 pp. 102–103

1. Yes   3. No   5. No

7.

9.

11.

13.

15.

17.
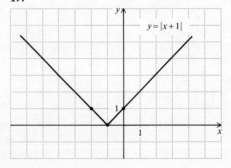

CHAPTER 3   EXERCISE SET 3.2   A-7

19.   **21.**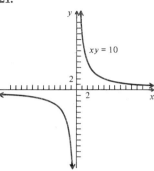

23. Same graphs   **25.** Domain, $\{x \mid 2 \leq x \leq 6\}$; range, $\{y \mid 1 \leq y \leq 5\}$ **27.** Horizontal line through $(0, 2)$ **29.** Line through $(0, 1)$ and $(-1, 0)$ **31.** Line through $(0, 0)$ and $(1, 2)$ **33.** See Exercise 11.

35.   **37.**

39.   **41.**   **43.**

Exercise Set 3.3 pp. 110–112

1. (b), (c), (d) **3.** (a) 0; (b) 1; (c) 57; (d) $5t^2 + 4t$; (e) $5t^2 - 6t + 1$; (f) $10a + 5h + 4$ **5.** (a) 5; (b) -2; (c) -4; (d) $4|y| + 6y$; (e) $2|a + h| + 3a + 3h$; (f) $\dfrac{2|a + h| + 3h - 2|a|}{h}$ **7.** (a) 3.14977; (b) 55.73147; (c) 3178.20675; (d) 1116.70323 **9.** (a) $\frac{2}{3}$; (b) $\frac{10}{9}$; (c) 0; (d) not possible **11.** All real numbers **13.** $\{x \mid x \neq 0\}$ **15.** $\{x \mid x \geq -\frac{4}{7}\}$ **17.** $\{x \mid x \neq 2, -2\}$ **19.** $\{x \mid x \neq -\frac{3}{4}, 2\}$ **21.** $f \circ g(x) = 12x^2 - 12x + 5$; $g \circ f(x) = 6x^2 + 3$ **23.** $f \circ g(x) = \dfrac{16}{x^2} - 1$; $g \circ f(x) = \dfrac{2}{4x^2 - 1}$ **25.** $f \circ g(x) = x^4 - 2x^2 + 2$; $g \circ f(x) = x^4 + 2x^2$ **27.** $0, -3, 3, 2$ **29.** $\dfrac{-1}{x(x + h)}$ **31.** $\dfrac{1}{\sqrt{x + h} + \sqrt{x}}$ **33.** $\{x \mid x \neq 2, -1 \text{ and } x \geq -3\}$ **35.** All real numbers **37.** Domain of $f \circ g$ is $\{x \mid x \neq 0\}$; domain of $g \circ f$ is $\{x \mid x \neq \frac{1}{2}, -\frac{1}{2}\}$

Exercise Set 3.4 p. 118

1. x-axis, no; y-axis, yes; origin, no **3.** x-axis, no; y-axis, yes; origin, no **5.** All yes **7.** All yes **9.** All no **11.** All no **13.** Yes **15.** Yes **17.** Yes **19.** Yes **21.** No **23.** No

25.

27.
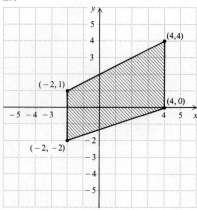

Exercise Set 3.5 pp. 124–126

1. and 3.
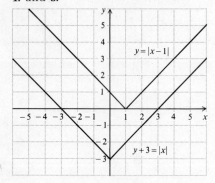

CHAPTER 3   EXERCISE SET 3.5   A-9

5. and 7.

9. and 11.

13.

15. and 17.

19.

21.

23.

25.

27.

29.

31.

33.

35.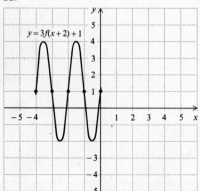

37. The graph is translated 1.8 units to the left; stretched vertically by a factor of $\sqrt{2}$; and reflected across the x-axis.

Exercise Set 3.6   pp. 132–135

1. (a) Even; (b) even; (c) neither; (d) neither   **3.** Neither   **5.** Even   **7.** Neither   **9.** Even   **11.** Odd   **13.** Neither   **15.** Odd   **17.** Even and odd   **19.** Even   **21.** (a) No; (b) yes; (c) yes; (d) no   **23.** 4   **25.** (a) $(-2, 2)$; (b) $(-5, -1)$; (c) $[c, d]$; (d) $[-5, 1)$ **27.** (a) $(-2, 4)$; (b) $(-\frac{1}{4}, \frac{1}{4})$; (c) $[7, 10\pi)$; (d) $[-9, -6]$ **29.** (a) Yes; (b) yes; (c) no; (d) yes; (e) yes   **31.** Where $x = -3$ and $x = 2$ **33.** (a) Increasing; (b) neither; (c) decreasing; (d) neither

35.

37.

39.
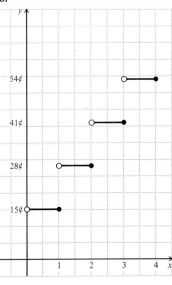

41. Increasing: $[0, 1]$; decreasing: $[-1, 0]$; many possible answers
43. Increasing: (a), (e); decreasing: (b); neither: (c), (d), (f)
45. (a) $[2, 3]$; (b) $(0, 9)$; (c) $(-6, 1)$

Exercise Set 3.7   pp. 141–142

1. $x = 4y - 5$ **3.** $y^2 - 3x^2 = 3$ **5.** $x = 3y^2 + 2$ **7.** $yx = 7$

9.

11.
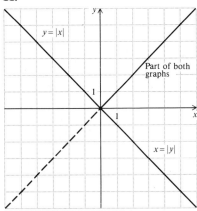

A-12 CHAPTER 3 ANSWERS EXERCISE SET 3.7 TO REVIEW

13. No **15.** Yes **17.** Yes **19.** Yes **21.** No **23.** Yes **25.** (a), (c) **27.** $f^{-1}(x) = \dfrac{x-5}{2}$ **29.** $f^{-1}(x) = x^2 - 1$
31. 3; -125 **33.** 12,053; $-17,243$ **35.** 1.8 **37.** x-axis: no; y-axis: yes; origin: no; y = x: no

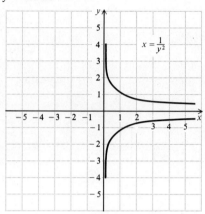

Chapter 3 Review pp. 143–146

1. {(1, 1), (1, 3), (1, 5), (1, 7), (3, 1), (3, 3), (3, 5), (3, 7), (5, 1), (5, 3), (5, 5), (5, 7), (7, 1), (7, 3), (7, 5), (7, 7)}

3.

5.

7.

9.

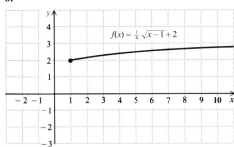

11. (b), (d) **13.** (b), (c), (d), (h) **15.** $x = 3y^2 + 2y - 1$ **17.** (b), (c) **19.** -3 **21.** $a^2 + 2ah + h^2 - a - h - 3$ **23.** 4
25. $f^{-1}(x) = (2x - 4)^2$ **27.** $\{x \mid x \leq \frac{7}{3}\}$ **29.** $\{x \mid x \neq 0, 3, -3\}$ **31.** $f \circ g(x) = \dfrac{4}{(3 - 2x)^2}$, $g \circ f(x) = 3 - \dfrac{8}{x^2}$ **33.** a
35. (a) $y = 1 + f(x)$; (b) $y = \frac{1}{2}f(x)$; (c) $y = f(x + 1)$

  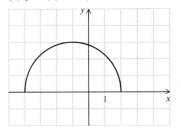

37. (e), (f) **39.** (a), (c), (d) **41.** (a) Yes; (b) no **43.** $(0, 1]$ **45.** (a)
47.

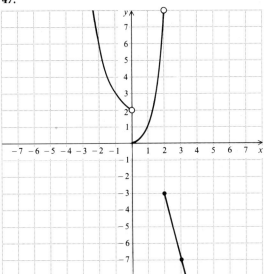

CHAPTER 4

Exercise Set 4.1 pp. 152–153

1. (a), (b), (d), (h) **3.** Line through $(0, -8)$ and $(3, 0)$ **5.** Line through $(0, 3)$ and $(-4, 0)$ **7.** Horizontal line through $(0, -2)$ **9.** Vertical line through $(3, 0)$ **11.** $\frac{1}{8}$ **13.** $\frac{1}{2}$ **15.** $\frac{1}{\pi}$ **17.** $y = 4x - 10$ **19.** $y = 2x - 5$ **21.** $y = \frac{x}{2} + \frac{7}{2}$ **23.** $m = 2; b = 3$ **25.** $m = -3; b = 5$ **27.** $m = \frac{3}{4}; b = -3$ **29.** $m = 0; b = -\frac{10}{3}$ **31.** $y = 3.516x - 13.1602$ **33.** $y = 1.2222x + 1.0949$ **35.** $f(x) = mx$ **37.** $f(x) = x + b$ **39.** False **41.** False **43.** Yes **45.** $\overline{AB}$, $\overline{DC}$: same slope; $\overline{BC}$, $\overline{AD}$: same slope **47.** $F = \frac{9}{5}C + 32$ **49.** $P = mQ + b, m \neq 0$. Then we can solve for Q: $Q = \frac{P}{m} - \frac{b}{m}$.

Exercise Set 4.2 pp. 158–159

1. Neither **3.** Perpendicular **5.** $y = 3x + 3$ **7.** $x = 3$ **9.** $y = -3$ **11.** $y = -\frac{2}{5}x - \frac{31}{5}$ **13.** $y = 3$ **15.** $x = -3$ **17.** $y = 0.6114x + 3.4094$ **19.** 5 **21.** $3\sqrt{2}$ **23.** $\sqrt{a^2 + 64}$ **25.** $\sqrt{a^2 + b^2}$ **27.** $2\sqrt{a}$ **29.** Yes **31.** $(-\frac{1}{2}, -1)$ **33.** $(a, 0)$ **35.** 18.8061 **37.** $(-0.4485, -0.2733)$ **39.** $y = -\frac{7}{3}x + \frac{22}{3}$ **41.** $(5, 0)$

Exercise Set 4.3 pp. 164–165

1. (a) $(0, 0)$; (b) $x = 0$; (c) 0 is a minimum
3. (a) $(\frac{1}{4}, 0)$; (b) $x = \frac{1}{4}$; (c) 0 is a minimum
5. (a) $(9, 0)$; (b) $x = 9$; (c) 0 is a maximum
7. (a) $(-\frac{1}{2}, 0)$; (b) $x = -\frac{1}{2}$; (c) 0 is a maximum
9. (a) $(4, 3)$; (b) $x = 4$; (c) 3 is a minimum
11. (a) $(-3, 5)$; (b) $x = -3$; (c) 5 is a maximum

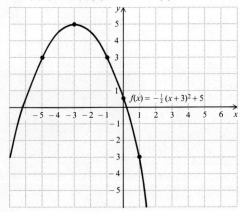

13. (a) $f(x) = -(x - 1)^2 + 4$; (b) $(1, 4)$; (c) $x = 1$; (d) 4 is a maximum
15. (a) $f(x) = (x + \frac{3}{2})^2 - \frac{9}{4}$; (b) $(-\frac{3}{2}, -\frac{9}{4})$; (c) $x = -\frac{3}{2}$; (d) $-\frac{9}{4}$ is a minimum
17. (a) $f(x) = -\frac{3}{4}(x - 4)^2 + 12$; (b) $(4, 12)$; (c) $x = 4$; (d) 12 is a maximum
19. (a) $f(x) = 3(x + \frac{1}{6})^2 - \frac{49}{12}$; (b) $(-\frac{1}{6}, -\frac{49}{12})$; (c) $x = -\frac{1}{6}$; (d) $-\frac{49}{12}$ is a minimum

21. $(3, 0), (-1, 0)$

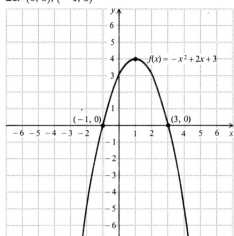

23. $(4 \pm \sqrt{11}, 0)$

25. No x-intercepts

27. $f(x) = 3\left(x + \dfrac{m}{6}\right)^2 + \dfrac{11m^2}{12}$

29.

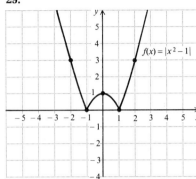

31. Minimum, -6.95

33. Minimum, -6.081; ± 2.466

Exercise Set 4.4 pp. 170–171

1. (a) $E = 0.15t + 72$; (b) 76.5, 77.25   **3.** (a) $R = -0.01t + 10.43$; (b) 9.79; 9.73; (c) 2063   **5.** (a) 1662.5 m after 15 sec; (b) after 33.4 sec   **7.** 10×10 **9.** 30   **11.** \$3.67   **13.** (a) $C(x) = 40x + 22{,}500$; (b) $R(x) = 85x$; (c) $P(x) = 45x - 22{,}500$; (d) Profit = \$112,500; (e) $x = 500$; (f) $x > 500$; (g) $x < 500$

Exercise Set 4.5 pp. 176–177

1. $\{3, 4, 5\}$ **3.** $\{0, 2, 4, 6, 8, 9\}$ **5.** $\{c\}$ **7.** Entire set of real numbers   **9.** $-\tfrac{1}{2}$ and all numbers between $-\tfrac{1}{2}$ and $\tfrac{1}{2}$
11. ⊶―――⊶   (−π, 0, π)   **13.** ●―――│   (−7, 0)
15. ∅   **17.** $\{x \mid -3 < x\}$ **19.** $\{x \mid x \geq -\tfrac{3}{2}\}$ **21.** $\{x \mid 4 < x\}$ **23.** ∅   **25.** $\{x \mid -3 \leq x < 3\}$ **27.** $\{x \mid 8 \leq x \leq 10\}$
29. $\{-7\}$ **31.** $\{x \mid -\tfrac{3}{2} < x < 2\}$ **33.** $\{x \mid 1 < x \leq 5\}$ **35.** $\{x \mid -\tfrac{11}{3} < x \leq \tfrac{13}{3}\}$ **37.** $\{x \mid x \leq -2 \text{ or } x > 1\}$
39. $\{x \mid x \leq -\tfrac{7}{2} \text{ or } x \geq \tfrac{1}{2}\}$ **41.** $\{x \mid x < 9.6 \text{ or } x > 10.4\}$ **43.** $\{x \mid x \leq -\tfrac{57}{4} \text{ or } x \geq -\tfrac{55}{4}\}$ **45.** $\{1\}$ **47.** $\{x \mid x > -\tfrac{1}{5}\}$
49. $\{x \mid x \geq -\tfrac{3}{2}\}$ **51.** $\{w \mid 4.885 \text{ cm} < w < 53.67 \text{ cm}\}$ **53.** $\{S \mid 97\% \leq S \leq 100\%\}$; yes   **55.** $\{x \mid x > 2\}$

Exercise Set 4.6 pp. 180–181

1. $\{-7, 7\}$ **3.** $\{x \mid -7 < x < 7\}$ **5.** $\{x \mid x \leq -\pi \text{ or } x \geq \pi\}$ **7.** $\{-3, 5\}$ **9.** $\{x \mid -17 < x < 1\}$
11. $\{x \mid x \leq -17 \text{ or } x \geq 1\}$ **13.** $\{x \mid -\tfrac{1}{4} < x < \tfrac{3}{4}\}$ **15.** $\{-\tfrac{1}{3}, \tfrac{1}{3}\}$ **17.** $\{-1, -\tfrac{1}{3}\}$ **19.** $\{x \mid -\tfrac{1}{3} < x < \tfrac{1}{3}\}$
21. $\{x \mid -6 \leq x \leq 3\}$ **23.** $\{x \mid x < 4.9 \text{ or } x > 5.1\}$ **25.** $\{x \mid -\tfrac{7}{3} \leq x \leq 1\}$ **27.** $\{x \mid -\tfrac{1}{2} \leq x \leq \tfrac{7}{2}\}$
29. $\{x \mid x < -8 \text{ or } x > 7\}$ **31.** $\{x \mid x < -\tfrac{7}{4} \text{ or } x > -\tfrac{3}{2}\}$ **33.** $\{x \mid \tfrac{3}{8} \leq x \leq \tfrac{9}{8}\}$ **35.** ∅   **37.** ∅
39. $\{x \mid 1.9234 < x < 2.1256\}$ **41.** $\{x \mid 0.98414 \leq x \leq 4.9808\}$ **43.** $\{2, \tfrac{4}{5}\}$ **45.** $\{-4, 4\}$ **47.** $\{x \mid x \leq \tfrac{3}{2}\}$
49. $\{x \mid -\tfrac{9}{2} < x < \tfrac{11}{2}\}$ **51.** $\{x \mid x < -\tfrac{8}{3} \text{ or } x > -2\}$

Exercise Set 4.7   pp. 184–185

1. $\{x \mid -1 < x < 2\}$ **3.** $\{x \mid x \leq -1 \text{ or } x \geq 1\}$ **5.** All real numbers **7.** $\{x \mid 3 - \sqrt{5} < x < 3 + \sqrt{5}\}$
9. $\{x \mid -2 \leq x \leq 10\}$ **11.** $\{x \mid x < -2 \text{ or } x > 4\}$ **13.** $\{x \mid -3 < x < \frac{5}{4}\}$ **15.** $\left\{x \mid x < \dfrac{-1 - \sqrt{41}}{4} \text{ or } x > \dfrac{-1 + \sqrt{41}}{4}\right\}$
17. $\{x \mid -1 < x < 0 \text{ or } x > 1\}$ **19.** $\{x \mid x < -3 \text{ or } -2 < x < 1\}$ **21.** $\{x \mid x > 4\}$ **23.** $\{x \mid x < -\frac{2}{3} \text{ or } x > 3\}$
25. $\{x \mid \frac{3}{2} < x \leq 4\}$ **27.** $\{x \mid x \leq -\frac{5}{2} \text{ or } x > -2\}$ **29.** $\{x \mid x < -\frac{11}{7}\}$ **31.** $\{x \mid x > 1\}$ **33.** $\{x \mid x > 0 \text{ and } x \neq 2\}$
35. $\{x \mid x < 0 \text{ or } x \geq 1\}$ **37.** $\{x \mid x < -3 \text{ or } -2 < x < -1 \text{ or } x > 2\}$ **39.** $\{x \mid x < \frac{5}{3} \text{ or } x > 11\}$ **41.** $\emptyset$
43. $\{x \mid x < -\frac{1}{4} \text{ or } x > \frac{1}{2}\}$ **45.** $\{x \mid x \neq 0\}$ **47.** $\{x \mid -4 < x < -2 \text{ or } -1 < x < 1\}$ **49.** $\{h \mid h > -2 + 2\sqrt{6} \text{ cm}\}$
51. (a) 10, 35; (b) $\{x \mid 10 < x < 35\}$; (c) $\{x \mid x < 10 \text{ or } x > 35\}$ **53.** (a) $\{k \mid k > 2 \text{ or } k < -2\}$; (b) $\{k \mid -2 < k < 2\}$
55. $\{x \mid -1 \leq x \leq 1\}$ **57.** $\{x \mid x \leq -3 \text{ or } x \geq 1\}$

Chapter 4 Review   pp. 185–187

1. -2 **3.** $y = 3x + 5$ **5.** $\sqrt{34}$ **7.** $y = -\frac{2}{3}x - \frac{1}{3}$ **9.** Parallel **11.** Perpendicular **13.** (a) $f(x) = -2(x + \frac{3}{4})^2 + \frac{57}{8}$;
(b) $(-\frac{3}{4}, \frac{57}{8})$; (c) $x = -\frac{3}{4}$; (d) $\frac{57}{8}$ is a maximum **15.** $\{x \mid 2 \leq x \leq 4\}$ **17.** $\{x \mid x > 0 \text{ or } x < -\frac{4}{3}\}$ **19.** $\{2, -7\}$
21. $\{x \mid x > 2 \text{ or } x < -\frac{1}{2}\}$ **23.** $\{x \mid -4 < x < 1 \text{ or } x > 2\}$ **25.** $\{x \mid x < -\frac{1}{2} \text{ or } x > \frac{1}{2}\}$ **27.** (a) $S = \frac{14}{15}d + 9\frac{2}{3}$; (b) 85
29. 15, 15 **31.** (a) 20; (b) $\{x \mid x > 20\}$; (c) $\{x \mid x < 20\}$ **33.** $\{x \mid \frac{1}{3} \leq x \leq 1\}$

CHAPTER 5

Exercise Set 5.1   p. 194

1. No **3.** $(-1, 3)$ **5.** $(\frac{39}{11}, -\frac{1}{11})$ **7.** $(-4, -2)$ **9.** $(-3, 0)$ **11.** $(10, 8)$ **13.** $(1, 1)$ **15.** $(-12, 0)$ **17.** $(0.924, -0.833)$
19. $(-\frac{1}{4}, -\frac{1}{2})$ **21.** $\{(5, 3), (-5, 3), (5, -3), (-5, -3)\}$

Exercise Set 5.2   p. 198

1. $(-4, 4)$ **3.** $\left(\dfrac{y + 5}{3}, y\right)$ or $(x, 3x - 5)$ **5.** $\emptyset$ **7.** Consistent: 1, 2, 3 **9.** $\left(\dfrac{5 - 2y}{3}, y\right)$ or $\left(x, \dfrac{5 - 3x}{2}\right)$ **11.** $\left(\dfrac{6y - 3}{4}, y\right)$ or
$\left(x, \dfrac{4x + 3}{6}\right)$ **13.** $-\frac{11}{2}, -\frac{9}{2}$ **15.** 20 km/h, 3 km/h **17.** 12.5 L, 7.5 L **19.** $\left(\dfrac{724y + 9160}{2013}, y\right)$ or $\left(x, \dfrac{2013x - 9160}{724}\right)$

Exercise Set 5.3   pp. 204–206

1. Yes **3.** $(3, -2, 1)$ **5.** $(-3, 2, 1)$ **7.** $(\frac{1}{2}, \frac{2}{3}, -\frac{5}{6})$ **9.** No solution **11.** $\left(\dfrac{10 + 11z}{9}, \dfrac{-11 + 5z}{9}, z\right)$; $(\frac{10}{9}, -\frac{11}{9}, 0)$, etc.
13. $(\frac{11}{9}z, \frac{5}{9}z, z)$; $(\frac{11}{9}, \frac{5}{9}, 1)$, etc. **15.** $(4z - 5, -3z + 2, z)$; $(-1, -1, 1)$, etc. **17.** $(0, 0, 0)$ **19.** $y = 2x^2 + 3x - 1$
21. (a) $E = -4t^2 + 40t + 2$; (b) \$98 **23.** $(-1, \frac{1}{5}, -\frac{1}{2})$ **25.** A: 4 hr; B: 6 hr; C: 12 hr **27.** $(-2, 4, -1, 1)$

Exercise Set 5.4   pp. 208–209

1. $(\frac{3}{2}, \frac{5}{2})$ **3.** $(-1, 2, -2)$ **5.** $(\frac{1}{2}, \frac{3}{2})$ **7.** $(\frac{3}{2}, -4, 3)$ **9.** $(r - 2, 3 - 2r, r)$ **11.** $(-3, -2, -1, 1)$ **13.** 4 dimes, 30 nickels
15. 10 nickels, 4 dimes, 8 quarters **17.** 5 lb of \$4.05, 10 lb of \$2.70 **19.** \$30,000 at $12\frac{1}{2}$%; \$40,000 at 13%

Exercise Set 5.5 p. 213

1. -11 3. $x^3 - 4x$ 5. -109 7. $-x^4 + x^2 - 5x$ 9. $(-\frac{25}{2}, -\frac{11}{2})$
11. $\left(\dfrac{4\pi - 5\sqrt{3}}{3 + \pi^2}, \dfrac{4\sqrt{3} + 5\pi}{-3 - \pi^2}\right)$ 13. $(\frac{3}{2}, \frac{13}{14}, \frac{33}{14})$ 15. $(\frac{1}{2}, \frac{2}{3}, -\frac{5}{6})$ 17. $2, -2$
19. $\{x \mid x \leq -\sqrt{3} \text{ or } x \geq \sqrt{3}\}$ 21. -34 23. 4
25. $\begin{vmatrix} L & -W \\ 2 & 2 \end{vmatrix}$ 27. $\begin{vmatrix} a & b \\ -b & a \end{vmatrix}$ 29. $\begin{vmatrix} 2\pi r & 2\pi r \\ -h & r \end{vmatrix}$

Exercise Set 5.6 p. 217

1.
3.
5.
7.
9.
11.
13.
15.
17.
19.
21.
23.

25.

27.

29.

31.

Exercise Set 5.7   pp. 220–221

1. Max 168, when $x = 0$, $y = 6$; min 0, when $x = 0$, $y = 0$ **3.** Max 152, when $x = 7$, $y = 0$; min 32, when $x = 0$, $y = 4$
5. Max 102, 8 of A, 10 of B   **7.** $7000 bank X, $15,000 bank Y, maximum income $1395   **9.** Max $192, 2 knits, 4 worsteds   **11.** Min $460, 30 P1 airplanes, 10 P2 airplanes

Chapter 5 Review   pp. 221–222

1. $(-2, -2)$ **3.** $(\frac{5}{18}, \frac{1}{7})$ **5.** $(13, 8, 2, -5)$ **7.** $1600 at 10%, $3400 at 10.5%   **9.** $(-3, 4, -2)$ **11.** $(1, -2, 3, -4)$
13. Inconsistent, independent   **15.** Consistent, independent   **17.** $10,000 at 12%, $12,000 at 13%, $18,000 at $14\frac{1}{2}$%
19. -13 **21.** -6 **23.** 0   **25.** $(a, 0)$ **27.** $(0, 9)$, $(2, 5)$, $(5, 1)$, $(8, 0)$ **29.** Type A: 0; Type B: 10; maximum score = 120 points   **31.**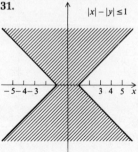

CHAPTER 6

Exercise Set 6.1   pp. 229–230

1. $\begin{bmatrix} -2 & 7 \\ 6 & 2 \end{bmatrix}$ **3.** $\begin{bmatrix} 1 & 3 \\ 2 & 6 \end{bmatrix}$ **5.** $\begin{bmatrix} 9 & 9 \\ -3 & -3 \end{bmatrix}$ **7.** $\begin{bmatrix} 11 & 13 \\ 5 & 3 \end{bmatrix}$ **9.** $\begin{bmatrix} -4 & 3 \\ -2 & -4 \end{bmatrix}$ **11.** $\begin{bmatrix} 17 & 9 \\ -2 & 1 \end{bmatrix}$ **13.** $\begin{bmatrix} 0 & 0 \\ 0 & 0 \end{bmatrix}$ **15.** $\begin{bmatrix} 1 & 2 \\ 4 & 3 \end{bmatrix}$ or A

17. $\begin{bmatrix} -5 & 4 & 3 \\ 5 & -9 & 4 \\ 7 & -18 & 17 \end{bmatrix}$ **19.** $\begin{bmatrix} -2 & 9 & 6 \\ -3 & 3 & 4 \\ 2 & -2 & 1 \end{bmatrix}$ or C **21.** $[-16]$ **23.** $[2 \quad -19]$ **25.** $\begin{bmatrix} 3 & -2 & 4 \\ 2 & 1 & -5 \end{bmatrix} \begin{bmatrix} x \\ y \\ z \end{bmatrix} = \begin{bmatrix} 17 \\ 13 \end{bmatrix}$

27. $\begin{bmatrix} 1 & -1 & 2 & -4 \\ 2 & -1 & -1 & 1 \\ 1 & 4 & -3 & -1 \\ 3 & 5 & -7 & 2 \end{bmatrix} \begin{bmatrix} x \\ y \\ z \\ w \end{bmatrix} = \begin{bmatrix} 12 \\ 0 \\ 1 \\ 9 \end{bmatrix}$ **29.** $\begin{bmatrix} -40.19 & 37.94 & 142.24 \\ -36.78 & 16.63 & 119.62 \\ -1.659 & 14.97 & 12.65 \end{bmatrix}$

31. $(A+B)(A-B) = \begin{bmatrix} -2 & 1 \\ 2 & -1 \end{bmatrix}$, $A^2 - B^2 = \begin{bmatrix} 0 & 3 \\ 0 & -3 \end{bmatrix}$

Exercise Set 6.2   pp. 234–235

1. $a_{11} = 7$, $a_{32} = 2$, $a_{22} = 0$ **3.** $M_{11} = 6$, $M_{32} = -9$, $M_{22} = -29$ **5.** $A_{11} = 6$, $A_{32} = 9$, $A_{22} = -29$ **7.** $|A| = -10$
9. $|A| = -10$ **11.** $M_{41} = -14$, $M_{33} = 20$ **13.** $A_{24} = 15$, $A_{43} = 30$ **15.** $|A| = 110$ **17.** 195

Exercise Set 6.3   pp. 239–241

1. $|A| = -10$, $|B| = 10$; $|A| = -|B|$ **3.** $|A| = 5$, $|B| = -5$; $|A| = -|B|$ **5.** $\begin{bmatrix} 3 & 7 & 3 \\ 1 & 5 & -3 \\ 4 & 9 & 1 \end{bmatrix}$ **7.** -70 **9.** -4

11. 9072   **13.** -153 **15.** 0   **17.** 0   **19.** $(x-y)(y-z)(x-z)$ **21.** $xyz(x-y)(y-z)(z-x)$ **23.** -4

Exercise Set 6.4   pp. 248–249

1. $A^{-1} = \begin{bmatrix} -3 & 2 \\ 5 & -3 \end{bmatrix}$ **3.** $A^{-1} = \begin{bmatrix} 2 & -3 \\ -7 & 11 \end{bmatrix}$ **5.** $A^{-1} = \begin{bmatrix} \frac{2}{11} & \frac{3}{11} \\ -\frac{1}{11} & \frac{4}{11} \end{bmatrix}$ **7.** $A^{-1} = \begin{bmatrix} \frac{3}{8} & -\frac{1}{4} & \frac{1}{8} \\ -\frac{1}{8} & \frac{3}{4} & -\frac{3}{8} \\ -\frac{1}{4} & \frac{1}{2} & \frac{1}{4} \end{bmatrix}$ **9.** $A^{-1} = \begin{bmatrix} \frac{1}{3} & 0 & \frac{1}{3} \\ -\frac{2}{5} & \frac{2}{5} & \frac{1}{5} \\ \frac{2}{15} & \frac{1}{5} & -\frac{1}{15} \end{bmatrix}$

11. A^{-1} does not exist.   **13.** $A^{-1} = \begin{bmatrix} 1 & -2 & 3 & 8 \\ 0 & 1 & -3 & 1 \\ 0 & 0 & 1 & -2 \\ 0 & 0 & 0 & -1 \end{bmatrix}$ **15.–27.** See Exercises 1–13.   **29.** $(-\frac{1}{39}, \frac{55}{39})$

31. $(3, -3, -2)$ **33.** Find AI and IA and compare with A.

35. A^{-1} exists if and only if $xy \neq 0$. $A^{-1} = \begin{bmatrix} x^{-1} & 0 \\ 0 & y^{-1} \end{bmatrix}$

37. A^{-1} exists if and only if $xyzw \neq 0$. $A^{-1} = \begin{bmatrix} \frac{1}{x} & -\frac{1}{xy} & -\frac{1}{xz} & -\frac{1}{xw} \\ 0 & \frac{1}{y} & 0 & 0 \\ 0 & 0 & \frac{1}{z} & 0 \\ 0 & 0 & 0 & \frac{1}{w} \end{bmatrix}$

Chapter 6 Review   pp. 249–250

1. $\begin{bmatrix} 0 & -1 & 6 \\ 3 & 1 & -2 \\ -2 & 1 & -2 \end{bmatrix}$ **3.** $\begin{bmatrix} -1 & 1 & 0 \\ -2 & -3 & 2 \\ 2 & 0 & -1 \end{bmatrix}$

5. Not possible   **7.** $\begin{bmatrix} 2 & -1 & -6 \\ 1 & 5 & -2 \\ -2 & -1 & 4 \end{bmatrix}$

9. $\begin{bmatrix} -\frac{1}{2} & 0 \\ \frac{1}{6} & \frac{1}{3} \end{bmatrix}$ **11.** $\begin{bmatrix} 1 & 0 & 0 & 0 \\ 0 & \frac{1}{9} & \frac{5}{18} & 0 \\ 0 & -\frac{1}{9} & \frac{2}{9} & 0 \\ 0 & 0 & 0 & 1 \end{bmatrix}$

13. -31 **15.** 0 **17.** If a matrix has all 0's below the main diagonal, then its determinant is the product of the elements on the main diagonal. Proof: Expand about the first column.   **19.** $(b-a)(c-b)(c-a)$
21. $(b-a)(c-a)(d-a)(c-b)(d-b)(d-c)$

CHAPTER 7

Exercise Set 7.1   pp. 256–257

1. (a) and (b)

(c)

3.

5.

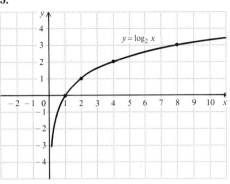

7. $2^5 = 32$ **9.** $10^{-2} = 0.01$ **11.** $6^1 = 6$ **13.** $\log_6 1 = 0$ **15.** $\log_{6/5} \frac{25}{36} = -2$ **17.** $\log_5 \frac{1}{25} = -2$ **19.** $\log_e 1.0833 = 0.08$
21. 10,000 **23.** $\frac{1}{2}$ **25.** 4 **27.** $\frac{1}{2}$ **29.** 2 **31.** 1 **33.** -3 **35.** $4x$ **37.** $\sqrt{5}$

39.

41.

43.

45.

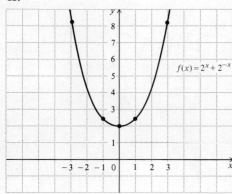

47. All real numbers
49. $\{x \mid x \neq 0\}$
51. $\{x \mid x > \frac{1}{3}\}$
53. $\{x \mid x < -3 \text{ or } x > 3\}$
55. $\{x \mid x \leq 0\}$
57. $\{x \mid x \geq 16\}$
59. $\{x \mid x > -3\}$
61. π^5

63.

Exercise Set 7.2   pp. 261–262

1. $2 \log_a x + 3 \log_a y + \log_a z$ **3.** $\log_b x + 2 \log_b y - 3 \log_b z$ **5.** $\log_a 4$ **7.** $\log_a \frac{2x^4}{y^3}$ **9.** $\log_a \frac{\sqrt{a}}{x}$ or $\frac{1}{2} - \log_a x$

11. $\log_a (x^2 - xy + y^2)$ **13.** $\frac{1}{2}[\log_a (1 - x) + \log_a (1 + x)]$ **15.** 0.602   **17.** 1.699   **19.** 1.778   **21.** -0.088 **23.** 1.954
25. -0.046 **27.** False   **29.** True   **31.** False   **33.** False   **35.** $\frac{1}{2}$ **37.** $\sqrt{7}$ **39.** $-2, 0$ **41.** $\{x \mid x > 0\}$

Exercise Set 7.3   pp. 268–269

1. 0.3909   **3.** 2.5403   **5.** 1.7202   **7.** 5.7952   **9.** $8.8463 - 10$ **11.** $6.3345 - 10$ **13.** 233   **15.** 0.018   **17.** 0.00000105
19. 25.2   **21.** 0.0973   **23.** 4.49   **25.** 0.0133   **27.** 190   **29.** 272   **31.** 8.77   **33.** 3.64   **35.** 25.7   **37.** 4.754264
39. -0.321371 **41.** 78,397,100   **43.** 0.000583

Exercise Set 7.4 p. 272

1. 1.6194 **3.** 0.4689 **5.** 2.8130 **7.** $9.1538 - 10$ **9.** $7.6291 - 10$ **11.** $9.2494 - 10$ **13.** 2.7786 **15.** $9.8445 - 10$
17. 224.5 **19.** 14.53 **21.** 70,030 **23.** 0.09245 **25.** 0.5343 **27.** 0.007295 **29.** 0.8268

Exercise Set 7.5 pp. 277–279

1. 5 **3.** $\frac{12}{5}$ **5.** $\frac{1}{2}, -3$ **7.** 2.7093 **9.** 10 **11.** 1 **13.** 5 **15.** 1; 100 **17.** 4 **19.** $x = \log_e (t + \sqrt{t^2 + 1})$
21. $x = \frac{1}{2} \log_5 \frac{t+1}{1-t}$ **23.** 11.9 years **25.** 65 db **27.** 140 db **29.** 6.7 **31.** $10^5 \cdot I_0$ **33.** (a) 82; (b) 68 **35.** 9 months
37. 4.2 **39.** 1; 10,000 **41.** ϕ **43.** $-9, 9$ **45.** $\frac{7}{4}$ **47.** $\log_x y - \log_x a$
49. $t = \dfrac{100(\log_e P - \log_e P_0)}{r}$
51. $t = -\dfrac{1}{k} \log \left[\dfrac{T - T_0}{T_1 - T_0} \right]$
53. $Q = a^b \cdot \sqrt[3]{y}$ **55.** 10; 100

Exercise Set 7.6 pp. 286–288

1. and 3.

5.

7. 0.6313 **9.** -3.9739 **11.** 0.7561 **13.** -1.5465 **15.** 8.4119 **17.** -7.4875 **19.** -2.5258 **21.** 13.7953 **23.** 4.6052
25. 4.0944 **27.** 2.3026 **29.** 140.67 **31.** $k = 0.0201$; $P = 1{,}243{,}000$ **33.** 0.4 gram **35.** 1.2 days **37.** (a) 3000 yr;
(b) 5100 yr **39.** 587 mb **41.** (a) $k = 0.061$, $P = \$100 e^{0.061t}$; (b) \$338.72; (c) 1978 **43.** 2.1610 **45.** -0.1544 **47.** 2.4849
49. $t = \dfrac{\ln P - \ln P_0}{k}$
51. By Theorem 7, $\ln x = \dfrac{\log x}{\log e} \approx \dfrac{\log x}{0.4343}$; by Table 2, $\approx 2.3026 \log x$
53. e^π **55.** 2; 2.25; 2.48832; 2.593742; 2.704814; 2.716924

Chapter 7 Review pp. 288–289

1.

3.

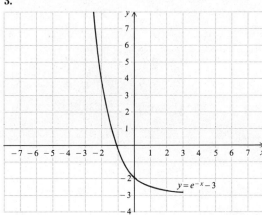

5. $\log_7 x = 2.3$ **7.** $\tfrac{1}{2}$ **9.** $3, -2$ **11.** 0.544   **13.** 0.2385   **15.** 1.4200   **17.** 73.9   **19.** 0.0276   **21.** 2.1861   **23.** -3.0748

25. $\tfrac{1}{5}$ **27.** $T = \dfrac{\log 2}{\log 1.13} \approx 5.7$ **29.** 6.2 g   **31.** 1   **33.** 3   **35.**   **37.** $\{x \mid x > e^{6/5}\}$

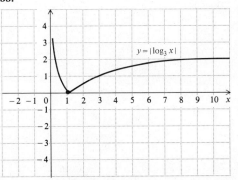

CHAPTER 8

Exercise Set 8.1 pp. 295–297

1. 52°   **3.** 17.5   **5.** $\sin\theta = \tfrac{5}{13}$, $\cos\theta = \tfrac{12}{13}$, $\tan\theta = \tfrac{5}{12}$, $\cot\theta = \tfrac{12}{5}$, $\sec\theta = \tfrac{13}{12}$, $\csc\theta = \tfrac{13}{5}$ **7.** $\sin\theta = 0.8788$, $\cos\theta = 0.4771$, $\tan\theta = 1.8419$, $\cot\theta = 0.5429$, $\sec\theta = 2.0958$, $\csc\theta = 1.1379$ **9.** (a) 53.5°; (b) 16.77; (c) 22.7   **11.** (a) 52.5; (b) 36.5; (c) 46.0

13. 3.3287   **15.** $\sin 45° = \dfrac{1}{\sqrt{2}}$, $\cos 45° = \dfrac{1}{\sqrt{2}}$, $\tan 45° = 1$

Exercise Set 8.2 pp. 299–301

1. (a) $(-5, -2)$; (b) $(5, 2)$; (c) $(5, -2)$; (d) yes, by symmetry   **3.** $M(0, 1)$; $N\left(-\dfrac{\sqrt{2}}{2}, \dfrac{\sqrt{2}}{2}\right)$; $P(0, -1)$; $Q\left(\dfrac{\sqrt{2}}{2}, -\dfrac{\sqrt{2}}{2}\right)$

5.

7.

9.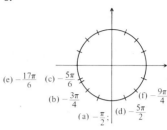

11. M: $\frac{2}{3}\pi, -\frac{4}{3}\pi$; N: $\frac{5}{6}\pi, -\frac{7}{6}\pi$; P: $\frac{5}{4}\pi, -\frac{3}{4}\pi$; Q: $\frac{11}{6}\pi, -\frac{\pi}{6}$ **13.** M: $\frac{8\pi}{3}$; N: $\frac{17\pi}{6}$; P: $\frac{13\pi}{4}$; Q: $\frac{23\pi}{6}$ **15.** $\pm\frac{\sqrt{3}}{2}$ **17.** $\left(\frac{\sqrt{3}}{2}, -\frac{1}{2}\right)$

19. ± 0.96649

Exercise Set 8.3 pp. 305-307

1. $y = \sin x$

3. (a) See Exercise 1; (b)

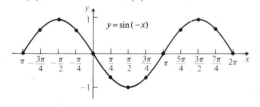

(c) same as (b); (d) same as (b)

5. (a) See Exercise 1; (b)

(c) same as (b); (d) the same

7. (a) See Exercise 2; (b)

(c) same as (b); (d) the same

9. $\cos x$ **11.** $-\sin x$ **13.** $-\cos x$ **15.** $\cos x$ **17.** $-\cos x$ **19.** (a) $\frac{\sqrt{2}}{2}$; (b) 0; (c) $\frac{1}{2}$; (d) $-\frac{\sqrt{2}}{2}$; (e) -1 (f) $\frac{1}{2}$; (g) $-\frac{\sqrt{2}}{2}$;

(h) 0; (j) $\frac{\sqrt{2}}{2}$; (k) 1; (m) $-\frac{\sqrt{3}}{2}$ **21.** (a) $\frac{\sqrt{2}}{2}$; (b) -1; (c) $-\frac{\sqrt{2}}{2}$; (d) $\frac{\sqrt{3}}{2}$; (e) 0; (f) $-\frac{\sqrt{3}}{2}$; (g) $\frac{\sqrt{2}}{2}$; (h) -1; (j) $-\frac{\sqrt{2}}{2}$; (k) 0;

(m) $\frac{1}{2}$ **23.** (a) $\frac{\pi}{2} + 2k\pi$, k any integer; (b) $\frac{3\pi}{2} + 2k\pi$, k any integer **25.** $x = k\pi$, k any integer

27. Domain: All reals; range: [−3, 3]; period: 2π; amplitude: 3   **29.**

31. $\frac{\sin x}{x}$ is between 1 and $\frac{2}{\pi}$. It approaches 1.   **33.** $\left(-\frac{3\pi}{4}, \frac{\pi}{4}\right)$ **35.** Solutions exist; cos x = $-\frac{2}{3}$. The range of the cosine function is [−1, 1].   **37.** (a) 0.8660; (b) 0.7071

Exercise Set 8.4   pp. 314–315

1.

	$\frac{\pi}{16}$	$\frac{\pi}{8}$	$\frac{\pi}{6}$	$\frac{\pi}{4}$	$\frac{3}{8}\pi$	$\frac{7}{16}\pi$
sin	0.19509	0.38268	0.5000	0.70711	0.92388	0.98079
cos	0.98079	0.92388	0.86603	0.70711	0.38268	0.19509
tan	0.19891	0.41421	0.57735	1.00000	2.41424	5.02737
cot	5.02737	2.41424	1.73206	1.00000	0.41421	0.19891
sec	1.01959	1.08239	1.15469	1.41421	2.61315	5.12584
csc	5.12584	2.61315	2.00000	1.41421	1.08239	1.01959

3. y = tan x   **5.** y = sec x

7. cos, sec   **9.** sin, cos, sec, cosec   **11.** Positive: I, III; negative: II, IV   **13.** Positive: I, IV; negative: II, III

15. $\sec(-x) \equiv \frac{1}{\cos(-x)} \equiv \frac{1}{\cos x} \equiv \sec x$ **17.** $\cot(x + \pi) \equiv \frac{\cos(x + \pi)}{\sin(x + \pi)} \equiv \frac{-\cos x}{-\sin x} \equiv \frac{\cos x}{\sin x} \equiv \cot x$

19. $\sec(x+\pi) \equiv \dfrac{1}{\cos(x+\pi)} \equiv -\dfrac{1}{\cos x} \equiv -\sec x$ **21.** Graph of $\sec(x-\pi)$ is like sec, moved π units to right. Graph of $-\sec x$ is sec reflected across x-axis. Graphs are identical.   **23.** If the graph of tan x were reflected across the y-axis and then translated to the right a distance of $\pi/2$, the graph of cot x would be obtained. There are other ways to describe the relation.   **25.** The sine and tangent functions; the cosine and cotangent functions.

27.

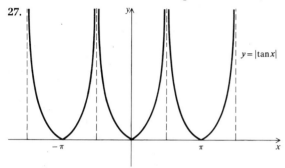

Exercise Set 8.5   p. 319

1. $\cot^2 x \equiv \csc^2 x - 1$; $\csc^2 x - \cot^2 x \equiv 1$ **3.** (a) $\csc x \equiv \pm\sqrt{1+\cot^2 x}$; (b) $\cot x \equiv \pm\sqrt{\csc^2 x - 1}$
5. $\tan\left(x - \dfrac{\pi}{2}\right) \equiv -\cot x$ **7.** $\sec\left(\dfrac{\pi}{2} - x\right) \equiv \csc x$ **9.** $\sin\left(x \pm \dfrac{\pi}{2}\right) \equiv \pm\cos x$; $\cos\left(x \pm \dfrac{\pi}{2}\right) \equiv \mp\sin x$; $\tan\left(x \pm \dfrac{\pi}{2}\right) \equiv -\cot x$; $\cot\left(x \pm \dfrac{\pi}{2}\right) \equiv -\tan x$; $\sec\left(x \pm \dfrac{\pi}{2}\right) \equiv \mp\csc x$; $\csc\left(x \pm \dfrac{\pi}{2}\right) \equiv \pm\sec x$ **11.** $\cos x = -0.9898$; $\tan x = -0.1440$; $\cot x = -6.9460$; $\sec x = -1.0103$; $\csc x = 7.0175$ **13.** $\sin x = 0.7987$; $\cos x = 0.6018$; $\tan x = 1.3273$; $\sec x = 1.6617$; $\csc x = 1.25210$ **15.** $\begin{vmatrix} \sin x & \cos x \\ -\cos x & \sin x \end{vmatrix} = 1$

Exercise Set 8.6   p. 324

1.   **3.**

5.   **7.**

9.   **11.**   **13.**

15. Amplitude, 3; period, $\frac{2\pi}{3}$; phase shift, $\frac{\pi}{6}$ **17.** Amplitude, 4; period, $\frac{2\pi}{5}$; phase shift, $-\frac{\pi}{10}$

19.   **21.**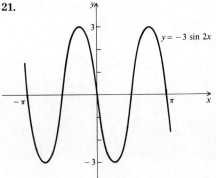

Exercise Set 8.7   pp. 327–328

1. $\sin^2 x - \cos^2 x$ **3.** $\sin x - \sec x$ **5.** $1 + 2\sin y \cos y$ **7.** $\sin^3 x + \csc^3 x$ **9.** $(\sin y - \cos y)(\sin y + \cos y)$
11. $\sin x(\sec x + 1)$ **13.** $3(\cot y + 1)^2$ **15.** $(\sin y + 3)(\sin^2 y - 3\sin y + 9)$ **17.** $\sin x(\tan y + \cot y)$ **19.** $\tan x$
21. $\cos x - 1$ **23.** $\cos x + 1$ **25.** $\cos x - 1$ **27.** $\frac{1}{2\cos x}$ **29.** $\frac{1}{3}\cot y$ **31.** $\frac{1 - 2\sin y + 2\cos y}{\sin^2 y - \cos^2 y}$ **33.** $\frac{5(\sin x - 3)}{3}$
35. $\sin x \cos x$ **37.** $\sin x + \cos x$ **39.** $1 - \sin y$ **41.** $\frac{\sqrt{\sin x \cos x}}{\cos x}$ **43.** $\frac{\cos x}{1 - \sin x}$ **45.** $3, -7$ **47.** $-10, 1$ **49.** $2, -3$
51. $3 \pm \sqrt{13}$

Chapter 8 Review   pp. 329–330

1. $(-3, -2)$ **3.** $(3, 2)$ **5. and 7.**

9. Set of all real numbers **11.** 2π **13.**

	$\dfrac{\pi}{6}$	$\dfrac{\pi}{4}$	$\dfrac{\pi}{3}$	$\dfrac{\pi}{2}$	$\dfrac{3\pi}{4}$	$\dfrac{5\pi}{4}$
sin x	$\dfrac{1}{2}$	$\dfrac{\sqrt{2}}{2}$	$\dfrac{\sqrt{3}}{2}$	1	$\dfrac{\sqrt{2}}{2}$	$-\dfrac{\sqrt{2}}{2}$
cos x	$\dfrac{\sqrt{3}}{2}$	$\dfrac{\sqrt{2}}{2}$	$\dfrac{1}{2}$	0	$-\dfrac{\sqrt{2}}{2}$	$-\dfrac{\sqrt{2}}{2}$

15. 2π **17.** I, IV **19.** 1 **21.** $-\sin x$ **23.** $-\cos x$ **25.** 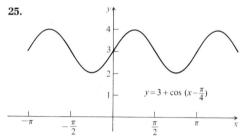 **27.** 2π **29.** $\dfrac{1}{\sin x}$

31. $\dfrac{\sqrt{\sec x \tan x}}{\sec x}$ **33.** No; $\sin x = \tfrac{7}{5}$, but sines are never greater than 1.

CHAPTER 9

Exercise Set 9.1 pp. 335–336

1. I **3.** III **5.** I **7.** II **9.** $\dfrac{\pi}{6}$ **11.** $\dfrac{\pi}{3}$ **13.** $\dfrac{5\pi}{12}$ **15.** 0.2095π **17.** 1.1922π **19.** 2.093 **21.** 5.582 **23.** 3.489
25. 2.0550 **27.** 0.0236 **29.** 57.32° **31.** 1440° **33.** 135° **35.** 74.69° **37.** 135.6° **39.** 1.1 radians; 63°
41. 5.233 radians **43.** 16 m **45.** $30° = \dfrac{\pi}{6}$; $60° = \dfrac{\pi}{3}$; $135° = \tfrac{3}{4}\pi$; $180° = \pi$; $225° = \tfrac{5}{4}\pi$; $270° = \tfrac{3}{2}\pi$; $315° = \tfrac{7}{4}\pi$ **47.** (a) 53.33;
(b) 170; (c) 25; (d) 142.86 **49.** 111.7 km; 69.81 mi **51.** $\tfrac{1}{30}$ radian

Exercise Set 9.2 pp. 338–339

1. 3150 cm/min **3.** 1047 mph **5.** 75.4 ft **7.** 563 radians **9.** 2.093 radians/sec **11.** (a) 395.35 rpm/sec;
(b) 41.4 radians/sec²

Exercise Set 9.3 pp. 345–346

1. $\sin\theta = \tfrac{5}{13}$, $\cos\theta = -\tfrac{12}{13}$, $\tan\theta = -\tfrac{5}{12}$, $\cot\theta = -\tfrac{12}{5}$, $\sec\theta = -\tfrac{13}{12}$, $\csc\theta = \tfrac{13}{5}$ **3.** $\sin\theta = -\tfrac{3}{4}$, $\cos\theta = -\dfrac{\sqrt{7}}{4}$, $\tan\theta = \dfrac{3}{\sqrt{7}}$,
$\cot\theta = \dfrac{\sqrt{7}}{3}$, $\sec\theta = -\dfrac{4}{\sqrt{7}}$, $\csc\theta = -\tfrac{4}{3}$ **5.** $\sin 30° = 0.5$, $\cos 30° = 0.866$, $\tan\theta = 0.577$, $\cot\theta = 1.732$, $\sec 30° = 1.155$,
$\csc 30° = 2$ **7.** -1 **9.** Does not exist. **11.** 1 **13.** -0.707 **15.** 0.5 **17.** 1.732 **19.** 1.414 **21.** 1 **23.** 0.5 **25.** 0.577
27. 0.707 **29.** -0.577

31.

33.

Exercise Set 9.4   p. 348

1. $\cos\theta = -\dfrac{2\sqrt{2}}{3}$, $\tan\theta = \dfrac{\sqrt{2}}{4}$, $\cot\theta = 2\sqrt{2}$, $\sec\theta = -\dfrac{3\sqrt{2}}{4}$, $\csc\theta = -3$ **3.** $\sin\theta = -\tfrac{4}{5}$, $\tan\theta = -\tfrac{4}{3}$, $\cot\theta = -\tfrac{3}{4}$, $\sec\theta = \tfrac{5}{3}$, $\csc\theta = -\tfrac{5}{4}$ **5.** $\sin\theta = -\dfrac{\sqrt{5}}{5}$, $\cos\theta = \dfrac{2\sqrt{5}}{5}$, $\tan\theta = -\tfrac{1}{2}$, $\sec\theta = \dfrac{\sqrt{5}}{2}$, $\csc\theta = -\sqrt{5}$ **7.** $\sin 25° = 0.4226$, $\cos 25° = 0.9063$, $\tan 25° = 0.4663$, $\cot 25° = 2.145$, $\sec 25° = 1.103$, $\csc 25° = 2.366$ **9.** (a) $\cos 27° = 0.89101$, $\tan 27° = 0.50952$, $\cot 27° = 1.96262$, $\sec 27° = 1.12232$, $\csc 27° = 2.20269$; (b) $\sin 63° = 0.89101$, $\cos 63° = 0.45399$, $\tan 63° = 1.96262$, $\cot 63° = 0.50952$, $\sec 63° = 2.202$; $\csc 63° = 1.2232$ **11.** $\sin 128° = 0.78801$, $\cos 128° = -0.61566$, $\tan 128° = -1.27994$, $\cot 128° = -0.78129$, $\sec 128° = -1.62427$, $\csc 128° = 1.26902$

Exercise Set 9.5   pp. 353–354

1. $2°30'$ **3.** $46°23'$ **5.** $67°50'$ **7.** $45.75°$ **9.** $76.88°$ **11.** 0.2306 **13.** 0.4176 **15.** 0.7824 **17.** -0.9228 **19.** -1.590 **21.** 0.4728 **23.** 0.5894 **25.** -0.1859 **27.** 0.2995 **29.** 0.2563 **31.** $13°40'$ **33.** $68°10'$ **35.** $69°17'$ **37.** 0.4392 **39.** 0.8145 **41.** $25°42'51''$; $61.63944°$ **43.** 0.48083 **45.** 8116 miles

Exercise Set 9.6   pp. 359–360

1. $\cos x = -0.9898$, $\tan x = -0.1440$, $\cot x = -6.9444$, $\sec x = -1.0103$, $\csc x = 7.0175$ **3.** $\sin x = 0.7987$, $\cos x = 0.6018$, $\tan x = 1.3273$, $\csc x = 1.2520$, $\sec x = 1.6618$ **5.** Odd   **7.**

$\sin\theta = \dfrac{-y}{1}$, $\sin-\theta = \dfrac{y}{1}$

9. Same as Exercise 31, Set 9.3   **11.** Domain: set of all real numbers; range: $[-1, 1]$; period; 2π **13.** Domain: all real numbers except odd multiples of $\pi/2$; range: set of all real numbers; period: π

15. $\cos\left(\dfrac{\pi}{2} - x\right) \equiv \sin x$ **17.** $\tan\left(x - \dfrac{\pi}{2}\right) \equiv -\cot x$

19. $\sec\left(x - \dfrac{\pi}{2}\right) \equiv \csc x$

21.

23.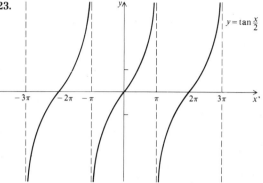

25. By similar triangles, $x/\cos s = y/\sin s = r/1$. Thus $\begin{cases} x = r\cos s, \\ y = r\sin s. \end{cases}$

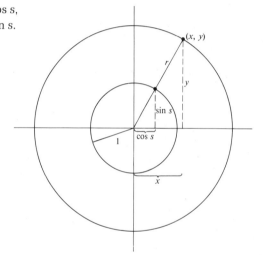

Chapter 9 Review   pp. 360–361

1. I; 0.483π; 1.52   **3.** I; 0.167π; 0.523   **5.** 270°   **7.** $\frac{7}{4}\pi$ or 5.5 cm   **9.** 1130 cm/min   **11.** $\sin\theta = \frac{3}{\sqrt{13}}$; $\cos\theta = -\frac{2}{\sqrt{13}}$; $\tan\theta = -\frac{3}{2}$; $\cot\theta = -\frac{2}{3}$; $\sec\theta = -\frac{\sqrt{13}}{2}$; $\csc\theta = \frac{\sqrt{13}}{3}$ **13.** $\frac{\sqrt{2}}{2}$ **15.** $\sqrt{3}$ **17.** $-\frac{2}{3}$ **19.** $\frac{\sqrt{5}}{2}$ **21.** $-\frac{3}{2}$ **23.** 0.6820   **25.** 0.9325   **27.** 1.3673   **29.** 47.55°   **31.** 0.5147   **33.** 16°10′

CHAPTER 10

Exercise Set 10.1   pp. 370–371

1. 0.9659   **3.** 0.2588   **5.** 0.9659   **7.** 3.7322   **9.** 1   **11.** 0   **13.** Undefined   **15.** 0.8448   **17.** sin 59°, or 0.8572   **19.** tan 52°, or 1.269   **21.** $2\sin\alpha\cos\beta$ **23.** $2\cos\alpha\cos\beta$ **25.** $\cos u$ **27.** $\frac{2\pi}{3}$ **29.** 0°   **31.** 86°40′   **33.** 135°   **35.** $2\sin\theta\cos\theta$

37. $\cot(\alpha+\beta) = \dfrac{\cos(\alpha+\beta)}{\sin(\alpha+\beta)} = \dfrac{\cos\alpha\cos\beta - \sin\alpha\sin\beta}{\sin\alpha\cos\beta + \cos\alpha\sin\beta} \cdot \dfrac{\frac{1}{\sin\alpha\sin\beta}}{\frac{1}{\sin\alpha\sin\beta}} = \dfrac{\frac{\cos\alpha\cos\beta}{\sin\alpha\sin\beta} - 1}{\frac{\sin\alpha\cos\beta}{\sin\alpha\sin\beta} + \frac{\cos\alpha\sin\beta}{\sin\alpha\sin\beta}} = \dfrac{\cot\alpha\cot\beta - 1}{\cot\alpha + \cot\beta}$

39. $\cot x$ **41.** $\frac{1}{7}$ **43.** $\frac{1}{6}$ **45.** 173°, approximately

Exercise Set 10.2 pp. 377–378

1. $\sin 2\theta = \frac{24}{25}$, $\cos 2\theta = -\frac{7}{25}$, $\tan 2\theta = -\frac{24}{7}$; 2θ in quadrant II **3.** $\cos 2\theta = \frac{7}{25}$, $\sin 2\theta = \frac{24}{25}$, $\tan 2\theta = \frac{24}{7}$; 2θ in quadrant I
5. $\sin 2\theta = \frac{24}{25}$, $\cos 2\theta = -\frac{7}{25}$, $\tan 2\theta = -\frac{24}{7}$; 2θ in quadrant II **7.** $8\sin\theta\cos^3\theta - 4\sin\theta\cos\theta$, or $4\sin\theta\cos^3\theta - 4\sin^3\theta\cos\theta$
or $4\sin\theta\cos\theta - 8\sin^3\theta\cos\theta$ **9.** $\dfrac{3 - 4\cos 2\theta + \cos 4\theta}{8}$ **11.** $\dfrac{\sqrt{2+\sqrt{3}}}{2}$ **13.** $2 + \sqrt{3}$ **15.** $\dfrac{\sqrt{2+\sqrt{2}}}{2}$ **17.** 0.6421
19. 0.9844 **21.** 0.1734 **23.** $\cos x$ **25.** $\cos x$ **27.** $\sin x$ **29.** $\cos x$ **31.** 1 **33.** 1 **35.** 8 **37.** $\sin 2x$
39. (i) $\frac{1}{2}[\sin(u+v) + \sin(u-v)] = \frac{1}{2}[\sin u\cos v + \cos u\sin v + \sin u\cos v - \cos u\sin v] = \sin u \cdot \cos v$. The other formulas follow in a similar manner.

Exercise Set 10.3 pp. 383–384

1.

$\csc x - \cos x \cot x$	$\sin x$
$\dfrac{1}{\sin x} - \cos x \dfrac{\cos x}{\sin x}$	$\sin x$
$\dfrac{1 - \cos^2 x}{\sin x}$	
$\dfrac{\sin^2 x}{\sin x}$	
$\sin x$	

3.

$\dfrac{1+\cos\theta}{\sin\theta} + \dfrac{\sin\theta}{\cos\theta}$	$\dfrac{\cos\theta + 1}{\sin\theta\cos\theta}$
$\dfrac{1+\cos\theta}{\sin\theta} \cdot \dfrac{\cos\theta}{\cos\theta} + \dfrac{\sin\theta \cdot \sin\theta}{\cos\theta\sin\theta}$	$\dfrac{1+\cos\theta}{\sin\theta\cos\theta}$
$\dfrac{\cos^2\theta + \cos\theta + \sin^2\theta}{\sin\theta\cos\theta}$	
$\dfrac{1+\cos\theta}{\sin\theta\cos\theta}$	

5.

$\dfrac{1 - \sin x}{\cos x}$	$\dfrac{\cos x}{1 + \sin x}$
$\dfrac{1-\sin x}{\cos x} \cdot \dfrac{\cos x}{\cos x}$	$\dfrac{\cos x}{1+\sin x} \cdot \dfrac{1-\sin x}{1-\sin x}$
$\dfrac{\cos x - \sin x\cos x}{\cos^2 x}$	$\dfrac{\cos x - \sin x\cos x}{1 - \sin^2 x}$
	$\dfrac{\cos x - \sin x\cos x}{\cos^2 x}$

7.

$\dfrac{1 + \tan\theta}{1 + \cot\theta}$	$\dfrac{\sec\theta}{\csc\theta}$
$\dfrac{1 + \dfrac{\sin\theta}{\cos\theta}}{1 + \dfrac{\cos\theta}{\sin\theta}}$	$\dfrac{\sin\theta}{\cos\theta}$
$\dfrac{\dfrac{\cos\theta + \sin\theta}{\cos\theta}}{\dfrac{\sin\theta + \cos\theta}{\sin\theta}}$	$\tan\theta$
$\dfrac{\cos\theta + \sin\theta}{\cos\theta} \cdot \dfrac{\sin\theta}{\sin\theta + \cos\theta}$	
$\dfrac{\sin\theta}{\cos\theta}$	
$\tan\theta$	

9.

$\dfrac{\sin x + \cos x}{\sec x + \csc x}$	$\dfrac{\sin x}{\sec x}$
$\dfrac{\sin x + \cos x}{\dfrac{1}{\cos x} + \dfrac{1}{\sin x}}$	$\dfrac{\sin x}{\sec x}$
$\dfrac{\sin x + \cos x}{\dfrac{\sin x + \cos x}{\sin x \cos x}}$	
$(\sin x + \cos x) \cdot \dfrac{\sin x \cos x}{\sin x + \cos x}$	
$\sin x \cos x$	
$\dfrac{\sin x}{\sec x}$	

11.

$\dfrac{1 + \tan \theta}{1 - \tan \theta} + \dfrac{1 + \cot \theta}{1 - \cot \theta}$	0
$\dfrac{1 + \dfrac{\sin \theta}{\cos \theta}}{1 - \dfrac{\sin \theta}{\cos \theta}} + \dfrac{1 + \dfrac{\cos \theta}{\sin \theta}}{1 - \dfrac{\cos \theta}{\sin \theta}}$	0
$\dfrac{\dfrac{\cos \theta + \sin \theta}{\cos \theta}}{\dfrac{\cos \theta - \sin \theta}{\cos \theta}} + \dfrac{\dfrac{\sin \theta + \cos \theta}{\sin \theta}}{\dfrac{\sin \theta - \cos \theta}{\sin \theta}}$	
$\dfrac{\cos \theta + \sin \theta}{\cos \theta} \cdot \dfrac{\cos \theta}{\cos \theta - \sin \theta} + \dfrac{\sin \theta + \cos \theta}{\sin \theta} \cdot \dfrac{\sin \theta}{\sin \theta - \cos \theta}$	
$\dfrac{\cos \theta + \sin \theta}{\cos \theta - \sin \theta} + \dfrac{\sin \theta + \cos \theta}{\sin \theta - \cos \theta}$	
$\dfrac{\cos \theta + \sin \theta}{\cos \theta - \sin \theta} - \dfrac{\cos \theta + \sin \theta}{\cos \theta - \sin \theta}$	
0	

13.

$\dfrac{1 + \cos 2\theta}{\sin 2\theta}$	$\cot \theta$
$\dfrac{1 + 2\cos^2 \theta - 1}{2 \sin \theta \cos \theta}$	$\dfrac{\cos \theta}{\sin \theta}$
$\dfrac{\cos \theta}{\sin \theta}$	

15.

$\sec 2\theta$	$\dfrac{\sec^2 \theta}{2 - \sec^2 \theta}$
$\dfrac{1}{\cos 2\theta}$	$\dfrac{1 + \tan^2 \theta}{2 - 1 - \tan^2 \theta}$
$\dfrac{1}{\cos^2 \theta - \sin^2 \theta}$	$\dfrac{1 + \tan^2 \theta}{1 - \tan^2 \theta}$
	$\dfrac{1 + \dfrac{\sin^2 \theta}{\cos^2 \theta}}{1 - \dfrac{\sin^2 \theta}{\cos^2 \theta}}$
	$\dfrac{\cos^2 \theta + \sin^2 \theta}{\cos^2 \theta - \sin^2 \theta}$
	$\dfrac{1}{\cos^2 \theta - \sin^2 \theta}$

17.

$\dfrac{\sin \alpha \cos \beta + \cos \alpha \sin \beta}{\cos \alpha \cos \beta}$	$\dfrac{\sin \alpha}{\cos \alpha} + \dfrac{\sin \beta}{\cos \beta}$
	$\dfrac{\sin \alpha \cos \beta + \cos \alpha \sin \beta}{\cos \alpha \cos \beta}$

19.

$1 - [\cos 5\theta \cos 3\theta + \sin 5\theta \sin 3\theta]$	$1 - \cos 2\theta$
$1 - \cos(5\theta - 3\theta)$	
$1 - \cos 2\theta$	

21.

$\dfrac{1}{2}\left[\dfrac{\dfrac{\sin\theta + \sin\theta\cos\theta}{\cos\theta}}{\dfrac{\sin\theta}{\cos\theta}}\right]$	$\dfrac{1+\cos\theta}{2}$
$\dfrac{1}{2} = \left[\dfrac{\sin\theta(1+\cos\theta)}{\cos\theta} \cdot \dfrac{\cos\theta}{\sin\theta}\right]$	
$\dfrac{1+\cos\theta}{2}$	

23.

$(\cos^2 x - \sin^2 x)(\cos^2 x + \sin^2 x)$	$\cos^2 x - \sin^2 x$
$\cos^2 x - \sin^2 x$	

25.

$\dfrac{\tan(3\theta - \theta)}{\tan 2\theta}$	$\tan 2\theta$

27.

$\dfrac{(\cos x - \sin x)(\cos^2 x + \cos x \sin x + \sin^2 x)}{\cos x - \sin x}$	$\dfrac{2 + 2\sin x \cos x}{2}$
$1 + \cos x \sin x$	$1 + \sin x \cos x$

29.

$(\sin\alpha\cos\beta + \cos\alpha\sin\beta)(\sin\alpha\cos\beta - \cos\alpha\sin\beta)$	$1 - \cos^2\alpha - (1-\cos^2\beta)$
$\sin^2\alpha\cos^2\beta - \cos^2\alpha\sin^2\beta$	$\cos^2\beta - \cos^2\alpha$
$\cos^2\beta(1-\cos^2\alpha) - \cos^2\alpha(1-\cos^2\beta)$	
$\cos^2\beta - \cos^2\beta\cos^2\alpha - \cos^2\alpha + \cos^2\alpha\cos^2\beta$	
$\cos^2\beta - \cos^2\alpha$	

31.

$\cos\alpha\cos\beta - \sin\alpha\sin\beta + \cos\alpha\cos\beta + \sin\alpha\sin\beta$	$2\cos\alpha\cos\beta$
$2\cos\alpha\cos\beta$	

33. $2\sin\left(2x + \dfrac{\pi}{3}\right)$ **35.** $5\sin(x+b)$, where $\cos b = \tfrac{4}{5}$, $\sin b = \tfrac{3}{5}$ **37.** $7.89\sin(0.374x + 31.2°)$

39. $y = \sqrt{2}\sin\left(2x - \dfrac{\pi}{4}\right)$

41. $\log(\cos x - \sin x) + \log(\cos x + \sin x) = \log[(\cos x - \sin x)(\cos x + \sin x)] = \log(\cos^2 x - \sin^2 x) = \log\cos 2x$

43. $\dfrac{1}{\omega C(\tan\theta + \tan\phi)} = \dfrac{1}{\omega C\left(\dfrac{\sin\theta}{\cos\theta} + \dfrac{\sin\phi}{\cos\phi}\right)} = \dfrac{1}{\omega C\left(\dfrac{\sin\theta\cos\phi + \sin\phi\cos\theta}{\cos\theta\cos\phi}\right)} = \dfrac{\cos\theta\cos\phi}{\omega C\sin(\theta+\phi)}$ **45.** By matrix multiplication, we obtain $\begin{bmatrix} \cos x\cos y - \sin x\sin y & \cos x\sin y + \sin x\cos y \\ -\sin x\cos y - \cos x\sin y & -\sin x\sin y + \cos x\cos y \end{bmatrix}$. The use of identities produces the desired result.

Exercise Set 10.4 pp. 389–390

1. $\frac{\pi}{4} + 2k\pi, \frac{3\pi}{4} + 2k\pi$ 3. $\frac{\pi}{4} + 2k\pi, -\frac{\pi}{4} + 2k\pi$ 5. $\frac{5\pi}{4} + 2k\pi, -\frac{\pi}{4} + 2k\pi$ 7. $\frac{3\pi}{4} + 2k\pi, \frac{5\pi}{4} + 2k\pi$ 9. $\frac{\pi}{3} + k\pi$
11. $\frac{\pi}{4} + k\pi$ 13. $\frac{5\pi}{6} + k\pi$ 15. $\frac{3\pi}{4} + k\pi$ 17. $0 + 2k\pi$ 19. $\frac{\pi}{2} + 2k\pi$ 21. $23° + k \cdot 360°, 157° + k \cdot 360°$
23. $39° + k \cdot 360°, 141° + k \cdot 360°$ 25. $36°58' + k \cdot 360°, 323°02' + k \cdot 360°$ 27. $21°25' + k \cdot 360°, 338°35' + k \cdot 360°$
29. $20°10' + k \cdot 180°$ 31. $38°20' + k \cdot 180°$ 33. $31° + k \cdot 360°, 329° + k \cdot 360°$ 35. $9°10' + k \cdot 360°, 170°50' + k \cdot 360°$
37. $\frac{\pi}{4}$ 39. $\frac{\pi}{3}$ 41. $-\frac{\pi}{3}$ 43. $\frac{3\pi}{4}$ 45. $-\frac{\pi}{6}$ 47. $\frac{2\pi}{3}$ 49. $13°30'$ 51. $-39°30'$ 53. $152°50'$ 55. $-22°10'$
57. $169°30'$ 59. $41.4167°$ 61. 0.2874 63. 0.7324 65. 2.0950 67. Arcsin: odd, Arctan: odd; Arcsin $(-x) \equiv -$Arcsin x; Arctan $(-x) \equiv -$Arctan x 69. 71. The limit exists and is 2.

Exercise Set 10.5 pp. 393–394

1. 0.3 3. -4.2 5. $\frac{\pi}{3}$ 7. $-\frac{\pi}{4}$ 9. $\frac{\pi}{5}$ 11. $-\frac{\pi}{3}$ 13. $\frac{\sqrt{3}}{2}$ 15. $\frac{1}{2}$ 17. 1 19. $\frac{\pi}{6}$ 21. $\frac{\pi}{3}$ 23. $\frac{\pi}{2}$ 25. $\frac{x}{\sqrt{x^2+4}}$
27. $\frac{3}{\sqrt{11}}$ 29. $\frac{1}{3\sqrt{11}}$ 31. $-\frac{1}{2\sqrt{6}}$ 33. $\frac{1}{\sqrt{1+y^2}}$ 35. $\frac{1}{2}$ 37. $\frac{\sqrt{2+\sqrt{3}}}{2}$ 39. $\frac{24}{25}$ 41. $\frac{119}{169}$ 43. $\frac{3+4\sqrt{3}}{10}$ 45. $\frac{-\sqrt{2}}{10}$
47. 0.9861

Exercise Set 10.6 pp. 397–398

1. $\frac{\pi}{3} + 2k\pi, \frac{2\pi}{3} + 2k\pi$ 3. $\frac{\pi}{4} + 2k\pi, \frac{-\pi}{4} + 2k\pi$ 5. $20°10' + k \cdot 360°, 159°50' + k \cdot 360°$ 7. $236°40', 123°20'$ 9. $\frac{4\pi}{3}, \frac{5\pi}{3}$
11. $123°40', 303°40'$ 13. $\frac{\pi}{6}, \frac{5\pi}{6}, \frac{7\pi}{6}, \frac{11\pi}{6}$ 15. $\frac{\pi}{6}, \frac{5\pi}{6}, \frac{7\pi}{6}, \frac{11\pi}{6}$ 17. $\frac{\pi}{6}, \frac{5\pi}{6}, \frac{3\pi}{2}$ 19. 0 21. $0, \frac{\pi}{6}, \frac{5\pi}{6}, \pi, \frac{7\pi}{6}, \frac{11\pi}{6}$ 23. $\frac{\pi}{6}, \frac{5\pi}{6}$
25. $109°30', 250°30', 120°, 240°$ 27. $\frac{\pi}{6}, \frac{5\pi}{6}, \pi$ 29. $0, \pi, \frac{\pi}{2}, \frac{3\pi}{2}$ 31. $60°, 120°, 240°, 300°$ 33. $60°, 240°$
35. $30°, 60°, 120°, 150°, 210°, 240°, 300°, 330°$ 37. $0, \pi$ 39. $0, \pi$ 41. $0, \pm 4.8, \pm 14.3, \pm 17.1$, etc.

Exercise Set 10.7 pp. 401–402

1. $0, \pi$ 3. $\frac{3\pi}{4}, \frac{7\pi}{4}$ 5. $\frac{\pi}{2}, \frac{3\pi}{2}, \frac{\pi}{6}, \frac{5\pi}{6}$ 7. $\frac{\pi}{2}, \frac{3\pi}{2}, \frac{\pi}{4}, \frac{3\pi}{4}, \frac{5\pi}{4}, \frac{7\pi}{4}$ 9. $0, \frac{\pi}{2}, \pi, \frac{3\pi}{2}$ 11. 0 13. $0, \frac{\pi}{2}, \pi, \frac{3\pi}{2}$ 15. $\frac{\pi}{6}, \frac{5\pi}{6}, \pi$
17. $\frac{\pi}{6}, \frac{5\pi}{6}, \frac{7\pi}{6}, \frac{11\pi}{6}$ 19. $63°30', 243°30', 101°20', 281°20'$ 21. $\frac{\pi}{6}, \frac{\pi}{2}, \frac{5\pi}{6}, \frac{7\pi}{6}, \frac{3\pi}{2}, \frac{11\pi}{6}$ 23. $\frac{2\pi}{3}, \frac{4\pi}{3}$ 25. $\frac{\pi}{4}, \frac{7\pi}{4}$
27. $\frac{\pi}{12}, \frac{5\pi}{12}, \frac{13\pi}{12}, \frac{17\pi}{12}$ 29. $\frac{\pi}{6}, \frac{3\pi}{2}$ 31. 1 33. $1.15, 5.65, -0.65$, etc.

Chapter 10 Review   pp. 402–403

1. $\sin x$ **3.** $\cos(27° - 16°)$ or $\cos 11°$ **5.** $\frac{1}{2}\sqrt{2 - \sqrt{2}}$ **7.**

9. $81° + k\,180°$ **11.** $\frac{7}{8}$ **13.**

[graph of $y + 1 = 2\cos^2 x$]

15. $0, \pi$

CHAPTER 11

Exercise Set 11.1   pp. 408–409

1. $B = 53°50'$, $b = 37.2$, $c = 46.1$ **3.** $A = 77°20'$, $a = 436$, $c = 447$ **5.** $B = 72°40'$, $a = 4.24$, $c = 14.2$ **7.** $A = 66°50'$, $b = 0.0148$, $c = 0.0375$ **9.** $B = 42°30'$, $a = 35.6$, $b = 32.6$ **11.** $A = 7°40'$, $a = 0.131$, $b = 0.973$ **13.** $c = 21.6$, $A = 33°40'$, $B = 56°20'$ **15.** $b = 12.0$, $A = 53°10'$, $B = 36°50'$ **17.** $a = 3.57$, $A = 63°20'$, $B = 26°40'$ **19.** 47.9 ft   **21.** $1°40'$ **23.** $30°10'$ **25.** 3.52 mi   **27.** 109 km   **29.** 23.9 km   **31.** 7.92 km   **33.** 3.45 km

Exercise Set 11.2   pp. 413–414

1. $C = 17°$, $a = 26.3$, $c = 10.5$ **3.** $A = 121°$, $a = 33.4$, $c = 14.0$ **5.** $B = 68°50'$, $a = 32.3$, $b = 32.3$ **7.** $B = 56°20'$, $C = 87°40'$, $c = 40.8$, and $B = 123°40'$, $C = 20°20'$, $c = 14.2$ **9.** $B = 19°$, $C = 44°40'$, $b = 6.25$ **11.** $A = 74°30'$, $B = 44°20'$, $a = 33.3$ **13.** 76.3 m   **15.** 50.8 ft   **17.** 1470 km   **19.** 10.6 km   **21.** [parallelogram figure] $A = bh$, $h = a\sin\theta$, so $A = ab\sin\theta$.

Exercise Set 11.3   pp. 416–418

1. $a = 14.9$, $B = 23°40'$, $C = 126°20'$ **3.** $a = 24.8$, $B = 20°40'$, $C = 26°20'$ **5.** $b = 41.9$, $A = 86°$, $C = 11°50'$ **7.** $A = 36°10'$, $B = 43°30'$, $C = 100°20'$ **9.** $A = 73°40'$, $B = 51°50'$, $C = 54°30'$ **11.** $A = 25°40'$, $B = 126°$, $C = 28°20'$ **13.** 28.2 nautical mi, S 55°20'E   **15.** 37 nautical mi   **17.** 59.4 ft   **19.** 68°, 68°, 44°   **21.** 42.6 ft   **23.** (a) 15.73 ft; (b) 120.8 ft²   **25.** 3424 yd²   **27.** 116.6 ft   **29.** $A = \frac{1}{2}a^2 \sin\theta$, when $\theta = 90°$.

Exercise Set 11.4   p. 420

1. 57, 38°   **3.** 18.4, 37°   **5.** 20.9, 58°30'   **7.** 68.3, 18°   **9.** 13 kg, 67°   **11.** 655 kg, 21°   **13.** 21.6 ft/sec, 33°40'   **15.** 726 lb, 47°   **17.** 174 nautical mi, S 15°E   **19.** An angle of 12° upstream

Exercise Set 11.5   p. 424

1. (5, 16)   **3.** (4.8, 13.7)   **5.** (−662, −426)   **7.** (5, 53°10′)   **9.** (18, 303°40′)   **11.** (5, 233°10′)   **13.** (18, 123°40′)   **15.** (3.46, 2)   **17.** (−5.74, −8.19)   **19.** (17.3, −10)   **21.** (70.7, −70.7)   **23.** Vertical 118, horizontal 92.4   **25.** S. 192 km/h, W. 161 km/h   **27.** 43 kg, S 35°30′W   **29.** (a) N. 28, W. 7; (b) 28.8, N. 14°W   **31.** (19, 36)   **33.** (14, 8)   **35.** 18   **37.** 17.89   **39.** 35.78   **41. 0**

Exercise Set 11.6   pp. 427–428

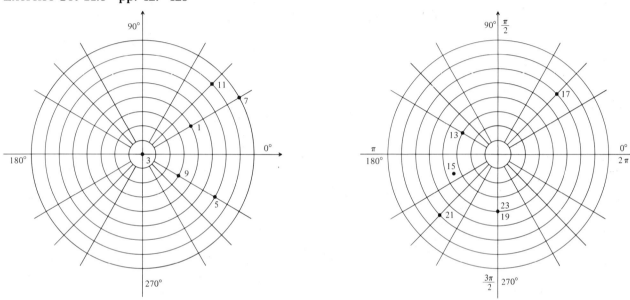

25. $(4\sqrt{2}, 45°)$ **27.** (5, 90°)   **29.** (4, 0°)   **31.** (6, 60°)   **33.** (2, 30°)   **35.** (6, 30°)   **37.** $\left(\dfrac{4}{\sqrt{2}}, \dfrac{4}{\sqrt{2}}\right)$ or (2.83, 2.83)   **39.** (0, 0)   **41.** $\left(\dfrac{-3}{\sqrt{2}}, \dfrac{-3}{\sqrt{2}}\right)$ **43.** $(3, -3\sqrt{3})$ **45.** $(5\sqrt{3}, 5)$ **47.** (4.33, −2.5)   **49.** $3r\cos\theta + 4r\sin\theta = 5$ **51.** $r\cos\theta = 5$ **53.** $r^2 = 36$ **55.** $r^2(\cos^2\theta - 4\sin^2\theta) = 4$ **57.** $x^2 + y^2 = 25$ **59.** $y = x$ **61.** $y = 2$ **63.** $x^2 + y^2 = 4x$ **65.** $x^2 - 4y = 4$ **67.** $x^2 - 2x + y^2 - 3y = 0$

Exercise Set 11.7   pp. 430–431

1. Cable 224-lb tension, boom 166-lb compression   **3.** Horizontal rod 168-kg tension, other rod 261-kg compression   **5.** Lift 2472 lb, drag 1315 lb   **7.** 60 kg   **9.** 80°20′   **11.** Horizontal rope 400 kg, other rope 566 kg   **13.** 2000 kg in each   **15.** 2242 kg on left, 1830 kg on right

Chapter 11 Review   p. 432

1. $A = 58°10′$, $B = 31°50′$, $c = 4.55$ **3.** 1748 cm   **5.** 420 cm   **7.** 13.95 ft   **9.** 75.8 mph; N 20°10′E   **11.** (−18.79, 6.84)   **13.** $r + 2\cos\theta - 3\sin\theta = 0$ **15.** (a) (25, 0); (b) $\sqrt{50}$ **17.** 50.52°, 129.48°

CHAPTER 12

Exercise Set 12.1 pp. 437–438

1. $i\sqrt{15}$ **3.** $4i$ **5.** $-2i\sqrt{3}$ **7.** $9i$ **9.** $i(\sqrt{7} - \sqrt{10})$ **11.** $-\sqrt{55}$ **13.** $2\sqrt{5}$ **15.** $\dfrac{\sqrt{5}}{\sqrt{2}}$ or $\dfrac{\sqrt{10}}{2}$ **17.** $\frac{3}{2}$ **19.** -2
21. $6 + 5i$ **23.** 8 **25.** $2 + 4i$ **27.** $-4 - i$ **29.** $-5 + 5i$ **31.** $7 - i$ **33.** $-6 + 12i$ **35.** $-5 + 12i$ **37.** i **39.** Yes
41. $x = -\frac{3}{2}, y = 7$ **43.** $-4 + 3i$ **45.** For example, $\sqrt{-1}\sqrt{-1} = i^2 = -1$ but $\sqrt{(-1)(-1)} = \sqrt{1} = 1$.
47. $\begin{bmatrix} -1 & -2 \\ 3 + 3i & 3 + 12i \end{bmatrix}$

Exercise Set 12.2 pp. 442–443

1. $\frac{4}{25} - \frac{3}{25}i$ **3.** $\frac{5}{29} + \frac{2}{29}i$ **5.** $-i$ **7.** $\dfrac{i}{4}$ **9.** $\frac{1}{2} + \frac{7}{2}i$ **11.** $\frac{1}{3} + \frac{2}{3}\sqrt{2}i$ **13.** $2 - 3i$ **15.** $\frac{1}{5} + \frac{2}{5}i$ **17.** $-\dfrac{1}{2} - \dfrac{i}{2}$ **19.** $\frac{28}{65} - \frac{29}{65}i$
21. $-\frac{1}{2} + \frac{3}{2}i$ **23.** $\frac{5}{2} + \frac{13}{2}i$ **25.** $3\bar{z}^5 - 4\bar{z}^2 + 3\bar{z} - 5$ **27.** $4\bar{z}^7 - 3\bar{z}^5 + 4\bar{z}$ **29.** $z = 1$ **31.** a **33.** $\dfrac{3 - i}{2 + i}$, or $1 - i$

Exercise Set 12.3 pp. 446–447

1. $x^2 + 4 = 0$ **3.** $x^2 - 2x + 2 = 0$ **5.** $x^2 - 4x + 13 = 0$ **7.** $x^2 - 3x - ix + 3i = 0$ **9.** $x^3 - x^2 + 9x - 9 = 0$
11. $x^3 - 2x^2i - 3x^2 + 5xi + x - 2i + 2 = 0$ **13.** $\frac{2}{5} + \frac{6}{5}i$ **15.** $\frac{8}{5} - \frac{9}{5}i$ **17.** $1 \pm 2i$ **19.** $-\dfrac{3}{2} \pm \dfrac{\sqrt{7}}{2}i$ **21.** $\dfrac{-1 + i \pm \sqrt{-6i}}{2}$
23. $\dfrac{-1 - 2i \pm \sqrt{-15 + 16i}}{6}$ **25.** $\sqrt{2} + \sqrt{2}i, -\sqrt{2} - \sqrt{2}i$ **27.** $2 + i, -2 - i$ **29.** $-1 \pm \sqrt{3}i, 2$ **31.** $x = 2 + i, y = 1 - 3i$

Exercise Set 12.4 pp. 451–452

9. $\dfrac{3\sqrt{3}}{2} + \dfrac{3}{2}i$ **11.** $-10i$ **13.** $2 + 2i$ **15.** $-2 - 2i$ **17.** $\sqrt{2}$ cis $315°$
19. 20 cis $330°$ **21.** 5 cis $180°$ **23.** 4 cis $0°$ or 4 **25.** 8 cis $120°$ **27.** cis $90°$ or i **29.** 2 cis $270°$ or $-2i$ **31.** $z = a + bi$,
$|z| = \sqrt{a^2 + b^2}$; $-z = -a - bi$, $|-z| = \sqrt{(-a)^2 + (-b)^2} = \sqrt{a^2 + b^2}$, $\therefore |z| = |-z|$
33. $|(a + bi)(a - bi)| = |a^2 + b^2| = a^2 + b^2$;
$|(a + bi)^2| = |a^2 + 2abi - b^2| = |a^2 - b^2 + 2abi| = \sqrt{(a^2 - b^2)^2 + (2ab)^2} = \sqrt{a^4 + 2a^2b^2 + b^4} = a^2 + b^2$

35. $z \cdot w = (r_1 \text{ cis } \theta_1)(r_2 \text{ cis } \theta_2) = r_1 r_2 \text{ cis } (\theta_1 + \theta_2)$, $|z \cdot w| = \sqrt{[r_1 r_2 \cos (\theta_1 + \theta_2)]^2 + [r_1 r_2 \sin (\theta_1 + \theta_2)]^2} = \sqrt{(r_1 r_2)^2} = |r_1 r_2|$, $|z| = \sqrt{(r_1 \cos \theta_1)^2 + (r_1 \sin \theta_1)^2} = \sqrt{r_1^2} = |r_1|$, $|w| = \sqrt{(r_2 \cos \theta_2)^2 + (r_2 \sin \theta_2)^2} = \sqrt{r_2^2} = |r_2|$. Then $|z| \cdot |w| = |r_1| \cdot |r_2| = |r_1 r_2| = |z \cdot w|$ **37.**

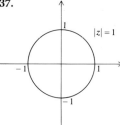

Exercise Set 12.5 p. 455

1. $8 \text{ cis } \pi$ **3.** $64 \text{ cis } \pi$ **5.** $8 \text{ cis } 270°$ **7.** $-8 - 8\sqrt{3}i$ **9.** $-8 - 8\sqrt{3}i$ **11.** i **13.** 1 **15.** $\sqrt{2} \text{ cis } 60°$, $\sqrt{2} \text{ cis } 240°$; or $\dfrac{\sqrt{2}}{2} + \dfrac{\sqrt{6}}{2}i, \dfrac{-\sqrt{2}}{2} - \dfrac{\sqrt{6}}{2}i$ **17.** $\text{cis } 30°, \text{cis } 150°, \text{cis } 270°$; or $\dfrac{\sqrt{3}}{2} + \dfrac{1}{2}i, \dfrac{-\sqrt{3}}{2} + \dfrac{1}{2}i, -i$ **19.** $2 \text{ cis } 0°, 2 \text{ cis } 90°, 2 \text{ cis } 180°, 2 \text{ cis } 270°$; or $2, 2i, -2, -2i$ **21.** $-1.366 + 1.366i, 0.366 - 0.366i$

Chapter 12 Review pp. 455–456

1. $14 + 2i$ **3.** $2 - i$ **5.** $x = 2, y = -4$ **7.** $x^2 - 2x + 5 = 0$ **9.** $\dfrac{(-3 \pm \sqrt{5})i}{2}$
11. 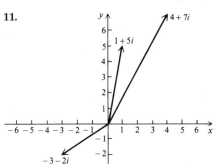 **13.** $\sqrt{2} \text{ cis } 45°$ **15.** $\sqrt[6]{2} \text{ cis } 15°, \sqrt[6]{2} \text{ cis } 135°, \sqrt[6]{2} \text{ cis } 255°$ **17.** $1 + 2i$
19. $3, \tfrac{3}{2}(-1 \pm \sqrt{3}i)$

CHAPTER 13

Exercise Set 13.1 pp. 460–461

1. 4 **3.** 1 **5.** 2 **7.** 0 **9.** 2 yes, 3 no, -1 no **11.** (a) yes; (b) no; (c) no **13.** $Q(x) = x^2 + 8x + 15, R(x) = 0$, $P(x) = (x - 2)(x^2 + 8x + 15) + 0$ **15.** $Q(x) = x^2 + 9x + 26, R(x) = 48, P(x) = (x - 3)(x^2 + 9x + 26) + 48$

17. $Q(x) = x^2 - 2x + 4$, $R(x) = -16$, $P(x) = (x + 2)(x^2 - 2x + 4) - 16$ **19.** $Q(x) = x^2 + 5$, $R(x) = 0$, $P(x) = (x^2 + 4)(x^2 + 5) + 0$ **21.** $P(x) = (2x^2 - x + 1)(\frac{5}{2}x^5 + \frac{5}{4}x^4 - \frac{5}{8}x^3 + \frac{39}{16}x^2 + \frac{49}{32}x + \frac{35}{64}) + \dfrac{-63x - 227}{64}$ **23.** (a) -32; (b) -32; (c) -65; (d) -65 **25.** (a) -2; (b) -2

Exercise Set 13.2 pp. 465–466

1. $Q(x) = 2x^3 + x^2 - 3x + 10$, $R(x) = -42$ **3.** $Q(x) = x^2 - 4x + 8$, $R(x) = -24$ **5.** $Q(x) = x^3 + x^2 + x + 1$, $R(x) = 0$ **7.** $Q(x) = 2x^3 + x^2 + \frac{7}{2}x + \frac{7}{4}$, $R(x) = -\frac{1}{8}$ **9.** $Q(x) = x^3 + x^2y + xy^2 + y^3$, $R(x) = 0$ **11.** $P(1) = 0$, $P(-2) = -60$, $P(3) = 0$ **13.** $P(20) = 5{,}935{,}988$, $P(-3) = -772$ **15.** -3 yes, 2 no **17.** -3 no, $\frac{1}{2}$ no **19.** $P(x) = (x - 1)(x + 2)(x + 3)$; $1, -2, -3$ **21.** $P(x) = (x - 2)(x - 5)(x + 1)$; $2, 5, -1$ **23.** $P(x) = (x - 2)(x - 3)(x + 4)$; $2, 3, -4$ **25.** $P(x) = (x - 1)(x - 2)(x - 3)(x + 5)$; $1, 2, 3, -5$ **27.** $-5 < x < 1$ or $x > 2$ **29.** $\frac{14}{3}$ **31.** $k = 0$

Exercise Set 13.3 pp. 470–471

1. $-3(m2), 1(m1)$ **3.** $0(m3), 1(m2), -4(m1)$ **5.** $x^3 - 6x^2 - x + 30$ **7.** $x^3 + 3x^2 + 4x + 12$ **9.** $x^3 - \sqrt{3}x^2 - 2x + 2\sqrt{3}$, no **11.** $-3 - 4i, 4 + \sqrt{5}$ **13.** $x^3 - 4x^2 + 6x - 4$ **15.** $x^3 - 5x^2 + 16x - 80$ **17.** $x^4 + 4x^2 - 45$ **19.** $i, 2, 3$ **21.** $1 + 2i, 1 - 2i$ **23.** $4, i, -i$ **25.** $i, -i, 1 \pm \sqrt{2}$ **27.** There is at least one complex value for sin x satisfying the equation (not necessarily a value of x).

Exercise Set 13.4 pp. 475–476

1. $1, -1$ **3.** $\pm(1, \frac{1}{3}, \frac{1}{5}, \frac{1}{15}, 2, \frac{2}{3}, \frac{2}{5}, \frac{2}{15})$ **5.** $-3, \sqrt{2}, -\sqrt{2}$ **7.** $-\frac{1}{5}, 1, 2i, -2i$ **9.** $-1, -2, 3 + \sqrt{13}, 3 - \sqrt{13}$ **11.** $1, -1, -3$ **13.** $-2, 1 \pm i\sqrt{3}$ **15.** $\frac{1}{2}, \dfrac{1 \pm \sqrt{5}}{2}$ **17.** None **19.** None **21.** None **23.** $-2, 1, 2$ **25.** 4 cm **27.** 3 cm, $\dfrac{7 - \sqrt{33}}{2}$ cm

Exercise Set 13.5 pp. 484–485

1. One or three **3.** None **5.** Two or none **7.** None **9.** Three or one **11.** Two or none **13.** 3 (Other answers are possible.) **15.** 4 **17.** -1 **19.** -3 **21.** 3 or 1 positive roots; 1 negative root; upper bound 2; lower bound -3 **23.** 1 positive, 1 negative root; upper bound 2; lower bound -2 **25.** 2 or 0 positive, 2 or 0 negative roots; upper bound 4; lower bound -3 **27.** No positive, no negative roots

Exercise Set 13.6 pp. 490–491

1.   **3.**

5.

7.
−1, 2

9.
2.2

11.
No real solution

13.
± 1.4
± 2

15.
−1
± 1.4

17. 0.75   **19.** −1.27

A-42 CHAPTER 13 ANSWERS   EXERCISE SET 13.7

Exercise Set 13.7   p. 498

1.

3.

5.

7.

9.

11.

13.

15.

17.

21.

25.

19.

23.

27.

29.

31.

33.

Exercise Set 13.8 pp. 502–503

1. $\dfrac{2}{x-3} + \dfrac{-1}{x+2}$ **3.** $\dfrac{-4}{3x-1} + \dfrac{5}{2x-1}$ **5.** $\dfrac{-3}{x-2} + \dfrac{2}{x+2} + \dfrac{4}{x+1}$ **7.** $\dfrac{-3}{(x+2)^2} + \dfrac{-1}{x+2} + \dfrac{1}{x-1}$ **9.** $\dfrac{3}{x-1} + \dfrac{-4}{2x-1}$

11. $x - 2 + \dfrac{-\frac{11}{4}}{(x+1)^2} + \dfrac{\frac{17}{16}}{x+1} + \dfrac{-\frac{17}{16}}{x-3}$ **13.** $\dfrac{-1}{x-3} + \dfrac{3x}{x^2+2x-5}$ **15.** $\dfrac{-2}{x+2} + \dfrac{10}{(x+2)^2} + \dfrac{3}{2x-1}$

17. $\dfrac{-\dfrac{x}{10a^2} - \dfrac{4}{5a}}{x^2 + a^2} + \dfrac{\dfrac{1}{4a^2}}{x-a} + \dfrac{\dfrac{1}{4a^2}}{x+a}$ **19.** $\dfrac{-1}{x+1} + \dfrac{4}{x+4}$

Chapter 13 Review pp. 503–504

1. 0 **3.** $x^3 - 3x^2 + 2x$ **5.** $(x-1)\left(x + \dfrac{1}{2} + i\dfrac{\sqrt{3}}{2}\right)\left(x + \dfrac{1}{2} - i\dfrac{\sqrt{3}}{2}\right)$ **7.** $(x-1)(x+3)(x+5)$; $1, -3, -5$ **9.** 7 **11.** 4

13. $\pm(1, 2, 3, 4, 6, \frac{1}{2}, \frac{3}{2})$ **15.** 4, 2, or none   **17.** -2 **19.** $-0.9, 1.3, 2.5$ **21.** $\dfrac{5}{x+1} - \dfrac{5}{x+2} - \dfrac{5}{(x+2)^2}$

CHAPTER 14

Exercise Set 14.1   pp. 510–511

1. The union of the graphs of $y = -x$ and $y = x$ **3.** The union of the graphs of $y = -x$ and $y = \frac{3}{2}x$ **5.** Just $(0, 0)$
7. $x^2 + y^2 = 49$ **9.** $C(-1, -3); r = 2$ **11.** $C(8, -3); r = 2\sqrt{10}$ **13.** $C(-4, 3); r = 2\sqrt{10}$ **15.** $C(2, 0); r = 2$
17. $C(-4.123, 3.174); r = 10.071$ **19.** $C(0, 0); r = \frac{1}{3}$ **21.** $x^2 + y^2 = 25$ **23.** $(x - 2)^2 + (y - 4)^2 = 16$
25. $(x - 1)^2 + (y - 2)^2 = 41$ **27.** $(x - h)(x - h) - (y - k)[-(y - k)] = r^2$, so $(x - h)^2 + (y - k)^2 = r^2$ **29.** Yes   **31.** No
33. Yes   **35.** No

Exercise Set 14.2   pp. 516–517

1. $C(0, 0); V: (2, 0), (-2, 0), (0, 1), (0, -1); F: (\sqrt{3}, 0), (-\sqrt{3}, 0)$ **3.** $C(1, 2); V: (3, 2), (-1, 2), (1, 3), (1, 1);$
$F: (1 + \sqrt{3}, 2), (1 - \sqrt{3}, 2)$ **5.** $C(-3, 2); V: (2, 2), (-8, 2), (-3, 6), (-3, -2); F: (0, 2), (-6, 2)$ **7.** $C(0, 0);$
$V: (-3, 0), (3, 0), (0, 4), (0, -4); F: (0, \sqrt{7}), (0, -\sqrt{7})$ **9.** $C(-2, 1); V: (-10, 1), (6, 1), (-2, 1 + 4\sqrt{3}), (-2, 1 - 4\sqrt{3});$
$F: (-6, 1), (2, 1)$ **11.** $C(0, 0); V: (-\sqrt{3}, 0), (\sqrt{3}, 0), (0, \sqrt{2}), (0, -\sqrt{2}); F: (-1, 0), (1, 0)$.   **13.** $C(0, 0);$
$V: (-\frac{1}{2}, 0), (\frac{1}{2}, 0), (0, \frac{1}{3}), (0, -\frac{1}{3}); F: \left(\dfrac{-\sqrt{5}}{6}, 0\right), \left(\dfrac{\sqrt{5}}{6}, 0\right)$ **15.** $C(2, -1); V: (-1, -1), (5, -1), (2, 1), (2, -3); F:$
$(2 - \sqrt{5}, -1), (2 + \sqrt{5}, -1)$ **17.** $C(1, 1); V: (0, 1), (2, 1), (1, 3), (1, -1); F: (1, 1 + \sqrt{3}), (1, 1 - \sqrt{3})$ **19.** $C(2.003125, -1.00513);$
$V: (5.0234302, -1.00515), (-1.0171802, -1.00515), (2.003125, -3.0186868), (2.003125, 1.0083868)$ **21.** $\dfrac{x^2}{4} + \dfrac{y^2}{9} = 1$
23. $\dfrac{(x - 3)^2}{4} + \dfrac{(y - 1)^2}{25} = 1$ **25.** $\dfrac{(x + 2)^2}{\frac{1}{4}} + \dfrac{(y - 3)^2}{4} = 1$
27.

(a) no; (b) $y = \pm 3\sqrt{1 - x^2}$; (c) yes, domain $\{x \mid -1 \leq x \leq 1\}$, range $\{y \mid 0 \leq y \leq 3\}$; (d) yes, domain $\{x \mid -1 \leq x \leq 1\}$, range $\{y \mid -3 \leq y \leq 0\}$

Exercise Set 14.3   pp. 522–523

1. $C(0, 0); V: (-3, 0), (3, 0); F: (-\sqrt{10}, 0), (\sqrt{10}, 0); A: y = \frac{1}{3}x, y = -\frac{1}{3}x$ **3.** $C(2, -5); V: (-1, -5), (5, -5); F: (2 - \sqrt{10}, -5),$
$(2 + \sqrt{10}, -5); A: y = -\dfrac{x}{3} - \dfrac{13}{3}, y = \dfrac{x}{3} - \dfrac{17}{3}$ **5.** $C(-1, -3); V: (-1, -1), (-1, -5); F: (-1, -3 + 2\sqrt{5}), (-1, -3 - 2\sqrt{5});$
$A: y = \frac{1}{2}x - \frac{5}{2}, y = -\frac{1}{2}x - \frac{7}{2}$ **7.** $C(0, 0); V: (-2, 0), (2, 0); F: (-\sqrt{5}, 0), (\sqrt{5}, 0); A: y = -\frac{1}{2}x, y = \frac{1}{2}x$ **9.** $C(0, 0);$
$V: (0, 1), (0, -1); F: (0, \sqrt{5}), (0, -\sqrt{5}); A: y = -\frac{1}{2}x, y = \frac{1}{2}x$ **11.** $C(0, 0); V: (-\sqrt{2}, 0), (\sqrt{2}, 0); F: (-2, 0), (2, 0); A: y = \pm x$

13. $C(0, 0)$; V: $(-\frac{1}{2}, 0)$, $(\frac{1}{2}, 0)$; F: $\left(-\frac{\sqrt{2}}{2}, 0\right), \left(\frac{\sqrt{2}}{2}, 0\right)$; A: $y = \pm x$ **15.** $C(1, -2)$; V: $(0, -2), (2, -2)$; F: $(1 - \sqrt{2}, -2), (1 + \sqrt{2}, -2)$; A: $y = -x - 1, y = x - 3$ **17.** $C(\frac{1}{3}, 3)$; V: $(-\frac{2}{3}, 3), (\frac{4}{3}, 3)$; F: $(\frac{1}{3} - \sqrt{37}, 3), (\frac{1}{3} + \sqrt{37}, 3)$; A: $y = 6x + 1, y = -6x + 5$ **19.** See text   **21.** See text   **23.** $C(1.023, -2.044)$; V: $(2.07, -2.044), (-0.024, -2.044)$; A: $y = x - 3.067, y = -x - 1.021$ **25.** $\frac{x^2}{4} - \frac{y^2}{9} = 1$

27. $\left(\frac{x-h}{a}\right)\left(\frac{x-h}{a}\right) - \left(\frac{y-k}{b}\right)\left(\frac{y-k}{b}\right) = 1 \therefore \frac{(x-h)^2}{a^2} - \frac{(y-k)^2}{b^2} = 1$

Exercise Set 14.4   pp. 526–527

1. $V(0, 0)$; $F(0, 2)$; D: $y = -2$ **3.** $V(0, 0)$; $F(-\frac{3}{2}, 0)$; D: $x = \frac{3}{2}$ **5.** $V(0, 0)$; $F(0, 1)$; D: $y = -1$ **7.** $V(0, 0)$; $F(0, \frac{1}{8})$; D: $y = -\frac{1}{8}$ **9.** $V(-2, 1)$; $F(-2, -\frac{1}{2})$; D: $y = \frac{5}{2}$ **11.** $V(-1, -3)$; $F(-1, -\frac{7}{2})$; D: $y = -\frac{5}{2}$ **13.** $V(0, -2)$; $F(0, -1\frac{3}{4})$; D: $y = -2\frac{1}{4}$ **15.** $V(-2, -1)$; $F(-2, -\frac{3}{4})$; D: $y = -1\frac{1}{4}$ **17.** $V(5\frac{3}{4}, \frac{1}{2})$; $F(6, \frac{1}{2})$; D: $x = -\frac{5}{2}$ **19.** $y^2 = 16x$ **21.** $y^2 = -4\sqrt{2}x$ **23.** $(y - 2)^2 = 14(x + \frac{1}{2})$ **25.** The graph of $x^2 - y^2 = 0$ is the union of the lines $y = x$, $y = -x$; the others are respectively a hyperbola, a circle, and a parabola   **27.** $V(0, 0)$; $F(0, 2014.0625)$; D: $y = -2014.0625$ **29.** $(x + 1)^2 = -4(y - 2)$ **31.** $(y - k)(y - k) - (x - h)(4p) = 0, (y - k)^2 = 4p(x - h)$

Exercise Set 14.5   pp. 530–531

1. $(-4, -3), (3, 4)$ **3.** $(0, -3), (4, 5)$ **5.** $(3, 0), (0, 2)$ **7.** $(-2, 1)$ **9.** $(3, 2), (4, \frac{3}{2})$
11. $\left(\frac{5 + \sqrt{70}}{3}, \frac{-1 + \sqrt{70}}{3}\right), \left(\frac{5 - \sqrt{70}}{3}, \frac{-1 - \sqrt{70}}{3}\right)$ **13.** $\left(\frac{15 + \sqrt{561}}{8}, \frac{11 - 3\sqrt{561}}{8}\right), \left(\frac{15 - \sqrt{561}}{8}, \frac{11 + 3\sqrt{561}}{8}\right)$
15. $9, 5$ **17.** $6\,\text{cm}, 8\,\text{cm}$ **19.** $4\,\text{in.}, 5\,\text{in.}$ **21.** $(0.965, 4402.33), (-0.965, -4402.33)$ **23.** $2(L + W) = P, L + W = \frac{P}{2}, LW = A$,
$L = \frac{P}{2} - W, W\left(\frac{P}{2} - W\right) = A, W^2 - \frac{WP}{2} + A = 0, W = \dfrac{\frac{P}{2} \pm \sqrt{\left(\frac{P}{2}\right)^2 - 4A}}{2} = \frac{P}{4} \pm \frac{\sqrt{P^2 - 16A}}{4} = \frac{1}{4}(P \pm \sqrt{P^2 - 16A})$
25. $(x - 2)^2 + (y - 3)^2 = 1$

Exercise Set 14.6   pp. 534–535

1. $(-5, 0), (4, 3), (4, -3)$ **3.** $(3, 0), (-3, 0)$ **5.** $(4, 3), (-4, -3), (3, 4), (-3, -4)$ **7.** No solution
9. $(\sqrt{2}, \sqrt{14}), (-\sqrt{2}, \sqrt{14}), (\sqrt{2}, -\sqrt{14}), (-\sqrt{2}, -\sqrt{14})$ **11.** $(1, 2), (-1, -2), (2, 1), (-2, -1)$
13. $(3, 2), (-3, -2), (2, 3), (-2, -3)$ **15.** $\left(\frac{5 - 9\sqrt{15}}{20}, \frac{-45 + 3\sqrt{15}}{20}\right), \left(\frac{5 + 9\sqrt{15}}{20}, \frac{-45 - 3\sqrt{15}}{20}\right)$
17. $(8.53, 2.53), (8.53, -2.53), (-8.53, 2.53), (-8.53, -2.53)$ **19.** $13, 12$ and $-13, -12$ **21.** $1\,\text{m}, \sqrt{3}\,\text{m}$ **23.** $16\,\text{ft}, 24\,\text{ft}$
25. $\left(x + \frac{5}{13}\right)^2 + \left(y - \frac{32}{13}\right)^2 = \frac{5365}{169}$

Chapter 14 Review p. 535

1.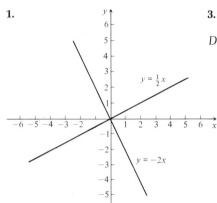

3. $(x-3)^2 + (y-4)^2 = 25$ **5.** $(x-2)^2 + (y-4)^2 = 26$ **7.** $F: (-3, 0)$; $V: (0, 0)$; $D: x = 3$ **9.** $\dfrac{x^2}{9} + \dfrac{y^2}{16} = 1$ **11.** $(-8\sqrt{2}, 8), (8\sqrt{2}, 8)$ **13.** 7, 4

CHAPTER 15

Exercise Set 15.1 pp. 540–542

1. 4, 7, 10, 13; 31; 46 **3.** $1, \frac{1}{2}, \frac{1}{3}, \frac{1}{4}; \frac{1}{10}; \frac{1}{15}$ **5.** $\frac{1}{2}, \frac{2}{3}, \frac{3}{4}, \frac{4}{5}; \frac{10}{11}; \frac{15}{16}$ **7.** $2n - 1$ **9.** $\dfrac{n+1}{n+2}$ **11.** $(\sqrt{3})^n$ **13.** $n - 1$ **15.** $4, 1\frac{1}{4}, 1\frac{4}{5}, 1\frac{5}{9}$
17. $64, 8, 2\sqrt{2}, \sqrt{2\sqrt{2}}$ **19.** 28 **21.** 30 **23.** $\frac{1}{2} + \frac{1}{4} + \frac{1}{6} + \frac{1}{8} + \frac{1}{10}$ **25.** $2^0 + 2^1 + 2^2 + 2^3 + 2^4 + 2^5$
27. $\log 7 + \log 8 + \log 9 + \log 10$ **29.** $\sum_{k=1}^{6} \dfrac{k}{k+1}$ **31.** $\sum_{k=1}^{4} \dfrac{1}{k^2}$ **33.** $\frac{3}{2}, \frac{3}{2}, \frac{3}{2}, \frac{3}{2}, \frac{3}{2}$ **35.** $1, 0, -1, 0, 1$ **37.** $\pi, 0, \pi, 0, \pi$
39. 1, 1, 2, 3, 5, 8 **41.** 2, 2.25, 2.37037, 2.441406, 2.488320, 2.521626 **43.** 2, 1.553774, 1.498834, 1.491398, 1.490379, 1.490238

Exercise Set 15.2 pp. 546–547

1. $a_{12} = 46$ **3.** 27th **5.** $a_{17} = 101$ **7.** $a_1 = 5$ **9.** $n = 28$ **11.** $a_1 = 8; d = -3; 8, 5, 2, -1, -4$ **13.** 670 **15.** 2500
17. 990 **19.** -264 **21.** 465 **23.** $5\frac{4}{5}, 7\frac{3}{5}, 9\frac{2}{5}, 11\frac{1}{5}$ **25.** n^2 **27.** $-10, -4, 2, 8$ **29.** Sides are $a, a + d, a + 2d$, and $a^2 + (a + d)^2 = (a + 2d)^2$. Solving, we get $d = a/3$. Thus sides are $a, 4a/3, 5a/3$, in the ratio $3:4:5$. **31.** 5200, 4687.50, 4175, 3662.5, 3150, 2637.5, 2125, 1612.5, 1100 **33.** 51, 679.65 **35.** Since we have an arithmetic sequence, $m = p + d$, $q = p + 2d$. Then $\dfrac{p+q}{2} = \dfrac{p+p+2d}{2} = p + d = m$.

Exercise Set 15.3 pp. 551–552

1. $a_{10} = \frac{81}{64}$ **3.** 1275 **5.** $\frac{547}{18}$ **7.** $\frac{63}{32}$ **9.** \$933.12 **11.** $\frac{1}{256}$ ft **13.** 5866.60 **15.** $\dfrac{a_n}{a_{n+1}} = r$, so $\dfrac{a_n^2}{a_{n+1}^2} = r^2$; hence $a_1^2, a_2^2, \ldots,$ is geometric, ratio r^2. **17.** $\dfrac{a_n}{a_{n+1}} = r$; $\ln(a_n) - \ln(a_{n+1}) = \ln r$; hence $\ln a_1, \ln a_2, \ldots,$ is arithmetic. **19.** $G = pr, q = pr^2$; $\sqrt{pq} = \sqrt{p^2 r^2} = pr = G$ **21.** 10,486 inches

Exercise Set 15.4   pp. 556–557

1. No   **3.** Yes   **5.** Yes   **7.** Yes   **9.** 8   **11.** 125   **13.** 2   **15.** $\frac{160}{9}$ **17.** 12,500   **19.** $\frac{7}{9}$ **21.** $\frac{7}{33}$ **23.** $\frac{170}{33}$ **25.** $\$5.33 \times 10^{10}$
27. 3.33×10^6; $66\frac{2}{3}\%$

Exercise Set 15.5   pp. 561–563

1. S_n: $1 + 2 + 3 + \cdots + n = \dfrac{n(n+1)}{2}$

S_1: $1 = \dfrac{1(1+1)}{2}$

S_k: $1 + 2 + 3 + \cdots + k = \dfrac{k(k+1)}{2}$

S_{k+1}: $1 + 2 + 3 + \cdots + k + (k+1) = \dfrac{(k+1)(k+2)}{2}$

1. Basis step: S_1 true by substitution.
2. Induction step: Assume S_k. Deduce S_{k+1}.
   Starting with the left side of S_{k+1}, we have
   $\underbrace{1 + 2 + 3 + \cdots + k}_{} + (k+1)$

   $= \dfrac{k(k+1)}{2} + (k+1)$ (by S_k)

   $= \dfrac{k(k+1) + 2(k+1)}{2}$ (adding)

   $= \dfrac{(k+1)(k+2)}{2}$ (distributive law)

3. S_n: $1 + 5 + 9 + \cdots + (4n - 3) = n(2n - 1)$
S_1: $1 = 1(2 \cdot 1 - 1)$
S_k: $1 + 5 + 9 + \cdots + (4k - 3) = k(2k - 1)$
S_{k+1}: $1 + 5 + 9 + \cdots + (4k - 3) + [4(k+1) - 3]$
$\qquad\qquad = (k+1)[2(k+1) - 1]$
$\qquad\qquad = (k+1)(2k+1)$

1. Basis step: S_1 true by substitution.
2. Induction step: Assume S_k. Deduce S_{k+1}.
   Starting with the left side of S_{k+1}, we have
   $\underbrace{1 + 5 + 9 + \cdots + (4k - 3)}_{} + [4(k+1) - 3]$

   $= k(2k - 1) + [4(k+1) - 3]$ (by S_k)
   $= 2k^2 - k + 4k + 4 - 3$
   $= (k+1)(2k+1)$

5. S_n: $\dfrac{1}{1 \cdot 2} + \dfrac{1}{2 \cdot 3} + \cdots + \dfrac{1}{n(n+1)} = \dfrac{n}{n+1}$

S_1: $\dfrac{1}{1 \cdot 2} = \dfrac{1}{1+1}$

S_k: $\dfrac{1}{1 \cdot 2} + \dfrac{1}{2 \cdot 3} + \cdots + \dfrac{1}{k(k+1)} = \dfrac{k}{k+1}$

S_{k+1}: $\dfrac{1}{1 \cdot 2} + \dfrac{1}{2 \cdot 3} + \cdots + \dfrac{1}{k(k+1)} + \dfrac{1}{(k+1)(k+2)}$

$\qquad\qquad = \dfrac{k+1}{k+2}$

7. 2. Induction step: Assume S_k. Deduce S_{k+1}. Now
$\qquad k < k + 1$ (by S_k)
$\qquad k + 1 < k + 1 + 1$ (adding 1)
$\therefore k + 1 < k + 2$

9. S_1: $3^1 < 3^{1+1}$
S_k: $3^k < 3^{k+1}$
$\qquad 3^k \cdot 3 < 3^{k+1} \cdot 3$
$\qquad 3^{k+1} < 3^{(k+1)+1}$

11. $S_1: 1^3 = \dfrac{1^2(1+1)^2}{4} = 1$

$S_k: 1^3 + 2^3 + \cdots + k^3 = \dfrac{k^2(k+1)^2}{4}$

$1^3 + 2^3 + \cdots + (k+1)^3 = \dfrac{k^2(k+1)^2}{4} + (k+1)^3$

$= \dfrac{(k+1)^2}{4}(k^2 + 4(k+1))$

$= \dfrac{(k+1)^2(k+2)^2}{4}$

13. $S_1: 1 + \dfrac{1}{1} = 1 + 1$

$S_k: \left(1 + \dfrac{1}{1}\right) \cdots \left(1 + \dfrac{1}{k}\right) = k + 1.$

Multiply by $\left(1 + \dfrac{1}{k+1}\right)$.

$\left(1 + \dfrac{1}{1}\right) \cdots \left(1 + \dfrac{1}{k+1}\right) = (k+1)\left(1 + \dfrac{1}{k+1}\right)$

$= (k+1)\left(\dfrac{k+1+1}{k+1}\right)$

$= (k+1) + 1$

15. $S_1: a_1 = \dfrac{a_1 - a_1 r}{1-r} = \dfrac{a_1(1-r)}{1-r} = a_1$

$S_k: a_1 + \cdots + a_1 r^{k-1} = \dfrac{a_1 - a_1 r^k}{1-r}$

Add $a_1 r^k$.

$a_1 + \cdots + a_1 r^k = \dfrac{a_1 - a_1 r^k}{1-r} + a_1 r^k \dfrac{1-r}{1-r} = \dfrac{a_1 - a_1 r^k + a_1 r^k - a_1 r^{k+1}}{1-r} = \dfrac{a_1 - a_1 r^{k+1}}{1-r}$

17. 2. Induction step: Assume S_k. Deduce S_{k+1}.
Starting with the left side of S_{k+1}, we have
$[r(\cos\theta + i\sin\theta)]^{k+1}$
$= [r(\cos\theta + i\sin\theta)]^k [r(\cos\theta + i\sin\theta)]$
$= r^k[\cos k\theta + i\sin k\theta][r(\cos\theta + i\sin\theta)]$
$= r^{k+1}(\cos k\theta + i\sin k\theta)(\cos\theta + i\sin\theta)$
$= r^{k+1}[\cos k\theta \cos\theta + i\sin k\theta \cos\theta + i\cos k\theta \sin\theta - \sin k\theta \sin\theta]$
$= r^{k+1}[(\cos k\theta \cos\theta - \sin k\theta \sin\theta) + i(\sin k\theta \cos\theta + \cos k\theta \sin\theta]$
$= r^{k+1}[\cos(k\theta + \theta) + i\sin(k\theta + \theta)]$
$= r^{k+1}[\cos(k+1)\theta + i\sin(k+1)\theta]$

19. 2. Induction step: Assume S_k. Deduce S_{k+1}.
We start with the left side of S_{k+1}.
$\underbrace{\cos x \cdot \cos 2x \cdots \cos 2^{k-1}x} \cdot \cos 2^k x$

$= \dfrac{\sin 2^k x}{2^k \sin x} \cdot \cos 2^k x = \dfrac{\sin 2^k x \cos 2^k x}{2^k \sin x} = \dfrac{\frac{1}{2}\sin 2(2^k x)}{2^k \sin x}$ (since $\sin 2\theta = 2\sin\theta\cos\theta$)

$= \dfrac{\sin 2^{k+1}x}{2^{k+1} \sin x}

21. S_n: $\left(1 - \frac{1}{2^2}\right)\left(1 - \frac{1}{3^2}\right) \cdots \left(1 - \frac{1}{n^2}\right) = \frac{n+1}{2n}$

S_2: $1 - \frac{1}{2^2} = \frac{2+1}{2 \cdot 2}$

S_k: $\left(1 - \frac{1}{2^2}\right)\left(1 - \frac{1}{3^2}\right) \cdots \left(1 - \frac{1}{k^2}\right) = \frac{k+1}{2k}$

S_{k+1}: $\left(1 - \frac{1}{2^2}\right)\left(1 - \frac{1}{3^2}\right) \cdots \left(1 - \frac{1}{k^2}\right)\left(1 - \frac{1}{(k+1)^2}\right) = \frac{k+2}{2(k+1)}$

1. Basis step: S_1 true by substitution.

2. Induction step: Assume S_k. Deduce S_{k+1}. Starting with the left side of S_{k+1} we have

$$\underbrace{\left(1 - \frac{1}{2^2}\right)\left(1 - \frac{1}{3^2}\right) \cdots \left(1 - \frac{1}{k^2}\right)}\left(1 - \frac{1}{(k+1)^2}\right)$$

$$= \frac{k+1}{2k}\left(1 - \frac{1}{(k+1)^2}\right)$$

$$= \frac{k+1}{2k} - \frac{1}{2k(k+1)} = \frac{(k+1)(k+1) - 1}{2k(k+1)}$$

$$= \frac{k^2 + 2k + 1 - 1}{2k(k+1)} = \frac{k^2 + 2k}{2k(k+1)}$$

$$= \frac{k(k+2)}{2k(k+1)} = \frac{k+2}{2(k+1)}.$$

23. S_2: $\overline{z_1 + z_2} = \overline{z_1} + \overline{z_2}$:

$\overline{(a + bi) + (c + di)} = \overline{(a + c) + (b + d)i} = (a + c) - (b + d)i$

$\overline{a + bi} + \overline{c + di} = a - bi + c - di = (a + c) - (b + d)i$

S_k: $\overline{z_1 + z_2 + \cdots + z_k} = \overline{z_1} + \overline{z_2} + \cdots + \overline{z_k}$

$\overline{(z_1 + z_2 + \cdots + z_k) + z_{k+1}} = \overline{z_1 + z_2 + \cdots + z_k} + \overline{z_{k+1}}$ by S_2;

$= \overline{z_1} + \overline{z_2} + \cdots + \overline{z_k} + \overline{z_{k+1}}$ by S_k.

25. S_1: i^1 is either i or -1 or $-i$ or 1.
S_k: i^k is either i or -1 or $-i$ or 1.
$i^{k+1} = i^k \cdot i$ is then $i \cdot i = -1$ or $-1 \cdot i = -i$ or $-i \cdot i = 1$ or $1 \cdot i = i$.

27. S_1: 2 is a factor of $1^2 + 1$.
S_k: 2 is a factor of $k^2 + k$.
$(k + 1)^2 + (k + 1) = k^2 + 2k + 1 + k + 1 = k^2 + k + 2(k + 1)$.
By S_k, 2 is a factor of $k^2 + k$, hence 2 is a factor of the right-hand side, and therefore a factor of $(k + 1)^2 + (k + 1)$.

29. S_1: 3 is a factor of $1(1 + 1)(1 + 2) = 6$.
S_k: 3 is a factor of $k(k + 1)(k + 2)$; i.e., $k(k + 1)(k + 2) = 3 \cdot m$ for some integer m, or $k^3 + 3k^2 + 2k = 3 \cdot m$.
S_{k+1}: $(k + 1)^3 + 3(k + 1)^2 + 2(k + 1) = 3p$ for some integer p.
The left-hand side is
$k^3 + 6k^2 + 11k + 6$ or
$(k^3 + 3k^2 + 2k) + (3k^2 + 9k + 6)$
$= 3 \cdot m + 3(k + 1)(k + 2) = 3 \cdot n$ for some integer n.

Chapter 15 Review   pp. 563–564

1. $3\frac{3}{4}$ **3.** 531   **5.** 11   **7.** $n = 6$, $S_6 = -126$ **9.** 0.27, 0.0027, 0.000027   **11.** $\frac{211}{99}$ **13.** $3\sqrt{5}$ **15.** \$7.38, \$1365.10
17. $\dfrac{16\sqrt{2}}{\sqrt{2}-1}$ in. or $32 + 16\sqrt{2}$ in.

19. S_n: $1 + 4 + 7 + \cdots + (3n - 2) = \dfrac{n(3n-1)}{2}$

S_1: $1 = \dfrac{1(3-1)}{2}$

S_k: $1 + 4 + 7 + \cdots + (3k - 2) = \dfrac{k(3k-1)}{2}$

S_{k+1}: $1 + 4 + 7 + \cdots + [3(k+1) - 2] =$

$1 + 4 + 7 + \cdots + (3k - 2) + (3k + 1) = \dfrac{(k+1)(3k+2)}{2}$

1. Basis step: $1 = \dfrac{2}{2} = \dfrac{1(3-1)}{2}$ is true.

2. Induction step: Assume S_k.

$1 + 4 + 7 + \cdots + (3k - 2) + (3k + 1)$

$= \dfrac{k(3k-1)}{2} + (3k + 1)$

$= \dfrac{k(3k-1)}{2} + \dfrac{2(3k+1)}{2}$

$= \dfrac{3k^2 - k + 6k + 2}{2}$

$= \dfrac{3k^2 + 5k + 2}{2}$

$= \dfrac{(k+1)(3k+2)}{2}.$

21. S_n: $\left(1 - \dfrac{1}{2}\right)\left(1 - \dfrac{1}{3}\right) \cdots \left(1 - \dfrac{1}{n}\right) = \dfrac{1}{n}$

S_2: $\left(1 - \dfrac{1}{2}\right) = \dfrac{1}{2}$

S_k: $\left(1 - \dfrac{1}{2}\right)\left(1 - \dfrac{1}{3}\right) \cdots \left(1 - \dfrac{1}{k}\right) = \dfrac{1}{k}$

S_{k+1}: $\left(1 - \dfrac{1}{2}\right)\left(1 - \dfrac{1}{3}\right) \cdots \left(1 - \dfrac{1}{k}\right)\left(1 - \dfrac{1}{k+1}\right) = \dfrac{1}{k+1}$

1. Basis step: S_2 is true by substitution.

2. Induction step: Assume S_k. Deduce S_{k+1}. Starting with the left side of S_{k+1}, we have

$\underbrace{\left(1 - \dfrac{1}{2}\right)\left(1 - \dfrac{1}{3}\right) \cdots \left(1 - \dfrac{1}{k}\right)}\left(1 - \dfrac{1}{k+1}\right)$

$= \dfrac{1}{k} \cdot \left(1 - \dfrac{1}{k+1}\right)$ (by S_k)

$= \dfrac{1}{k} \cdot \left(\dfrac{k+1-1}{k+1}\right)$

$= \dfrac{1}{k} \cdot \dfrac{k}{k+1}$

$= \dfrac{1}{k+1}$ (simplifying)

23. $\dfrac{a_{k+1}}{a_k} = r_1$, $\dfrac{b_{k+1}}{b_k} = r_2$, so $\dfrac{a_{k+1}b_{k+1}}{a_k b_k} = r_1 r_2$ (const.)

25. (a) a_n all positive or all negative; (b) always; (c) always; (d) never; (e) $a_n = 0$; (f) $a_n = 0$

CHAPTER 16

Exercise Set 16.1   pp. 572–573

1. $4 \cdot 3 \cdot 2$, or 24   **3.** $_{10}P_7 = 10 \cdot 9 \cdot 8 \cdot 7 \cdot 6 \cdot 5 \cdot 4$, or 604,800   **5.** 120; 3125   **7.** 120; 24   **9.** $\dfrac{5!}{2!\,1!\,1!\,1!} = 5 \cdot 4 \cdot 3 = 60$
11. $9 \cdot 9 \cdot 8 \cdot 7 \cdot 6 \cdot 5 \cdot 4$, or 544,320   **13.** $\dfrac{9!}{2!\,3!\,4!} = 1260$ **15.** 12!, or 479,001,600   **17.** (a) 120; (b) 3840
19. $52 \cdot 51 \cdot 50 \cdot 49 = 6{,}497{,}400$ **21.** $\dfrac{24!}{3!\,5!\,9!\,4!\,3!} = 16{,}491{,}024{,}950{,}400$ **23.** $4! = 24$, $8! \div 3! = 6720$, $\dfrac{13!}{2!\,2!\,2!\,2!\,2!} = 194{,}594{,}400$
25. $80 \cdot 26 \cdot 9999 = 20{,}797{,}920$ **27.** 11   **29.** 9

Exercise Set 16.2 p. 575

1. 126 **3.** 1225 **5.** $\dfrac{n(n-1)(n-2)}{3!}$ **7.** 8855 **9.** 210 **11.** $\binom{8}{2} = 28$, $\binom{8}{3} = 56$ **13.** $\binom{10}{7} \cdot \binom{5}{3} = 1200$ **15.** $\binom{58}{6} \cdot \binom{42}{4}$
17. $\binom{4}{3} \cdot \binom{48}{2} = 4512$ **19.** 2,598,960 **21.** $\binom{8}{3} = 56$ **23.** $\binom{5}{2}\binom{8}{2} = 280$ **25.** 5 **27.** 7

Exercise Set 16.3 pp. 579–580

1. $15a^4b^2$ **3.** $-745{,}472a^3$ **5.** $1120x^{12}y^2$ **7.** $-1{,}959{,}552u^5v^{10}$ **9.** $m^5 + 5m^4n + 10m^3n^2 + 10m^2n^3 + 5mn^4 + n^5$
11. $x^{10} - 15x^8y + 90x^6y^2 - 270x^4y^3 + 405x^2y^4 - 243y^5$ **13.** $x^{-8} + 4x^{-4} + 6 + 4x^4 + x^8$
15. $\binom{n}{0} - \binom{n}{1} + \binom{n}{2} - \binom{n}{3} + \cdots + (-1)^n\binom{n}{n}$ **17.** $140\sqrt{2}$ **19.** $9 - 12\sqrt{3}t + 18t^2 - 4\sqrt{3}t^3 + t^4$ **21.** 128 **23.** 2^{26}, or 67,108,864 **25.** $-7 - 4\sqrt{2}i$ **27.** $\sin^7 t - 7\sin^5 t + 21\sin^3 t - 35\sin t + 35\csc t - 21\csc^3 t + 7\csc^5 t - \csc^7 t$
29. $\sum_{r=0}^{n} \binom{n}{r}(-1)^r a^{n-r} b^r$ **31.** -3 **33.** 5 **35.** $2^7 - 1$, or 127

Exercise Set 16.4 pp. 585–586

1. 52 **3.** $\frac{1}{4}$ **5.** $\frac{1}{13}$ **7.** $\frac{1}{2}$ **9.** $\frac{2}{13}$ **11.** $\frac{5}{7}$ **13.** 0 **15.** $\frac{11}{4165}$ **17.** $\frac{28}{65}$ **19.** $\frac{1}{18}$ **21.** $\frac{1}{36}$ **23.** $\frac{30}{323}$ **25.** $\frac{9}{19}$ **27.** $\frac{1}{38}$ **29.** $\frac{1}{19}$
31. 2,598,960 **33.** (a) 36; (b) 1.23×10^{-5}

Chapter 16 Review pp. 586–587

1. $6! = 720$ **3.** $\binom{15}{8} = 6435$ **5.** $\dfrac{9!}{1!\,4!\,2!\,2!} = 3780$ **7.** (a) $P(6,5) = 720$; (b) $6^5 = 7776$; (c) $P(5,4) = 120$; (d) $P(3,2) = 6$
9. 36 **11.** $220a^9x^3$ **13.** $m^7 + 7m^6n + 21m^5n^2 + 35m^4n^3 + 35m^3n^4 + 21m^2n^5 + 7mn^6 + n^7$ **15.** $-6624 + 16{,}280i$
17. $\left(\log \dfrac{x}{y}\right)^{10}$ **19.** $\frac{1}{4}$

Index

Abscissa, 99
Absolute value, 3, 37–39, 177–180
  complex numbers, 448
Addition, 3, 4
  fractional expressions, 32
  matrices, 224
  polynomials, 17
  vectors, 418
  vectors using components, 421
Addition method, 191–193
Addition principle, 25, 56
Additive identity
  matrices, 225
  zero, 3
Additive inverses, 4, 5, 7
  matrices, 225
  polynomials, 17
Algebra, fundamental theorem of, 454, 466
Algebraic manipulations, 324
Amplitude, 303, 304
Angle of depression, 407
Angle of elevation, 407
Angles
  between lines, 368–370
  central, 334
  measure of, 332–335
  reference, 344, 352
  rotations and, 332
  trigonometric functions of, 339–344, 350
Angular speed, 336–338
Antilogarithms, 263, 266, 267
  interpolation, 271, 272
Applied problems, 62–69, 77–79, 86, 89–92, 166–169, 196, 197, 203, 204, 275–277, 282–285, 292, 294, 295, 337, 338, 406–408, 419, 421, 428–430, 530, 534
Arc length, 334
Area problems, 66
Argument, 448
Arithmetic means, 545
Arithmetic sequences, 542
  common difference of, 542
  recursive definition of, 539
Arithmetic series, 543–545
Associativity
  addition, 3
  multiplication, 6
Asymptotes, 308, 492–495
  horizontal, 494
  of a hyperbola, 518
  vertical, 493
Augmented matrix, 245
Axes, 99
  symmetry with respect to, 113–115
Axis
  conjugate, 518
  major, 513
  minor, 513
  polar, 425
  of symmetry, 113
  transverse, 518

Basis step, 559
Bearings, 408
Bels, 275
Bernoulli's inequality, 562
Binomial coefficient, 577
Binomial notation, 449
  to polar notation, 449
Binomial series, 576–579
Binomial theorem, 576
Binomials, 16
  product of, 19
  square of, 20

Calculators, 9, 352, 389, 407
Canceling, 31
Cartesian coordinate system, 99, 425
Cartesian products, 96, 97
Cartesian square, 96
Center
  of circle, 509, 510
  of ellipse, 513
  of hyperbola, 517
Central angles, 334
Change of base, 285, 286
Changing the sign, 5
Characteristic, 265
Checking solutions, 57, 59, 82
Circles, 508–510
  center, 509
  equation of, 509
  graphs, 510
  radius, 509
  stretched, as ellipse, 513
  unit, 297–299, 354
Circular arrangements, 571
Circular functions, 292, 301–311
  signs of, 311, 342, 343
Clearing of fractions, 59
Closed interval, 129
Coefficients, 16, 458
  binomial, 577
  leading, 458
Cofactor method, 243–245
Cofactors, 232
Cofunction identities, 316, 317, 347, 358, 359

Coincident lines, 154
Combinations, 573-574
Combinatorics, 566
Common difference of arithmetic
    sequences, 542
Common logarithms, 262-268
Common ratio of geometric
    sequences, 547
Commutativity
    addition, 3
    multiplication, 6
Complements, 347
Completing the square, 72, 73, 162, 163
Complex fractional expressions, 33, 34
Complex numbers, 435-437
    absolute value of, 448
    conjugates of, 438
    division of, 438, 439, 451
    equality of, 436
    equations involving, 402-446
    graphs of, 447, 448
    multiplication of, 450
    polar notation, 448
    properties of, 436
    reciprocals of, 439
    roots of, 453-455
Components of vectors, 420
    in addition, 421
Composition of functions, 108, 109
Compound interest problems, 64-66
Compound sentence, 173
Cones, 506
    nappes of, 506
    right circular, 506
Conic sections, 506
    circles, 508-510
    ellipses, 512-515
    hyperbolas, 517-521
    parabolas, 523-526
Conical surface, 506
Conjugate axis of hyperbola, 518
Conjugates
    of complex numbers, 438, 439
    of polynomials, 442
    properties of, 439-442

Conjunction, 173, 215
Constant functions, 106
Constraints, 218
Consumer price index, 287
Continuous functions, 130, 302, 304, 308-311
Convex polygon, 218
Coordinates
    first, 99
    polar, 425-427
    second, 99
Cosecant function, 293, 311, 340
Cosine function, 293, 303-305, 320, 340
Cosines, law of, 415
Cotangent function, 293, 309, 310, 340
Counting principle, fundamental, 566
Cramer's rule, 211-213
Cubic polynomials, 486

Decibels, 276
Decimals
    repeating, 2
    terminating, 2
Decreasing functions, 130, 131
Degree
    of polynomials, 16, 458
    of terms, 16
Degree measure, 332, 350
    to radian measure, 333
DeMoivre's theorem, 453
Dependent equations, 195, 196, 201, 211
Depression, angle of, 407
Descartes' rule of signs, 476
    corollary, 478
Determinants, 209-213
    evaluating, using cofactors, 233
    expanding, 233
    higher-order, 231-234
    minor, 231
    properties of, 235-239
Diagonals, of a polygon, 80
Diagram, set, 171
Difference formulas, 364-368
Dimension symbols, 49, 50

Direct variation, 88, 89
Directrix of a parabola, 523
Discriminant, 75, 164, 444
Disjunction, 174
Displacement, 418
Distance formula, 156
Distributivity
    over addition, 6
    over subtraction, 6
Division, 8
    of complex numbers, 438, 451
    of fractional expressions, 29-31
    of polynomials, 459, 460
    synthetic, 462-464
Domain, 97, 102
    of functions, 106, 107
Doubling time, 278, 288

e, 279, 280
Earthquake magnitude, 276
Economic multiplier, 556
Elevation, angle of, 407
Ellipse, 506, 512-515
    center, 513
    construction, 512
    equations, 513
    foci, 512
    graphs, 513, 514
    major axis, 513
    minor axis, 513
    as stretched circle, 513
    vertices, 513
Empty set, 26
Equations
    with absolute value, 177-180
    of circles, 509
    and complex numbers, 442-446
    dependent, 195, 196, 201, 211
    of ellipses, 513, 514
    equivalent, 56
    exponential, 255, 273, 274, 282
    fractional, 59
    graphs of, 99-102
    homogeneous, 201-203
    of hyperbolas, 518-520
    inconsistent, 195, 200, 211
    linear, 148

logarithmic, 254, 274, 275
matrix, 228
of parabolas, 524, 525
polar to rectangular, 426
quadratic, 71–75, 444
radical, 82, 83
rectangular to polar, 426
reducible to quadratic, 84–87
of second degree, 71, 506–510
solutions of, 24, 99
solving, 24–27, 56–59, 445
solving systems of, using matrices, 206–208
  addition method, 191–193
  Cramer's rule, 211–213
  substitution method for, 190
systems of equations, 190–193, 196, 197, 198–200
  first-degree and second-degree, 528–530
  second-degree, 530–534
trigonometric, 394–397
writing from solutions, 75, 444
Equilibrium, forces in, 428–430
Equivalent equations, 56
Even functions, 126, 127, 304, 356
Exponential equations, 255, 273, 274, 282
Exponential functions, 252–254
Exponential notation, 11–14, 252
Exponents, 11–13
  fractional, 46, 47
  irrational, 252
  rational, 46, 47
Expressions
  complex fractional, 33, 34
  fractional, 29–34
  radical, 36, 37

Factor theorem, 464
Factorial notation, 567
Factoring
  differences of cubes, 23
  differences of squares, 22
  by grouping, 22
  polynomials, 22, 23
  sums of cubes, 23

trinomial squares, 23
Fibonacci sequence, 541
Field, 6
Foci
  of an ellipse, 512
  of a hyperbola, 517
Focus of parabola, 523
FOIL, 19
Force, 418
Forces in equilibrium, 428–430
Forgetting, 277
Formulas, 61, 77
  distance, 156
  midpoint, 157
  quadratic, 74, 443
Fourth-degree polynomials, 486
Fractional equations, 59
Fractional exponents, 46, 47
Fractional expressions, 29–34
  addition of, 32
  complex, 33
  decompose, 499–502
  division of, 29–31
  multiplication of, 29–31
  simplifying, 29–31
  subtraction of, 33
Fractions, partial, 499–502
Function machine, 107
Function values, trigonometric, 340–344, 350
  relationships among, 346, 347
Functions, 104–109
  circular, 292
  composition of, 108, 109
  constant, 106
  continuous, 130, 302, 304, 308, 309–311
  cosecant, 311
  cosine, 303–305
  cotangent, 309, 310
  decreasing, 130, 131
  defined piecewise, 131, 132
  domains of, 106, 107
  even, 126, 127, 304, 356
  exponential, 252–254
  greatest integer, 112
  increasing, 130, 131

inputs of, 108
inverses of, 138–141
linear, 152
logarithmic, 254–256
as mappings, 107
as mathematical models, 166–169
notation for, 105
odd, 127, 302, 356
outputs of, 108
periodic, 128, 302, 304, 308–311
polynomial, 458
postage, 135
quadratic, 159–164
rational, 491–498
secant, 310, 311
sine, 301–303
special classes of, 126–132
tangent, 307–309
trigonometric, 293, 301–311
vertical line test for, 105
Fundamental counting principle, 566
Fundamental theorem of algebra, 454, 466

Gauss-Jordan method, 245–248
Geometric progressions, 547
Geometric sequences, 547
  common ratio, 547
  infinite, 554
Geometric series, 549, 550
  repeating decimals as, 554
Grad, 336
Graphs
  circles, 510
  complex numbers, 447, 448
  cosecant function, 311
  cosine function, 304, 357
  contangent function, 309, 357
  ellipses, 513, 514
  equations, 98–102
  functions, 104, 105
  hyperbolas, 519–521
  inequalities, 171–175, 214–216
  inverse relations, 136
  linear equations, 148
  parabolas, 524, 525
  polar coordinates, 425

polynomial functions, 485–489
polynomials, 485–489
quadratic functions, 160, 161
rational functions, 492–498
secant function, 310
sine and cosine, variations of, 320–322, 383
sine function, 302, 357
solving equations by, 489, 490
sums, 323
systems of equations, 190
systems of inequalities, 215, 216
tangent function, 309, 357
Greatest integer function, 112

Half-angle identities, 374–376
Half-open interval, 129
Half-plane, 214
Homogeneous equations, 201–203
Horizontal asymptotes, 494
Horizontal shrinkings, 123, 124
Horizontal stretchings, 123, 124
Horizontal translations, 120, 121
Hyperbola, 506, 517–521
  asymptotes, 518
  center, 517
  conjugate axis, 518
  equations, 518–520
  foci, 517
  graphs, 519–521
  transverse axis, 518
  vertices, 518
Hypotenuse, 292

i, 434
  powers of, 434
Identities
  additive, 3
    matrix, 225
  multiplicative, 6
    matrix, 241
  proving, 380, 380–382
  in solving trigonometric equations, 398–401
  trigonometric, 303, 305, 312, 313, 315–319, 354–359, 364–368, 371–382, 391

Imaginary axis, 447
Imaginary numbers, 434
Inconsistent equations, 195, 200, 211
Increasing functions, 130, 131
Index, 36
Induction, mathematical, 557–561
Induction step, 559
Inequalities
  with absolute value, 177–180
  graphing, 171–175
  polynomial, 182, 183
  quadratic, 181, 182
  rational, 183, 184
  solving, 27, 173–175
  systems of, 215, 216
Initial side, 332
Inputs, 108
Integers, 2
  as exponents, 11–13
Intercepts, 148
  x, 163
  y, 151
Interpolation, 269–272, 350
  linear, 270
Intersection of sets, 171, 173
Interval notation, 129
Intervals
  closed, 129
  half-open, 129
  open, 129
Inverse variation, 89, 90
Inverses
  additive, 4, 5, 7, 17
    of matrices, 225
  of functions, 138–141
  of matrices, 242
  multiplicative, 6
  of relations, 136
  of trigonometric functions, 385–392
Irrational exponents, 252

Joint variation, 91

Law of cosines, 415
Law of sines, 410
Least common multiple, 32, 59

Life expectancy, 170
Line
  of reflection, 112
  of symmetry, 112, 160–163, 518
  symmetry with respect to a, 112, 113
Linear equations, 148
  graphs of, 148
Linear functions, 152
Linear interpolation, 270
Linear programming, 217–220
Linear speed, 337
Linear transformations, 125, 126
Lines, 148
  coincident, 154
  parallel, 154
  perpendicular, 154
  point-slope equation of, 150
  slope of, 148–150
  slope-intercept equation of, 151
  two-point equation of, 151
  vertical, 154
Logarithmic equations, 254, 274, 275
  solving, 255
Logarithmic functions, 254–256
  properties of, 257–261
Logarithms
  anti-, 263, 266–267
  calculations with, 267, 268
  change of base, 285, 286
  characteristic of, 265
  common, 262–268
  in computation, 262–264
  interpolation, 269–272
  mantissa of, 265
  natural, 281, 282
  tables of, 264
Loudness of sound, 275
Lower bounds on roots, 480–483

Main diagonal, 208
Major axis of an ellipse, 513
Manipulations
  algebraic, 324
  trigonometric, 325–327
Mantissa, 265
Mappings, 107

Mathematical induction, 557–561
   principle of, 559
Mathematical models, 166–169, 203, 204
Matrices
   addition of, 224
   additive identity, 225
   additive inverses of, 225
   augmented, 245
   column, 226
   columns of, 224
   determinants of, 209–213
   dimensions of, 224
   inverses of, 242–248
   main diagonal of, 208
   multiplying, 226–229
   operations on, 224–229
   row, 226
   row-equivalent, 208
   rows of, 224
   square, 209, 224
   subtraction of, 225
   transpose, 243
   zero, 224
Matrices and systems of equations, 206–208
Matrix equations, 228
Matrix inverses, 242–248
   cofactor method, 243–245
   Gauss-Jordan method, 245–248
Maximum, 160–163, 169, 170, 218
   linear programming, 217–220
Measure
   of angles, 332
   degree, 332, 350
   radian, 332–335
   of rotations, 332
Midpoint formula, 157
Mil, 336
Minimum, 160–163
   linear programming, 217–220
Minor axis of an ellipse, 513
Minors, 231
Models, mathematical, 166–169, 203, 204
Monomials, 16
Mortgage payment, 15

Motion problems, 67, 68
Multiplication, 5, 6
   complex numbers, 450
   of fractional expressions, 29–31
   of matrices, 226–229
   of polynomials, 19–21
   scalar, 423
Multiplication principle, 25, 57
Multiplicative identity, matrix, 241
Multiplicative inverses, 6
Multiplying by one, 333

Nappes of cones, 506
Natural logarithms, 281, 282
Natural numbers, 2
Nondistinguishable objects, permutations of, 570
Nonsensible replacements, 29–31, 37
Notation
   binomial, 449
   exponential, 11–14, 252
   factorial, 567
   function, 105
   interval, 129
   inverse function, 138
   polar, 422, 448
   radical, 36, 37
   rectangular, 449
   scientific, 13, 14
   set-builder, 27
   sigma, 540
   trigonometric, 448
Numbers
   complex, 435–437
   imaginary, 434
   integers, 2
   irrational, 2, 3
   natural, 2
   rational, 2
   real, 2–9
   whole, 2

Odd functions, 127, 302, 356
One, as an exponent, 11
Open interval, 129
Order, 8, 9
Ordered pairs, 96

Ordinate, 99
Origin, 99
   symmetry with respect to, 116–118
Outcomes, 581
Outputs, 108

Parabola, 159, 506, 523–526
   directrix, 523
   equations, 524, 525
   focus, 523
   graphs, 524, 525
Parallel lines, 154
Partial fractions, 499–502
Percent problems, 63
Periodic functions, 128, 302, 304, 308, 309, 310, 311
   amplitude, 303, 304
Permutations, 567–571
   circular arrangements, 571
   of n objects, r at a time, 568
   of nondistinguishable objects, 570
Perpendicular lines, 154
Phase shift, 322
Pi, π, 3
Point of symmetry, 116
Point-slope equation, 150
Poker hands, 585
Polar axis, 425
Polar coordinate system, 425
Polar coordinates, 425–427
Polar equations, 426
Polar form, for vectors, 422
Polar notation
   to binomial notation, 449
   for complex numbers, 448
   for vectors, 422
Pole, 425
Polygon, convex, 218
Polynomial functions, 458
   graphs of, 485–489
Polynomial inequalities, 182, 183
Polynomials
   addition of, 17
   conjugates of, 442
   cubic, 486
   definition, 16, 458

degree of, 16, 458
division of, 459, 460
factoring, 22, 23
fourth-degree, 486
graphs of, 485–489
with integer coefficients, 471
multiplication of, 19–21
quartic, 486
with rational coefficients, 469, 475
rational roots of, 471–475
with real coefficients, 468
roots of, 458, 466–484
subtraction of, 17, 18
third-degree, 486
zeros of, 458
Population growth, 282
Postage function, 135
Powers, principle of, 82
Precision of angle measure, 406
Principal values of inverses of trigonometric functions, 387
Principle
addition, 25, 26
fundamental counting, 566
of mathematical induction, 559
multiplication, 25, 57
of powers, 82
of zero products, 26, 58, 444, 451
Probability, 580–585
experimental, 580
origin, 584
theoretical, 580
Product
Cartesian, 96, 97
scalar, 226
Profit and loss, 168, 171
Programming, linear, 217–220
Projectile problem, 169, 170
Properties
of complex numbers, 436
of conjugates, 439–442
of determinants, 235–239
of exponents, 12, 13
of a field, 6
of logarithmic functions, 257–261
of radicals, 39–41
of real numbers, 3–9

of vectors, 423
Proportion, 88
Pythagorean identities, 315, 316

Quadrants, 99
Quadratic equations, 71–75, 84–87
with real coefficients, 444
Quadratic formula, 73, 74, 443
Quadratic functions, 159–164
discriminant of, 164
Quadratic inequalities, 181, 182
Quartic polynomials, 486

Radian, 332
Radian measure, 332–335
to degree measure, 333
Radical
properties of, 39–41
symbol, 36
Radical equations, 82, 83
Radical expressions, 39–44
Radical notation, 36, 37
Radicand, 36
Radioactive decay, 283
Radiocarbon dating, 287
Radius, 509
Raising a power to a power, 13
Range, 97, 102
Rational exponents, 46, 47
Rational functions, 491–498
graphs of, 492–498
roots of, 495
zeros of, 495
Rational inequalities, 183, 184
Rational numbers, 2
Rational roots theorem, 472
Rationalizing
denominator, 41, 43, 44
numerator, 44
Real axis, 447
Real numbers, 2–9
properties of, 3–9
Reciprocals, 6, 8
of complex numbers, 439
Rectangular coordinate system, 425
Rectangular equations, 426
Rectangular form for vectors, 422

Recursive definition, 539
Reference angle, 344, 352
Reference triangle, 344
Reflections, 112
line of, 112
Relations, 97, 98
domain, 97, 102
inverses of, 136
range, 97, 102
Remainder theorem, 461
Removing a factor of 1, 30, 31
Repeating decimals, 2
to fractional notation, 555
as geometric series, 554
Right triangles, 292
solving, 406–408
Roots
of complex numbers, 453–455
cube, 36
even, 36
lower bounds on, 480–483
negative, 477
odd, 36
of polynomials, 458, 466–484
principal, 36
rational, 471–475
rational functions, 495
real, 476–478
square, 36
upper bounds on, 478–480, 483
Rotations, 332
measures of, 332
Row-equivalent matrices, 208
Row-equivalent operations, 208
Rule of signs, Descartes', 476
corollary, 478

Scalar multiplication, 423
Scalar product, 226
Scientific notation, 13, 14
Secant function, 293, 310, 311, 340
Second-degree equations, 71–75
systems of, 530–534
systems of first degree and, 528–530
Second-degree polynomials, 159
Sensible replacements, 29–31, 37

Sequences, 538
   arithmetic, 542
      sum of first n terms, 544, 545
   geometric, 547
   infinite, 552
   terms of, 538
Series, 540
   arithmetic, 543-545
   binomial, 576-579
   geometric, 549, 550
   infinite, 553
Set diagrams, 171
Set-builder notation, 27
Sets
   intersection of, 171, 173
   solution, 24
   union of, 172
Shift, phase, 322
Shrinkings
   horizontal, 123, 124
   vertical, 121-123
Side
   initial, 332
   terminal, 332
Sigma, 540
   notation, 540
Similar terms, 16
Similar triangles, 292
Simplifying
   fractional expressions, 29-31, 32-34
   radical expressions, 39-47
Sine function, 293, 301-303, 320-322, 339
Sine, law of, 410
Slope, 148-150
Slope-intercept equation, 151
Small angles, function values of, 350
Solutions
   of equations, 24, 99
   of inequalities, 214
Solving equations, 24-27, 56-59, 445
   by graphing, 489, 490
Solving inequalities, 27
Solving oblique triangles, 409
Solving right triangles, 406-408
   applications of, 406-408

Solving triangles, 406-416
   oblique, 409
   right, 406-408
   AAS, 410
   SAS, 415
   SSA, 411
   SSS, 416
Speed, 48, 67
   angular, 336-338
   linear, 337
Square matrices, 209
Standard form
   equation of circle in, 509
   quadratic equations in, 73, 74
Standard position of vectors, 422
Step
   basis, 559
   induction, 559
Stretchings
   horizontal, 123, 124
   vertical, 121-123
Subsets, 578
Substitution method, 190, 191
Subtraction, 6, 7
   of fractional expressions, 33
   of matrices, 225
   of polynomials, 17, 18
Sum formulas, 364-368
Sum-product identities, 376-377
Sums, graphs of, 323
Surface element, 506
Symmetry, 112-118
   axis of, 113
   line of, 112, 160-163
   point of, 116
   with respect to axes, 113-115
   with respect to a line, 112, 113
   with respect to the origin, 116-118
   with respect to a point, 116
   with respect to $y = x$, 137
Synthetic division, 462-464
Systems of equations
   dependent, 195, 196, 201
   of first-degree and second-degree, 528-530
   inconsistent, 195, 200
   of second-degree, 530-534

   in three variables, 198-200
   in two variables, 190-193, 196, 197
Systems of inequalities, 215-220

Tables
   common, 264
   logarithm, 264
   natural, 281
   trigonometric functions, 293, 349, 352
Tangent function, 293, 307-309, 340
Terminal side, 332
Terminating decimal, 2
Terms, 16
   degree of, 16
   similar, 16
Theorem
   of algebra, fundamental, the, 454, 466
   binomial, 576
   factor, 464
   rational roots, 472
   remainder, 461
Third-degree polynomials, 486
Transformations, 119-124
   linear, 125
Translations, 119
   horizontal, 120, 121
   vertical, 119, 120
Tranpose, 243
Transverse axis of hyperbola, 518
Triangles
   reference, 344
   right, 292
   similar, 292
   solving, 406-416
      AAS, 410
      SAS, 415
      SSA, 411
      SSS, 416
Trigonometric equations, 394-397
   using identities in solving, 398-401
Trigonometric functions, 293, 301-311
   of angles, 339-344
   cosecant, 311, 340

cosine, 303–305, 340, 357
cotangent, 309, 310, 340, 357
graphs of, 357
inverses of, 385–389
and their inverses, 390–392
relationships among, 346, 347, 354, 355
secant, 310, 311, 340
signs of, 311, 342, 343
sine, 301–303, 339, 357
for small angles, 350
tables of, 293, 349, 352
tangent, 307–309, 340, 357
Trigonometric identities, 303, 305, 312, 313, 315–319, 354–359, 364–368, 371–382, 391
cofunction, 316, 317, 347
double-angle, 372–374
half-angle, 374–376
proving, 380–382
Pythagorean, 315, 316, 355, 356
in solving equations, 398–401
sum-product, 376, 377
Trigonometric manipulations, 325–327
Trigonometric notation, 448
Trigonometric ratios, 293
Trinomials, 16

Tripling time, 278
Two-point equation, 151

Union, of sets, 172, 174
Unit circle, 297–299, 354
distances on, 298, 299
Units, changes of, 48, 49
Upper bounds on roots, 478–480, 483

Variation, 88–92
direct, 88, 89
inverse, 89, 90
joint, 91
Variation constant, 88, 89
Vector quantity, 418
Vectors, 418, 419
addition of, 418, 421
components of, 420
and coordinates, 420–423
polar form, 422
properties of, 423
rectangular form, 422
resultant of, 419
standard position, 422
Velocity, 418
Velocity, initial, 79
Vertex, 159–164, 217, 506

Vertical asymptotes, 493
Vertical-line test, 105
Vertical lines, 155
Vertical shrinkings, 121–123
Vertical stretchings, 121–123
Vertical translations, 119, 120
Vertices
of ellipse, 513
of hyperbola, 518

Whispering gallery, 515
Whole numbers, 2
Work problems, 69

x-intercepts, 148, 163, 164

y-intercepts, 148, 151, 154
$y = x$, symmetry with respect to, 137

Zero, 2
as an exponent, 11
Zero matrix, 224
Zero products, principle of, 26, 58, 444, 451
Zeros
of polynomials, 458
of rational functions, 495